Electrochemistry

The Past Thirty and the Next Thirty Years

J. O'M. Bockris

A VOLUME IN HONOR OF J. O'M. BOCKRIS

Electrochemistry
The Past Thirty and the Next Thirty Years

Edited by
Harry Bloom
The University of Tasmania
Hobart, Tasmania, Australia

and
Felix Gutmann
Macquarie University
Sydney, New South Wales, Australia

Plenum Press · New York and London

Library of Congress Cataloging in Publication Data

Main entry under title:

Electrochemistry, the past thirty and the next thirty years.

Proceedings of a symposium held at the Imperial College, London, in 1975.
Includes bibliographical references.
1. Electrochemistry–Congresses. 2. Bockris, John O'M. I. Bockris, John O'M. II.
Bloom, Harry, 1921-   III. Gutmann, Felix, 1908-
QD551.E54   541'.372   76-49951
ISBN 0-306-30921-1

Proceedings of an International Symposium on Electrochemistry–The Past
Thirty and the Next Thirty Years held at the Imperial College
in London, England, April 3–4, 1975

© 1977 Plenum Press, New York
A Division of Plenum Publishing Corporation
227 West 17th Street, New York, N.Y. 10011

Printed in the United States of America

SYMPOSIUM COMMITTEE

The overall direction was in the hands of an International Committee consisting of:

D. Inman *(Imperial College, London)* Chairman
F. Gutmann *(Macquarie University, Sydney)* Hon. Secretary/Treasurer
H. Bloom *(University of Tasmania, Hobart)*
B. E. Conway *(University of Ottawa)*
A. R. Despic *(University of Beograd)*
J. D. Mackenzie *(University of California, Los Angeles)*
I. A. Menzies *(University of Technology, Loughborough)*

while the organization was in charge of an Australian Sub-Committee consisting of:

H. Bloom *(University of Tasmania, Hobart)* Chairman
F. Gutmann *(Macquarie University, Sydney)* Hon. Secretary/Treasurer
R. B. Hunter *(University of Sydney, Sydney)*
D. F. A. Koch *(Commonwealth Industrial and Scientific Research Organization, Melbourne)*
E. C. Potter *(Minerals Research Laboratory, Sydney)*
B. J. Welch *(University of New South Wales, Sydney)*

LIST OF DONORS

Financial support from the donors listed below is gratefully acknowledged.

British Iron and Steel Corporation, London, England
Comalco Limited, Melbourne, Victoria, Australia
Energy Research Corporation, Bethel, Connecticut
IBM (United Kingdom) Ltd., Southampton, England
Kaiser Aluminum and Chemical Corporation, Permanente, California
Metallurgie Hoboken-Overpelt, Hoboken, Belgium
Nalco Chemical Company, Chicago, Illinois
Philips Laboratory, Briarcliff Manor, New York
Reynolds Metals Company, Listerhill, Alabama
Siemens A. G., Erlangen, West Germany
Swiss Aluminum Ltd., Chippis, Switzerland
West Australian Breweries Ltd., Perth, Western Australia

Preface

Some time ago a group of present and former collaborators of Professor John O'M. Bockris, following a suggestion by Professor J. D. Mackenzie (Los Angeles), conceived the idea of an International Symposium devoted to reviewing the active and developing aspects of the science of electrochemistry. From this beginning has sprung the "Electrochemistry Symposium—The Past Thirty and the Next Thirty Years," which took place at Imperial College, London, from April 3-6, 1975.

The plan for this symposium is unusual, since it features pairs of invited addresses, one to summarize the "state of the art" and the other to suggest directions for future research in particular aspects of electrochemistry.

This volume of proceedings gives these papers in their final, considered, and fully referenced form, arranged in the sequence of their delivery at the symposium. Also included are introductory addresses given by Professor Ubbelohde, Professor Frumkin, Dr. Egan, and Dr. Inman.

Both aspects of nearly every topic, plus the discussions, are integrated in a Report or Summary. A synopsis of the matters raised at the symposium and prepared by Professor John O'M. Bockris closes this volume.

The cooperation of Plenum Press, New York, is gratefully acknowledged.

H. Bloom

Hobart

F. Gutmann

Sydney

Preface

Contributors

A. J. Appleby, Laboratoire d'Electrolyse, C.N.R.S. Bellevue, 92-Bellevue, France

J. O'M. Bockris, School of Physical Sciences, The Flinders University of South Australia, Bedford Park, South Australia, 5042

G. L. Cahen, Jr., Department of Material Science, University of Virginia, Charlottesville, Virginia 22901

B. E. Conway, Chemistry Department, University of Ottawa, Ottawa, Canada

A. Damjanovic, Xerox Corporation, Webster Research Center, Webster, New York 14580

A. R. Despić, Faculty of Technology and Metallurgy, University of Beograd, and the Institute of Electrochemistry ICTM, Beograd, Yugoslavia

G. Eckert, Sydney Hospital, Sydney, NSW, Australia

Thomas W. Healy, Department of Physical Chemistry, University of Melbourne, Parkville 3052 Victoria, Australia

Karl H. Hauffe, Institute of Physical Chemistry, University of Göttingen, West Germany

D. Inman, Department of Metallurgy and Materials Science, Imperial College of Science and Technology, London SW7 2BP, England

James W. Johnson, Chemical Engineering Department, University of Missouri, Rolla, Missouri

Hendrik Keyzer, Chemistry Department, California State University, 5151 State University Drive, Los Angeles, California 90032

Hideaki Kita, Department of Chemistry, Faculty of Science, Hokkaido University, Sapporo, Japan

D. F. A. Koch, Division of Mineral Chemistry, CSIRO, Melbourne, Australia

A. T. Kuhn, Department of Chemistry and Applied Chemistry, University of Salford, Salford, M5 4WT, Lancashire, England

John D. Mackenzie, School of Engineering and Applied Science, University of California, Los Angeles, California 90024

Einar Mattsson, Director of Research, Swedish Corrosion Institute, Drottning Kristinas väg 48, S-114 28 Stockholm, Sweden

James McBreen, Electrochemistry Department, Research Laboratories, General Motors Corporation, Warren, Michigan 48090

E. C. Potter, CSIRO, Division of Process Technology, North Ryde, Australia 2113

A. K. N. Reddy, Department of Inorganic and Physical Chemistry, Indian Institute of Science, Bangalore 560012, India

Nolan E. Richards, Reduction Research Division, Reynolds Metals Company, Sheffield, Alabama 35660

Eric Sheldon, Department of Physics, University of Lowell, Lowell, Massachusetts

M. A. Slifkin, Department of Pure and Applied Physics, University of Salford, Salford M5 4WT, Lancashire, England

S. Srinivasan, Department of Applied Science, Brookhaven National Laboratory, Upton, New York 11973

G. E. Stoner, Department of Material Science, University of Virginia, Charlottesville, Virginia 22901

J. W. Tomlinson, Professor of Physical Chemistry, Victoria University of Wellington, New Zealand

A. T. Ward, Xerox Corporation, Webster Research Center, Webster, New York 14580

B. J. Welch, School of Chemical Technology, University of New South Wales, Australia

S. H. White, Department of Metallurgy, Imperial College, London SW7, England

D. E. Williams, Department of Metallurgy and Materials Science, Imperial College of Science and Technology, London SW7 2BP, England

Contents

Introductory Address

A. R. Ubbelohde

A provocative list of electrochemical themes has been prepared, and their discussion is likely to yield new inspiration in science and engineering, in more than one direction.

It is the privilege of the person opening the symposium to select those parts of the program for comment which appeal to him particularly, without having to confess ignorance about some of the other themes on the bill of fare. I don't want to take up this privilege, so I will start with some of my favorites, but will then go on to humble confessions.

Quite definitely, electro-organic chemistry is my favorite. I consider that the physical chemistry and in particular the bond-reaction chemistry of organic ions to be pregnant with all kinds of unexplored innovations. When an organic ion is discharged at an electrode, we all know the rather elementary formulation of Kolbe reactions leading mainly to radical pairing. But in fact, electrode discharge is one of the most versatile means for producing an enormous variety of organic free radicals. The behavior of these radicals immediately before or immediately after the ion is discharged at an electrode must in principle permit various kinds of bond rearrangements, if the molecular structure is at all complex. Furthermore, although the electrode material imposes a constraint on the course of the reactions, one should be able to present other reaction alternatives to the free radical, which may prefer these to Kolbe pairing with a neighbor. For example, it should be interesting to "flood" an electrode with a gas capable of reacting with free radicals, such as ethylene, acetylene, oxygen, or a halogen. This could become a route to reaction products quite different from the usual Kolbe-paired radicals.

Quite possibly, some of the more interesting possibilities can only be realized at elevated temperatures, above the boiling point of electrolytes in conventional solutions. In this sense, developments of organic electrolysis in molten salts, particularly organic salts of low melting point such as the groups of substances we have recently been studying in my laboratories, considerably extend as well as raise the range of working temperatures available.

I should add that in addition to yielding a diversity of organic products after bond rearrangement, some free radicals produced by the discharge of organic cations will amalgamate with mercury, forming the so-called "ammonium amalgams" in which they dissolve and are stabilized after losing an electron by transfer to the electron band of the liquid metal.

These fascinating avenues for further inquiry depend in part on technological innovations in the study of what happens at an electrode, i.e., of *electrode processes*. Our ignorance about the detailed physical chemistry of electrode processes is partly due to our comparative ignorance of solid-state reaction chemistry. The surface of most conventional electrodes is (geometrically) very restricted, which means that only those reactions can be conveniently studied for which the electrode surface is "self-clearing." Quite possibly use of fluidized-bed electrodes, or dispersed liquid electrodes, will offer much more favor-

---

A. R. Ubbelohde, C. B. E., F. R. S. • Chemical Engineering Department, Imperial College, London, England.

able conditions for investigating and indeed for using novel electrode processes on the large scale.

This brings me back to two other principal themes due to be discussed near the beginning of this symposium. The subject of electrocatalysis is full of intriguing possibilities since: (1) The electrode surface can be continually renewed by exchange with the electrolyte. (2) More than a single reactive molecular species can be brought up to the electrode/electrolyte interface during electrolysis. The activating regime prevailing there during electrolysis may well catalyze chemical reactions between molecules impinging on it, because of its "freshness," being kept free from poisons as it were.

Some years ago I invented a special case of "electrocatalysis" in which catalytic reactions were studied at the back of a thin septum electrode during active electrolysis on its other face. "Septum hydrogenation," which was the principal reaction I studied, did show certain interesting possibilities. Again these might be more rewarding at higher temperatures, with molten salt electrolytes, than with the aqueous electrolytes I used.

Electrode processes in relation to solid-state reactions are in my view also very much wrapped up with electrochemical problems of *energy conversion*, at any rate from the particular aspect of storage of electrical energy in a secondary cell. Presumably, the reason the lead accumulator has had such a long and successful history is because the solid-state reactions at the electrolyte/electrode interface can occur in both charge and discharge directions, with little or no irreversible hindrances. But the heavy weight per unit charge of this secondary cell militates greatly against its wider uses. There *must* be other electrode/electrolyte interface reactions less objectionable from this point of view; quite possibly, working at somewhat higher temperatures using molten salts will offer possibilities of secondary cells not available with solutions of electrolytes.

Obviously, like every physical chemist who has done research on cooperative hydrogen bonds in solids, I am tempted to stick my neck out concerning *bioelectrochemistry*. But perhaps I had better not venture too far except to wish well for the discussion of this very provocative theme.

I also note with pleasure substantial themes of longstanding importance, such as corrosion, electrowinning of aluminum and of other metals, and mineral processing. I may just mention the very recent founding of a new Kodak Chair of Interface Science at Imperial College in my department. All kinds of electrochemical problems are basically interface problems, and we can confidently have hope for progress in this direction too for the years ahead.

Finally, in wishing well for the days of discussion ahead you may desire an electrochemical comment even about the symposium dinner. Unfortunately I cannot be present at this so I will publish my contribution beforehand. I may term it the Ubbelohde electrochemical couple, measured by our lips when we are drinking wine or beer. If you think of it, when you bring a glass to your lips you cannot help forming an electrochemical couple; indeed, wine or beer tastes quite different from cups of conducting materials such as gold, silver, and pewter compared with nonconductors such as glass or porcelain. My theory requires that the electrochemical couple is different in each case. Note that I will not classify milk or tea as electrolytes—although perhaps I must compromise over Coca-Cola!

Chairman's Opening Remarks

Douglas Inman

On a practical note, let me first welcome you to Imperial College, and apologize for the organizational difficulties which during the last few months have made your successful arrival here as difficult as possible!

I wish to say very little at this stage as, certainly, at least with regard to electrochemical science, it is all going to be said during the course of our proceedings. The second report to the Club of Rome, which has just appeared, is entitled: "Mankind at the Turning Point," and it is of course our contention that electrochemical science will make a major and necessary contribution to man's existence on earth in the years which lie ahead. Although electrochemists are often more concerned with interphases rather than interfaces, the year 1975 is a well-defined landmark of the second type as it is now 30 years after the finish of World War II and the year in which Professor Bockris obtained his Ph.D., whereas 30 years from now we shall be 5 years into the much-heralded 21st century.

I can distinguish four main levels of technology. These are the perhaps illusory, blissful state of no technology, the state of labor-intensive intermediate technology which is perhaps most suitable for some of the undeveloped regions of the world, the state of Western technology from the early 20th century to the present day which we should be busily exporting to the underdeveloped nations to allow them to expand economically, and advanced technology which will be necessary in the looped spaceship Earth economy which the developed nations will have to pursue if they and indeed the rest of the world are to survive. This advanced technology, which will necessarily have to concentrate less on throughput and more on a materials- and energy-conserving and pollution-minimizing economy, will without doubt contain a large contribution of the electrochemical science which we are to discuss during the next four days.

Douglas Inman • Nuffield Research Group, Imperial College of Science, London SW7, England.

Symposium Dinner Speech*

Harold Egan

Prof. Bockris, Mrs. Bockris, ladies and gentlemen, I would like first of all, if it is not too late, to say welcome to Britain, welcome to London, welcome to Imperial College; welcome back to Imperial College, with a special welcome back to Prof. John Bockris and to Mrs. Lili Bockris and all other guests of the evening. Our subject tonight is to enjoy a pleasant, informal evening against a background of electrochemistry over the past 30 years and the next 30 years. Let me say straightaway that my own contribution has been to the former era and is, like the slide of an earlier speaker, slightly out of focus. I suppose I am a sort of a paleoelectrochemist.

I would like to take this opportunity to express thanks on behalf of all of us to a number of individuals: to the International Committee which under the chairmanship of Douglas Inman organized the symposium, to Harry Bloom (unfortunately unwell and unable to come to London) who has also acted as chairman of the Australian Subcommittee, to Douglas Mackenzie (who had the original idea for the whole symposium), and to Alex Despić, Ian Menzies, my old friend Prof. Brian Conway, and to Felix Gutmann, who has been honorary secretary both of the Organizing Committee and the Australian Subcommittee. We should not forget other members of the subcommittee: Dr. Hunter, Dr. Welch, and Dr. Edmund Potter. To the whole Committee, to the Australian Subcommittee, to Harry Bloom, Edmund Potter, Felix Gutmann, and to Douglas Inman, we are especially grateful. And we are also most grateful to Imperial College for having us here in London and to everyone behind the scenes in so many activities.

It is now my pleasure to convey to the assembly a message of greeting which has been received from Prof. A. N. Frumkin of the Institute of Electrochemistry, Soviet Academy of Sciences in Moscow (p. 7).

Like Eddie Potter, I first met John Bockris in 1943. We sat the same London External BSc examination that summer, in the old Imperial Institute building about 100 m to the southeast of where we are now. Like Eddie Potter, I have no record in my 1943 pocket diary of our first encounter. But there is a note the following year, under September 7, 1944, which says simply "JOMB." And there is an abrupt change in my diary at that time. On September 6 it said (among other things) "Jam, kippers, bread, shirts, tie, tyres, inner tubes." On September 7 it said "JOMB," and on September 8 "Phone Bockris." It is certainly not for me tonight to anticipate the words of appreciation which Prof. Conway will be delivering tomorrow. However, someone else is with us tonight, Dr. T. P.

*The symposium dinner was held at Imperial College Union on the evening of April 5. The chair was taken by Dr. Harold Egan, who was introduced by Dr. Inman as one of Prof. Bockris' earliest coworkers in 1946–1948 and who is now Government Chemist of the United Kingdom. During the course of the dinner Dr. Egan proposed the health of HM The Queen (coupled with the Heads of State of all of those countries represented at the symposium), to Prof. Bockris and to Mrs. Lili Bockris, and all the other guests present. The dinner closed with presentations to Prof. Bockris by Dr. Egan of greetings signed by all present, a special commemorative cartoon drawn specially for the occasion by Prof. H. Keyzer, and a presentation by Prof. A. Despić of an inscribed souvenir copper electrode. Following the dinner Dr. Inman showed a short film taken on laboratory location in Philadelphia in 1958. [Ed.]

Harold Egan • U. K. Government Chemist, London SE1, England.

Hoar, Reader in Chemical Metallurgy at the University of Cambridge, who has been in electrochemistry longer than everyone else in this room. Sam Hoar is well known in Britain for, among other things, the report of the official Committee on Corrosion and Protection (of which he was the chairman) which was published in 1971. This report made the first realistic estimate of the cost of metallic corrosion in the United Kingdom, on the basis of the avoidable (as opposed to total) costs to the community. But his contribution to electrochemistry goes back to the early 1930s.

Message of Greeting

A. N. Frumkin

I think that it was a good idea to link the symposium "Marking the Contributions of John O'M. Bockris" with the discussion of the past 30 and the next 30 years of electrochemistry. Indeed, there is hardly any important branch of electrochemistry to which Prof. Bockris has not made a fundamental contribution, whether it be the adsorption theory of organic substances, electric double-layer structure, passivity, ellipsometry, electrocrystallization of metals, growth of dendrites, kinetics of oxygen reduction and iron dissolution, hydrogen embrittlement, structure of molten salts, electrocatalysis, bioelectrochemistry, and last but not least the quantum theory of the elementary act. So much for Prof. Bockris's activity in the past 30 years. As regards the next 30 years of electrochemistry, I do not think there exists another scientist who has done so much for the propaganda of its future possibilities, the more so in connection with dramatic problems of energy conversion and cleaner environments. Prof. Bockris said once that electrochemistry is "an underdeveloped science." This means that electrochemistry could yield much more than it does at present, if more funds were invested in its development. I think that Prof. Bockris is quite right. I would also add that electrochemistry is an underestimated science and that what electrochemists have achieved has not been duly appreciated. I hope that this symposium will remedy the situation.

I would like to add another item to the list of Prof. Bockris's merits. I remember very well that during the sad period of the Cold War, I received a letter from London from the British–Soviet Friendship Society, which said that a young British electrochemist, John Bockris, wished to establish scientific contacts with us. Bockris's initiative was very helpful in restoring the relations between Soviet and Western electrochemists broken soon after the war. True enough, it hasn't always been plain sailing between Bockris and Soviet electrochemists. Sometimes there have been difficulties, but they have been only lovers' quarrels. The division of electrochemistry into two parts, "ionics" and "electrodics," proposed by Prof. Bockris, is becoming very popular everywhere in our country as well. I must say that in Russian these terms are even nicer: "ionica" and "electrodica." These sound like sweet feminine names, especially "electrodica," quite like Everedica (the Russian pronunciation of Euridice). In conclusion, I wish great success to this sensational symposium and express my hope that in 20 years a symposium "The Past Fifty and the Next Fifty Years of Electrochemistry," marking the contributions of John O'M. Bockris and associates will be held. The participants will come to the meeting in fuel-cell-driven electromobiles, as by that time the adverse effects of the great Nernstian hiatus will have been at last overcome.

A. N. Frumkin • Institute of Electrochemistry, Academy of Sciences of the USSR, Moscow.

1

The Problems of Energy Conversion— Experience of the Past

A. R. Despić

1. Dynamics of Energy Conversion

The use of energy is one of the basic characteristics of the development of mankind. The per capita consumption of energy is a good measure of the level of physical civilization reached by society, and the trend toward increased consumption still shows no tendency to level off. Thus, according to some estimates,[1] during the last million years mankind has used a total of about 13 Q* of energy, most of it (10.5 Q) coming from the combustion of wood and similar materials. The present annual rate of consumption amounts to 0.15 Q, 90% of which is obtained by the combustion of fossil fuels. The projected consumption around the year 2050 should be about 5 Q per annum.

It is interesting to note that from the very dawn of civilization till some 30 years ago there was only one basic source of energy used on the earth and that was the sun. All other forms have been derived from it through more or less involved processes of energy conversion, mapped schematically in the energy conversion chart shown in Figure 1. Thus, the vital energy on which all life is based is formed in a complex biochemical conversion process in plants, in which materials brought to a higher state of organization (lower entropy) and free energy with the help of photochemical synthesis eventually degrade to the materials from which they are formed, i.e., CO_2 and H_2O. Mechanical energy, needed for the propulsion of ships, has been obtained by the conversion of solar energy into that of the wind. In this way, the sun helped man to discover the globe and establish worldwide communication. Water circulation caused by the absorption of solar energy has long been used for creating mechanical power in mills and other mechanical devices, and ever since the mastering of electrodynamics, it has been a channel for the flow of solar energy to all types of energy uses for human needs.

By far the largest store of solar energy has, however, been relatively recently introduced into massive use. Those are the fossil fuels created from billions of years of accumulation of residues of life, with some help from geothermal energy in shaping up their present form. Although in the year 1850 fossil fuels provided only 10% of energy, by 1950 they became the main secondary source of energy on which modern civilization is operating.

It was not by chance that carbon has emerged as the basic constituent of fossil fuels and the most important element for storage of solar energy in high-energy-content materials on the earth. This is due to the particular qualities of its electronic shell—the number of electrons per atom which are ready for interaction with other atoms as well as the selection of orbitals at hand (s, p, sp, sp^2, sp^3) to suit the particular situation. That helped the development of the complex path of biochemical photosynthesis which leads to

---

*1 Q of energy equals 10^{18} Btu and can be obtained by combustion of 38.10^9 tons of coal.

A. R. Despić • Faculty of Technology and Metallurgy, University of Beograd, and the Institute of Electrochemistry ICTM, Beograd, Yugoslavia.

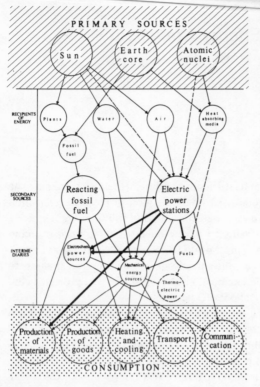

PRIMARY SOURCES

Sun   Earth core   Atomic nuclei

RECIPIENTS OF ENERGY   Plants   Water   Air   Heat absorbing media

Fossil fuel

SECONDARY SOURCES   Reacting fossil fuel   Electric power stations

INTERME- DIARIES   Electrochemical power sources   Fuels

Mechanical energy sources

Thermo- electric power

Production of materials   Production of goods   Heating and cooling   Transport   Communi- cation

CONSUMPTION

Fig. 1. Energy conversion chart— energy sources and lines of flow. Thick line—conversion based on electrochemical process; dashed line—recently developed process still in experimental stage or in very limited use.

splitting carbon from oxygen, the reverse process returning all the energy, so far mostly in the form of combustion heat.

The transformation of liquid or solid carbon-containing material in that process into gaseous products facilitated the waste-disposal problems of both the living world and of the technological world created by man.

If humanity had to go on relying on the energy of the sun as the only primary source of satisfying its growing needs, it is quite likely that carbon would remain the most important energy transmitter.

However, the reserves of fossil fuels are found to be limited, i.e., expendable, at the present and projected consumption rate within a very short period on the order of 100–200 years. Thus, the reviews of fossil fuels reserves of the United States[2,3] show that a total of 130 Q could hopefully be available for the needs of that country. However, only 28 Q are likely to be available at a reasonable cost (consisting of 21 Q in coal, 6 Q in crude oil and shale oil, and 1 Q in natural gas). Considering the estimated trend of consumption, this amount of energy is likely to be spent by the year 2080. The situation is not unlike that elsewhere in the world. The estimates in the U.S.S.R. range between an optimistic 150-year limit and a pessimistic one of only 70 years.

After this supply is exhausted, mankind would have to turn to the primary energy sources only, the remainder of the energy conversion scheme being only stages, with larger or smaller storage capacity, in the conversion process from source to sink.

It can easily be shown that lines of conversion of solar energy are of too low a cross section to allow for all of the energy to be taken from that source only. The strongest line, photochemical biosynthesis, would require that a significant portion of the land be devoted solely to this use. The total energy delivery from the sun amounts to 6.6×10^{-6} Q/km^2 per year. Hence, the yearly consumption of 5 Q attained by the year 2050 could be covered by harvesting $750,000 \text{ km}^2$ at 100% collection efficiency. Of course, a much larger territory is needed in reality as the collection efficiency is bound to be much lower.

Thirty years ago, man started mastering the use of another basic energy source—the atomic nucleus—and has already reached a level at which this process of acquisition of energy in a basic form could favorably compete with the tedious collection of solar energy. This extended the energy conversion pattern (Figure 1) and made it even more dynamic, some lines receiving increasing parts of the total energy flow and others decreasing in importance with the likelihood of being eventually eliminated for all practical purposes. Besides the obvious reason of exhaustion of some energy transmitters essential for the maintenance of a line, as in the case of fossil fuels, the economy of the whole sequence of conversions along a path, reflected in the price of energy at the point of consumption, is the major factor affecting these dynamics. It is illustrative that heat should be obtainable in the near future from nuclear reactors by using practically inexhaustible supplies of deuterium in sea water at a cost of $0.001/MBtu, while that from reactors running on a 0.5% ore of uranium or thorium would cost as little as $0.003/MBtu. This should be compared with $0.20/MBtu for coal or even $0.40/MBtu for oil.

Hence, nuclear energy is likely not only to close the gap arising from the exhaustion of fossil fuels but also to take over increasing parts of energy supply even before that time. Geothermal energy has found little use in the past. Its main contribution was in modifying fossil sediments, increasing their energy content. It is difficult to forecast at this stage the extent to which its role will increase in the future. If it does, it will be just an alternative to atomic power stations supplying high-quality heat.

It is conceivable, of course, that man is unaware of the existence of other primary energy sources in nature and that some entirely new form of energy may become available in the future. However, the likelihood of new discoveries in this direction is considerably less now than it was even at the time H. Becquerel noticed radioactivity. It is highly unlikely that with all the scientific potential existing at present in the world a similar phenomenon of fundamental importance could have gone unnoticed. The change of the basic source of energy is bound to strongly affect the energy conversion lines all the way down to the points of consumption. In particular, production of materials, heating, and transport will undergo some major changes.

At first sight, it would appear that the likely initial step in nuclear heat conversion would be into electrical energy and that the energy would flow in that form all the way to most points of consumption. This concept is being questioned lately on several grounds: (1) The losses of energy in the process of transport between the source and the consumer are by no means negligible or, at least, they can be made smaller if some other concept is applied. (2) The rate of consumption fluctuates so much that the needs at the peaks of demand can be several-fold larger than the average demand; having the power installed to match the peak demands would be irrational and uneconomical. (3) Supplying electricity from ground stations to mobile consumers is rather impractical.

Hence, there are good reasons to believe that a chemical energy storage in the form of a "fuel"* should be interpolated at different stages along the conversion paths:

1. Heat from nuclear energy can be used directly for the production of such a fuel.
2. Electrical energy obtained in nuclear power plants can also be stored in the form of high-energy-content materials capable of undergoing a stoichiometrically reversible electrochemical reaction.

---

*It should be noted that the word "fuel" is taken here conditionally. Thus, liquid air can be considered as a fuel if it is used to provide a thermoelectric power generator with frigories (negative heat).

3. When stored in such a way, the energy can also be transported in that form to the point of consumption. This can be more practical than direct supply of electrical energy, especially when mobile consumers are concerned. The latter are presently using 20% of the total energy flux with good prospects of their use increasing to 25%.

Since at least one step along the energy conversion path should thus involve the transformation of electrical energy into chemical energy or vice versa, of all chemical processes this gives particular importance to the electrochemical ones, equal to the importance of processes converting heat into electricity or electricity into mechanical work.

Electrochemical lines are stressed in the energy conversion chart. The likelihood of their use depends (1) on the availability of the energy transmitter upon the use of which a particular line of electrochemical energy conversion is based, (2) on the inherent qualities of the transmitter (energy capacity, reactivity, etc.), and (3) on the state of development of the electrochemical processes involving the transmitter.

Electrochemical conversion has so far been popularly associated with primary and secondary chemical energy sources (batteries and accumulators). No doubt any consideration of future development should start by analyzing those experiences gained from past development in that field. However, one should not overlook the fact that this line of energy conversion has so far carried a negligible portion of the total flux of electrochemical energy.

If lead-acid batteries used as automobile starters are eliminated from consideration (since energy conversion there is of secondary importance), the rest of the battery world covers demands using comparatively very small amounts of energy (communication, reserve power, minor portion of traction, etc.).

Far larger and indeed significant amounts of energy have already been flowing for quite some time through electrochemical processes of production of materials. Thus, chlorine and caustics occupy a significant portion of the volume of production of the chemical industry. If the recent figure of the total annual world production of chlorine[9] of 25×10^6 tons is taken, one can calculate that this is equivalent to converting 87.5×10^9 kWh of electrical energy into electrochemically active materials of a very high free-energy content. Metals obtained by electrowinning represent a significant portion of the production of metallurgy. If only aluminum is taken into account, the annual world production of 12×10^6 tons[10] converts another 161×10^9 kWh of electrical energy into the chemical one.

Hence, when considering the future in which a much larger part of the total energy flux will flow through electrochemical lines than is the case today, one should review *ab initio* the possible main transmitters of energy on which these lines are likely to operate.

One could first argue that if electrochemical processes become an essential part of future energy conversion and if the choice of the energy transmitter is not restricted by considerations prevailing in the past (solar energy as the only source), the classical transmitter—carbon—no longer represents the most suitable element:

1. A number of elements in the upper left corner of the periodic chart of elements can store larger amounts of energy per unit of weight, recoverable by combustion of electrochemical oxidation, than carbon, as shown in Figure 2.

2. Inasmuch as it could be useful when the individual consumer is concerned, the gaseous nature of the reaction product of carbon oxidation may prove a serious problem on an increased level of energy consumption, as pointed out by Bockris,[4] Plass,[5] and others.

H			
H_2O - 33.000 Wh/kg 1.23 V			
Li	**Be**	**B**	**C**
Li_2O -11.950 Wh/kg 3.09 V $LiOH \cdot H_2O$ -27.600 Wh/kg 7.15 V	BeO -18.000 Wh/kg 3.01 V $BeO \cdot H_2O$ -26.800 Wh/kg 4.83V	B_2O_3 -15.200 Wh/kg 2.04 V $B_2O_3 3H_2O$ -24.800 Wh/kg 3.21 V	CO_2 - 9.140 Wh/kg 1.02 V
Na	**Mg**	**Al**	**Si**
Na_2O - 2.270 Wh/kg 1.95 V $NaOH_{aq}$ - 5.060 Wh/kg 4.34 V	MgO -6.530 Wh/kg 2.95 V $Mg(OH)_2$ -9.550 Wh/kg 4.17V	Al_2O_3- **8.140 Wh/kg** **2.72v** $Al_2O_3 3H_2O$ **3.82v** **11.800 Wh/kg**	SiO_2 -7.910 Wh/kg 2.06 V
K	**Ca**	**Ga**	**Ge**
K_2O - 1.285 Wh/kg 1.87 V KOH_{aq}- 1.715 Wh/kg 4.81 V	CaO -4.210 Wh/kg 3.14 V $Ca(OH)_2$ -6.260 Wh/kg 4.48V	Ga_2O_3 - 2.160 Wh/kg 1.87 V $Ga(OH)_3$-3.335 Wh/kg 2.89 V	GeO_2 -2.050 Wh/kg 1.39 V $GeO_2 \cdot H_2O$-3.185 Wh/kg 2.16 V

Fig. 2. Energy of combustion of elements in the upper left corner of the periodic chart of elements.

3. Finally, electrochemical reactions of substances involving carbon have so far proven thermodynamically highly irreversible. Hence, too high a portion of energy contained in them cannot be obtained in the form of electrical energy.

This has oriented attention towards other energy transmitters and, in particular, to hydrogen which is considered (Figure 2) to be at the top of the list as far as energy content is concerned. It seems to have many other advantages as well: (1) the raw material from which it is obtained (water) is readily available everywhere except deserts; (2) it has remarkable fluidity and hence can be transported through pipes at a lower loss of energy than electricity; and (3) its electrochemical reaction is fairly reversible and has the prospect of being made even more so with the further development of electrocatalysis.

All this led to the development of the concept of hydrogen economy. One of the most active advocates of this, J. O'M. Bockris, showed in many of his writings[4,6,7] how the discoveries in the field of electrochemistry in the near past laid good foundations for the application of that concept in the future, and initiated a detailed analysis of its perspectives.[8]

However, one should not overlook at least two serious shortcomings of hydrogen as a competitor in energy transmission: Its handling, away from pipelines, is cumbersome and not without hazards; its storage represents a considerable problem.

So far, four ways of storing hydrogen are envisaged:

1. Low-pressure storage in enormous reservoirs made available underground by the exhaustion of oil or natural gas
2. High-pressure storage in tanks
3. Liquefaction and cryogenic storage
4. Reversible chemical bonding to some carrier (hydrides)

None of these ways is entirely satisfactory. The first can be used only in large energy production and distribution networks; the second involves weight of tanks as an unavoidable part of the weight content of the system and increases the price; the third requires a rather expensive and energy-consuming cooling process; and the fourth has the disadvantage of the same kind as the second unless the other constitutent is a high-energy-capacity fuel as well.

These disadvantages leave room for investigating the potentialities of other elements on the list. Lithium, magnesium, and aluminum seem to be the promising ones.

Hence, one could visualize economies based on the use of these metals as well, as energy storers, or at least take them into account in forecasting the directions which future energy conversion may take.

One should also note at this point that electrochemists tend to overlook the competition electrochemical conversion may have in the future from new energy conversion lines. Thus, thermoelectricity seems to have obtained a significant chance with a new alloy being born in Sweden of exceptional thermoelectric coefficient and with the idea to use liquid air as the energy transmitter, i.e., to use frigories rather than calories in the conversion of atomic heat into electricity. Also, internal combustion engines can extend their usefulness into the unlimited future if they are adapted to new synthetic fuels (e.g., hydrogen).

Hence, electrochemistry is likely to face serious challenges on the way to conquering more room in the field of energy conversion rather than leisurely occupying it when left by the withdrawal of classical ways of conversion via fossil fuels.

The above considerations seem to set up the ground for establishing the areas worth further consideration, the history of which should be reviewed and analyzed in order to draw some conclusions which could help establish their potential to fill future needs. These are:

1. Chemical energy sources, in the usual sense of the word, which have been successful in satisfying past and present demands placed upon the limited part of the field of energy conversion which has so far been covered by electrochemical processes.

2. Substances which are potential energy transmitters but so far have practically not been used for this purpose.

3. Recent developments in understanding the fundamentals of processes involved in electrochemical energy conversion which could be of substantial help for solving many serious problems involved in the large-scale application of electrochemical systems in the field of energy conversion.

2. The Morals of the Story

2.1. The Chemical Energy Sources

If Newton had discovered the laws of electrochemistry instead of those of mechanics, it is quite likely that today's traffic would be running on electricity produced by electrochemical storers of high specific energy rather than on internal combustion engines powered by gasoline. Electrochemistry was late in entering the history of science. It took a good deal of learning about both electricity and inorganic chemistry before the two could start being systematically combined. This, as well as the relative delay in the application of knowledge developed in branches of physical chemistry other than chemical thermodynamics, made electrochemistry an underdeveloped science[11] up to the present time, with all the natural consequences of such a situation reflecting on its application in the field of energy conversion. The relatively young age of that science also caused, as a result, too little interaction between its "pure" and "applied" side: between the fundamentalist and the practitioner, the scientist and the inventor, characteristic of the relationship between, e.g., early Carnot followers and the steam engine engineers of the time.

It is hardly an exaggeration to say that those people inventing the most important electrochemical devices and technologies and developing them to a very high level of productivity—mostly mechanical and electrical engineers, used barely anything more of the fundamental understanding of electrochemistry than Faraday's laws. All optimization achieved was based on professional common-sense considerations, as, e.g., bringing electrodes closer together or using concentrated electrolyte to reduce cell resistance, as the basic parameter which they could recognize.

Thus, the most important electrochemical technologies (lead–acid batteries, electrolysis of brine, electrowinning of aluminum, etc.) have remained practically unchanged for 100 years as far as their electrochemistry goes, and any improvements that were introduced were made purely on the engineering side.

To be truthful, until 30 years ago there was not much fundamental electrochemistry could offer to a practitioner as theoretical guidance. The number of great names electrochemistry could cite prior to World War II have struggled with the very essentials of what is governing events at electrified interfaces. Hence, their results were of little help to any practical problem. The most a learned engineer could gain from reading an electrochemistry textbook of the time was to become aware that reversible potential is not the only criterion indicating if an electrochemical process will go and at what cell voltage.

This situation also obtained throughout most of the 30 years of development we are trying to analyze. It is true that this virgin land for application of the achievements, which fundamental electrochemistry had accumulated, is already under heavy attack, bringing positive results. However, the effects on the bulk of traditional electrochemical technologies have so far been so small that the phenomena of our interest can well be analyzed as though there had been no such influence.

What are those phenomena?

2.1.1. The Rule of the Leclanché Cell

Ever since 1800, when Volta found out[12] that electricity could be drawn from a pile, like one shown in Figure 3, in much larger quantities than from a Leyden jar, an enormous spectrum of combinations constituting electrochemical systems awaited the inventors. There is little doubt that many more have indeed been tried than the history of science has recorded.

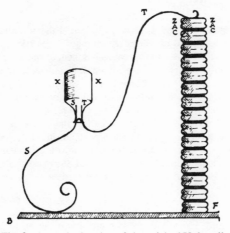

Fig. 3. An early drawing of the original Volta pile.

The Leclanché cell, designed by G. Leclanché in France as early as 1865[13] and developed only 20 years later into the weird package of manganese dioxide wrapped up around a graphite rod, wetted, and encased into a zinc can, overwhelmed the use of any other system to such an extent that, according to present-day statistics on dry-battery production, all the others could well be neglected without much damage to that line of energy conversion.

The use of dry batteries exhibited a considerable rise in recent years with the introduction of portable electronics, and this rise is likely to continue without a tendency to level off. So far, however, the Leclanché cell has had little competition, although new sources (and especially the small portable secondary batteries) are appearing in increasing number. According to latest figures,[14] out of 10 billion pieces of dry batteries produced in the world in 1974, 9 billion are the Leclanché cells, consuming 200,000 tons of zinc and 560,000 tons of MnO_2. It appears that this wasteful battery (using 12–15% of zinc only) can be driven off the market only by a lack of raw materials (exhaustion forecast: 80 years and 94 years, respectively), or by the steady increase in their price, something which has already been observed for some time.

In the past 30 years, the mass-produced version of the cell has experienced little change. That which has occurred has been in the direction of increasing reliability and extending the range of conditions under which they can operate. This is achieved by better sealing (preventing spilling out of the electrolyte and increasing internal resistance due to drying out) and by better metallurgy of the zinc can, as well as by a better control of its corrosion.

The voltage of the Leclanché cell at the very beginning of discharge, 1.5 V, represents in itself a rather good portion of the theoretical voltage of the $Zn–MnO_2$ couple. However, it decreases rather sharply with operation. Also, inadequate use of the active substances makes the overall energy efficiency very poor.

In the field of primary batteries, there are three other combinations that should be considered since they have taken over a part of the market in recent years.

These are the mercury (Ruben–Mallory) cell, the alkaline cell, and the zinc-air cell. It is striking to note, and hardly by chance, that zinc has remained the only electrochemically active material used for negative plates of all those primary sources.

2.1.2. The Phenomenon of the Lead–Acid Battery

In 1859 G. Planté wrote[15]:

> In a note already presented to the Academy, I indicated the advantages of substituting platinum with lead for the use of secondary currents in electric telegraphy, proposed recently by M. Jacobi. Special studies I have undertaken, of those currents revealed that the counter electromotive force, obtained from lead immersed in acidified water, is about $2\frac{1}{2}$ times larger than that obtained from electrodes of platinized platinum, and about $6\frac{1}{2}$ times larger than that obtained from ordinary platinum. This electromotive force, although produced by plates of one and the same metal, is also very much superior to that of the elements of Grove or Bunsen, because of the large affinity of lead peroxide for hydrogen. . . . This secondary pile is very easy to construct . . . taking lead in sufficiently thin foils, one can place a very large surface area into a small space. The 9 elements I constructed, are contained in a square box of 36 cm each side; . . . it can be preserved in a physical cabinet always charged and ready to serve . . . giving potent discharges of dynamic electricity.

With this invention and Faure's improvement made as early as 1881, the common car battery was born in more or less its present form.

The production of lead-acid batteries quite closely followed the increase of production in the automotive industry, reaching a volume of 100 million pieces worth 1 billion dollars in 1974. An estimate is made[14] that with the 300 million lead-acid batteries that are presently active in the world, the total power contained in this source equals the total power of all electric power stations. They are, however, not used as an energy conversion line but are idle most of the time, waiting for moments of engine starting. The present production consumes about 200 million tons of lead per annum, threatening to exhaust the world reserves in 40–65 years.

The lead-acid battery has traditionally been used for a limited amount of electrically powered traction as well. The classical traction battery, however, has too low an energy density (about 20 Wh/kg) to pretend to power anything but locally employed slowly moving vehicles, typified by the English milk delivery van. Still, contrary to the dry battery, the lead-acid battery has undergone considerable improvement in recent years in energy density with little loss of other good qualities (large number of cycles and low price). Most of it is due to the replacement of the heavy hard-rubber casing by a considerably lighter one of polypropylene, but improvements in the qualities of plates should not be underrated.

Hence, the newest types of batteries intended for traction developed in some countries (Japan, U.S.A., England, West Germany) have more than doubled the energy density, while the city of Moscow is expected soon to have a fleet of 5000 taxis running on batteries with 50 Wh/kg energy density at 2 hr rate of discharge.

The energy efficiency of the lead-acid battery can be calculated by using the ratio of the voltages at the plateaus of discharge to that at the plateaus of charging, corrected for the faradaic yield of the two processes. When this is done, a figure of about 61% is obtained.

2.1.3. Competitive Capacity of Other Secondary Batteries

The field of application of secondary batteries, being considerably broader than that of the primary ones, did not give the lead-acid batteries the same degree of dominance attained by the Leclanché cell, particularly when true energy conversion applications are considered.

The alkaline iron–nickel oxide system selected by Edison around 1900, after a thorough and systematic review of all possible chemical systems which might be used as a storage battery, and developed in 1908 to a state in which it could find practical use, could never push the lead-acid battery out of the market. However, because of its exceptional sturdiness and resistance to abuse, as well as long life, it is still being produced in Poland, France, and U.S.S.R., in spite of a 4–5 times higher price per unit of electrical energy than that of the lead-acid battery. This is due mainly to the steel casing as well as to the addition of more expensive metals (50% copper in French batteries and 20% cadmium in Polish ones) to the negative electrode. Energy efficiency of the Edison battery, calculated in the same manner as above, comes to about 52%.

The close relatives of this battery are those in which iron has been replaced by cadmium and zinc. The first one was born about the same time as the Edison battery (W. Jungner and K. L. Berg, 1893–1909). Better performance and lower self-discharge rate could make it competitive in the same applications, in spite of considerably higher cost. However, this battery secured its position in the world only after a major improvement was recently made by the introduction of sintered electrodes. This made it the best battery presently available in terms of all qualities except price. Its energy efficiency comes to about 62%.

It is interesting to note that zinc, which found exclusive use in primary cells, has so far been unsuccessful in rechargeable batteries. In combination with the nickel oxide positive electrode, it could not survive in spite of considerable effort put into its development in Ireland. The somewhat better fate of the silver oxide–zinc battery developed relatively recently by H. G. André is due to one quality alone—exceptionally high power density. However, because of its prohibitively high price (with the price of silver increasing daily) and its very limited life due to poor rechargeability of the zinc electrode, its use is limited (military).

A similar experience with the zinc–air cells still in the development stage (combined with additional difficulties of the air electrode suffering from recharging) left the general impression in the battery world that little hope should be placed in this metal for a rechargeable negative plate. The attitude taken recently in quite a few large research establishments in the field range from the complete abandoning of further work to shifting it to the development of a hybrid between the primary and the secondary cell—the chargeable-plate battery.

2.1.4. The Phenomenon of the Fuel Cell

Astounding confidence has been given to an entirely new development—the hydrogen–oxygen fuel cell: The whole success of the Apollo mission relies upon this system. This was the coronation of an idea born almost as early as the oldest chemical energy sources and earlier than the lead–acid battery, back in 1839 by R. W. Grove. Although cherished by electrochemists, the idea remained without much support or strong devotion for more than a century. It was only after World War II that interest was revived and that workable solutions started appearing with Bacon leading the field in England in the 1950s.

The new solutions were due primarily to the fact that, for the first time, the ends of fundamental and applied electrochemistry started meeting. For a while, there has been considerable public and government support to research and development in the field because of increasing awareness of impending energy crises and already acute environmental problems with conventional energy conversion.

The system proved, however, to be much more difficult to master than any chemical energy source made before, and where good progress has been made, it rendered solutions whose economy could not match those already existing in energy conversion. The Apollo battery (and, quite likely, a similar Soviet development) was so far the only practical result of all the effort. That had a negative effect on public enthusiasm, and support reached its low several years ago. Now, the interest seems to be reviving again, especially with the prospects of introduction of the hydrogen economy, but the future prospects of this type of chemical energy source are outside the scope of this analysis.

The purpose of the above review was to create a basis for analyzing the relationship between the properties of a chemical energy source and its success as a product of practical use. Several such properties are of equal interest for understanding both the past and the future of chemical energy sources and hence the relevance of this analysis:

2.1.4.1. Simplicity. All the chemical energy sources that have found application in the past, even to a limited extent (such as the nickel–iron battery) are characterized by great simplicity. Even relatively simple cells, but with two electrodes in two different electrolytes, like the Daniel cell, have disappeared during the course of time from all but (for some queer reason) the electrochemistry textbooks. One of the main reasons for the slow development of fuel cells lies in the complexity of the system affecting its reliability. The problems of delicate balance at the three-phase boundary in the body of the electrodes, as well as those of heat and mass transfer and water balance, are probably even more challenging than those connected with valves, mobile parts, etc., since the latter have already been overcome in mechanical energy-conversion devices. These facts are to remain a permanent handicap of fuel cells compared to conventional energy sources and, to some extent, this even applies to hybrids using the air electrode.

Hence, those oriented toward the development of secondary chemical energy sources with two solid electrodes in an aqueous solution at ambient temperature rechargeable by electricity have an advantage, gained before the start, compared to researchers working on more complex systems—not only fuel cells but even simple cells require elevated temperature, exclusion of water, prevention of carbonization, etc.

2.1.4.2. Price. Seems to be the major factor in deciding on the mass use of a particular chemical energy source. This is certainly the main cause for the commercial success of the Leclanché cell vs. all other types of dry batteries. Moreover, their cost seemed to be so acceptable for whatever application they found that little stimulus seemed to exist for improving the use of active materials. Although those constitute a relatively small portion of the cost of production (e.g., 10–12% for zinc), there is sufficient room in that improvement to help the manufacturers reduce the price if endangered by the appearance of a competitive source.

So far, consumers have been interested primarily in the initial price of the source rather than in the cost of unit of energy obtained by the battery. Hence, the small rechargeable nickel–cadmium battery, although giving cheaper energy in the long run, has still not been able to make a significant penetration into the market of batteries for portable communication devices.

The lead–acid battery is another example where price has proven more important than life expectancy and reliability, since much higher qualities in the latter two features in, e.g., nickel–iron or nickel–cadmium battery, did not win over the market. One should recognize the fact, however, that the present-day lead–acid battery is of sufficiently long life for most applications (e.g., almost matching the active life of an average passenger car expected by the manufacturer). Hence, this battery is also an example of the right balance between several important properties.

2.1.4.3. Life Expectancy and Reliability. These seem to be closely related through the fact that the number of cycles attained by the average battery and the probability of failure are drawn from the same statistics. The nickel–iron cell has reached the ideal as far as those two qualities are concerned. Let this anecdote serve as an example:

The University of Belgrade had a stationary battery of nickel–iron cells built in the 1930s for the needs of the electrochemistry laboratory. At the outset of the war, the building was bombed and the battery in the cellar was buried under piles of debris and flooded till the end of the war. During the reconstruction of the building, it was dug out, rinsed of mud and dirt, refilled, and taken apart to be used as individual cells for student exercises. They survived another 30 years of mistreatment, and some are still in operation.

Still, for most terrestrial applications, battery failure often leads to unpleasant but hardly ever hazardous consequences. Hence, these qualities are not rated sufficiently high to compensate for high price. It is quite likely that an average car driver would not mind having to change his battery every six months if the price were, e.g., ten times lower. It is the corresponding decrease of reliability which would become a nuisance in that case. Hence, life expectancy alone is important only in combination with the price, and although long life eventually leads to cheap energy, at this moment the preference seems to be for shorter life but lower investment.

2.1.4.4. Power Density. This is of considerable importance in applications where there are large differences between peak and average power consumption. Although the lead–acid battery is quite a good performer in that respect as far as car-starting requirements are concerned, in some other cases the needs are so stringent that opportunity appeared for the development of the sintered nickel–cadmium battery. Even the silver oxide–zinc battery, which is at the negative end of examples with respect to the other two qualities cited above, was left some chance for survival.

However, as far as traction is concerned, a recent invention reduced the importance of this quality: the hybridization of an "energy" battery with a "power" battery, supplying average and peak power consumptions, respectively. Hence, this is becoming a sufficiently desirable but not essential quality as to allow for sacrifices in other more important ones.

2.1.4.5. Energy Density. This is of particular importance for energy conversion in transport. Considering the participation of transport in the total energy conversion (20–25%) and the ambitions for electrochemical conversions in that field, this is emerging as a quality of primary importance for the mass use of a battery in the future. So far, however, no chemical energy source presently in production is near to matching the needs, except where local traffic (delivery vehicles, commuter cars, etc.) is concerned.

It appears that a total neglect of that quality was present even among those early researchers in the field who, like Edison, did try to make a systematic estimate of the merits of different electrochemical systems. Hence, theoretical energy density, which seems to be the starting point of today's researchers, did not seem to have any effect upon that selection. It was later development which tried to correct this. As was seen, some impressive results had been achieved which, however, could not eliminate the inherent shortcomings of the selected systems. Thus, the silver oxide–zinc battery is so far the only one that has approached the lower limit of medium-energy-density region, 100 Wh/kg. However, one can roughly estimate the needs of an average car by assuming that the number of kilometers driven between chargings equals double the energy density in Wh/kg. Hence, the medium-density battery is indeed essential for successful penetration of the electric car into the market. Since the silver oxide–zinc battery is out of competition from any other respect, the field is as yet entirely open. This is where the lead–acid battery is likely to finally give way to other, more reasonable developments.

2.1.4.6. Energy Efficiency. This has obviously also not been of much concern since no relationship can be found between this quality and the present popularity of a chemical energy source. This is not unexpected, particularly when one has in mind that the efficiency of electrochemical conversion has, in general, been much higher than that of most other types of conversion. Thus, car drivers are presently satisfied with 8–12% efficiency of using the free energy of combustion of fossil fuels. Efficiency has its objective merits in the short period of history when the fossil fuels and other materials are approaching exhaustion and other energy sources (particularly atomic) have not yet been developed to cover the demands. In the situation of abundant energy[1] the merits of energy efficiency should be taken relative to all other merits in its economic value. Hence, it is difficult to say a priori what weight this consideration will have in estimating the overall competitive capacity of an energy source. It is only evident that this weight is bound to decrease with the decrease in the cost of electric energy. This, of course, should not discourage efforts to increase energy efficiency as this always brings very obvious returns. One should only match the expense of these efforts with the returns which they are likely to bring in the case of success.

2.2. Great Electrochemical Technologies

Electrolysis of brine and electrowinning of zinc and aluminum are by far the largest electrochemical technologies of present times in terms of volume of production and con-

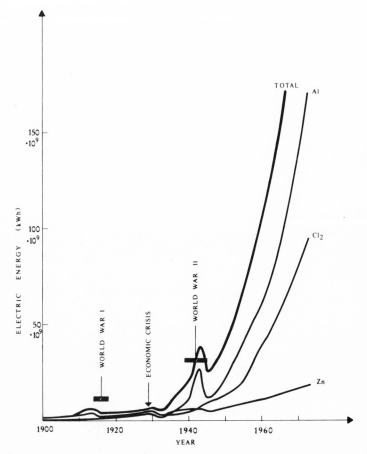

Fig. 4. Energy converted into chemical energy by the main electrochemical processes.

sumption of energy. The total amount of electrical energy converted into chemical energy in these three processes alone is seen in Figure 4. It appears that today they consume 3.6% of all the electrical energy produced in the world.

Although the knowledge that chlorine can be obtained by electrolysis has been present ever since 1800 (W. Cruikshank), the industrial era for that process started with the introduction of mass production of electricity by dynamo. Since the very beginning, two different processes competed with each other: the diaphragm-cell process and the mercury-cell process. This competition extended to present times because of a fairly good balance between the virtues and vices of both processes. For a while, the mercury process dominated because of lower overall price of product. However, with the increasing price of mercury and the growing concern with pollution by mercury in effluents found to accumulate eventually in animal life used as food, the situation is likely to change in the future.

Brine electrolysis experienced a major development in recent years. This is due to the introduction of the so-called "dimensionally stable anode" of titanium, catalyzed at present mostly by ruthenium oxide or a platinum–iridium alloy, as a replacement for graphite anodes which have been used up in the process. The achievement is illustrated in Figure 5; it shows significantly better polarization characteristics of titanium anodes.

Hence, not only constant adjustment and replacement of anodes is avoided, but also space and quantity of mercury are saved by increasing the operating current density by

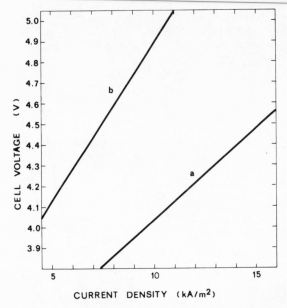

Fig. 5. Current–voltage relationship for mercury cell using anodes of graphite (a) and of catalyzed titanium (b) (310 g/liter NaCl; 70–80°C; distance between the electrodes, 3 mm).

about 30%. Moreover, energy consumption was also lowered from 3900 kWh to 3500 kWh/ton Cl_2.

The world need for chlorine and caustics seems to have a stable tendency of rise, attaining a present volume of 25×10^6 tons of chlorine and 28×10^6 tons of sodium hydroxide. When converted into energy used up in the process, the growth of this energy conversion line is seen (Figure 4) to attain presently the amount of 95 billion kWh per annum.

The energy efficiency of the process can be estimated if the energy consumed in the process is compared with the energy which can be obtained by reversing the process under ideally reversible conditions. The theoretical cell voltage for the process, obtained from the standard free-energy change of formation of NaCl and water from chlorine, hydrogen, and sodium hydroxide, is 2.3 V. The practically required cell voltage of the diaphragm cells is 4.0 V with an average faradaic efficiency of 92%; that of the mercury-cell process is 4.5 V with 96% efficiency. This yields energy efficiencies of 53 and 49%, respectively.

With the first industrial plant erected by Anaconda Co. in the United States as late as 1915, zinc became a relative newcomer in the electrolytic processes, fighting its way through different varieties of classical thermal processes. However, it has already succeeded in attaining a volume of production in 1973 of 5×10^6 tons and thus contributes 70% to total production of zinc.

In a modern electrolytic zinc plant, zinc is obtained at a rate of 300–1000 A/m², with 3.1–3.8 V of the cell voltage, and a faradaic efficiency varying between 85 and 90%. This implies an energy consumption of 3200–3300 kWh/ton.

Since the theoretical cell voltage of the zinc–oxygen couple is 2.0 V, the energy efficiency of that process calculated from those parameters is about 47.5–55%. Also, when the free energy of formation of zinc sulfate and water from zinc and sulfuric acid is taken as that obtainable by reversing the process, and amounting to 1640 kWh/ton, the conversion efficiency of 50–51% is obtained.

Aluminum is the most abundant metal at the surface of the earth, constituting 8.13% of the earth's crust. However, its recovery in the metallic state is of relatively recent origin (H. C. Oersted, 1825), compared to metals constituting the basis of the development of civilization (copper and iron), because of its high affinity to oxygen, i.e., the necessity to apply a potent reducing agent unavailable in early history. Even this could be done at first only with one aluminum-containing mineral—bauxite. Nevertheless, the recovery once being mastered, its use is increasing at the high pace of 10% per annum (Figure 6) to reach 12.7×10^6 tons in 1973.

Most of the major developments in the technology of electrowinning of aluminum took place more than 30 years ago. As seen in Figure 6, this resulted in a considerable decrease in specific energy consumption up till the outbreak of World War II. The lower rate of improvement since then has led to a present energy-consumption figure of 13,400 kWh/ton.

This specific energy, combined with a large volume of production, results in the participation of the process in the total electrochemical energy conversion of about 60%.

With an average cell voltage of 3.90 V, compared to recoverable cell voltage of 2.72 for the formation of Al_2O_3 at room temperature and a faradaic efficiency of 87%, the energy efficiency of the process amounts to 61% with the prospects of being further increased in future developments[16] to about 67%. When calculated on the basis of the recoverable free energy, amounting to 11,800 kWh/ton for the formation of the hydrated form of Al_2O_3 in aqueous solutions, the energy efficiency rises to 88%. The relatively high energy efficiency is due mostly to the fact that the process of electrowinning takes place at a very elevated temperature (955–980°C), which makes the reversible cell voltage rather low (1.14 V) and even leaves room for voltage losses to reach the theoretical cell voltage at room temperature.

What conclusions can be drawn from the development and present status of the great electrochemical technologies?

Electrochemistry has taken up that part of the chemical industry where substances of high chemical potential (with respect to products of reactions they could participate in)

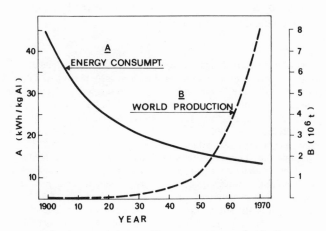

Fig. 6. Production of aluminum and the specific energy consumption in the process of recovery.

are produced. Thus, all the three products discussed above are at the far ends of the scale of electrode potentials. All of these substances can and have been produced chemically (Cl_2 by MnO_2 and Al by Na) or thermally (Zn from ZnO), but the electrochemical processes exhibited clear advantages.

All the three technologies are relatively simple, especially in the electrochemical stage, and run smoothly, requiring relatively little maintenance and hence low manpower per unit of production volume.

The conversion efficiency of the processes is at present relatively high compared to most energy-conversion processes, except those involving interconversion of mechanical and electrical energy. This is due to steady improvements mainly of an engineering nature, as shown in Figure 6 for aluminum. In the chlorine industry, the major development cited above increased the conversion efficiency as well. However, this is still far from satisfactory when we know that the interconversion of chemical and electrical energy can, in principle, be complete. Hence, considerable room is left for further improvements.*

Because of high chemical potential, all three substances should in principle be suitable for energy transmission in the energy conversion lines (reversible acceptance and release of energy), providing a high concentration of energy during storage. However, only zinc has been used so far in that capacity, dominating, as was shown earlier, in the field of the primary chemical energy sources. Having in mind that zinc is primarily produced today by the electrochemical process, those sources differ in fact from the secondary ones only in that the input and output of electrical energy are spatially separated.

Chlorine has not been used as an energy transmitter, although some recent attempts have been made (the lithium–chlorine battery[20]). In fact, oxygen is so much better than chlorine as a reactant on the positive side of the energy storer, in any respect but electrochemical reactivity, that there is little wonder why chlorine is left somewhat aside.

The case of aluminum is different. It has many qualities superior to those of zinc (availability, energy density, price) so that it is a real wonder that more effort has not been put into its use in chemical energy conversion. This is an example of how the lack of interaction between fundamental and applied science, and the tendency of practical engineers to take materials as they are, led the development away from optimal direction and made aluminum exclusively a construction material.

It is our intention to show, later on in this text, how the knowledge accumulated in electrochemistry could help change this situation.

---

*One such improvement can be offered as an example in support of this statement, although it has not yet found its place in industrial production. In the presently used mercury-cell process of brine electrolysis sodium amalgam is decomposed in separate decomposers where pieces of graphite are left to float on mercury in contact with sodium hydroxide solutions. Because of relatively low hydrogen overvoltage compared to that at mercury, this is evolved on graphite corroding sodium from the amalgam. The rate of hydrogen evolution determines the size of the decomposer and the concentration of hydroxide which could be attained. Recently, it was shown[18,19] that considerable depolarization of graphite, i.e., increase in the rate of hydrogen evolution, can be achieved if a combination of alkali molybdate and cobalt ions in very small quantities is added to the solution. They are both deposited at graphite. However, cobalt goes first, enabling reduction of molybdate into a black nonmetallic sponge of extremely high activity. The increased rate of hydrogen evolution permits considerable reduction in the size of the decomposers and the increase of the concentration of hydroxide in the solution from 50% to 65–70%. Taking into account the world production of hydroxide, this should save expenses for evaporation of $9-16 \times 10^6$ tons of water in the solidification process.

3. The Launching Pad

It has only been in the last 30 years that the understanding of many electrochemical problems has zoomed to heights at which it could start being useful in solving problems of electrochemical practice. Marked contributions to that development have been made by John O'Mara Bockris and his co-workers, to the acknowledgement of which this meeting is devoted. An attempt is made here to demonstrate the relevance of the achievements made in electrochemistry during that period for the foundations of future energy-conversion practice.

However, a detailed systematic analysis of the field from that point of view would far transgress the scope of this contribution. Instead, the reader is directed toward review articles on each particular problem, and the space below is used for a few examples arising from the work of the author's own laboratory. The excuse for doing so is found in the fact that this work is of a rather recent origin and hence not yet so well known that its reconsideration would be dull to a learned electrochemist.

3.1. Mechanisms of Electrode Reactions

Electrode polarization is interesting not only because it ranks low from the point of view of energy efficiency of an electrode process. A far more important aspect is its effect on the competitive capacity of the desired electrode process with respect to other electrode processes which are occurring at a given potential. A good example of the importance of that aspect in practice is the difficulty of electrowinning of aluminum from aqueous solutions. Electrochemists of the classical school, relying on thermodynamic considerations only, would consider this to be a very obvious case because of the large difference in the potential of evolution of hydrogen from water and the reversible potential of aluminum in the aqueous solution of its salt. However, alkali metals are in the same or even worse situation, and yet they can all be deposited on mercury with practically 100% efficiency. It is obvious that for hydrogen evolution the overvoltage at those potentials is too low to drive it at a significant rate. However, finite aluminum deposition rates require highly negative potentials (high overpotentials) where the rate of hydrogen evolution is much higher than that of the former. Obviously, one must learn much more about hydrogen evolution kinetics and aluminum deposition kinetics in order to be able to reverse the situation.

To these ends, studies of reaction mechanisms could be of considerable use. It became generally realized in the early postwar days that the Tafel relationship is quite a general one applying to any activation-controlled electrochemical process. However, it took quite some time to build a complete picture of how the value of the transfer coefficient and reaction orders reflect the complexities of the reaction mechanism.[22]

Once this issue was clarified, powerful diagnostic criteria were obtained for discussing what can be, and in some cases even what should be, the path of an electrode reaction. The stepwise nature of many reactions of reduction of metal ions to metal has been realized. Recognition should be accorded to the stimulating role played by Bockris and co-workers [22-24] in their work on copper deposition, as well as to the theoretical arguments of Bockris and Conway,[25] for the application of this concept to the reduction of practically all polyvalent cations. Today, it appears almost justified to generalize that all but the alkali metals deposit after a stepwise reduction involving a certain sequence of electrochemical and chemical steps exemplified by the following reaction path pertaining to deposition of zinc from zincate.[26-29]

$$Zn(II)Q \underset{}{\overset{K'}{\rightleftharpoons}} Zn(II)X + 2OH^- + 2H_2O$$
$$\text{chemical}$$

$$Zn(II)X + e \underset{}{\overset{K'}{\rightleftharpoons}} Zn(I)X$$
$$\text{electrochemical}$$

$$\left.\begin{array}{l} Zn(I)X \underset{}{\overset{K}{\rightleftharpoons}} Zn(I)Y + OH^- \\ \quad\quad\quad \text{chemical} \\[6pt] Zn(I)Y + e \underset{}{\overset{K''}{\rightleftharpoons}} Zn + OH^- \\ \quad\quad\quad \text{electrochemical} \end{array}\right\} 2H_2O$$

where Q is quite likely $(OH)_4 \cdot 4H_2O$, the structure of other species following from the stoichiometry. The first electrochemical step is rate determining in the case of solid zinc,[27] while the second chemical step is the rate-determining step in the case of zinc amalgam.[29]

3.2. Understanding the Possible Effects of the Electrical Double Layer

The double-layer structure has been given considerable attention from the very outset of the modern approach to electrochemistry. This resulted in the accumulation of a large amount of experimental material and the development of detailed theoretical conclusions concerning the structure of the interface, its thermodynamics, and the kinetics of structural changes.[30] The importance of the potential of zero charge has been stressed by Antropov[31] and has since been advocated particularly strongly by Frumkin and co-workers.[32] This enables us to know the actual sign of the charge on an electrode surface at a given relative potential and what its concentration is, i.e., how high a field is across the interface. These affect electrode process kinetics in a direct way (the Frumkin effect) if other conditions exist for the development of the diffuse part of the double layer. In addition to that, there are important indirect effects resulting from the influence of the charge on adsorption of ions and neutral substances, acting catalytically or inhibitively on electrode processes.

Thus, one could know in which range of positive electrode potential one cannot expect adsorption of cations to play any role or what negative potentials of an electrode exclude the presence of anions at the electrode surface.

The adsorption of neutral molecules is also indirectly governed by the field as it determines the limits above which strong binding of water molecules drives neutral molecules away from the surface (the Bockris–Devanathan–Müller theory[33]).

Even the changes in the position of a molecule with respect to the surface with the increase of field have been recorded. A review of the complex problem of the adsorption of organics at electrodes and the effect of the field (or charge) summarized in detail the present level of understanding of that problem.[34]

3.3. The Role of Adsorption at an Electrode Surface

Purely electrostatic or specific adsorption of ionic substances as such can have a very beneficial effect on the rates of some electrochemical reactions.[35]

It was known for some time,[36,37] but only recently better understood, that the deposition of monolayers of metal adatoms or partially charged adions occurs on foreign substrates at potentials far more positive than the reversible potential (underpotential deposition)—a situation which was unthinkable only a few years ago. Yet, it is so logical when one realizes that adhesion forces (or adsorption energy on foreign substrate) need not necessarily be weaker than the cohesion forces (adsorption energy on a like substrate). Such monolayers can also have electrocatalytic effects, as will be shown below.

So far, however, adsorption has been used more often to inhibit electrode reactions than to catalyze them. This applies particularly to reactions underlying corrosion of materials. In addition to the many investigations reported in the literature,[38] it would be interesting to cite here an example where both inhibition and catalysis change places in time in one and the same system. This is taken from our work on the effect of some amines on the hydrogen-evolution reaction on zinc in alkaline solutions.[39]

Corrosion of zinc has been followed by a volumetric determination of the rate of hydrogen evolution. The difficulty in such a system is that in concentrated alkalies substituted amines are sparingly soluble even when sulfonated, probably because of a salting-out effect. On the other hand, this helps their attachment to a solid–liquid interface, once they have been put there. Hence, different substituted amines were adsorbed at the surface of zinc by first immersing it into alcohol or the aqueous solution of the amine and then transferring the sample into concentrated alkalies for investigation. The corrosion rate was followed with time. A typical dependence of the ratio of rates in the absence and presence of the sulfonated dibenzylamine, defining the degree of inhibition, is shown in Figure 7. It exhibited an unexpected change after about 100 hr of contact between the sample and the electrolyte. Indeed, the change could be followed even visually, the clearly observable evolution of hydrogen from the surface suddenly ceasing. After another 30 hr, the hydrogen started evolving again, and the rate returned to such values as to make the substance present at the surface an activator rather than an inhibitor of corrosion. The logical way to explain this phenomenon is to assume that there is a slow chemical transformation of the substance present at the surface, and acting as an activator, into an un-

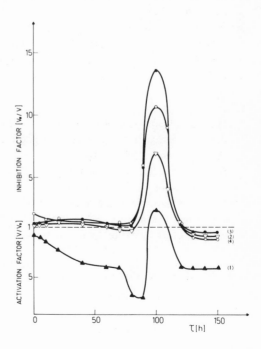

Fig. 7. Time dependence of the inhibitive effect of sulfonated dibenzylamine on hydrogen corrosion of zinc in concentrated alkali hydroxide. Concentrations of KOH: 1 M(1), 3 M (2), 6 M (3), and 10 M (4).

known but rather efficient corrosion inhibitor. Unfortunately, this substance is not stable either, and it transforms further into a new activator of hydrogen evolution. This example is given to show how the field of corrosion inhibition by adsorption is open to discoveries which could be of considerable practical significance. If the unstable intermediate could be analyzed in order to learn about its structure, stabilized by some modification of the latter (which would not harm its quality), and synthesized, a new corrosion inhibitor of considerable use would be born.

3.4. Electrocatalysis

Significant progress has been made in our understanding of what governs the rates of electrochemical reactions and how they can be increased. Besides the well-known review of this problem by Bockris,[40] several excellent studies of quite recent origin are available.[41-43]

One should stress here the importance of the development of methods of studying surfaces with ultrahigh sensitivity.

It has only recently become possible to detect the presence of as little as 10^{-12} moles of a substance at the surface by using either optical methods (ellipsometry,[44] reflectance spectroscopy,[45] etc.) or some electrochemical scanning techniques.[46] This being made possible, one can investigate the effect of such a substance on a given electrode process. To demonstrate this, one could cite some recent results[47,48] on the catalytic effect of adatoms of metals deposited at underpotential.

Figure 8A shows the potential sweep diagram for platinum immersed into a perchlorate solution containing lead ions. One can see waves due to adsorption–desorption of possibly partially discharged lead starting at potentials almost 1 V more positive that those of the reversible deposition and dissolution. By integrating the currents, one could calculate the degree of coverage of the surface by lead. For example, at a potential of 550 mV vs. NHE, it amounts to 0.31. The second potential sweep diagram (Figure 8B) shows the oxidation of formic acid on platinum in the same electrolyte with and without the addition of lead ions. It is seen that in the presence of the latter, the rate of oxidation of the formic acid, reflected in the anodic current at the sweep going positive, has undergone an increase of almost two orders of magnitude at the peak, e.g., at 550 mV vs. NHE. The differences in the potential sweep patterns indicate that there could be some changes, caused by the presence of lead, in the reaction mechanism or at least in the rate-determining step.

Thus, a field of electrocatalysis by trace amounts of foreign substances at the surface has been opened to investigation with good prospects of rendering some entirely new phenomena of considerable practical interest.

At the same time these results point to the importance of surface states of electrocatalysts. It seems that the moment has come to launch a decisive assault on establishing a better understanding of the solid phase of an electrochemical system. The use of various optical techniques *in situ* and possibly *ex situ* (LEED, Anger, ESCA) could provide an atomic level of understanding of various electrocatalytic reactions.

3.5. Electrocrystallization

As already mentioned, zinc electrodes in secondary chemical energy sources have been considered so difficult to master that, in many instances, research on sources using zinc has been abandoned. This only shows that the knowledge already accumulated in the

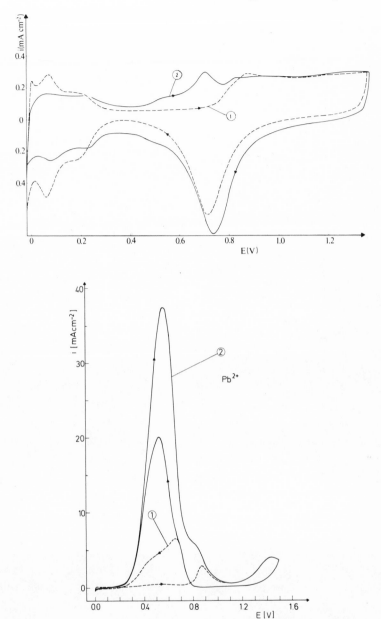

Fig. 8. Potential sweep diagrams for platinum in 2 M perchloric acid solution in the absence of Pb^{2+} ions in solution (1) and in their presence (10^{-3} moles) (2) without (a) and with addition of formic acid (b).

field of electrocrystallization has not penetrated sufficiently into applied research and development.

In the case of zinc in alkaline solutions, detailed morphological studies[49, 50] have pretty well defined those conditions under which zinc plates out in four different ways. These are shown in the electronmicrographs in Figure 9.

Dendritic growth is one of the deterrents limiting the life of the zinc electrode. The causes of this growth have now been quite well established, the leading role in this having

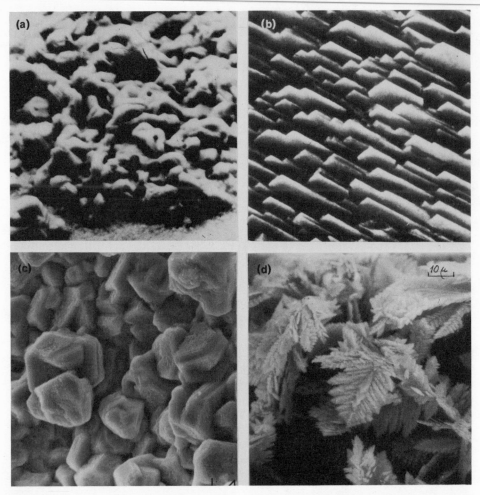

Fig. 9. Electron micrographs of different types of crystal growth of zinc from alkaline solution: (a) mossy deposit, (b) epitaxial layer growth, (c) boulders, (d) dendrites.

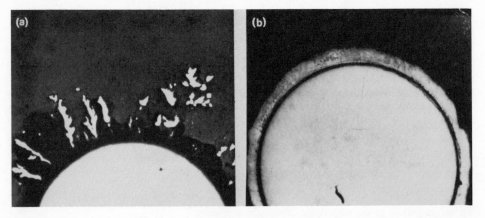

Fig. 10. Zinc deposit obtained from alkaline zincate solution on platinum, using d.c. (a) and square pulses of potential at a frequency of 2 kHz (b).

been played by the work of Bockris and Barton[51] in 1962. Subsequently, those ideas have been applied to the very case of zinc in alkalies and extended further to cover some new phenomena observed in that system.[52, 53]

Finally, it has been shown how the increase in surface roughness and dendritic growth can be prevented if electrodeposition is carried out by a pulsating current rather than by d.c.[54] Figure 10 shows how the pulsating regime permitted a very smooth zinc deposit at the same rate at which the d.c. caused deposition of rough and loose zinc.

Hence, the use of appropriately designed pulsating regimes for charging zinc plates should remove at least one major problem in the use of zinc in rechargeable energy sources.

3.6. Effect of Minor Constituents on Electrochemical Properties of Metals

Too little work has been done as yet in attempting to improve electrochemical properties of metals which are potential energy transmitters by alloying them with minor quantities of other elements. A traditional example is that of reducing zinc corrosion in dry batteries by adding mercury to the system. However, little has been done in making similar intervention on aluminum and thus make it a usable material in chemical energy sources. Its use is prevented by the fact that under normal conditions it is covered by a very tough protective layer of oxide. This makes it an excellent construction material but, if immersed into a neutral electrolyte like the one used in the Leclanché cell, its rest potential is much more positive than that of zinc so that most of its inherently high free-energy content is lost. Moreover, when polarized anodically, the rate of hydrogen evolution reaction causing its corrosion is increased ("negative differential effect") instead of decreased, as would be expected by normal electrochemical reasoning. The increase is proportional to the anodic dissolution current. Hence, about 16–20% of aluminum is lost during dissolution by simultaneous hydrogen corrosion.

Work on this system was started in our laboratory in order to see if this situation could be improved. Several considerations were taken as a guideline in the search for an appropriate additive:

1. The additive should increase hydrogen overvoltage.

2. The additive should be intimately mixed with aluminum; hence, solid solutions are needed rather than separate phases.

3. For the additive to be effective, its concentration should be high at the spot where corrosion takes place; if this were to be throughout the metal, this would require a high percentage of additive in the alloy. Hence, surface-active additives are needed which would accumulate at the grain boundaries during crystallization and cooling.

4. If the additive dissolved from the boundary does not redeposit when a grain starts dissolving, the grain would attain the properties of pure aluminum; hence, the additive should have a more positive reversible potential.

5. The additive accumulated at the surface should either prevent the formation of the tough oxide film or should modify its properties so as to change the negative differential effect into a positive one or at least reduce its value.

It was found that three elements could be expected to fulfill these requirements. Those are indium, gallium, and thallium.

Indeed, minor additions of these elements exhibited a significant effect.[56] For example, as little as 0.025% of indium (1 atom per 20,000 atoms of aluminum) rendered a negative shift of the rest potential by 200 mV and a reduction of the negative differential effect to 6%. Figure 11 shows the effect of the addition of 0.055% of indium on the

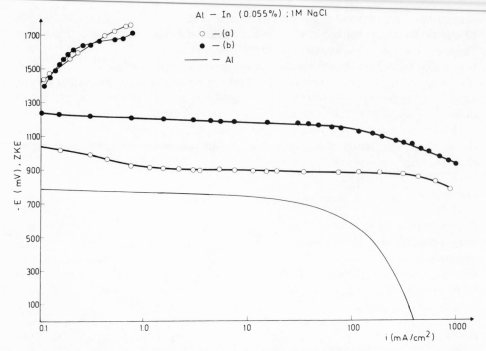

Fig. 11. Polarization curves for anodic dissolution of aluminum into 1 M NaCl solution when alloyed with 0.055% (w/w) of indium and quenched (a) or annealed (b). Solid line represents that for pure aluminum.

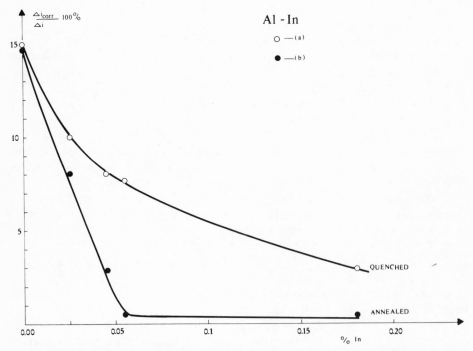

Fig. 12. The dependence of the negative differential effect on the content of indium in the alloy. Curves (a) and (b) pertain to the alloys obtained by quenching and annealing, respectively.

polarization curve of the alloy as compared to that of pure aluminum in 1 M NaCl. Figure 12 shows the dependence of the negative differential effect on the content of indium. Thermal treatment (annealing) led to further improvement of the qualities of the alloy. It is seen that the alloy exhibits a satisfactory negative rest potential (more negative than zinc under corresponding conditions), low polarizability up to current densities of the order of 1 A/cm^2, and the rate of corrosion decreased to only 0.5%. Other elements showed a similar effect.

Thus, such additions make aluminum a good electrochemically active material, superior in all qualities, but the rechargeability, i.e., reducibility from aqueous solutions, to zinc. They open the way for its use as an energy transmitter in primary batteries or exchangeable plate secondary ones, as well as in the protection of materials against corrosion wherever zinc could be used for that purpose.

It is hoped that the few examples cited above have shown that the developments in fundamental electrochemistry during the past 30 years have created a solid platform for the propulsion of its application into a higher level of historical significance. Energy conversion is likely to gain some new electrochemical channels of considerably higher quality than those available nowadays. This, in turn, will give even greater importance to the work which devoted electrochemists are presently carrying out.

References

1. R. Phillip Hammond, in: *Electrochemistry of Cleaner Environment* (J. O'M. Bockris, ed.), Chapter 7, Plenum Press, New York (1972).
2. A. B. Cambel, *Energy Research and Development and National Progress*, U.S. Government Printing Office, Washington, D.C. (1964).
3. M. K. Hubert, *Energy Resources*, National Academy of Science, Publ. 1000-D, Washington, D.C. (1962).
4. J. O'M. Bockris, in: *Electrochemistry of Cleaner Environment* (J. O'M. Bockris, ed.), Chapter 1, Plenum Press, New York (1972).
5. G. N. Plass, in: *Electrochemistry of Cleaner Environment* (J. O'M. Bockris, ed.), Chapter 2, Plenum Press, New York (1972).
6. J. O'M. Bockris and D. Drazić, *Electrochemical Science*, Taylor & Francis, London (1972).
7. J. O'M. Bockris, N. Bonciocat, and F. Gutmann, *An Introduction to Electrochemical Science*, Wykeham Publications (London) Ltd., London (1974).
8. D. P. Gregory, D. Y. C. Ng, and G. M. Long, in: *Electrochemistry of Cleaner Environment* (J. O'M. Bockris, ed.), Chapter 8, Plenum Press, New York (1972).
9. Extrapolated figure from N.Ibl, Plenary lecture, Symp. on Electrochem. Eng., Newcastle 1971, *Inst. Chem. Eng., Symp. Ser. No. 37* (1973).
10. Metallgeselschaft, Metal Statistics 1964–1974, Frankfurt (1975).
11. J. O'M. Bockris, *Electrochemistry—the Underdeveloped Science.*
12. A. Volta, *Trans. R. Soc. 90*, 403 (1800).
13. G. Leclanché, *Les Mondes 16*, 532 (1868).
14. N. S. Lidorenko, report, at the V Allsoviet Meeting of Electrochemistry, Moscow, (January 1975).
15. G. Planté, *Compt. Rend. 50*, 640 (1860).
16. P. Gallone, *Trattato di Ingegneria Elettrochimica*, Tamburini Editore, Milano (1973).
17. C. D. Hodgman, ed., *Handbook of Chemistry and Physics*, Chemical Rubber Publ. Co., Cleveland, Ohio (1959).
18. M. M. Jakšić, D. R. Jovanović, and J. M. Czonka, *Electrochim. Acta 13*, 2077, (1968).
19. M. M. Jakšić, V. Komnenić, and R. Atanasoski, Yug. Pat. Appl., YU-755-P-122/75.
20. D. A. Y. Swinkels, in: *Advances in Molten Salt Chemistry* (Y. Braunstein, ed.), Vol. 1, Plenum Press, New York (1971).
21. J. O'M. Bockris and A. K. Reddy, *Modern Electrochemistry*, Plenum Press, New York (1970).
22. E. Mattson and J. O'M. Bockris, *Trans. Faraday Soc. 55*, 1586 (1959).
23. J. O'M. Bockris and M. Enyo, *Trans. Faraday Soc. 58*, 1187 (1962).

24. J. O'M. Bockris and H. Kita, *J. Electrochem. Soc. 109*, 928 (1962).
25. B. E. Conway and J. O'M Bockris, *Proc. R. Soc. A248*, 394 (1958).
26. H. Gerischer, *Z. Phys. Chem. 202*, 302, (1953); *Z. Electrochem. 57*, 604, (1953); *Angew. Chem. 68*, 20 (1956).
27. J. O'M. Bockris, Z. Nagy, and A. Damjanović, *J. Electrochem. Soc. 119*, 285 (1972).
28. D. A. Payne and A. J. Bard, *J. Electrochem. Soc. 119*, 1665 (1972).
29. A. R. Despić, Dj. Jovanović, and T. M. Rakić, *Electrochim. Acta* (submitted).
30. P. Delahay, *Double Layer and Electrode Kinetics*, Interscience, New York (1965).
31. L. I. Antropov, *Zh. Fiz. Khim. 25*, 1494 (1951).
32. A. N. Frumkin, *J. Electroanal. Chem. 9*, 173 (1965).
33. J. O'M. Bockris, M. A. V. Devanathan, and K. Müller, *Proc. R. Soc. A274*, 55 (1963).
34. B. B. Damaskin, O. A. Petrii, and V. V. Batrakov, *Adsorption of Organic Compounds on Electrodes*, Plenum Press, New York (1971).
35. L. Gierst, E. Nikolas, and L. Tytgat-Vandenberghen, in: *Solid Liquid Interfaces (Proceedings of the International Summer School, Dubrovnik-Cavtat 1969)*, Zagreb (1971), p. 37.
36. M. Haissinsky, *Experientia 8*, 125 (1952).
37. L. B. Rogers, *Rec. Chem. Prog. 16*, 197 (1955).
38. H. Kaesche and N. Hackerman, *J. Electrochem. Soc. 105*, 191 (1958).
39. L. Ž. Vorkapić, A. R. Despić, and D. M. Dražić, *Bull. Soc. Chim. (Belgrade)*, 7–8 (1974).
40. J. O'M. Bockris and H. Wroblowa, *J. Electroanal. Chem. 7*, 428 (1964).
41. R. Parsons, *Surface Sci. 18*, 28 (1969).
42. S. Trasatti, *J. Electroanal. Chem. 39*, 163 (1972).
43. H. Kita and T. Kurisu, *J. Res. Inst. Catalysis, Hokkaido Univ. 21*, 200 (1973).
44. J. Kruger, in: *Advances in Electrochemistry and Electrochemical Engineering*, Vol. 9, (R. H. Müller, ed.), Wiley, New York (1973).
45. J. D. E. McIntyre, in: *Advances in Electrochemistry and Electrochemical Engineering*, Vol. 9, (R. H. Müller, ed.), Wiley, New York (1973).
46. E. Yeager and J. Kuty, *Physical Chemistry*, (H. Eyring, ed.), Vol. 9a, Academic Press, New York (1970).
47. R. R. Adzić and A. R. Despić, *J. Chem. Phys., 61*, 3482 (1974).
48. R. R. Adzić, D. N. Simić, D. M. Dražić, and A. R. Despić, *J. Electroanal. Chem.* (submitted).
49. J. O'M. Bockris, L. Nagy, and D. Dražić, *J. Electrochem. Soc. 120*, 30 (1973).
50. I. N. Justinijanović, J. N. Jovićević, and A. R. Despić, *J. Appl. Electrochem. 3*, 193 (1973).
51. J. L. Barton and J. O'M. Bockris, *Proc. R. Soc. A268*, 485 (1962).
52. J. W. Diggle, A. R. Despić, and J. O'M. Bockris, *J. Electrochem. Soc. 116*, 1503 (1969).
53. A. R. Despić and M. M. Purenović, *J. Electrochem. Soc. 121*, 329 (1974).
54. Cf. ref. in: A. R. Despić and K. J. Popov, *Modern Aspects of Electrochemistry* (B. E. Conway and J. O'M. Bockris, eds.), Vol. 7, p. 199, Plenum Press, New York (1972).
55. A. R. Despić and K. J. Popov. *J. Appl. Electrochem. 1*, 275 (1971).
56. M. M. Purenović, A. R. Despić, and D. M. Dražić, *Elektrokhimiya* (in press).
57. M. M. Purenović, A. R. Despić, and D. M. Dražić. Yug. Pat. Appl.

2

The Future of Electrochemical Energy Conversion—The Next Thirty Years

James McBreen

1. Introduction

The future of electrochemical energy conversion cannot be discussed in a meaningful way without consideration of the future primary sources of energy. On a long time scale there are numerous possible ways in which energy can be generated and utilized. There are many variables, and it has recently become voguish to arrange these into "scenarios" for the future. These planning games are fascinating intellectual exercises. This paper, however, makes no pretense at outlining a "scenario" for the future but rather considers what the alternative sources of energy are and discusses how electrochemistry in general and electrochemical energy conversion in particular can fit into future energy systems. The present development of fuel cells is discussed, and suggested areas of future work are outlined.

2. Energy in the Past

Figure 1 is a plot of the annual world energy production of individual energy sources from 1860 to the present day.[1] The salient thing about this plot is that new fuels have always come in with a steep slope, generally not displacing but rather adding to the contribution of established fuels. The overall increase in energy consumption has been due in part to the increase in world population. However, in industrialized nations, there has been an increase in the per capita energy consumption in this period.[2] For instance, the per capita energy consumption for the United States increased from 1.3×10^8 kJ to 3.3×10^8 kJ in this period. A general correlation can be made between energy consumption and gross national product for the various nations (Figure 2).[3] The degree of causality between the two factors is difficult to assess with certainty, but every indication is that the economy and standard of living of industrialized nations are strongly dependent on a large supply of energy and a high use of energy per capita.

The end manner in which the primary sources of energy have been used has changed over the years. The fastest growing type of energy is electricity. In recent times the use of electricity in the United States has been doubling every decade.[4]

None of the primary sources of energy used to date have been suitable for direct use in fuel cells. Bauer's attempt to use coal directly in fuel cells was unsuccessful.[5] The direct use of hydrocarbons in fuel cells has presented both technical and economic difficulties.[6]

3. Energy in the Future

World energy production in the past has been almost exclusively from fossil fuels (Figure 1). In the next 30 years this picture will change. Most projections indicate that by

James McBreen • Electrochemistry Department, Research Laboratories, General Motors Corporation, Warren, Michigan 48090.

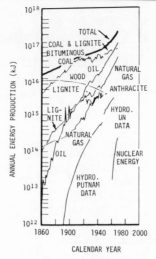

Fig. 1. Annual world energy production of individual energy sources.

2005 virtually all of the oil in the world will have been discovered and we will be well past the point of maximum reserves and maximum production of petroleum.[1,7] Thirty years later (2035), according to some authors,[7] the maximum annual production of all fossil fuels will be reached (Figure 3). As a historical event the fossil fuel era will be shorter than such former human institutions as the Spanish Empire.

Some authors have suggested that the industrialized nations shift to an all-electric economy.[8] An economy tied to massive electrical networks would lack the flexibility we now have. Other problems are the transmission of large quantities of electric power

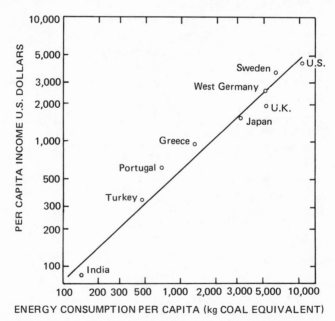

Fig. 2. Energy consumption and per capita income for various nations. Line is a least-square fit for 80 nations.[3] Some representative countries are shown.

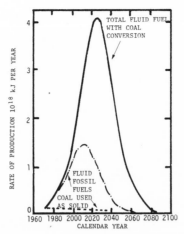

Fig. 3. Fluid fuels from fossil fuels in the future with and without coal conversion.[7] Total energy, 214.4×10^{18} kJ; energy from coal, 159.7×10^{18} kJ; energy from oil and gas, 54.7×10^{18} kJ.

and the storage of large quantities of electrical energy near consumer sites for peak-shaving purposes. In the past, fluid fuels have been a very effective medium for the storage and transmission of energy. Liquid fuels will be indispensable for large aircraft, flexible long-range transportation, and agriculture. There will also be a need for fluid fuels in the processing industries. In the future the end use of energy will be both in the form of fluid fuels and electricity. The things to consider then are how the electricity will be generated, what the fluid fuels will be, in what manner such fluid fuels will be used, and how electrochemistry can fit into such a scheme.

4. Future Production of Electricity

In all countries, even those with large fossil fuel reserves (U.S. and U.K.), there is a shift to a greater use of electricity.[4,10] In the near term (1975–1985) the primary energy sources will be fossil fuels and fissionable nuclear materials.

4.1. Fossil Fuels

The general consensus is that there will be a shift away from petroleum to coal.[9] Near-term efforts will be focused on removing sulfur from coal and increasing the efficiency with which coal produces electricity. The latter could be accomplished by the use of combined-cycle powerplants (i.e., combined gas turbine–steam turbine plants), magnetohydrodynamics (MHD), or fuel cells.[9]

Removal of undesirable components (e.g., Pb, As, Zn, Hg, Cd, Ni, Be, Se, and S) from coal will necessitate the conversion of solid coal to a gas or liquid.[11] Coal could be converted to power gas (a mixture of CO and H_2) and used at the minehead or upgraded to methane and piped for use elsewhere.[12] Methanol is one of the more likely liquid fuels to be produced from coal.[13]

Power gas could be used as a fuel for molten carbonate fuel cells that operate at 500–750°C.[6] Another possibility is the use of solid electrolyte fuel cells which operate at 1000°C.

In the near future (by 1980) the combined-cycle process[9] is the most likely advanced method in which coal will be used to generate electricity. Success in the control of the corrosion of ceramic materials at high temperatures could result in the use of MHD and high-temperature fuel cells by 1985.

If processes to economically produce methane and methanol from coal are developed, fuel cells would be an ideal method to upgrade these fuels to electricity. The indirect use of methane in a fuel cell has already been demonstrated,[14] and methanol is a more suitable fuel than hydrocarbons for direct use in fuel cells.

Alternative fossil fuel sources that have been considered in North America include the tar sands of Alberta's Athabasca deposit and the oil shale deposits of Colorado, Utah, and Wyoming.[12] The atomic hydrogen/carbon ratios of tar sands and oil shale are higher than that of coal (approx. 1.5 for the tar bitumen, 1.6 for the oil shale kerogen, and 0.72–0.92 for coal). Because of this, less water and less sophisticated methods are needed to produce oil from these sources as compared to coal.[15] These sources, if developed, would more likely be used to produce gasoline-like fuels which would not be suitable for direct use in fuel cells.

4.2. Fissionable Nuclear Fuels

4.2.1. Convertor Reactors

Several kinds of convertor reactors, based on the fission of the scarce isotope of uranium ^{235}U, have been developed. These reactors vary in the type of heat-exchange system, the moderator used, and the type of the fuel elements. A tabulation of the salient features of some of these reactors is given in Table 1. Lightwater reactors have been predominantly used in the United States while gas-cooled reactors have been used in the United Kingdom. The low operating temperature of convertor reactors limits their use to the generation of steam for steam turbines. From the chemist's point of view, the important parameter is the temperature of operation since this determines what chemical processes might be carried out using nuclear reactors.

At the end of 1973 nuclear power provided only 5% of the electricity in the United States. In 1974 there were about 52 reactors in operation in the United States. The nuclear industry expects to build 950 more within the next 25 years.[16] France plans to build only nuclear power plants from now on.[17] All indications are that the present trend toward convertor reactors will gather momentum and that nuclear power will continue to capture a larger share of the electrical generating capacity. By 1995 nuclear power will constitute more than 50% of the electrical generating capacity in the United States and the United Kingdom.[18,19]

The ^{235}U which is used in convertor reactors constitutes only 0.7% of naturally occurring uranium. The exact reserves of uranium are not known with certainty, but the general consensus is that there is enough uranium for convertor reactor requirements for the next 30 years without any great increase in fuel price.[15]

In looking beyond 1985, one can consider many other energy sources. These are breeder reactors, nuclear fusion, geothermal energy, tidal power, indirect solar energy, and direct solar energy.

4.2.2. Breeder Reactors

Convertor reactors will continue to be the mainstay of nuclear power until the end of the century. However, extensive work is going on in the United States, Russia, Japan, and several European countries on liquid-metal-cooled fast-breeder reactors (LMFBR).[20,21] The hope is that breeders will eliminate the need for gaseous diffusion plants and ensure a much larger supply of nuclear fuel.

Table 1. Convertor Reactor Options and Parameters

Reactor type	Water cooled				Gas cooled				Na cooled
Acronym[a]	SCHWR	BWR	PWR	HWR	AGR	Magnox	HTR	GBR	
Nominal output (MWe)	625	1,110	850	750	625	550	1,320	1,320	1,250
Fuel	UO_2	UO_2	UO_2	UO_2	UO_2	U	UO_2	UO_2	U/PuO_2
Fuel cladding	Zircaloy	Zircaloy	Zircaloy	Zircaloy	SS[b]	Magnesium alloy	SiC	SiC	SS[b]
Coolant	H_2O	H_2O	H_2O	D_2O	CO_2	CO_2	He	CO_2	Na
Inlet temperature (°C)	271	278	288	261	287	247	273	260	340
Outlet temperature (°C)	282	286	321	299	651	414	749	650	540
Thermal efficiency (%)	32	31	32	32	41	32	38	35	42

[a]SCHWR, steam-generating heavy-water-moderated reactor; BWR, boiling-water reactor; PWR, pressurized-water reactor; HWR, heavy-water reactor; AGR, Advanced gas-cooled reactor; HTR, high-temperature gas-cooled reactor; GBR, Gas-cooled breeder reactor.
[b]Stainless steel.

Fast-breeder reactors use unmoderated, high-energy neutrons (approx. 2 MeV) and operate at higher power densities than convertor reactors. In the breeder reactor practically all of the heat comes from the kinetic energy of the splitting nucleus. The neutrons perpetuate the reaction and convert the ^{238}U and ^{232}Th into fissionable ^{239}Pu and ^{233}U, respectively. Thus, the available fissionable fuel is increased by a factor of 140.

High neutron fluxes (approx. 1×10^{23} neutrons/cm^2) have caused materials problems, such as swelling of stainless steel in the reactor core.[9] In a breeder reactor the fluxes will be higher (approx. 3×10^{23} neutrons/cm^2). It is not yet known how this will affect the core materials. Plutonium is one of the most toxic substances known,[22] and the radioactive isotope ^{239}Pu has a half-life of 24,400 years. It has been projected that by the year 2000 as much as 2.7×10^{10} C of radioactive wastes could accumulate. The maximum proposed body burden limit for ^{239}Pu varies between 1.4×10^{-13} C[23] and 4.0×10^{-8} C.[24] These materials would have to be stored for thousands of years under virtually constant surveillance.[25] This presumes a degree of permanence for human institutions that has no historical precedent. Doubts have been expressed about the wisdom of a commitment to breeder nuclear reactors.[11,26] If breeder reactors are not developed to everybody's satisfaction by the end of the century, nuclear fission will not play much of a role in nuclear generation beyond that time.[15] Although the technology for nuclear fission reactors is now reasonably well assured, there are still reservations about its widespread use.[11]

4.3. Nuclear Fusion

Nuclear fusion, the combination of the nuclei of light elements into heavier ones, is the energy source of the sun and stars. If useful energy could be generated from nuclear fusion then there would be a virtually inexhaustible energy source. However, fusion power does not exist yet. To achieve fusion the fuel must reach a sufficiently high kinetic temperature ($>10^4$ eV) and the product of the particle density (n) and the confinement time (τ) must exceed the Lawson criterion ($n\tau > 10^{14}$ cm^{-3} s).[27] Two approaches that have been used to achieve this criterion are tokomak magnetic containment devices and converging neodymium doped glass lasers.[9,27] It has been estimated that the tokomak fusion reactors would be large (500–2000 MW) and that laser fusion reactors could be made small (50–200 MW). Thus, laser fusion reactors offer greater flexibility for use.

As opposed to fission, most of the heat from fusion reactors would come from the kinetic energy of the neutrons. These high-energy neutrons (14 MeV) can activate the components of a fusion reactor. However, the major radiation hazard from fusion reactors comes from ^{3}H.[20] This isotope has a half-life of 12 years and can get incorporated into the environment. Indications are that the biological hazard potential of fusion reactors is two or more orders of magnitude less than LMFBR reactors.[24] Fission reactors generate heat even after they have been shut down and thus require continuous cooling; fusion reactors do not and hence have a decided advantage. Although technical feasibility has yet to be demonstrated, there is a good possibility that the commercialization of fusion plants could begin by the year 2000.

4.4. Geothermal Energy

Geothermal energy has not received serious consideration until recently.[28] By the end of the century geothermal energy could be used extensively for power generation, particularly in countries on the rim of the Pacific Ocean.

4.5. Tidal Power

Tidal power has been harnessed and a 240-MW plant was built on the estuary of the Rance river on the coast of Brittany in France.[9] However, like hydroelectric power, there are only a limited number of sites available for the use of this source of power.

4.6. Indirect Solar Energy

The sun radiates 1.7×10^{17} W of energy onto the earth continuously. This energy is the source of the energy of the wind and the waves and the heat of ocean thermal gradients.

4.6.1. Ocean Thermal Gradients

The technical feasibility of the use of ocean thermal gradients to generate electricity was demonstrated 45 years ago when a 22-kW powerplant, using the Claude cycle, was constructed off the coast of Cuba.[9] Future systems would probably use a secondary working fluid such as ammonia or freon to turn a turbine. Corrosion of installations in seawater is a problem, and electrochemistry could have important contributions to make here. If the economics are favorable and if no major environmental problems are uncovered, ocean thermal gradients could contribute to electricity generation by 1990.

4.6.2. Wind Power

Wind power has been used extensively, particularly on farms in the western United States, to generate electricity and to pump water. A wind machine that generated 1.25 MW of electricity was operated at Grandpa's Knob in Vermont from 1941-45.[29] It has been estimated that as much as 5.4×10^{15} kJ/yr could be generated from wind in the United States.[9] Other favorable sites for wind generation are the British Isles, Norway, Japan, New Zealand, and the vicinity of the Bay of Biscay.[30] The technology for electricity generation is within reach; the main problem is storage of energy to compensate for the variable nature of the wind.

4.6.3. Wave Power

Wave power has been considered for electricity generation. In special locations such as the Hebrides it is conceivable that such power could be harnessed by the end of the century.[31]

4.7. Direct Solar Energy

At the top of the earth's atmosphere the energy flux from the sun is 1.3 kW/m^2. Absorption and scattering reduce this to about 0.9 kW/m^2 at the earth's surface.[32] The average annual insolation in the United States is 5.4 GJ/m^2 and about 3.6 GJ/m^2 in the United Kingdom. Although this vast inexhaustible source has been so obvious, little interest has been shown in its development until recently. In the past few years there has been a growing interest in solar energy. The techniques one uses to harness solar energy depend on the type of sunlight. In areas with no cloud cover, such as the American Southwest, Australia, Africa, and India, focusing collectors can be used. In countries such as

Great Britain where cloud cover diffuses more than half the incident radiation, flat-plate collectors would have to be used since diffused light cannot be focused.

At present the emphasis is on the development of small solar energy gadgets, such as rooftop solar collectors for use in homes. The solar energy would be used for water heating, space heating, and air conditioning. Energy storage is a problem and intriguing methods of storage, such as rocks, magnetite, flywheels, Glauber's salt,[15] and metal hydrides,[33] have been proposed. Homebuilding is an extremely fragmented and conservative industry and although some of these solar energy devices are already on the market, rapid implementation of the use of solar energy on a local scale may be difficult.

An interesting approach to the use of solar energy on a large scale has been proposed by Aden and Marjorie Meinel of the University of Arizona.[34] They suggest a focusing collector using Fresnel lenses to heat a pipe containing a heat-exchange fluid at 500°C. The pipe is in an evacuated glass tube to reduce conductive and convective losses and is coated with a material that permits only 5–10% of the energy absorbed to be emitted. The heat is used to run a steam turbine. The concept of using focusing collectors to run a heat engine is not new. A 74.5-kW solar-powered steam engine was built and operated in Egypt in 1912.[9] As with most direct or indirect uses of solar energy, provisions have to be made for energy storage. Large areas of land are required; the scheme proposed above would require 1.3×10^{10} m^2 for a 1.0-TW (10^{12} W) plant.

In 1839, the same year as Grove discovered the fuel cell, Becquerel first observed the photovoltaic effect.[35] Photovoltaic solar energy conversion by means of silicon solar cells has provided power for spacecraft with missions exceeding 2–3 weeks. These cells have proved to be a very reliable source of power which converts solar energy to electricity with an efficiency of 14%. To be attractive for terrestrial applications the current price of solar cells would have to be reduced by at least a factor of 100.

A concept for the use of photovoltaic cells on a large scale has been proposed.[36] Arrays of solar cells (approx. 30 km^2) would be set up in space in synchronous orbit. The energy would be beamed to earth by 10-cm microwaves using a transmitting dish antenna (approx. 2 km in diameter) and a receiving antenna (3–7 km in diameter) on the earth's surface. This station would yield about 20 GW of power on the ground. Construction of such an array is predicated on the existence of an inexpensive space shuttle. Energy storage would not be needed; two solar arrays could be used and be positioned so that at least one of the arrays would receive radiation at all times. The most logical location for the receiving antenna is at sea, but transmission of energy to land would be a problem.

A United States National Science Foundation report[37] has contended that solar energy could provide 35% of the total building heating and cooling load and 20% of the electrical energy requirements by the year 2020. Australia, with plentiful sunshine and no firm commitment to nuclear energy,[38] could conceivably receive 50% of its low-grade heat from solar energy by the end of the century.[32] Japan has long used solar water heaters and is now planning a new solar energy program. Interest in solar energy is rapidly developing. If the economic problems can be solved, then the end of the century should see a rapid expansion in the use of this clean, infinite energy source.

5. Transmission of Electricity

Because of the problems of pollution from fossil fuel plants and the danger of radiation leakage from nuclear plants, a growing number of electrical powerplants are likely to be built at large distances from urban centers. This will require enormous amounts of

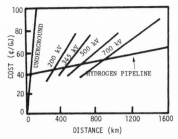

Fig. 4. Comparison of costs of electricity transmission and hydrogen transmission in high-pressure pipe lines. Data for underground electricity transmission and various high-voltage transmission lines are shown.[44]

electrical transmission. Electrical transmission is the most expensive method of energy transportation.[9] These costs are bound to rise even more since overhead lines use 3×10^4 m^2 of land per kilometer of length.

Some of the schemes, discussed above, for producing electricity are located in the ocean and many nuclear plants will probably be built in the ocean so that seawater can be used for cooling. Transmission of large quantities of electricity in the ocean is a problem. It has been suggested that the electricity be converted into hydrogen[39] and that the hydrogen be transported in pipelines, tankers, or tank cars to the point of use. A comparison of the cost of electricity transmission in high-voltage lines and hydrogen transportation in high-pressure pipelines is given in Figure 4.[39] For long distances the savings are significant using hydrogen as the medium of energy. Hydrogen pipelines would solve the problem of energy transport from powerplants located in the ocean.

The possible role for electrochemistry in a scheme where hydrogen is used as the medium of energy is enormous. Electrolysis could be used to generate the hydrogen, and fuel cells are an ideal way to upgrade hydrogen to electricity.

6. Energy Storage and Load Leveling

In any scheme of power production, the incorporation of an energy storage system can improve the system performance or reduce the cost or both. In most schemes where solar energy is used, either directly or indirectly, energy storage is an absolute necessity.

The peak-power problem is of concern to electric utilities. Figure 5 is a typical daily load profile of a utility. In the past, utilities have used their highly efficient modern large

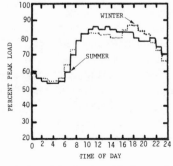

Fig. 5. A typical weekday load profile for an electric utility.

plants to deliver the base load and have switched their smaller, less efficient, old plants into the grid to meet the increase in power demand. In the future the "old" plants will be large and will be difficult to switch in and out of the grid. At present the peak-power problem is temporarily resolved by a patchwork of diesel generators and gas turbines with low fuel efficiencies. Nuclear powerplants cannot be switched off and on in short periods, so in the future a more permanent satisfactory solution must be obtained for the peak-power problem.

For large electrical networks the options for energy storage are:

1. Pumped hydroelectric storage
2. Compressed air storage
3. Hydrogen generation and use as a fuel
4. Electrically rechargeable batteries

Pumped hydroelectric storage is the only significant energy storage system in use at the present time. Existing facilities and presently proposed facilities represent less than 3% of present United States generating capacity. Site limitations and environmental considerations will limit the growth of hydroelectric storage.

Compressed air can increase the fuel efficiency of a gas turbine by as much as a factor of 2-3 and has been considered for energy storage.[40,41] The first underground reservoirs for compressed air are being dug in West Germany by a Hamburg utility.[42] These reservoirs are at a depth of 600 m in a salt formation; they expect to be producing electric power commercially by 1977. These developments are of interest to the electrochemist since this technology, if successful, would be an ideal method to store hydrogen and oxygen for fuel cells.

From an environmental point of view, the combination of an electrolyzer, a hydrogen and oxygen storage system, and a fuel cell is a desirable method of energy storage. Present electrolyzers have efficiencies as high as 85% [43] and hydrogen–oxygen fuel cells have efficiencies of 60%. Thus the overall efficiency for such a system is 51%. It has been pointed out that hydrogen–oxygen gas turbines would yield a similar overall efficiency.[44] However, this is only true if the gas turbine is operated at full power. The efficiency of a fuel cell remains high over an operating range approximately 25-125% of the nominal power rating of the powerplant.[9] Thermal efficiencies of 100% appear to be feasible for electrolyzers,[44] and fuel-cell efficiencies will likely reach 70%. This would yield an overall efficiency of 70% for such a storage system. Energy storage systems of this type could be made in sizes from tens of watts to tens of megawatts. They would be pollution free, reject all their heat into the air, and would have a noise level of about that of the kitchen refrigerator. Preliminary work is going on in the development of these systems.[45] There is every likelihood that this work will expand, and in 30 years this type of energy storage should have widespread use.

Batteries are a simple method of energy storage. It is estimated that about 18×10^{10} kJ of electricity are now stored in automobile batteries.[46] Batteries have the advantages that they can be easily modularized, have no emissions, and no special siting requirements. Another advantage is that of instant start-up. Batteries store electricity with an overall efficiency of 70%. The main problems are life and cost. Two systems which promise to meet the cost requirements are the Li/FeS_2[47] and Na/S batteries.[48] Work is being done on these systems, and economics and performance will ultimately determine which electrochemical system, batteries or fuel cells, will capture a share of the energy storage market.

7. Fluid Fuels of the Future

Apart from the fluid fuels that have been discussed before (Section 4), several other schemes have been proposed for the synthesis of fluid fuels.

It has been estimated that nonfossil production of carbon by photosynthesis is about 1.3×10^{14} kg/year.[49] Various schemes have been proposed to convert some of this carbon to methane. The major drawback is the massive areas required for the production of methane. However, one report suggests the possibility of providing 30% of the gaseous fuel and 10% of the liquid fuel for the United States in this manner by 2020.[37] Another suggested source of methane is organic wastes.[50] Fuel cells could be used to upgrade the methane to electricity.

Hydrogen has been proposed as a synthetic fuel to replace fossil fuels.[44,51,52] Hydrogen could be used to directly reduce the ores of iron and aluminum. These processes would be cleaner than those now used and would result in more uniform products. A plentiful supply of hydrogen would alleviate the present world fertilizer shortage. An internal combustion engine operating on hydrogen fuel has already been demonstrated.[53] The problem of a simple, lightweight, economical means of hydrogen storage remains to be solved. Hydrogen is the ideal fuel-cell fuel, and the scope for fuel cells in a scheme where hydrogen is readily available is enormous.

Except in locations where cheap hydroelectric power is available, hydrogen is now produced by various steam-reforming processes using hydrocarbons as the energy sources. Several complex chemical schemes have been proposed for the direct formation of hydrogen from thermal energy.[54,55] A typical reaction sequence is

$$CaBr_2 + 2H_2O \longrightarrow Ca(OH)_2 + 2HBr \qquad \text{at } 730°C$$
$$2HBr + Hg \longrightarrow HgBr_2 + H_2 \qquad \text{at } 250°C$$
$$HgBr_2 + Ca(OH)_2 \longrightarrow CaBr_2 + HgO + H_2O \qquad \text{at } 100°C$$
$$HgO \longrightarrow Hg + 1/2O_2 \qquad \text{at } 600°C$$

The incentive for using these processes is the hope of a more efficient use of nuclear reactor heat to produce hydrogen directly rather than using electrolysis. The temperature of water-cooled reactors is too low, and the temperature of gas-cooled reactors is borderline (Table 1). All of the processes require large amounts of chemicals and present formidable corrosion problems. A more fundamental shortcoming is the slow kinetics of the heterogeneous reaction steps. At the temperature of the high-temperature reactor (HTR) these reactions would be displaced far from equilibrium, and the net effect is a reduction in the real efficiency of 45%, if equilibrium is assumed,[56] to an efficiency of closer to 35%.[57] Electricity generation with a thermal efficiency of about 40% and electrolysis with a thermal efficiency of 100% would be a more efficient method of hydrogen generation.

Electrochemical photolysis of water has been suggested as a method for hydrogen generation. The idea is to use n-type semiconductor electrodes for oxygen evolution and p-type semiconductor electrodes for hydrogen evolution and irradiate both electrodes with sunlight.[58-60] At present, the costs look prohibitive;[61] however, it has been pointed out that unlike solar cells, the fabrication of p–n junctions is unnecessary and the possibility of the use of inexpensive polycrystalline materials exists.[62]

Another method that has been proposed for hydrogen generation is the use of photochemical reaction steps in conjunction with either thermochemical or electrochemical steps.[63] A typical photocatalyst is I_2; it can be used to produce HI photochemically, and HI can be thermochemically decomposed to H_2 and I_2.

Work on electrolyzers is being carried out, and efficiencies of 85% at 0.5 A/cm² have been reported.[43] High-temperature electrolyzers at 1100°C using solid electrolytes have been proposed. Other proposed methods are the use of radioactive compounds to reduce the oxygen overvoltage or the use of photochemical assists to reduce the overall electrolyzer overvoltage.[64] Progress on electrolyzers is being made. At present there is no shortage of ideas for improvement, and the prospects are good that thermal efficiencies of 100% can be achieved.

Biophotolysis of water to produce hydrogen gas has been proposed.[37] It has been suggested that the best method for the large-scale utilization of solar energy may perhaps result from the conversion of sunlight directly into hydrogen gas.[65]

The major problem facing the hydrogen economy is one of implementation.[66] The problem of substituting hydrogen in the present fluid-fuel distribution networks we now have is formidable. Another problem is the storage of hydrogen on vehicles. Table 2 gives the energies per unit volume and weight for various fuels. Liquid hydrogen looks promising from the point of view of weight if methods for handling this liquid can be successfully developed. Jet aircraft need a high-specific-energy fuel, and it is highly likely that some aircraft will use liquid hydrogen by the end of the century.[67] The massive changes required in the supportive infrastructure of other sectors of transportation militate against rapid widespread use of hydrogen in the next 30 years.

Other synthetic fuels that have been proposed are methanol, ethanol, ammonia, and hydrazine.[50,68] These fuels are more suitable for direct use in fuel cells than hydrocarbons, and fuel cells could play a major role in upgrading these fuels to electricity. Methanol is an attractive fuel, and it has been suggested that it would be easier to introduce methanol into the present liquid-fuel distribution system than liquid hydrogen. The idea was that methanol could be mixed with gasoline and gradually replace the latter.[69] However, recent studies indicate that phase separation and other problems may preclude such an arrangement[70] and slow down the rate at which methanol will become readily available.

No overall complete assessment of the future primary energy sources, on a worldwide basis, has been made. Most assessments for the United States ignore the contribution that solar energy will probably make. For this reason the author thinks that the projections of Wolf[71] give the most accurate picture of the energy source mix for the United States (Figure 6). Like most recent projections, these data are based on a projected overall growth rate of 3.4% per year. Recently, projections have been made using lower growth rates.[11] As to how these lower growth rates would affect the energy source mix is not

Table 2. Energetics of Various Fuels

| Fuel | Heat of combustion | | Fuel-cell energetics | |
	kJ/kg ($\times 10^4$)	kJ/m³ ($\times 10^7$)	$E°$ (volts)	kJ/kg ($\times 10^3$)
Liquid hydrogen	12.00	0.85	1.23	13.10
Ammonia	2.13	1.74	1.12	7.92
Hydrazine	3.02	3.06	1.56	9.38
Methanol	2.33	1.85	1.18	8.55
Ethanol	2.86	2.26		
Gasoline	4.83	3.56		
FeTiH	0.12	0.70		
LaNi$_5$H$_6$	0.17	1.10		

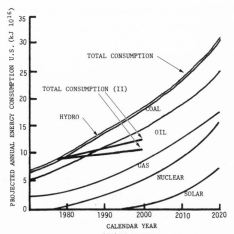

Fig. 6. Future energy source mix for the U.S.[71] Two lower estimates for total energy use are also shown.[11]

known, but the indications are that the fluid fuels of the future will be more suitable for fuel cells than the primary energy sources of the past.

8. Current Development of Fuel Cells

In recent history, the development of fuel cells has swayed between euphoria and disillusionment. In the late fifties, the United States government started funding the development of fuel cells for the space program; this funding reached a high of nearly $16 million in 1963. By 1970 the funding had dropped to about $3 million. In the United States the development was focused on hardware for space applications where cost was no object. This work culminated in the eminently successful Apollo fuel cells.[72] Little or no effort was made, using this money, to develop a commercially viable fuel cell. In Europe, on the other hand, there has always been a low-level effort to develop low-cost fuel cells.[73,74]

So far, the fuel cell has not had widespread use in the consumer sector of the economy. The main reason is the cost of the electrocatalyst. Hydrogen, the ideal fuel-cell fuel, has not been available, and other available fuels, such as hydrocarbons, can be directly used only with large amounts of platinum catalysts.[6] Low specific power has precluded mobile applications of fuel cells.[75] Operation of alkaline fuel cells using air as an oxidizer has been a problem because of CO_2 removal.[76] Acid electrolyte fuel cells obviate the problem; however, at present these fuel cells have only about one third the specific power of alkaline fuel cells. In the past, electrode life has been a problem; however, great strides have been made in improving electrode life.[76,77]

The commercial fuel cell did not come as the promised "spin off" from the space program. However, recent developments indicate that a serious effort is under way to make a commercial fuel cell.[72,77] In 1967 various utilities and Pratt and Whitney Aircraft started a development program on fuel cells. Gas utilities funded the development of a 12.5-kW fuel-cell system using methane in conjunction with a steam reformer. This program has resulted in the limited production of 12.5-kW units which have been tested at 37 locations in North America.[9,78] An improved 12.5-kW unit has been designed with specifications shown in Table 3. If this fuel cell is produced, it will represent the first genuine commercial fuel cell in the United States.

**Table 3. Provisional Specifications for a
Reformed Methane–Air Fuel Cell**

Continuous rated power	12.5 kW
Volume	0.93 m^3
Weight	2200 kg
Life	2000 h
Specific power (continuous)	5.68 W/kg
Power density (continuous)	13.4 Kw/m^3

More recently electric utilities have funded a program at Pratt and Whitney for development of a 26-MW fuel-cell system using 16-kW modules. This fuel cell would use reformed liquid hydrocarbons, such as No. 2 fuel oil or jet fuel. Tests have been carried out on a 37.5-kW unit of this type.[14] If this fuel cell is developed, then dispersed electric power generation by fuel cells could rapidly become a reality. The incentives are reduced transmission cost, reduced capital costs, short lead times, and compatibility with the environment.

Another fuel cell that is under development is the methanol–air fuel cell.[79,80] The goal here appears to be the development of a compact high-specific-power fuel cell for remote power and emergency power applications.

In Germany an alkaline hydrogen–oxygen fuel cell will be test-marketed soon. These units are for emergency power and remote power applications.[81]

The cost of the catalyst has always been a problem with fuel cells. However, promising results have been reported using low-cost catalysts. Tungsten carbide has been found to be a satisfactory anode catalyst for acid-electrolyte fuel cells.[82] Ammonia-activated graphited charcoal has shown promise as a cathode catalyst. Penovskite compounds, such as doped lanthanum cobaltite, have shown superior performance to platinum catalysts at 180°C in alkaline fuel cells.[83]

The technical feasibility of the fuel cell has been demonstrated, and work is under way to reduce its cost and demonstrate economic feasibility. In the United States great strides have been made in reducing the quantity of platinum electrocatalysts, and in acid fuel cells all-carbon support structures have been used instead of gold-plated tantalum screens.[84] In Europe the focus is on the replacement of platinoid catalysts.[72] Present developments indicate that the fuel cell will be first used commercially for remote power applications. The next major area for fuel cells is for upgrading methane to electricity. A later development will be the use of fuel cells for dispersed production of electricity. These three uses of fuel cells do not require any changes in the present fluid-fuel distribution network and could be readily implemented.

9. Future Work

If fuel cells are to play an important role in future energy systems, much more research and development than is currently under way is needed. The key problem area is the electrocatalyst. At present, even when only a few milligrams of platinoid catalysts per square centimeter are used, the catalyst cost for the fuel cell is $300–400/kW.[85] The study of electrocatalysts is not a simple matter and will have to be tackled on a broad front.

It is necessary to carry out both fundamental and applied work on electrocatalysis. Fundamental theoretical studies to elucidate the nature of the electron-transfer step are necessary. Techniques for *in situ* observations of electrode surfaces must be developed. Other important areas are the determination of zero charge potentials on metals and semiconductors and the elucidation of the role of absorbed intermediates in electrochemical reactions. It might be profitable for the fundamental electrochemist to abandon his vertical thinking and use a horizontal approach and learn from other fields, such as biology. It might be possible to use molecular engineering to produce enzyme-like catalysts for oxygen reduction and methanol oxidation.

There is a need for electrochemical engineering on a microscale to properly locate electrocatalysts so that the minimum amount may be more effectively used. The engineer will have to optimize mass, heat, and momentum transfer in the cell. If the fuel cell is ever to be used for mobile applications, such as automotive propulsion, the specific power of the system has to be increased.[75,85] This can be done by reducing electrode overvoltage, decreasing electrolyte weight, and clever engineering of auxiliary fuel-cell hardware.

Materials research must be carried out on a much more sophisticated level than has been done in the past. In the case of high-temperature fuel cells, it would be well to forego construction of large test units and instead focus on the materials problems that have to be solved if these systems are to be used for electricity generation from our remaining fossil fuels. Materials are also important in lower-temperature fuel cells. The goal of the effort should be the improvement of electrode life and the development of lightweight, durable fuel-cell stacks.

Parallel efforts will be necessary in acid-, alkaline-, and neutral-electrolyte fuel cells. Acid-electrolyte fuel cells are ideal for the reformed hydrocarbon–air systems that are currently under development. This type of fuel cell would be a suitable hydrogen–air fuel cell if hydrogen becomes readily available. Alkaline fuel cells would be more suitable for hydrogen–oxygen storage systems and for hydrazine. If enzyme-type catalysts were developed, it would be necessary to use a neutral electrolyte.

More efficient electrolyzers are necessary if the fuel cell is to be used in energy-storage systems. The work currently under way in this area should be expanded.

Fuel cells will not succeed without an adequate supply of suitable fuels. Apart from hydrogen generation, it is necessary to carry out work on hydrogen-storage systems. Hydrazine synthesis is another area where the electrochemist can contribute. However, the question of hydrazine toxicity has to be settled before work is carried out in this area.[86] In the long run, the best method of fuel synthesis may be the conversion of solar energy into hydrogen. Fundamental studies should be done in this area. The electrochemist can contribute to solar energy studies in general by developing processes for the low-cost production of solar cells. An important way in which the physical chemist can contribute is by developing imaginative ways to use oxygen. Large new markets for oxygen would accelerate implementation of the hydrogen economy.

The opportunities for electrochemical energy conversion in future energy systems are good. Timely preparations must be made if these hopes are to be realized. The electrocatalyst problem and the materials problems will not be solved overnight. The problems of one-shot devices, such as primary batteries or nuclear weapons, can be solved in short periods of time by large task forces. However, the problems of devices, where lifetime is a factor, such as rechargeable batteries, fuel cells, or nuclear reactors, do not readily yield to large teams in a short period of time. If fuel cells are to play an important role in the next 30 years, a commitment to the solution of these problems must be made now.

References

1. J. D. Parent and H. R. Linden, *Energy World* p. 3 (January 1974).
2. M. Altman, M. Telkes, and M. Wolf, *Energy Convers. 12*, 53 (1972).
3. W. N. Peach, *The Energy Outlook for the 1980's*, U.S. Government Printing Office, Washington, D.C. (1973), p. 5.

4. R. D. Doctor, *The Growing Demand for Energy*, A Report from Rand Corp., National Technical Information Service, U.S. Department of Commerce, Springfield, Virginia, (January 1972).

5. E. Bauer and J. Tobler, *Z. Elecktrochem. 39*, 169 (1933).

6. H. A. Liebhafsky and E. J. Cairns, *Fuel Cells and Fuel Batteries*, Wiley, New York (1968).

7. M. A. Elliott and N. C. Turner, *J. Inst. Fuel 47*, 55 (1974).

8. P. N. Ross, in: *Proceedings of the Miami Hydrogen Energy Conference* (T. N. Veziorglu, ed.), School of Engineering and Environmental Design, University of Miami, Coral Gables, Florida (March 1974), pp. S15–79.

9. A. L. Hamond, W. D. Metz, and T. H. Maugh II, *Energy and the Future*, American Association for the Advancement of Science, Washington, D.C. (1973).

10. L. R. Shepherd, *Nature 249*, 697 (1974).

11. P. Baldwin *et al.*, Exploring Energy Choices, A Preliminary Report of the Ford Foundation Energy Policy Project, The Ford Foundation, Washington, D.C. (1974).

12. S. H. Schurr, Energy Research Needs, A Report to the National Science Foundation, from Resources for the Future Inc., Washington, D.C., U.S. Government Printing Office, Washington, D.C. (October 1971).

13. P. Soedjanto and F. W. Schaffert, *Oil Gas J.* p. 88 (June 11, 1973).

14. W. Podolny, Presentation to the Subcommittee on Energy R&D Policy of the House Space and Astronautics Committee (July 1972).

15. H. C. Hottel and J. B. Howard, *New Energy Technology—Some Facts and Assessments*, The MIT Press, Cambridge, Massachusetts (1971).

16. A.E.C. Stresses Reactor's Value, *New York Times*, Dec. 29, 1974, p. 25.

17. A. J. Appleby, private communication (September 30, 1974).

18. W. G. Dupree, Jr., and J. A. West, *U.S. Energy Through the Year 2000*, U.S. Department of the Interior (December 1972), p. 20.

19. G. R. Bainbridge, *Nature 249*, 733 (1974).

20. W. Hafele and C. Starr, *J. Br. Nucl. Energy Soc. 13*, 131 (1974).

21. R. Balent and R. J. Beeley, *Energy Sources 1*, 189 (1974).

22. W. J. Blair and R. C. Thompson, *Science 183*, 715 (1974).

23. A. R. Tamplin and T. B. Cochran, *Radiation Standards for Hot Particles*, Natural Resources Defense Council, Washington, D.C., (1974).

24. D. J. Rose, *Science 184*, 351 (1974).

25. A. M. Weinberg, *Science 177*, 27 (1972).

26. D. P. Geesaman, in: *The Energy Crisis*, (R. S. Lewis and B. J. Spinard, eds.), p. 57, The Educational Foundation for Nuclear Science, Chicago, Illinois (1972).

27. R. F. Post and F. L. Ribe, *Science 186*, 397 (1974).

28. L. J. Carter, *Science 186*, 811 (1974).

29. B. E. Smith, in: *Wind Energy Conversion Systems*, (J. M. Savino, ed.), p. 5, National Technical Information Service, Washington, D.C. (June 1973).

30. A. H. Stodhard, in: *Wind Energy Conversion Systems* (J. M. Savino, ed.), p. 65, National Technical Information Service, Washington, D.C. (June 1973).

31. S. H. Salter, *Nature 249*, 720 (1974).

32. B. J. Brinkworth, *Nature 249*, 726 (1974).

33. G. G. Libowitz, *Proceedings 9th Intersociety Energy Conversion Engineering Conference*, San Francisco, California, August 1974, American Society of Mechanical Engineers, New York (1974), p. 322.

34. A. B. Meinel and M. J. Meinel, *Phys. Today 25*, 44 (1972).

35. H. Becquerel, *Compt. Rend. 9*, 561 (1839).

36. P. E. Glaser, *Science 162*, 857 (1968).

37. P. Donovan, W. Woodward, W. R. Cherry, F. H. Morse, and L. O. Herwig, An Assessment of Solar Energy as a National Energy Resource, NSF/NASA Solar Energy Panel (December 1972).

38. D. G. Evans, *J. Inst. Fuel 47*, 129 (1974).

39. D. P. Gregory, A Hydrogen-Energy System, American Gas Association, Catalogue No. L21173 (August 1972).

40. G. C. Szego, in: *Wind Energy Conversion Systems* (J. M. Savino, ed.), p. 152, National Technical Information Service, Washington, D.C. (June 1973).

41. M. L. Kyle, E. J. Cairns, and D. S. Webster, Lithium/Sulfur Batteries for Off-Peak Energy Storage, A Report to the U.S. AEC, Contract No. W-31-109-Eng-38, Report No. ANL-7958, National Technical Information Service (March 1973).

42. L. I. Ricci, *Chem. Eng.* p. 24 (February 3, 1975).

43. W. A. Titterington and J. F. Austin, *Extended Abstracts*, Vol. 74-2, Abstract No. 233, Electrochemical Society Inc. (October 1974).
44. D. P. Gregory, *Extended Abstracts*, Vol. 74-2, Abstract No. 226, Electrochemical Society Inc. (October 1974).
45. J. M. Burger, B. A. Lewis, R. J. Isler, F. J. Salzano, and J. M. King, *Proceedings 9th Intersociety Energy Conversion Engineering Conference*, San Francisco, California, August 1974, American Society of Mechanical Engineers, New York (1974) p. 428.
46. H. J. Schwartz, in: *Wind Energy Conversion Systems* (J. M. Savino, ed.), p. 146, National Technical Information Service, Washington, D.C. (June 1973).
47. D. R. Vissers, Z. Tomczuk, and R. K. Steunenberg, *J. Electrochem Soc. 121*, 665 (1974).
48. R. P. Hamlen, E. L. Simons, and F. G. Will, *Extended Abstracts*, Vol. 72-2, Abstract No. 35, Electrochemical Society Inc. (October 1972).
49. D. L. Klass, *Chem. Technol.*, 161 (1974).
50. M. Celvin, *Science 184*, 375 (1974).
51. J. O'M. Bockris, *Science 176*, 1323 (1972).
52. W. E. Wirsche, K. C. Hoffmann, and F. J. Salzano, *Science 80*, 1325 (1973).
53. R. E. Billings, *Extended Abstracts*, Vol. 74-2, Abstract No. 240, Electrochemical Society Inc. (October 1974).
54. J. E. Funk and R. M. Reinstrom, *System Study of Hydrogen Generation by Thermal Energy*, Vol. 2, Supplement A, Energy Depot Electrolysis System Study, Final Report TID 20441, Allison Division, General Motors Corporation, Indianapolis, Report EDR-3714, U.S. AEC, Washington, D.C. (June 1964).
55. G. de Beni and C. Marchetti, *Eur. Spectra 9*, 46 (1970).
56. J. B. Pangborn and J. C. Slater, in: *Proceedings of the Miami Hydrogen Energy Conference* (T. N. Veziorglu, ed.), School of Engineering and Environmental Design, University of Miami, Coral Gables, Florida (March 1974), pp. S11–35.
57. A. J. Appleby, *Extended Abstracts*, Vol. 74-2, Abstract No. 230, Electrochemical Society Inc. (October 1974).
58. A. Fujishima, K. Honda, and S. Kikuchi, *Kogyo Kagaku Zasshi 72*, 108 (1969).
59. A. Fujishima and K. Honda, *Bull. Chem. Soc. Jpn. 44*, 1148 (1971).
60. A. Fujishima and K. Honda, *Nature 238*, 37 (1972).
61. S. N. Paleocrossas, in: *Proceedings of The Miami Hydrogen Energy Conference* (T. N. Veziorglu, ed.), School of Engineering and Environmental Design, University of Miami, Coral Gables, Florida (March 1974), p. S7–33.
62. T. S. Jayadevaiah, *Applied Physics Letters 25*, 399 (1974).
63. T. Ohta and M. Kimaya, *Proceedings 9th Intersociety Energy Conversion Engineering Conference*, San Francisco, California, August 1974, American Society of Mechanical Engineers, New York (1974), p. 317.
64. J. O'M. Bockris, in: *Proceedings of The Miami Hydrogen Energy Conference* (T. N. Veziorglu, ed.), School of Engineering and Environmental Design, University of Miami, Coral Gables, Florida (March 1974), p. S9–1.
65. G. Seaborg, Finding a New Approach to Energetics—Fast, *Saturday Review—World* (December 14 1974), pp. 44–48.
66. E. M. Dickson, T. J. Logothetti, and J. W. Ryan, in: *Proceedings of The Miami Hydrogen Energy Conference* (T. N. Veziorglu, ed.), School of Engineering and Environmental Design, University of Miami, Coral Gables, Florida (March 1974), pp. S8–1.
67. G. D. Brewer, *Astronaut. Aeronaut. 12-5*, 40 (1974).
68. J. O'M. Bockris, *Ambio 3*, 15 (1974).
69. T. B. Reed and R. M. Lerner, *Science 182*, 1299 (1973).
70. E. E. Wigg, *Science 186*, 785 (1974).
71. M. Wolf, *Energy Convers. 14*, 9 (1974).
72. F. T. Bacon and T. M. Fry, *Proc. R. Soc. 334*, 427 (1973).
73. W. Vielstich, *Brennstoftelemente*, Verlag Chemie Gmbh, Weinheim/Bergstr. (1965).
74. E. Justi and A. W. Winsel, *Kalte-Verbrennung*, Franz Steiner Verlag, Wiesbaden (1962).
75. E. H. Hietbrink, J. McBreen, S. M. Selis, S. B. Tricklebank, and R. R. Witherspoon, Electrochemical Sources for Vehicle Propulsion, in: *The Electrochemistry of Cleaner Environments*, (J. O'M. Bockris, ed.), Plenum Press, New York (1972).
76. K. V. Kordesch, Fuel cells and their future prospects, Paper No. 70CP213-PWR, The Institute of Electrical and Electronics Engineers Inc., New York (December 1969).
77. A. D. S. Tantram, *Energy Policy* p. 55 (March 1974).

78. R. W. Benedict, Little Black Box, A discussion of the TARGET Program, *Wall Street Journal* (May 19, 1971).
79. Fuel Cell Development Set by French Firm, Jersey Standard Unit, *Wall Street Journal* (December 7, 1970), p. 3.
80. B. Warzawski, B. Verger, and J. C. Dumas, *J. Marine Tech. Soc. 5*, 28 (1971).
81. F. von Sturm, private communication (October 1974).
82. L. Baudendistel, H. Bohn, J. Heffler, G. Louis, and F. A. Pohl, *Proceedings 7th Intersociety Energy Conversion Engineering Conference*, San Diego, California September 1972, American Chemical Society, Washington (1972), p. 20.
83. A. C. C. Tseung and H. L. Bevan, *J. Electroanal. Chem. Interfacial Electrochem. 45*, 429 (1973).
84. G. H. Dews and R. W. Vine, U.S. Pat. 3,801,374.
85. E. J. Cairns, in: *Physics and the Energy Problem—1974*, (M. D. Fiske and W. W. Havens, Jr., eds.), p. 160, American Institute of Physics, New York (1974).
86. *Evaluation of Carcinogenic Risk of Chemicals to Man*, Vol. 1, p. 95, International Agency for Research on Cancer, World Health Organisation, Geneva (1972).

3

Summary: Energy Conversion—The Past and the Future

A. J. Appleby

Future society, after the year 2000 or 2030, will be essentially based on nonfossil energy. We can certainly expect that oil and gas reserves will be largely depleted by that time if energy growth continues at recent rates. Even if massive conservation measures are implemented (which appears currently improbable), the peak of production will have been passed in 25 years. While large reserves of coal will still be found in many countries, particularly in China, the Soviet Union, and the United States, they will be concentrated in remote locations. The coal will consequently be very difficult to mine and transport without enormous expenditure of effort in the form of capital, labor, and intrinsic energy, which will result in a low overall efficiency as an energy source. Environmental considerations, necessitating sulfur removal and preparation (at low efficiency) of synthetic fuels, will further degrade its value as an energy resource. Production of coal (and coal-based synthetic fuel) on a scale to supply the U.S. industrial society will be rendered improbable by these considerations, by the problems of geography, and by the difficulties of finding the required labor on the free market. In consequence, coal-based synthetic fuels and coal-based production of electricity will be as costly as the nuclear or solar alternatives. Shale oil, another possible fossil-fuel source, will have little impact on the energy economy since its extraction requires more energy than coal and will be restricted by the limitations of water supply in the regions in which it is found. Underground gasification of coal, which, if developed, represents the most attractive method for coal use, will be similarly restricted by the problem of water supply: lignite gasification may in consequence be more attractive.

A nonfossil energy basis for the industrial world may be provided by nuclear (fission, fusion), solar (direct or indirect), tidal, and geothermal energy. Future energy will be either in the form of electricity or of fluid fuels, according to best end-use optimization and to transportation considerations. Since this nonfossil energy will be very capital-intensive, it will be very much more costly than today's cheap oil if conventional accounting methods are used. Its development promises to completely change the structure of industrial society, in that the necessary investment for energy will compete with that for less necessary industries. As the total available investment will be limited by lower growth rates resulting from more costly energy, it will inevitably be those industrial areas that are the largest energy consumers (particularly transportation-based industries, industries dedicated to planned obsolescence) which will be squeezed out of the economy. This will require a rethinking of the whole structure of industry and will necessitate a change of many of the recently acquired habits of our society. In the long term, it will release the pressure on valuable natural resources.

To ensure that society will be able to afford this nonfossil-fuel energy base, calculated both from the viewpoint of total investment needed (both financial and energetic) and from the energy requirements of industry, measures will have to be instituted for the conservation of energy, both active (increased efficiency of end use and of conversion)

A. J. Appleby • Laboratoire d'Electrolyse, C.N.R.S. Bellevue, France.

and passive (use of waste heat, restructuring of the GNP). Low-grade heat (including solar heat and thermal waste, perhaps using heat pumps) will find widespread use both in the domestic and industrial sectors. The GNP will be restructured to eliminate those industries that are most energy-consuming, particularly those that require high-grade energy in the form of work. This is particularly true of transportation, currently an area of exceedingly poor end-use efficiency and of throwaway products.

Electrochemistry will have a large part to play in transportation (long-lasting, low-maintenance electric vehicles with high efficiencies) and in stretching fossil-fuel reserves, where appropriate, by the use of fuel cells. The latter will first be introduced for use in remote areas and in peak-shaving applications, since they promise quite high efficiencies even in small units. As they consume reformed carbonaceous fuels, low-temperature fuel cells (acid type) currently have catalysis and poisoning, hence cost and lifetime, problems. Direct hydrocarbon fuel cells are still a long way off and may not arrive in time to have much impact on the energy economy. High-temperature (molten carbonate) systems are now beginning to show a good deal of promise. Research is clearly needed in the areas of low-temperature electrocatalysis and in the problem of materials at high temperature.

In the nonfossil-fuel economy of the future, the fuel cell will be of further importance as a convertor of stored hydrogen for transportation, commercial, and domestic use. Catalytic problems are much less acute than with the reformed hydrocarbon low-temperature systems, and it is ironic that today's technology will be perfectly applicable. The remoteness of the locations of tomorrow's nonfossil energy-producing centers from population concentrations will force the development of cheap methods of energy storage, transport, and reconversion for appropriate end use. Electricity is unstorable, and its transport is costly and involves high energy losses. The distances involved for transportation become clear when it is realized siting problems (cooling requirements, possible pollution dangers) necessitate the location of primary and breeder fission plants on remote sea coasts or in the oceans. Plants operating on thermal ocean temperature gradients will be immersed in the tropical seas. Plants using wave and wind power will be located in the northern oceans and remote sea coasts or mountains, respectively. Solar plants will be located in clear-sky areas, particularly deserts. Small domestic-size solar or wind units for local use will avoid the problem of transportation but will involve year-round storage.

The most appropriate storage and transportation medium is hydrogen produced by decomposing water. The technology for its storage and transmission by pipeline will be of low cost and can be extrapolated from current natural-gas practice. In addition, the use of hydrogen will be nonpolluting. Other materials may be rejected as energy vectors on account of their low specific density (e.g., the water gas–methane "chemical heat pipe," aluminum metal), especially when it is considered that they will involve a two-way transport of energy vector and its residue for reprocessing, whereas hydrogen requires only one-way transport. However, I should remark that the use of aluminum as an energy vector in a cycle over limited distances is not to be rejected out of hand: Dr. Despić's interesting results on the very low self-discharge of Al–In alloys point the way to a possible medium-efficiency (perhaps 40%) Al–air battery that could serve as a stop-gap system for vehicle propulsion until the hydrogen fuel cell is ready.

The great question facing future technology will be the highest efficiency and lowest investment cost method for the manufacture of nonfossil hydrogen. Two rival technologies exist—namely electrolysis and thermochemical cycles. The object of the latter is to attempt to produce hydrogen without the use of free energy (work). This is, however, impossible, since irreversible losses within the system require inevitably a Carnot-limited work input. While irreversible losses resulting from thermal gradients have generally been considered in estimating efficiencies of model cycles, they have been normally neglected for the chemical kinetic and separation steps, which have been assumed to operate under reversible conditions. Approximate calculations show that when all such factors are considered for published cycles working up to 850°C (using high-temperature reactors), overall high-heating-value efficiencies will rarely be over 30–35%, and up to 50% of the heat input will be required for making electricity. The plants will be expensive, will have a heavy chemical inventory and turnover, and are likely to suffer from substantial materials problems. By contrast, direct low-temperature electrolysis (KOH, 130°C) using HTR electricity produced at 42% efficiency should be able to produce hydrogen at 39% high-heating-value efficiency, with problems that can be anticipated now and solved relatively cheaply. It is only when very high-temperature heat sources are available (over 1200°C) from special reactors or from focused solar radiation that thermochemical cycles may have a definite edge over electrolysis, since electricity production making efficient use of such temperatures will involve very costly topping cycles. Even then, the most efficient cycles may involve combined thermal and electrochemical steps. One may therefore conclude that electrochemistry will have a dominant part to play in the future energy economy.

One final and very important point is the question of costs, either in terms of investment per unit output or in terms of the final cost of energy. Investment per unit of output is meaningless unless the time of amortization is considered and unless certain maintenance costs are allowed for. It appears that the current generation of acid fuel cells have acceptable investment costs, but, according to Prof. Bockris's comment, insufficient lifetime. Again, these costs seem high when it is considered that the units (apart from their low catalyst loadings) consist of the simple packing of identical parts made from very cheap materials. The costs may therefore not consider the future scale of production that would be desirable. Their current expense inhibits market penetration, without which they will always be expensive. These points will have to be solved by future policy decisions. Energy supply to a society in the future will be much too important to be left to the vagaries of the capital marketplace. It will require vast amounts of capital, which must be invested to ensure the very survival of society. It will be at least as important as defense expenditure, for which conventional bookkeeping does not apply. In any case, if conventional cost calculations are carried out for the nonfossil energy of the future, using the current interest rates (which may be assumed to be a permanent fact in the future demand-inflated economy), we find that society could not afford such energy at anything resembling the present growth rate. Primary energy to fuel food production for the expanding world population could not be found. Energy production, like defense or health, education, and welfare, must have a budget that is either a percentage of the GNP, or is provided in the form of government-guaranteed low-interest (i.e., inflation-free) loans, with energy costs indexed to the cost of living. This must be combined with a conservation program, as outlined at the beginning of this summary.

4

Electrochemistry in the Biomedical Sciences

S. Srinivasan, G. L. Cahen, Jr., and G. E. Stoner

1. Introduction

One of the most advanced fuel-cell systems is the human body in which chemical energy liberated during enzymatic reactions of organic compounds and oxygen is transformed into mechanical energy probably via electrical energy. Physiological and biochemical processes (e.g., ion transport, energy metabolism, nerve and muscle conduction, bone growth and healing, blood clotting. cancer, etc.) involve electrochemical mechanisms (ion fluxes, adsorption at solid–electrolyte interfaces, proton transfer, electron transfer, electrokinetic effects, etc.), but it was only during the last 30 years that such electrochemical mechanisms of biological processes have even begun to be elucidated.

At the first Symposium on Electrochemistry in Biology and Medicine in 1953,[1] Shedlovsky in his introductory remarks stated that, "Living matter or a living cell is not a mere assembly of chemical compounds. It is an oriented, dynamic system of complex materials in constant interaction with its environment, a complex chemical laboratory manufacturing many compounds no chemist has yet been able to synthesize, and electrochemical in many if not perhaps all of its functions." Since the early sixties, there has been a growing interest in the interdisciplinary field of bioelectrochemistry and an increasing number of physicists, chemists, mathematicians, and engineers are engaged in active research, either in collaboration with biological and medical scientists or independently, trying to understand and in some cases control these complex physiological and biochemical processes. Within the last four years there have been three symposia on bioelectrochemistry,[2-4] and two more are scheduled this year.[5,6]

Bioelectrochemistry is a very extensive field. An attempt is made in Table 1 to present a number of areas of active research at the present time. They are classified under the broad headings of electrophysiology; energy metabolism; cellular chemistry; clinical medicine, surgery, and dentistry; and biomedical engineering. This table by no means covers all the topics in bioelectrochemistry. Due to the extensive nature of bioelectrochemistry, the present article only deals with some of the topics selected from Table 1.

2. Electrochemical Potentials of Living Cells

2.1. General Phenomenology of Electrical Events at Cell Boundaries

One of the most widely investigated and controversial fields in electrophysiology is the origin of the rest potentials across cell membranes. Before presenting the different viewpoints, it is worthwhile illustrating the general phenomenology of electrical events in cell boundaries (Figure 1) as presented by Teorell.[7] According to this view, the cell membrane is considered as an organized architecture of interlinked negatively charged groups,

S. Srinivasan • Department of Applied Science, Brookhaven National Laboratory, Upton, New York 11973. *G. L. Cahen, Jr., and G. E. Stoner* • Department of Material Science, University of Virginia, Charlottesville, Virginia 22901.

Table 1. Electrochemistry in the Biomedical Sciences—Some Areas of Active Research

Electrophysiology	Energy metabolism	Cellular chemistry	Clinical medicine, surgery, and dentistry	Biomedical engineering
1. Rest potentials across membranes	1. Electron transfer and energy conservation	1. Contributions of chemical groups to surface charge	1. Intravascular thrombosis	1. Inactivation of pathogens
2. Action potentials	2. Phases, membranes, and electron transfer	2. Antigen–antibody reactions	2. Bone healing and growth	2. Microbial detection
3. Transport of ions and neutral molecules across membranes	3. Oxidative phosphorylation		3. Acupuncture	3. Blood-gas determinations
4. Bilayer lipid membranes			4. Iontophoresis	4. Specific ion electrodes
5. Nerve conduction, excitation, and impulse transmission			5. Cancer	5. Enzyme electrodes
6. Synaptic and ephatic transmission				6. Biochemical fuel cells
7. Neuromuscular functions				7. Energy sources for pacemakers and artificial hearts
8. Neuroskeletal phenomena				

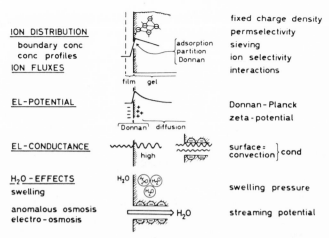

ION DISTRIBUTION
boundary conc
conc profiles
ION FLUXES
film   gel

fixed charge density
permselectivity
sieving
ion selectivity
interactions

adsorption
partition
Donnan

EL-POTENTIAL
Donnan diffusion

Donnan-Planck
zeta-potential

EL-CONDUCTANCE
high

surface=
convection } cond

H₂O-EFFECTS
swelling
anomalous osmosis
electro-osmosis

swelling pressure
streaming potential

Fig. 1. Scheme of the general phenomenology at a fixed charged membrane. [From Teorell.[7]]

which are immobilized. Carboxyl groups from proteins and phosphate groups from lipoids contribute to this negative charge. Thus, as in the case of an ion-exchange membrane, one may expect the same type of electrochemical phenomena on a microscale as are encountered on the macroscale. There are three important characteristics: (1) a discontinuity of the concentration distribution of free ions; (2) the electrical boundary potential, i.e., the Donnan potential; and (3) an osmotic pressure effect.

The rest potential across a cell membrane, according to Teorell, is the sum of a two-phase boundary and an internal intramembrane potential related to diffusion phenomena. Figure 2 presents a summary of electrical events at cell boundaries.[7] The transport processes of charged particles such as ions or colloid electrolytes across cell boundaries are electrokinetic in origin. As seen in Figure 2, the cell mosaic structure is composed of negatively charged and noncharged pores. The cell membrane dynamics proposed by Teorell led to a model for excitability of cells.

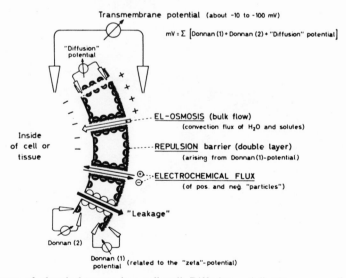

Fig. 2. Summary of electrical events in a cell wall. Diffusion and "active processes" are omitted. [From Teorell.[7]]

Table 2. Theories of Origin of Resting (Maintained or Standing) EMF in Biological Systems[8]

1. Diffusion potentials[73-79]
   a. Chemical chain
   b. Concentration chain
2. Oxidation–reduction potentials[80-86]
3. Phase boundary[87]
4. Metabolic influences on (1) above[88]
5. Donnan effects on (1) or (3) above
6. Na or K pumps[89-91]
7. Adsorption or fixed charge[92-98]
8. Association capacitor[99]

2.2. Diffusion Theory for Membrane Potentials

There are more than two dozen theories of the origin of rest potentials across cell membranes. Jahn[8] has listed them under eight categories (Table 2). One of the most widely accepted theories is based on the diffusion model.[9,10] According to Hodgkin and Katz,[11] the rest potential (ψ) may be expressed by the equation

$$\psi = \frac{RT}{F} \ \ln \ \frac{P_{K^+}[K^+]_{in} + P_{Na^+}[Na^+]_{in} + P_{Cl^-}[Cl^-]_{in}}{P_{K^+}[K^+]_{out} + P_{Na^+}[Na^+]_{out} + P_{Cl^-}[Cl^-]_{out}} \tag{1}$$

where P_{K^+}, P_{Na^+}, and P_{Cl^-} are the permeability constants of the cell membrane for the indicated ions. Hodgkin and Katz found that if $P_{K^+}:P_{Na^+}:P_{Cl^-} = 1:0.04:0.45$, the calculated and experimental ψ values are in reasonably good agreement. One of the questions which arises is how the cell maintains its ionic composition and its internal negative potential. Hodgkin and Keynes[12] proposed that the maintenance of the unequal distribution of ions and of the potential difference is a result of a metabolic process supplying energy to expel sodium ions which have leaked into the cell, thus accumulating potassium in the interior. Evidence for a Na^+/K^+ exchange pump was obtained when it

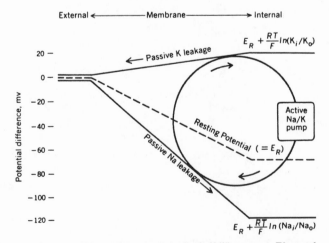

Fig. 3. Electrochemical (downhill) gradients and ionic (uphill) pumps. Electrochemical potentials cause K^+ ions to leak outward and Na^+ ions to leak inward through the cell membrane. Ionic distribution is maintained by an active secretory process, requiring continuous supply of energy. The electrical potential gradient across the membrane is indicated by broken line. [From Katz.[10]]

Table 3. Anti-K_i^+/K_0^+ [after Jahn[14]] [a] Evidence Against the Diffusion Potential Theory

I. Varied K_0^+ and did not obtain the expected results
  1. Ling, various old data
  2. Stampfli, node of Ranvier
  3. Tasaki, nerve fibers
  4. Schoffeniels, electroplax
II. Varied K_i^+ and did not obtain the expected result
  5. Grundfest, nerve fibers
  6. Baker, Hodgkin, and Shaw; Baker, Hodgkin, and Meves, nerve fibers
  7. Tasaki, Watanabe and Takenaka; Tasaki and Shimamura; Tasaki and Takenaka; Tasaki, nerve fibers
  8. Falk and Gerard, muscle fiber
  9. Ling, soaked muscle
III. Natural variation of K_i^+/K_0^+
  10. Blinks, marine algae, *Valonia* and *Halicystis*
IV. Varied Ca_0^{2+} and other cations. All lower the emf
  11. Naitoh and Eckert
V. K_i^+ is bound to cytoplasmic material
  12. Nasonov; Aleksandrov; Tarusov; Troshin; Ling; Eisenman; Fischer and Hooker; Jahn; Dunham and Child; Cope; Ling

[a]References for original papers may be obtained from the paper by Jahn.

was found that the same metabolic poisons which inhibit the efflux of Na^+ ions also decrease the transport of K^+ ions into the cell. Further, a large part of active efflux of sodium ions from the cell ceases when K^+ ions are removed from the external medium. A single coupled transport mechanism, as shown in Figure 3, was inferred from these experimental findings.

There are some researchers who have strongly opposed the diffusion potential theory. Jahn[8] has tabulated evidence against this theory (Table 3). As seen in Table 3, when the potassium concentration outside the cell (K_0^+) was varied, the experimental rest potentials differed from those calculated using the Goldman equation. Likewise, when the potassium concentration within the cell was altered, a similar discrepancy between calculated and experimental results was observed.

2.3. Association Induction Theory for Membrane Potentials

In light of these anomalies, it is worthwhile to examine some of the alternate theories. Calculations by Ling revealed that more than three times the maximum available energy of a resting frog muscle is necessary for the ionic pumps proposed in the widely accepted diffusion theory. Ling[13] thus suggested a theory, "association–induction hypothesis," for the selective ionic distribution and rest potential based on equilibrium phenomena due to adsorption on fixed ionic sites on cellular proteins. According to the association–induction hypothesis, the living cell constitutes a proteinaceous fixed-charge system (Figure 4) containing a three-dimensional network bearing ionic groups and H-bonding groups which polarize and orient the solvent water molecules in multilayers such that the motions (particularly the number of rotational ones) of multicationic solvents and solute molecules (e.g., sugars) or hydrated ions become greatly reduced, comparable to Traubes hydrated glass surfaces, collodion membranes, and exchange-resin membranes.

In the living cell, the solvent molecules and ions are associated with the protein matrix. Thus, they have a lower entropy within the cell, which affects the distribution of these entities between the intra- and extracellular phases. In addition to association of ions and neutral molecules on proteins, there are inductive effects through intervening polypeptide

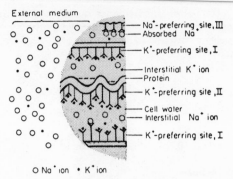

External medium

— Na$^+$-preferring site,III
— Absorbed Na$^+$
— K$^+$-preferring site,I
— Interstitial K$^+$ ion
— Protein
— K$^+$-preferring site,II
— Cell water
— Interstitial Na$^+$ ion
— K$^+$-preferring site,I

O Na$^+$ ion   • K$^+$ ion

Fig. 4. Diagrammatic illustrations of a living cell. Stipled area represents space filled with water in polarized multilayers. [From Ling.[13]]

chains. This cooperative behavior is responsible for the propagated action potential of muscle and nerve in muscle contraction and in other similar cell activities. The consequences of this model are:

1. Due to the lower rotational entropy of cellular water as compared with normal water, it has a lesser affinity for neutral molecules or ions.
2. The proteins offer the adsorption sites for ions and neutral molecules.
3. The concentrations of adsorbed ions or neutral molecules depend on steric factors and on the electron density of the ionic or hydrogen bonding sites.
4. The affinities of the ions or neutral molecules can be altered by inductive effects.

In contrast with the membrane theory which explains electrochemical phenomena in a macroscopic way, the association–induction hypothesis deals with cellular potentials, ionic permeability, and other aspects of cell physiology in molecular terms.

2.4. Association–Capacitance Theory for Membrane Potentials

A model closely allied with the "association–induction hypothesis" is the "association–capacitance theory" proposed by Jahn.[8,14] The principle of this theory is shown in Figure 5. The negative charge of the membranes is assumed to be due to carboxyl ions. The anionic field strengths are such that the inside is selective for potassium ions while the outside is so for sodium ions. The potential across the membrane is the algebraic sum of the potentials across the two interfaces. Since the sodium ions are bound at a greater distance than the potassium ions, the potential at the exterior interface is greater than across the interior interface. This would indicate a positive rest potential. Action potentials are observed when the stimulating potential removes the associated Na$^+$ ions from a small section of the surface. When a decrease in potential greater than the normal membrane potential occurs, the latter is reversed in sign. Ions from adjacent areas fill the vacancy causing a self-propagating action potential, transmitted at a more or less constant amplitude and velocity.

2.5. Oxidation–Reduction Potential Theories for Membrane Potentials

The widely accepted diffusion theory for membrane potentials which originates from the early work of Bernstein[15] is based on nineteenth-century electrochemical concepts.

RESTING POTENTIAL

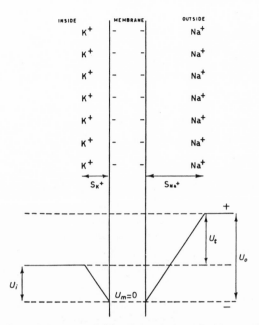

Fig. 5. Representation of association–capacitance theory for membrane potentials. The average separation of anions and cations is shown by S values. U_m, the voltage drop within the membrane, is assumed to be zero. U_i and U_0 are the "inside and outside" potentials with respect to the membrane. U_t is the "membrane potential." [From Jahn.[14]]

At the present time, it is well known that redox reactions occur at biological interfaces, and it is more probable that the observed potentials across these interfaces are a result of these redox processes rather than due to the thermodynamic reversibility with respect to the alkali ions. Some workers, including Lund,[16] Marsh,[17] Korr,[18] and Lorenti de No,[19] made suggestions along these lines over 25 years ago. More recently, Bockris[20] proposed that the potential difference across biological cells is a mixed potential of two electron-transfer reactions occurring at different points along the membrane, e.g., at different groups of enzymes adsorbed on the membrane. The reducible species is probably oxygen, and the oxidizable one is an organic compound. This model is similar to the one proposed by Del duca and Fuscoe.[21] Further support for this view comes from the work of Cope,[22] who showed that the kinetics of decay of photogenerated free radicals in eye melanin particles illustrates an exponential dependence of the rate of the reaction on potential across the membrane of the cell. Thus, Cope[22] suggested that there is a rate-determining interfacial electron-transfer reaction in the overall reaction. The mechanism of electronic conductivity between the sites at which the anodic and cathodic reactions occur was interpreted by Bockris[23] on the basis of electron tunneling over short distances within the membrane. However, in the functioning of a biological membrane, there is also the associated transport of alkali metal ions through ionically conducting pores in the membrane. It would not, however, be the cause but a result of interfacial reactions in micro-fuel cells[23] which occur to transform the chemical energy of biochemical oxidation reactions to the electrochemical energy of biological cells.

3. Ion Transport through Membranes

Biological membranes are complex in structure and difficult to use for experimental investigations. Mueller *et al.*[24,25] introduced the idea of using an artificial phospholipid bilayer membrane to study transport processes across membranes. In many respects, the properties of the artificial bilayer lipid membranes resemble those of the lipid part of natural cell membranes. To increase the electrical conductivity of the artificial membranes (which are in the region of 10^{-9}–10^{-7} ohm$^-$ cm^{-1}), it is necessary to add uncouplers for oxidative phosphorylation, e.g., dinitrophenol, pentachlorphenol, valinomycin, or iodine.

It may now be worthwhile to examine how carriers help in the transport of ions across membranes. This subject has been covered in an excellent review article by McLaughlin and Eisenberg.[26] The antibiotics valinomycin and actin have been mostly used as neutral carriers. The antibiotics bind alkali metal cations (K^+ preferentially over Na^+) and make them soluble in bulk hydrocarbon phases, which then helps their transport across the membrane. Without a carrier the alkali metal ions as well as chlorides are insoluble in the hydrocarbon phase and thus cannot be transported across the membrane. An illustration of the currently accepted model for transport of ions across bilayer lipid membranes is shown in Figure 6. An ion (I^+) is captured at one interface by the carrier (S), transported to the other interface where the carrier discharges the ion, ($I^+S^+ \rightarrow I^+ + S$) and diffuses back to the first interface. For valinomycin, the selectivity of the alkali metal ions is in the following ratio:

$$Rb:K:Cs:Na:Li \equiv 2:1:8 \times 10^{-1}:4 \times 10^{-5}:5 \times 10^{-6} \qquad (2)$$

whereas for actin it is

$$K:Rb:Cs:Na:Li \equiv 1:6 \times 10^{-1}:4 \times 10^{-2}:7 \times 10^{-3}:4 \times 10^{-4} \qquad (3)$$

A charge pulse technique was developed by Feldberg and Kissel[27] to determine the kinetics of carrier-mediated transport of ions across bilayer lipid membranes. In this method, the bilayer lipid membrane is charged within a microsecond. Since the charge can only decay through the membrane's conductive mechanism, the resulting voltage decay reflects membrane and boundary layer phenomena. As the membrane capacitance is independent of voltage, this method appears to be more useful for studying phenomena across membranes than those across electrode–solution interfaces. Feldberg and Kissel[27] carried out an analysis of the voltage–time transients obtained on actin (non-, mon-, din-, and trin-) actin-mediated transport of ammonium ions and valinomycin-mediated transport of cesium and potassium through glycerol mono-oleate bilayer membranes. In

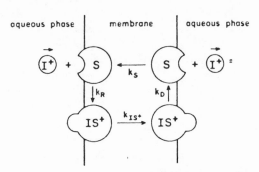

Fig. 6. Diagram of carrier mediated ion transport across a membrane defining rate constants. [From McLaughlin and Eisenberg.[26]]

Table 4. Rate Constants for Carrier-Mediated Ion Transport Across Bilayer Lipid Membranes[27]

Constant	Trinactin-NH_4^+		Valinomycin-K^+	
	GDO[a]	(GMO)[b]	(GDO)	(GMO)
k_{is}^*	1.7×10^4	1.3×10^4	9.8×10^4	7.5×10^5
k_s	4.4×10^4	$>1.5 \times 10^5$	9.0×10^4	1.6×10^5
k_1	6.6×10^7	1.2×10^8	7.4×10^7	1.7×10^9
k_{-1}	9.2×10^3	1.2×10^4	7.4×10^4	4.7×10^5
K_1	7.2×10^3	1.0×10^4	1.0×10^3	3×10^3
k_1/k_s	1.5×10^3	$<8 \times 10^2$	8.2×10^2	1.1×10^4

[a] Values on glycerol dioleate (GDO) membranes obtained by Laprade *et al.*[28]
[b] Values on glycerol mono-oleate (GMO) membranes obtained by Feldberg and Kissel.[27]

Table 4 the results of this work are compared with the transport studies carried out on glycerol dioleate bilayer membranes using the voltage-clamp (potentiostatic) method.[28] The rate constant for the formation of the ion–antibiotic complex is represented by k_1 while k_{-1} refers to the dissociation constant, k_s is the rate constant for the transport of free carrier across the membrane, and k_{is}^* is the rate constant for the transport of ion complex halfway across the membrane. There is good agreement between the two sets of data for actin-mediated ammonium ion transport, but this is not so for valinomycin-mediated potassium ion transport.

There are some antibiotics which facilitate the transport of ions by another mechanism—they cause an increase in the size of the pores of the membrane. Using the voltage-clamp technique, one finds that these antibiotics (e.g., gramicidin, alamethicin) produce discrete conduction steps across the membrane.[29,30] Such a phenomenon is also found when "excitability-inducing material" (EIM, a protein extracted from *Aerobacter cloacae*) is added to black lipid membranes.

4. Electron-Transfer Reactions of Hemoprotein

Biological electron transfer invariably occurs between two redox couples as shown:

$$\text{Red}_1 + \text{Ox}_2 \rightleftharpoons \text{Ox}_1 + \text{Red}_2 \tag{4}$$

An analogous reaction involving inorganic species is:

$$Fe^{2+} + Ce^{4+} \rightleftharpoons Fe^{3+} + Ce^{3+} \tag{5}$$

The field of electron transfer in energy metabolism is too vast to be covered in this paper. As a model, however, electron-transfer reactions of hemoproteins, as investigated by Jordan, is briefly dealt with in this section.[31] In Figure 7 is represented the "redox potential cascade" for the terminal oxidation chain of cellular respiration. The overall "cross-reaction":

$$
\begin{array}{l}
\text{COO}^- \\
| \\
\text{CH}_2 \\
| \\
\text{HCOH} \\
| \\
\text{COO}^-
\end{array}
+ \tfrac{1}{2}O_2 \longrightarrow
\begin{array}{l}
\text{COO}^- \\
| \\
\text{CH}_2 \\
| \\
\text{CO} \\
| \\
\text{COO}^-
\end{array}
+ H_2O \tag{6}
$$

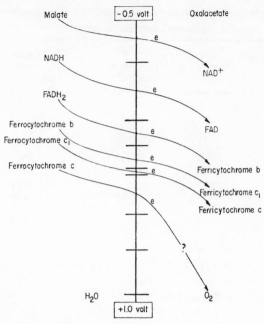

Fig. 7. Terminal oxidation chain of cellular respiration. [From Jordan.[31]]

is a result of the two half-cell reactions:

$$
\begin{array}{ccc}
\text{COO}^- & & \text{COO}^- \\
| & & | \\
\text{CH}_2 & \longrightarrow & \text{CH}_2 \\
| & & | \\
\text{HCOH} & & \text{CO} \\
| & & | \\
\text{COO}^- & & \text{COO}^-
\end{array}
+ 2\text{H}^+ + 2e_0^-
\tag{7}
$$

and

$$
\tfrac{1}{2}\text{O}_2 + 2\text{H}^+ + 2e_0^- \longrightarrow \text{H}_2\text{O}
\tag{8}
$$

The reversible potentials for these two half-cell reactions differ by over 1 V, but still there is no appreciable oxidation if a malate solution is exposed to oxygen. The oxidation of malate to oxalacetate occurs rapidly via the intermediate steps represented in Figure 7. Even though the redox potential differences for the successive steps are small, the electron transfer rates for these steps are quite high.

Table 5 shows the rate constants and free energies of activation for the electron-transfer reactions of heme and hemoproteins as obtained by Jordan using electrode kinetic techniques. The rate constants for the electron-transfer reactions of heme dimers and monomers, hemochromes, hemoglobin, and myoglobin are quite low. Due to the fact that the axial ligands in these compounds vary widely, Jordan concludes that electron transfer occurs via the plane of the porphyrin. The rate constant for the electron transfer reaction involving cytochrome c is considerably higher, and it is very probable that the mechanism of this reaction is not the same as for the other compounds.

Table 5. Rate Constants and Free Energies of Activation for
Some Biological Electron-Transfer Reactions[31]

Redox couple		$k°$ (cm/sec at 25°C)	$\Delta G^{\ddagger}$ (kcal/mole)
Ferriheme	Ferroheme (monomers in ethanol)	0.2	7
Ferridicyano hernichrome	Ferrodicyano hemochrome	4	5
Ferriheme Ferridimer	Ferriheme } Ferrodimer }	0.8	6
Ferrimyoglobin	Ferromyoglobin	2	5
Ferrihemoglobin	Ferrohemoglobin	0.6	6
Ferricytochrome c	Ferrocytochrome c	1000	1

5. Chemistry of Cell Surfaces

Many phenomena (e.g., cell adhesion, aggregation, immunology, etc.) depend on the nature of the cell surface. For a better understanding of the processes which occur across cell–solution interfaces, it is essential to determine the chemical nature of cell surfaces at a molecular level. By using cell electrophoresis techniques, Mehrishi[32] has been able to gain information on intact cell surfaces without causing any damage to the cell. Most mammalian cell surfaces carry a negative surface charge.[33] A modification of the cellular surface, which is carried out with specific agents, leads to alterations in the surface charge density. By comparing the surface charge density with the controls (i.e., without adding the chemicals), it is possible to draw inferences on the nature of the chemical groups on cell surfaces.

The positively charged amino groups on human lymphocytes and platelets can be blocked using 2,3-dimethyl maleic anhydride. A positively charged amino group is replaced by a negatively charged carboxyl group, which means that there is an increase of 2 negative charges per amino group which is blocked. By lowering the pH from 6.8 to 6, the adduct is removed.

Sulfhydro groups on the surface of human platelets and erythrocytes are trapped by 6,6-dithiodinicotinic acid (UV absorption peak 290 nm). An ionizable carboxyl group replaces a sulfhydro group, which corresponds to an increase of electron charge of one per sulfhydro group blocked. A thione absorbing at 344 nm is a product, and its identification serves as a check of the electrokinetic method.

Enzymes may be used for chemical modification of cells which enables characterizations of chemical groups from electrokinetic studies. For example, neuraminidase blocks α-carboxyl groups of N-acetyl neuraminic acid (NANA). Similarly, ribonuclease acts on the phosphate groups of human lymphocyte surfaces while alkaline phosphatase alters phosphate groups on human platelets.

It may be seen from Table 6 that there are negatively charged carboxyl as well as phosphate groups which are identifiable on the surfaces of platelets and lymphocytes. The remainder of the negative charges have not been identified but include weakly acidic groups of pK about 3.8. On the surfaces of erythrocytes, only two types of negative charges are present—α-carboxyl of NANA and weakly acidic groups of pK about 3.8.

Mehrishi[32] has proposed hexagonal and cubic lattice packing for the groups on cell surfaces. Figure 8 illustrates the two types of packing on the surface of human lympho-

Table 6. Surface-Charge Characteristics and Nature of Chemical Groups Contributing to Surface Charge per Cell Surface Area for Platelets, Lymphocytes, and Erythrocytes[32]

Cell type	Anodic electrophoretic mobility (μm/sec/V/cm)	Apparent electron charges (×10^6)	Corrected electron charges (×10^6)	Amino groups (+) (×10^5)	Negatively charged groups				
					NANA α-carboxyl (pK$_a$ 2.6)[a] (×10^5)	Phosphate susceptible to[b] RNase[c] (×10^5)	Alk. phos.[d] (×10^5)	Weak acidic (pK ~ 4) (×10^5)	–SH groups (×10^6)
Platelets[c]	0.85 ± 0.04	1.8	2.04	2.42	8.9	–	5.0	6.5	0.28
Lymphocytes[d]	1.09 ± 0.08	9.34	10.29	9.5	29.0	8.7[f]		55.5	1.98
Erythrocytes[e]	1.08 ± 0.03	10.29	10.29	Not detected	62.0	–	–	40.9	Not detected

a Surface areas of cells: 28.27 μm²; 113 μm²; 163 μm². Electrokinetic data obtained in physiological saline (0.145 M NaCl; pH 7.2 at 25°C).
b May be present but <5 × 10^5 per cell.
c Thromb. Diath. Haemorrh. Suppl. 26, 53 (1967); 26, 370 (1971).
d Int. Arch. Allergy 42, 69 (1972).
e Arch. Biochem. Biophys. 135, 356 (1969).
f Exp. Cell Res. 50, 441 (1968).

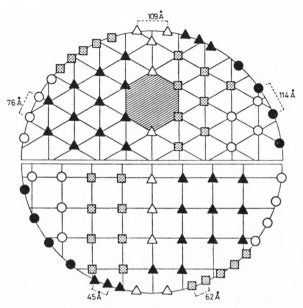

Fig. 8. The human lymphocyte is assumed to have a spherical shape with a diameter of 6 μm and a surface area of 113 μm². The groups are arranged at the corners of squares or equilateral triangles. Only half of the cell surface area illustrating the two different types of packing is shown; the upper half shows hexagonal packing and the lower half cubic lattice packing. ⊡, α-carboxyl groups of neuraminidase-susceptible *N*-acetyl neuraminic acid; ○, sulfhydro groups; ●, phosphate groups; △, amino groups; ▲, unidentified negatively charged groups. [From Mehrishi.[32]]

cytes. Both arrangements are shown. At the present time, it is not possible to identify the preferred arrangement.

By using combined electrokinetic–electron microscopic techniques on erythrocytes incubated with neuraminic acidase, fixed with gluteraldehyde and treated with colloidal ferric oxide particles, it was found by Weiss *et al.*[34] that the sialic acid groups are arranged in clusters having a charge density more than expected on the basis of the total charge when it is uniformly distributed. Mehrishi[32] concludes that the membrane surface is organized as a mosaic structure of macromolecular complexes, functionally discrete and biochemically heterogeneous. The receptor sites probably participate in specific cellular interactions.

6. Electrochemistry in Immunology

Immunology deals with the study of antigen–antibody reactions. An immunologic topic of interest to electrochemists is immunoelectroadsorption, which was initiated by Mathot, Rothen, and Cassals.[35] In this method, an antigen is first adsorbed on a metal slide. The slide is then dipped in a solution of a metal serum. Due to the immunologic interaction, there is an adsorption of antibodies from the solution on the metal slide. The change in thickness of the films on the metal slide, measured by ellipsometry, gives the extent of the immunologic reaction. In dilute solutions of antigens or antibodies this method is not sufficiently sensitive. However, when the potential across the metal-solution interface is altered, specific interactions of antigens with antibodies can be detected at concentrations as low as 10^{-14}-10^{-15} g/ml.[36]

The initial studies were conducted with chromium-coated slides. Electron-diffraction patterns, obtained by reflection from the surface, revealed that the structure of the metallic surface plays an important part in determining the sensitivity of the method. An amorphous metallic layer is relatively inactive (concentrations as high as 10^{-8}-10^{-7} g/ml of antigen are needed for the detection of the immune reactions). With the metal film in a crystalline state and having a partial magnetic orientation, an antigen concentration of only 10^{-14} g/ml is necessary for detection of the immunologic reaction.

Rothen[36] then carried out studies on slides coated with nickel, a ferromagnetic metal. The nickel-coated slides were inactive. However, by placing them in a magnetic field with the lines of force perpendicular to the surface and the metallized surface facing the south pole, they became active. If the polarity was reversed, the slides reverted to the inactive state. By heating active slides to 70-80°C or on long-term standing, they lose their activity. Rothen proposed that the long-range order of nickel atoms, produced by the strong influence of the magnetic field, permits the long-range order of well-separated antigen molecules, a *sine qua non* condition for detecting the immunologic reaction. Under these conditions, a large number of antibodies are combined nonstoichiometrically with the antigen (Rothen refers to these combinations as Berthollides).

In another investigation, Rothen and Mathot[37] determined the kinetics of adsorption of antibodies on slides coated with antigens. Human γ globulin was first adsorbed on a chromium-coated glass slide for a 3-min period from a veronal solution (10^{-3} g/ml of globulin). The thickness of the antigen film was 30 Å. The adsorption of anti-human γ globulin was then recorded as a function of time, using ellipsometric techniques. In Figure 9 curve (a) shows the adsorption–time relation when the antigen-coated slide was introduced into the antibody-containing solution in a wet condition. Curve (b), which shows a lower adsorption, was obtained when the dry slide was immersed in the solution. Curve (c) reflects the low adsorption of antibody on a slide coated with a heterologous antigen–rabbit globulin. The data were further analyzed by Rothen and Mathot.[37] It

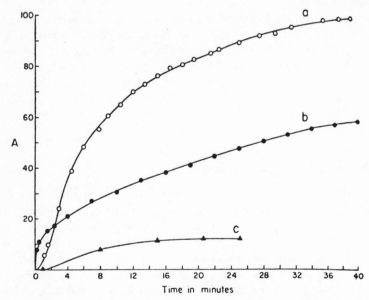

Fig. 9. The adsorption (A) of anti-human γ globulin from an antiserum diluted 1/10 in veronal buffer on a chromium slide covered with human γ globulin as a function of time as measured ellipsometrically. (a) slide introduced wet; (b) slide introduced dry; (c) slide introduced into veronal solution containing rabbit anti γ globulin. [From Rothen and Mathot.[37]]

was found that a plot of the square of the thickness vs. time showed a two-section linear relation for case (a) but a single line for case (b). The linear relations were taken as evidence of a diffusion-controlled process for the adsorption of antibody on the antigen. Rothen and Mathot concluded that the reason for the break in case (a) is probably due to different energy levels involved in the adsorption process.

7. Intravascular Thrombosis—An Electrochemical View

7.1. Definition of Terms Thrombosis, Thromboembolic Disease

Intravascular thrombosis is a reaction at the blood vessel wall or prosthetic material–blood interface. The term heart attack is more commonly used than thrombosis. The heart attack is the end result. Thrombosis is the real cause of the heart attack. Thrombosis may be defined as the formation of a coalescent or agglutinated solid mass of blood components in the blood stream. Thrombus deposits produce occlusions in arteries and veins, inhibiting blood flow. Obstruction to blood flow usually occurs at the site at which the thrombus first deposits. Sometimes, thrombi break loose, travel through the blood stream and cause obstruction at some distal point of narrowing elsewhere. Such thrombi are called "emboli." The two phenomena are commonly referred to as "thromboembolic disorders."

7.2. Evidence for an Electrochemical Mechanism of Thrombosis

The adsorption of blood proteins on the blood vessel wall or on implanted vascular or heart-valve prostheses, their reactions at the interface with blood, the adhesion of blood cells on these surfaces, the aggregation of blood cells either directly or via bridges (e.g., fibrin strands) are intermediate steps in the overall thrombosis or clotting reaction. Since the intermediate enzymatic reactions which lead to thrombosis are interfacial in nature and occur in electrolyte environments, one may expect many of them to be electrochemical in origin. Sawyer, Srinivasan, and co-workers have accumulated considerable evidence[38-44] for an electrochemical mechanism of thrombosis. It may be summarized as follows:

1. Under normal conditions, the surfaces of the blood vessel wall and of blood cells are negatively charged.
2. Injury or atherosclerosis make the blood vessel wall less negatively charged or even positively charged. These situations are more prone to thrombosis.
3. A decrease in pH makes the surface charge of the blood vessel wall and of blood cells less negative. The isoelectric point for the vascular components is at a pH of about 4.5.
4. Anticoagulant drugs increase or maintain the negative surface charge of the blood vessel wall and of blood cells while procoagulants have the opposite effect and quite often even cause a reversal in the sign of the surface charge.
5. Of electronically conducting materials, all those which maintain negative potentials in blood vs. the normal hydrogen electrode (NHE) tend to be antithrombogenic while those which register positive potentials vs. NHE are invariably thrombogenic.
6. Of insulator materials, the uniformly negatively charged ones tend to be antithrombogenic.
7. Blood coagulation factors take part in adsorption and charge transfer reactions at

metal–solution interfaces. At least some of these reactions seem to be relevant in the "thrombosis cascade."

Certain aspects of this work will be dealt with in some detail in the following subsections.

7.3. Correlations between Thrombogenic or Antithrombogenic Properties of Materials and Positions of Metals in Electromotive Series

A systematic study of the thrombogenic or antithrombogenic properties of some metals, covering a wide range in the electromotive series (Table 7), was determined in a relatively simple manner.[45] Clean wires of the metals were inserted into canine carotid and femoral arteries through side branches and left in position for 30–40 min. The potentials of these electrodes in contact with flowing blood were measured with respect to a standard calomel electrode. At termination, the vessels were clamped above and below the electrodes. The dogs were sacrificed, after injection of formalin into the vessels, for fixation of thrombus deposits. The vessels were slit open and the electrodes examined. The results proved to be quite interesting (Table 7). Electrodes of metals establishing negative potentials in blood vs. the NHE were antithrombogenic while those exhibiting positive potentials were thrombogenic. Subsequent to this investigation, several long-term studies with vascular or heart-valve prostheses fabricated with metals and alloys confirmed the potential dependence of thrombus deposition on metallic implants.[46]

7.4. Antithrombogenic Characteristics of Cathodically Polarized Copper Prostheses

From the results summarized in the preceding section, it would appear that only the corrodible metals and alloys are nonthrombogenic while the more noble ones are thrombogenic. Blood-compatible metals and alloys are essential for certain types of medical devices. One obvious method of maintaining a negative potential on a metallic prosthesis is by cathodic polarization. Experiments were thus carried out to test the antithrombogenic characteristics of cathodically polarized copper prostheses.[47] One of the best sites for such testing is the canine thoracic inferior vena cava (TIVC). A copper prosthesis (a tube 5 cm long, 1 cm i.d., 1 mm wall thickness) was implanted in the canine TIVC. It was

Table 7. Dependence of Thrombus Deposition at Metal Electrodes on Position Metal in Electromotive Series[45]

Metal	M/M^{n+} standard electrode potential (volts vs. NHE)	Resting potential at metal–blood interface (volts vs. NHE)	Occurrence (+) or nonoccurrence (X) of thrombus deposition
Mg	−2.375	−1.360	X
Al	−1.670	−0.750	X
Cd	−0.402	−0.050	X
Cu	+0.346	−0.025	+
Ni	−0.230	+0.029	+
Au	+1.420	+0.120	+
Pt	+1.200	+0.125	+

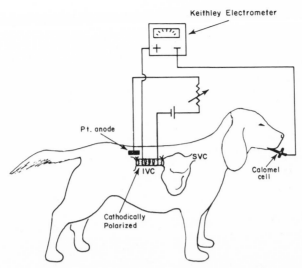

Fig. 10. Experimental arrangement for maintaining a copper prosthesis, implanted in the canine thoracic inferior vena cava, at negative potentials. [From Gileadi *et al.*[47]]

maintained at a cathodic potential using a suitable polarizing circuit (Figure 10). Three experiments were conducted in which the potentials of the copper tubes ranged from −160 to −60 mV/NHE. The tubes were left in position for 6, 8, and 14 days. All tubes were free of thrombus deposits. In two control experiments, with no polarizing circuit, the copper tubes registered positive potentials and were found to be occluded in 8 and 14 days. These experiments confirm that the potential across the metal–blood interface is a more basic parameter than the chemical nature of the surface in determining its blood-compatible characteristic.

7.5. Correlations between Surface Charge Characteristics and Pro- or Anticoagulant Properties of Insulator Materials

Insulator materials are frequently used in the fabrication of vascular and heart-valve prostheses. There are also cases in which insulator materials are used for coating prostheses constructed with metals. From the interfacial nature of the clotting reaction, one can expect the surface-charge property of the material to play a role in determining its pro- or antithrombogenic activity. This subject has been treated extensively in recent review articles.[44,48] It is worthwhile to present some of the interesting correlations. Heparin is the most potent anticoagulant known at the present time. Several attempts have been made to bond heparin to plastic surfaces. When heparin is bonded ionically to a plastic surface[49] via a bridge, e.g., a polyquaternary ammonium salt, the zeta potential across the coated plastic material–electrolyte interface (as determined from streaming potential measurements) is more negative than for the uncoated material. The heparinized material was also antithrombogenic. Unfortunately, because of the ionic nature of the bonds in the coating of the material with heparin, there was elution of the heparin, from the surface, in blood—an electrolyte environment. To overcome this difficulty, Lagagren and Eriksson[50] "cross-linked" the heparin on the surface with gluteraldehyde. In short-term implantation studies, these surfaces were antithrombogenic. Attempts also were made to bond heparin covalently to surfaces. Although the heparin thus bonded was more stable than the ionically bonded heparin in electrolyte environments, there was some loss in heparin activity. The negative-charge effect of heparin is due to a total of six negative charges per tetrasaccharide unit. Carboxylate and sulfonate groups contribute to this negative charge. Sulfonated and carboxylated Teflon surfaces were more stable than heparinized surfaces and were also found to be antithrombogenic.[51] Negatively charged electrets were more thromboresistant than nonelectrified ones.[52]

7.6. Correlations between Effects of Drugs on the Surface-Charge Characteristics of the Vascular System and Their Pro- or Anticoagulant Properties

A measurement of the alteration in surface-charge characteristics of the vascular components produced by chemicals, either *in vivo* or *in vitro*, using electrokinetic techniques (streaming potential, electro-osmosis, electrophoresis), has been one of the most promising areas in screening the pro- or antithrombogenic activities of these compounds.[44,53] Anticoagulant drugs increase or maintain the negative charge densities of the blood vessel wall or blood cells. Procoagulants have the opposite effect. Heparin shows a pronounced charge effect but could be reversed by protamine. There is a strong urge to find oral anticoagulants. Aspirin may fall under this class, but it is not as potent as heparin. Flavonoids are also similar to aspirin in this respect, but from long-term studies it appears that their breakdown phenolic products have a stronger anticoagulant action. The electrokinetic studies confirm that the oral contraceptives are mildly procoagulant and their charge effects on the vascular wall may be the cause for their inducing thromboembolic disorders in some women on the "pill." Introduction of a negatively charged substituent in the steroid ring may reduce this effect. Alternatively, it may be necessary to include an anticoagulant (e.g., aspirin) in the "pill." Figure 11 illustrates schematically the regions of stability, reversible aggregation, and irreversible aggregation of blood cells, as may be inferred from electrophoretic measurements on blood cells.

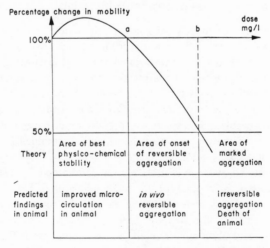

Fig. 11. Qualitative representation of percentage change in electrophoretic mobility as a function of chemical concentration, showing regions of stability or aggregation. [From Stoltz *et al.*[54]]

7.7. Electrochemical Reactions of Blood Coagulation Factors— Their Role in the "Thrombosis Cascade"

Using electrochemical (electrocapillary, differential capacity, potentiodynamic, and potentiostatic), ellipsometric, electron microscopic, and hematologic methods, the adsorption and charge transfer reactions, as well as, where possible, their role in blood coagulation, were investigated at solid (mostly metal)–solution interfaces.[41] Studies have been conducted on amino acids, peptides, and blood coagulation factors. The results of these studies may be summarized as follows: (1) adsorption of amino acids, peptides and proteins on metals is a potential-dependent phenomenon; (2) blood coagulation factors (fibrinogen, prothrombin, thrombin, factors V, VIII, and IX) take part in electron-transfer reactions; (3) fibrinogen is electropolymerized to a fibrin-like product at anodic potentials; (4) fibrinogen clotting times are increased by its prior exposure to an electrode maintained at a potential of –800 mV; (5) electrochemical activation of prothrombin at positive potentials yields a product with thrombin activity; and (6) heparin inhibits the adsorption of fibrinogen on freshly cleaved mica surfaces. These results are most valuable in the elucidation of the electrochemical mechanism of thrombosis. Several of the enzyme reactions in blood coagulation appear to occur by electrochemical pathways.

8. Automated Microbial Detection and Quantification

The detection of microorganisms is important in many areas of medical, environmental, and industrial activity. For example, representative situations, in which knowledge of the presence and/or quantity of microbes is desired, are as follows: (1) infectious disease (measurements and typing of organisms in blood, urine, cerebrospinal fluid, sputum, etc.), (2) food quality control, (3) fermentation process in industry, (4) environmental quality control (air, water), and (5) pharmaceutical manufacture.

Frequently, the method of "culture growth and observation" is used to establish both the presence of microbes and the quantitative information. Growth rates are observed by cultures derived from multiple dilutions of the same biological sample. An alternative approach to microbial analysis is based on the fact that a certain concentration of living cells causes a reproducible change in the potential of an electrode (e.g., platinum) with respect to a reference electrode (e.g., calomel) immersed in the aerobic or anaerobic culture suspension. This concentration or minimum detection level is in the range 10^4–10^6 viable cells/ml for a variety of organisms.[55,56] In the initial region, there is no change of potential with time. It is then followed by a sinusoid potential–time behavior which is characteristic of the test organism. In Figure 12 are shown the potential–time relationships of organisms which do or do not produce hydrogen. Examples of the former type are *Escherichia coli, Enterobacter aerogenes, Serratia marcescens, Protens mirabilis, Citrobacter intermedium, Salmonella, Klebsiella*, and of the latter type are *Staphylococcus aureus, Staphylococcus epidermidis, Streptococcus, Listeri monocylogenes, Pseudomonas aeruginosa, Moraxella*, and *Shigella alkalescens*.

In order to interpret this type of potential–time behavior, cells from a culture, in which a signal was detected, were first concentrated by centrifugation. The cells were then re-suspended in culture media or physiological saline at different concentrations and showed the following behavior: (1) the supernatant solution produced no potential–time signal; (2) the voltage signal produced showed a concentration dependence, very similar to an adsorption isotherm (Figure 13); (3) no voltage signal was produced if the cells were

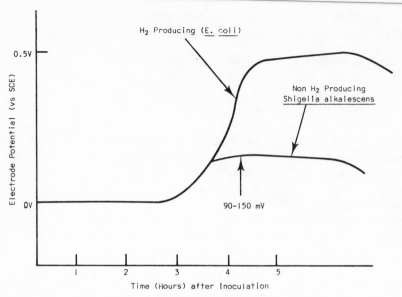

Fig. 12. Comparison of the growth portion of the voltage–time curves for H$_2$- and non-H$_2$-producing bacteria. The same "lag time" is observed when the initial inoculum and growth rates are the same (Pt–calomel electrodes). [From Stoner *et al.*[56]]

thermally destroyed; (4) in old or weak cultures, containing a comparable number of dead and viable cells, a lower potential change was observed; a curve of the same general shape as in Figure 13 was obtained when cells are suspended in a nonnutrient medium (electrolyte only). These results indicate that it is the presence of the living organisms

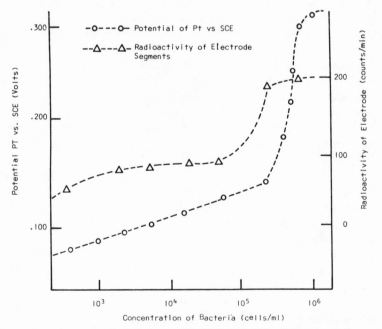

Fig. 13. Comparison between adsorption–concentration and potential–concentration curves for *Shigolla alkalescens* in TSB media.

rather than any metabolic product which accounts for the observed relations between population and potential. The results thus suggest microbial interaction with an electrode surface.

To further examine the hypothesis, radioactively labeled bacteria were placed in cultures at various concentrations and small segments of electrode were removed and analyzed for radioactivity (presumably a measure of adsorbed bacteria). Figure 13 also shows a correspondence between potential change and radioactivity as a function of cell concentration. A similar adsorption behavior was also observed on Ni, Ta, W, Mo, Cu, Al, and stainless steel. Some of these metals or their alloys may be useful for the development of disposable electrode sensors.

A clinical evaluation of this method was carried out at the Hôtel Dieu Medical Center, Rouen, France. Blood cultures were monitored using platinum–calomel or tungsten–molybdenum electrode systems. Approximately 25 cultures inoculated with blood from healthy donors were incubated for 5 days at 35°C with no false positives indicated by either electrode combination. The same cultures were then inoculated with organisms from patients known to have both pure and mixed bacterial infections. In all cases, both electrode systems indicated the same lag time response at least 18 hr before the first visible growth was detected on agar plates.

The feasibility of using the microbial electrode system is being currently examined for antibody screening, antigen–antibody reactions, and bacterial typing. Perhaps of more fundamental interest, however, is the fact that electrochemical sensors are able to detect and distinguish a population of microorganisms; the nature of this detection is dependent on the viability of the organism.

9. Electrochemical Inactivation of Pathogens

Water is generally disinfected by chlorination to reduce the population of indicator organisms such as *Escherichia coli* to acceptable levels. In some cases, however, chlorination is either undesirable or insufficient as a water-disinfection process. Further, pathogenic viruses, especially the nonlipid picorno viruses such as polio, rhino, entero, and the encephalomyocarditis viruses, have been shown to survive chlorination as well as conventional water-treatment methods which function effectively for removing bacterial contamination. In addition, there has been recent interest, for a variety of environmental and health reasons, to obtain a water-disinfection process other than chlorination.

An electrochemical process[57,58] has been developed which is capable of destroying the above-mentioned viruses, as well as bacteria and larger organisms like protozoa and flukes. With this process, the water is passed between electrodes, cycled between potentials such that a current density of about 10–15 mA/cm^2 at a frequency of around 1 Hz is obtained. Under these conditions, with either triangular or square waveforms, water disinfection is achieved without the production of chlorine. The mechanisms of this process are twofold. The first is achieved if the organisms, by means of good mass-transfer conditions, are brought in contact with the electrode. This is a sufficient condition for disinfection as under this condition all organisms are destroyed. However, some organisms are also destroyed away from the electrode, or downstream from the reactor. The disinfection agent in this second case is not defined and is much shorter-lived than any chlorine residue. The first mechanism is more straightforward and involves electrochemical destruction of the protective coat of the organism during the anodic cycle of the process—in short, electro-oxidation or denaturation. During the cathodic part of the cycle, the organism is dispelled giving rise to a self-cleaning electrode process.

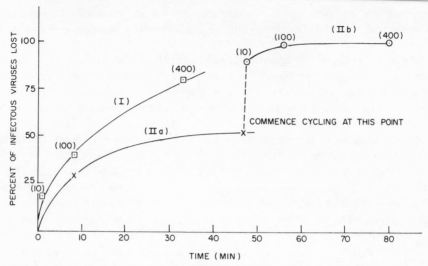

Fig. 14. Electrochemical destruction of pathogens—percent of infectious viruses lost as a function of time. ⊡, with cycling (adsorption and electrolysis); X, no cycling (only adsorption); ⊙, cycling (after absorption).

The first test for feasibility of this process was carried out on an osmotic-shock-resistant strain of T_4 bacteriophage. Figure 14 shows the results of this test using carbon rods (2 cm^2) as electrodes and cycling between ±1 Volt vs. standard calomel electrode (SCE), with a magnetic stirrer used for mixing.

Since this early work, more experiments have been carried out on a variety of organisms and flow-through reactors of variable design. Waters ranging from unsettled raw sewage to distilled water have been disinfected by the process. At the present time, it is economically competitive with other disinfection techniques on the small to medium scale, such as individual dwelling water treatment, swimming pools, fish tanks, and the like, but needs to be improved before it will be competitive with chlorination on the municipal level.

In summary, another area of bioelectrochemistry, microbe–electrode interaction, has provided us with possible new technology of electrochemical disinfection.

10. An Electrochemical Method for Measuring Blood Coagulation

The latter stages of blood clotting can be thought of as a two-stage polymerization reaction, the first being the formation of non-cross-linked polymer strands and the second being the insolubilization and cross-linking to produce a pure fibrin clot. Both of these processes have been studied[59,60] using biochemical and enzymatic techniques.

Various instrumentation methods currently exist to measure the first polymerization process. They are based on changes in optical density or viscosity of the blood during coagulation. However, if one measures the potential of a gold electrode with a small steady-state current in contact with blood or plasma during coagulation,[61,62] it turns out that both the polymerization (step one) and cross-linking (step two) can be resolved. This resolution manifests itself in the form of two sigmoid curves (inflections) on a potential-time scale. It turns out that only the positive electrode is sensitive to coagulation while the negative electrode remains relatively constant during the process (Figure 15).

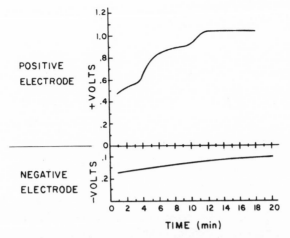

Fig. 15. Potential vs. time relationship of two electrodes applied to blood plasma under constant d.c. conditions. While the negative electrode potential is seen to change but little, and that linearly with time, two reaction steps can be distinguished on the anodic potential curve. The first step is considered to be linked to the polymerization reaction while the second step is thought to be associated with the cross-linking process.

It has been shown[61,62] that if one blocks the cross-linking process, but not the polymerization process, then only the first inflection is obtained. Furthermore, patients with a deficiency in the enzyme (factor XIII) which promotes this cross-linking only display the first inflection. One of the particular points regarding this process is the fact that only gold electrodes seem to detect both stages whereas platinum and several other metals detect just the first stage. One possible explanation for the mechanism of the method comes from the fact that several fragments (or peptides) are released during each stage of the process and the gold electrode may be acting as a polarographic detector for these peptides. Consistent with this hypothesis is the fact that the peptides are negatively charged and the detection is predominantly at the positive electrode.

In addition to being of potential use in the area of clinical hematology, the method has shown promise in areas of drug or chemotherapy for cancer patients. In these situations, the drugs interfere with the coagulation mechanism, and this can be measured by this technique.[63]

11. Electrochemical Destruction of Tumors

Electric shock stress, if given for 3 days after virus inoculation to mice, results in tumors which are 50% smaller.[64] A small electrically caused lesion at a particular site on the dorsal hypothalamus in rabbits caused a complete suppression of complement-fixing antibodies to primary immunization with horse serum. The same hypothalamic region can be electrically stimulated to cause an enhanced immunologic response by electrodes placed on either side of it. Recently it was shown by Rosenberg and Van Camp[65,66] that when an electric current was passed between two platinum electrodes in a culture of *Escherichia coli*, they did not divide but continued to grow and form long filaments. These workers pointed out that platinum is not completely inert and platinum ions combine with the culture medium to form a compound which prevents bacterial cells from dividing. On this assumption, the compound bisdichlorodiamine platinum(II) was administered to tumor-bearing Swiss white mice. The tumor was found to regress completely in every animal so treated.

Recently, it was shown that electrochemical methods can be used to destroy viruses (cf. Section 9). When rats, with intrahepatic implantation of positively and negatively polarized platinum electrodes, were injected intraportally with tumor cells, the cells were attracted to the positive but not the negative electrode.[67] It is not known whether the tumor cells did reach the cathode and were subsequently destroyed.

With this background information, some preliminary experiments were conducted on the response of tumors to electric currents.[68] Tumors (Walker's carcinosarcoma) were produced in Sprague–Dawley rats. A platinum-foil electrode (area 3 cm^2) was placed over the tumor (diameter about 5 cm). Two or three stainless-steel electrodes were inserted into the tumor and served as the anode. An electric current of about 5 mA was passed between the platinum-foil cathodes and stainless-steel anodes for three periods of 2 hr each. There was a 2-hr rest period between current passages. The animals were followed in respect to tumor size and infection. In all six animals with such treatment the tumor growth was inhibited and there was a gradual disappearance of the tumor with time. In one animal, there was complete disappearance of the tumor with no evidence of recurrence for over at least three months of observation. One of the problems encountered in these experiments was the large amount of tumor necrosis on account of the current passage. This problem was greatly reduced by decreasing the current to 1–2 mA. In these experiments, it was necessary to pass the electric currents for a longer time, 4–6 hr.

Tumor growth is a surface reaction, probably involving adsorption and charge-transfer reactions. The fact that radiation therapy works best under aerobic conditions suggests the formation of the peroxide ion (HO_2^-) is an essential step in tumor destruction.[69, 70] This ion is formed more easily by the electrochemical reduction of oxygen. Even if this were not the mechanism, it is very probable that there are other charge-transfer reactions which may cause destruction of tumors. Heptinization of tumor antigen by the process of electrochemical destruction of tumor cells and the release of the antigen into systemic circulation may set up actual immunization against tumor by the host.

12. Biogalvanic and Biofuel Cells

In the past decade the scope of bioelectrochemistry was further extended to develop implantable power sources. In this process the body itself is forced to play an active role in delivering the electrochemical power. Efforts are now underway to generate bioelectrochemical power inside the human body[71] to drive such implantable devices as cardiac pacemakers. For this purpose two different systems have received particular attention.[72]

The biogalvanic cell, for instance, consists of a sacrificial anode like aluminum and oxygen cathodes. The cell utilizes body fluids as electrolyte and the dissolved oxygen of the body fluids for cathodic reaction. The oxidation rate and the amount of the metal determine the life of the cell. By a careful choice a life period up to 15 years can be projected at a steady power output of 150 μW at 0.7 V.

A special type of biogalvanic cell is schematically represented in Figure 16. The anode is placed between two oxygen cathodes to form a chamber cell. The chamber space between the electrodes is filled with physiological solution. The compartments on either side of the anode also serve to hold back the resulting anodic oxidation products. The interior and exterior surfaces of the cathodes are covered with hydrophilic and hydrophobic membranes, respectively. In the former case the electrolytic contact is maintained between the electrodes, and in the latter case the direct contact of body fluids

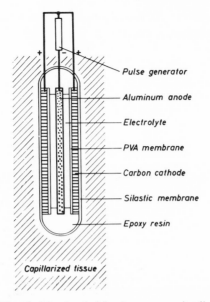

Fig. 16. Biogalvanic (aluminum–oxygen) cell.

with the electrodes is prevented. Only oxygen is transported to the electrode surface in gaseous phase. Animal tests extending over a year indicated that an electrical power of 100 μW at 0.6 V can be realized with this type of a cell.

The other system is based on fuel cell principles. Body fluids contain sufficient quantities of reactants such as glucose and dissolved oxygen. With appropriate electro-catalysts these substances can be converted to derive electrochemical energy. However, selective catalysts are required for this purpose. Siemens has developed an implantable biofuel cell taking the physiological and electrochemical factors into consideration.[71, 72]

In the schematic diagram of the cell, as shown in Figure 17, the anode catalyst (glucose electrode) is sandwiched between two selective and porous cathode catalysts (oxygen electrodes). Thin hydrophilic membranes are inserted between the electrodes to prevent electrode short-circuiting. The outer surfaces of the cathodes are also coated with

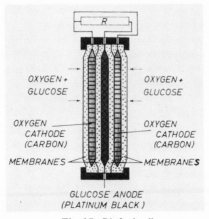

Fig. 17. Biofuel cell.

body-compatible hydrophilic membranes. The entire cell unit is packed tightly. During operation, oxygen is consumed by the cathodes and glucose alone diffuses to the central anode. Under physiological conditions a long range *in vitro* power of 100-150 μW at 0.5-0.6 V could be derived with this cell. The life of the biofuel cell is dependent on the constant supply of reactants and the activity of the electrocatalysts. If these factors are ensured, cell life will be indefinitely long. The initial animal tests have to show the feasibility of biofuel cells as implantable power sources.

Acknowledgments

The authors wish to thank Dr. M. Somasundaram of the Department of Neurology, State University of New York, Downstate Medical Center, for helpful discussions and suggestions in the field of electrophysiology.

References

1. T. Shedlovsky, ed., *Electrochemistry in Biology and Medicine*, Wiley, New York (1955).
2. First International Symposium on Bioelectrochemistry,
3. A Workshop in Bioelectrochemistry, Princeton, New Jersey, October 10–14, 1971, sponsored by National Science Foundation, Washington, D.C. and ESB Inc. Technology Center, Yardley, Pennsylvania.
4. Second International Symposium on Bioelectrochemistry, Pont à Mousson, France, October 1–5, 1973, *Proc. Bioelectrochem. Bioenerget. 1*, 1–565 (1974).
5. Bioelectrochemistry—Interfacial Phenomena in Biological Systems, Joint Symposium of Organic and Biological Electrochemistry, Corrosion, and Physical Electrochemistry Divisions of Electrochemical Society, Spring Meeting of Electrochemical Society, Toronto, Canada, May 11–16, 1975.
6. Third International Symposium on Bioelectrochemistry, Munich, Germany, Oct. 1975.
7. T. Teorell, in: *Biophysical Mechanisms in Vascular Homeostasis and Intravascular Thrombosis* (P. N. Sawyer, ed.), pp. 19–29, Appleton-Century-Crofts, New York (1965).
8. T. L. Jahn, in: *Proceedings of a Workshop in Bioelectrochemistry*, Princeton, New Jersey, October 10–14, 1971 (A. A. Pilla, ed.), pp. 225–267, National Science Foundation, Washington, D.C. and ESB Inc. Technology Center, Yardley, Pennsylvania.
9. A. L. Hodgkin and A. F. Huxley, *J. Physiol. 116*, 473 (1952).
10. B. Katz, *Nerve, Muscle and Synapse*, McGraw-Hill, New York (1966).
11. A. L. Hodgkin and B. Katz, *J. Physiol. 108*, 37 (1949).
12. A. L. Hodgkin and R. D. Keynes, *J. Physiol. 128*, 28 (1955).
13. G. N. Ling, Int. Rev. Cytol. *26*, 1 (1969).
14. T. L. Jahn, *Bioelectrochem. Bioenerget. 1*, 441 (1974).
15. J. Bernstein, *Arch. Ges. Physiol. Pfluegers 92*, 521 (1902).
16. E. J. Lund, *J. Exp. Zool. 51*, 265 (1928).
17. G. Marsh, *J. Plant Physiol. 10*, 681 (1935).
18. J. M. Korr, *Cold Spring Harbor Symp. Quant. Biol. 7*, 74 (1939).
19. R. Lorenti de No, *Stud. Rockefeller Inst. 131*, (1947).
20. J. O'M. Bockris, *Nature 224*, 775 (1969).
21. M. G. Del duca, J. M. Fuscoe, and R. W. Zurilla, *Dev. Ind. Microbiol. 4*, 81 (1962).
22. F. W. Cope, *Arch. Biochem. Biophys. 103*, 352 (1963); *J. Biol. Phys. 3*, 1 (1975).
23. J. O'M. Bockris and S. Srinivasan, *Nature 215*, 197 (1967).
24. P. Mueller, D. O. Rudin, H. Tian, and W. Westcott, *Nature 226*, 1204 (1962).
25. P. Mueller and D. O. Rudin, in: *Current Topics in Bioenergetics*, (D. R. Sanadi, ed.), pp. 157–242, Academic Press, New York (1969).
26. S. McLaughlin and M. Eisenberg, *Annu. Rev. Biophys. Bioeng., 4*, 335–366 (1975).
27. S. Feldberg and G. Kissel, *J. Membr. Biol. 20*, 269–300 (1975).
28. R. Laprade, S. M. Ciani, G. Eisenman, and G. Szabo, in: *Membranes, A Series of Advances*, (A. Eisenman, ed.), Vol. 3, Marcel Dekker (1974).

29. S. B. Hladky and D. A. Haydon, *Nature 225*, 451 (1970).

30. J. Eisenberg, J. E. Hall, and C. A. Mead, *J. Membr. Biol. 14*, 143 (1973).

31. J. Jordan, in: *Proceedings of a Workshop in Bioelectrochemistry*, Princeton, New Jersey, October 10-14, 1971 (A. A. Pilla, ed.), pp. 103-122, National Science Foundation, Washington, D.C. and ESB Inc. Technology Center, Yardley, Pennsylvania.

32. J. N. Mehrishi, *Bibl. Anat. 11*, 260 (1973).

33. H. A. Abramson, L. S. Moyer, and M. H. Gorin, *Electrophoresis of Proteins*, Hafner Publishing, New York (1964).

34. L. Weiss, R. Zeigel, O. S. Jung, and I. D. J. Bross, *Exp. Cell Res. 70*, 57 (1972).

35. C. Mathot, A. Rothen, and J. Cassals, *Nature 202*, 1181 (1964).

36. A. Rothen, *Chimica 25*, 411 (1971).

37. A. Rothen and C. Mathot, *Helv. Chim. Acta 54*, 1208 (1971).

38. P. N. Sawyer, ed., *Biophysical Mechanisms in Vascular Homeostasis and Intravascular Thrombosis*, Appleton-Century-Crofts, New York (1965).

39. S. Srinivasan and P. N. Sawyer, *J. Assoc. Adv. Med. Instrum. 3*, 116 (1969).

40. S. Srinivasan and P. N. Sawyer, *J. Colloid. Interface Sci. 32*, 456 (1970).

41. S. Srinivasan, L. Duic, N. Ramasamy, P. N. Sawyer, and G. E. Stoner, *Ber. Bunsen Ges. Phys. Chem. 77*, 798 (1973).

42. P. N. Sawyer and S. Srinivasan, *Bibl. Anat. 12*, 106 (1973).

43. N. Ramasamy, J. T. Keates, S. Srinivasan, and P. N. Sawyer, *Bioelectrochem. Bioenerget. 1*, 244 (1974).

44. S. Srinivasan and B. R. Weiss, in: *Colloid Dispersions and Micellar Behavior* (K. L. Mittal, ed.), Chapter 24, ACS Symposium Series No. 9, Washington, D.C. (1975).

45. P. S. Chopra, S. Srinivasan, T. R. Lucas, and P. N. Sawyer, *Nature 215*, 1494 (1967).

46. P. N. Sawyer and S. Srinivasan, in: *Medical Engineering* (C. D. Ray, ed.), Chapter 82, Year Book Publishers, Chicago (1973).

47. E. Gileadi, B. Stanczewski, A. Parmeggiani, T. R. Lucas, M. Ranganathan, S. Srinivasan, and P. N. Sawyer, *J. Biomed. Mat. Res. 6*, 489 (1972).

48. P. N. Sawyer, R. K. Aaron, and S. Srinivasan, in: *Medical Engineering* (C. D. Ray, ed.), Chapter 83, Year Book Publishers, Chicago (1973).

49. G. A. Grode, S. J. Anderson, H. M. Grotte, and R. D. Falb, *Trans. Am. Soc. Artif. Intern. Organs 15*, 1 (1969).

50. H. R. Lagagren and J. C. Eriksson, *Trans. Am. Soc. Artif. Intern. Organs 17*, 10 (1971).

51. J. A. Martin, A. Afshar, M. J. Kaplitt, P. S. Chopra, S. Srinivasan, and P. N. Sawyer, *Trans. Am. Soc. Artif. Intern. Organs 14*, 78 (1968).

52. P. V. Murphy, A. Lacroix, S. Merchant, and W. Bernhard, *J. Biomed. Mat. Res., Symp. 1*, 59 (1971).

53. S. Srinivasan and P. N. Sawyer, in: *Electrochemical Bioscience and Bioengineering* (H. T. Silverman, I. F. Miller, and A. J. Salkind, eds.), pp. 17-36, Electrochemical Society, Princeton, New Jersey, 1973.

54. J. F. Stoltz, M. Stoltz, and A. Larcan, *Bibl. Anat. 10*, 474 (1969).

55. J. R. Wilkins, G. E. Stoner, and E. H. Boykin, *Appl. Microbiol. 25*, 951 (1974).

56. G. E. Stoner, J. R. Wilkins, and J. F. Lemeland, presented at the Spring Meeting of the Electrochemical Society, Toronto, Canada, May 1975; *J. Electrochem. Soc. 122*, 109C (1975).

57. G. E. Stoner, U.S. Pat. 3,725,226.

58. G. E. Stoner, G. L. Cahen, Jr., and J. Parcells, presented at the Spring Meeting of the Electrochemical Society, Toronto, Canada, May 1975; *J. Electrochem. Soc. 122*, 106C (1975).

59. A. Henschen and B. Blomback, *Arkiv. Kemi. 22*, 355 (1964).

60. B. Blomback, M. Blomback, N. J. Grondhal, and E. Holmberg, *Arkiv. Kemi. 25*, 411 (1966).

61. T. H. Boyd, III, and G. E. Stoner, *Thrombosis Res. 3*, 209 (1973).

62. T. H. Boyd, III, and G. E. Stoner, U.S. Pat. 3,840,806.

63. D. M. Komp, R. J. Lyles, T. H. Boyd, III, B. Cox, and G. E. Stoner, *Pediatr. Res. 8*, 75 (1974).

64. *Medical Tribune 16* (May 15, 1971).

65. *Medical World News 40U* (May 7, 1971).

66. B. Rosenberg and L. Van Camp, private communication (1975).

67. B. Gardener, A. Sterling, and J. Brown, *Surgery 62*, 361 (1965).

68. S. Srinivasan, A. Ofodile, T. R. Lucas, E. C. Ateyeh, E. Ahrens, S. Jacobson, and R. Redner, unpublished.

69. H. L. Gray, *Am. J. Roentgenol. 85*, 803 (1961).

70. T. Alper, *Organic Peroxides in Radiobiology*, Pergamon Press, New York (1958).
71. J. R. Rao and G. Richter, *Naturwissenschaften 61*, 200–206 (1974).
72. J. R. Rao, G. Richter, F. von Sturm, E. Weidlich, and M. Wenzel, *Biomed. Eng. 9* (3), 98–103 (1974).
73. W. Ostwald, Elektrische Eigenschäden halbdurchlässiger Scheidewände, *Z. für Physik. Chemie 6*, 71 (1890).
74. J. Bernstein, *Arch. Ges. Physiol. Pfluegers. 92*, 521 (1902).
75. L. Michaelis, Contribution to the theory of permeability of membranes for electrolytes, *J. Gen. Physiol. 8*, 33–60 (1925).
76. L. Michaelis, Molecular sieve membranes, *Bull. Nat. Res. Council, 69*, 119–145 (1929).
77. W. J. V. Osterhout, Physiological studies of single plant cells, *Biol. Rev. 6*, 369–411 (1931).
78. R. Höber, Membrane permeability to solutes in its relations to cellular physiology, *Physiol. Rev. 16*, 52–102 (1936).
79. R. Höber, *Physical Chemistry of Cells and Tissues*, Blakiston, Philadelphia (1945).
80. E. J. Lund, *J. Exp. Zool. 51*, 265 (1928).
81. E. J. Lund *et al.*, *Bioelectric Fields and Growth*, University of Texas Press, Austin (1947).
82. G. Marsh, *J. Plant Physiol. 10*, 681 (1935).
83. T. L. Jahn, *J. Cell. Comp. Physiol. 60*, 217 (1962).
84. F. W. Cope, *Arch. Biochem. Biophys. 103*, 352 (1963).
85. F. W. Cope, in: *Oxidases and Related Redox Systems*, Vols. 1 and 2 (T. E. King, H. S. Mason, and M. Morrison, eds.), pp. 51–52, John Wiley & Sons, Inc., New York (1965).
86. J. O'M. Bockris, *Nature 224*, 775 (1969).
87. R. Beutner, *Physical Chemistry of Living Tissues and Life Processes*, Williams & Wilkins, Baltimore (1933).
88. L. R. Blinks, The relations of bioelectric phenomena to ionic permeability and to metabolism in large plant cells, *CSHSOB 8*, 204–215 (1940).
89. H. H. Ussing, in: *Metabolic Aspects of Transport Across Cell Membranes* (Q. R. Murphy, ed.), University of Wisconsin Press, Madison (1957).
90. H. H. Ussing and K. Zehrahn, Active transport of sodium as the source of electric current in the short-circuited isolated frog skin, *Acta Physiol. Scand. 23*, 110–127 (1951).
91. G. E. Briggs, A. B. Hope, and R. N. Robertson, *Electrolytes and Plant Cells*, Blackwell Scientific Publications, Oxford, England (1961).
92. T. Teorell, An attempt to formulate a quantitative theory of membrane permeability, *Soc. Exp. Biol. Med. 33*, 282–285 (1935).
93. T. Teorell, Transport processes and electrical phenomena in ionic membranes, *Prog. Biophys. 3*, 305–369 (1953).
94. K. H. Meyer, and J. F. Sievers, La perméabilité des membranes I. Théorie de la perméabilité ionique, *Helv. Chem. Acta 19*, 649–664 (1936).
95. G. N. Ling, *A Physical Theory of the Living State*, pp. 134–138, 336, Blaisdell Publishing Co., New York (1962).
96. G. Eisenman, *Boletin Inst. Estud. Med. Mex. 21*, 155 (1963).
97. G. Eisenman, in: *The Glass Electrode*, Interscience, New York (1965).
98. G. Eisenman, in: *Glass Electrodes for Hydrogen and Other Cations*, (G. Eisenman, ed.), Marcel Dekker, Inc., New York (1967).
99. T. L. Jahn, *Bioelectrochemistry and Bioenergetics 1*, 441 (1974).

5

Summing-Up: Bioelectrochemistry

G. Eckert

Dr. Srinivasan has thoroughly summed up the field in his tabulated list under the headings of electrophysiology, energy metabolism, cellular chemistry, clinical medicine, and surgery and biomedical engineering. I wish to reinforce the previous speakers' arguments and add some thoughts on the possible future of the field. By necessity, the bias of this presentation will be toward the clinical.

1. Blood Coagulation

We have heard of the interesting work which has been carried out on the electro-chemistry of coagulation, including possible mechanisms for the anticoagulant heparin. Why does blood coagulation matter? This can perhaps be illustrated by the statement that in a group of people such as this audience, drawn from the affluent, rather sedentary, professionals, almost half of the males will die from an unwanted, inappropriate, blood clot. Myocardial infarction, caused by coronary thrombosis, is the "great killer" of the adult members of the Western world. Each time then that a physician gives an injection of heparin to prevent intravascular coagulation, he is in fact carrying out an electrochemical procedure. Each time he reverses the effect of the anionic heparin with cationic prot-amine, he is also performing a process with an electrochemical basis. Application of the knowledge of the underlying mechanism to the prevention of unwanted clot formation is obviously of the greatest importance.

2. Neurophysiology

Much experimental work has been carried out on the process of the transmission of the nerve impulse. The Nobel Prize has been awarded to a number of investigators in this field. As pointed out by Prof. Bockris, however, the explanation of the underlying mechanism has been retarded by unsound application of classical thermodynamics to this kinetic situation. It will be the role of the electrochemist to put the theories of the mechanisms underlying the observations of the neurophysiology experimentalists into a sound practical framework.

This example illustrates what I believe will be one of the important problems in the future of the interdisciplinary field of bioelectrochemistry, namely, the cooperation be-tween the biologists and clinicians on one hand, and the electrochemists on the other. As a clinician, I would like to observe that my colleagues in this field have much to offer because they are in the position to know where the practical problems lie. A number of them have the expertise to make the observations and carry out the experiments. How-ever, it is also true that we have often been guilty of retarding the progress of this and other subjects (or at least of running into dead ends) through an unfortunate inability to communicate and cooperate with the other members of the possible team, including the electrochemists in this case.

G. Eckert • Sydney Hospital, Sydney, NSW, Australia.

3. Energy Metabolism

Although great progress has been made in the understanding of the metabolic processes at a molecular level, allowing the reaction involved to be set out in "clearly defined pathways," remarkably little progress has been made toward the elucidation of the mechanisms underlying the couple of chemical processes for energy utilization in the living organism. Oxidative phosphorylation remains an incomplete chapter in the story of basic biochemistry. The classical approach of biochemistry has apparently broken down at this point. Szent-György has compared the traditional approaches of biochemistry to energy considerations with attempting to study the functions of a sensitive piece of apparatus by first dissolving it in nitric acid. Expressed in another way, we have, in the history of theory of biological and pathological phenomena, seen the progress from a humoral to a subcellular to a molecular approach. The step to a submolecular approach remains to be accomplished. It is here that the electrochemist has a lot to offer.

4. Control of Cell Division, Tissue Growth, and Healing and the Understanding of Cancer

Dr. Pilla has commented that "one of the most challenging problems in electrochemistry today has been pointed out by the recent successful attempts to electrically control certain basic, biological growth and healing processes. Thus considerations of the concepts of electrochemical information transfer at living cell membranes have led from basic membrane and cellular studies to clinical application in the control of certain bone disorders. Evidence is already available that many other disorders may be influenced in a similar manner once the particular information code is found." The practical clinical significance that will result from any success in this area is obvious.

Szent-György has suggested that the redox potential of the environment is vital in the control of cell division and therefore in the control of cancer. Much experimental work is at the moment being carried out to test this hypothesis.

The experiments of Rosenberg and others on the use of transition element complexes on cell division opens up the possibilities of using these agents both as anticancer agents and as experimental tools in the selective switching off of certain steps in the sequence of reactions. The possibility of an electrochemical mechanism in the understanding of these processes has yet to be considered.

5. Noninvasive Clinical Investigations

Electrocardiography and electroencephalography are two of the well established, noninvasive, clinical investigations which have an electrochemical basis. The extension and refinement of such techniques is generally considered a worthwhile field for future research.

6. Intracellular Chemical Analysis

In routine clinical chemistry, it is customary to perform analyses on blood serum, but *intra*cellular concentrations are usually regarded as too difficult for determination. The information on intracellular concentrations, however, may be important because with certain species, such as potassium ions, there may be gross abnormality within the cells, while the *extra*cellular concentration remains relatively normal. Ion-specific microelectrodes and other electrochemical techniques may be applicable to this problem. As a word of warning, however, it should be noted that pathological investigations already account for a considerable proportion of the available financial resources. Clearly, a limit will soon be reached and priorities will have to be specified with respect to any further expenses.

7. Drug Interactions

A considerable literature has been developed on the interaction between drugs and tissues and drugs and drugs. Much of this literature is of questionable clinical significance. The significance of many reported phenomena to basic understanding of drug actions has not been thoroughly evaluated. Conductometric titration is one technique which can be used in the *in vitro* investigation of possible drug interactions. The physical measurements should always be correlated with the appropriate clinical investigations. An example of the application of this technique is the following which was published by Dr. F. Gutmann and the author:

Heparin, which acts as an anticoagulant in the circulating blood itself, relies on its anionic charge for its activity. The anticoagulant action is therefore reversed by certain cationic agents such as protamine. References on incompatibilities and interactions have therefore listed other cationic drugs such as lignocaine, a drug which is used as an antiarrhythmic and required for patients who may also require heparinization. When examined critically, these references are not based on any documented experiments, but are based on the "it might be expected that" type of argument. Conductometric titration failed to detect any interaction, and appropriate hematological investigations showed no impairment of detectable anticoagulant activity in the presence of the other drug. Many of the common statements in clinical references require careful physicochemical evaluation.

The interaction between certain drugs and dyes exhibits the phenomena metachromasia, which is well recognized by the histologists. The spectroscopic shift involved has been explained in terms of charge-transfer interactions, and this type of interaction may be more important in clinical phenomena than has so far been generally recognized.

8. Medical Engineering

The development of a glucose-driven pacemaker and improvements in renal dialysis have been described by the previous speaker. The electrochemical principles involved in the problem of choice of prosthesis (e.g., in a hip replacement) have also been noted. Many improvements could be effected by a cooperative interdisciplinary approach.

Further, for a number of reasons, it seems likely that the "golden age" of the development of new drugs is over and the resources, that were in the past devoted to designing new chemical substances, will be applied to improved methods of delivery of the drug to the patient. Electrochemists can be expected to have a role in these developments.

6

Some Aspects of Electrocatalysis of Oxygen Reactions at Bare and Oxide-Covered Electrodes with a View on the Future

A. Damjanovic and A. T. Ward

1. Background

It has been clearly demonstrated, for the first time in the Electrochemistry Laboratory of Prof. John O'M. Bockris at the University of Pennsylvania, that thin surface oxide films anodically formed at noble-metal catalysts profoundly affect not only catalytic activity of a given reaction—for instance of the oxygen reduction reaction[1,2] or the electro-oxidation of ethylene[3,4]—but also the mechanism of the reactions, as in the case of the oxygen reduction reaction. This discovery is significant in its own right, but it becomes even more significant when one realizes that at that time, the early 1960s, many electrochemists, including some leading ones, leaned heavily to the adsorption rather than to phase oxide theory for the anodic films at platinum and other noble-metal and alloy electrodes.

Figure 1 illustrates these differences in behavior between oxide-free electrodes and electrodes covered with a thin anodic oxide film. It is evident that the oxide-free electrode is a better catalyst than the oxide-covered surface by about 100 times, at least at 0.85 V; in fact, the ratio of the activity at these two surfaces increases with decreasing potential. The $\partial V/\partial \log i$ slope, which is indicative of the reaction mechanism, is close to 2.3 RT/F at oxide-free electrodes, but 2.3($2RT/F$) at oxide-covered electrodes. This change in the mechanism is, in fact, reflected in the catalytic activity for the reaction.

The work just described has been followed by extensive research and study of the oxygen reaction both in the laboratory of Prof. Bockris and elsewhere. The research was frequently centered around such questions as: Why does the oxygen reduction reaction at oxide-free electrodes extend only up to 1.0 V, whereas at oxide-covered electrodes it extends up to the reversible potential for the reaction (1.23 V)? Why, at oxide-free electrodes, is a rest potential established close to 1.0 V, and not the reversible potential, and what is the significance of this rest potential? What are the factors that affect reaction mechanism and kinetics at oxide-free and at oxide-covered electrodes? What is the nature of the anodic oxide film? What is the mechanism of charge transport (electronic or ionic) through the oxide film? And, of course, the question: how can a catalyst be tailored to enhance the catalytic activity of such a technologically important reaction as the oxygen reaction?

The progress made in these studies has been well reviewed.[5-11] In this article, emphasis will be placed on the effect of unpaired d electrons on the kinetics and mechanism of the oxygen reaction. Also, it will be shown how thin anodic oxide films can affect the mechanism and catalytic activity of this reaction. Finally, on the basis of the progress gained in this area, future directions in the research will be speculated upon.

A. Damjanovic and A. T. Ward • Xerox Corporation, Webster Research Center, Webster, New York 14580. A. D.'s present address is: Allied Chemical Corp., Corporate Engineering Dept., Morristown, N.J. 07960.

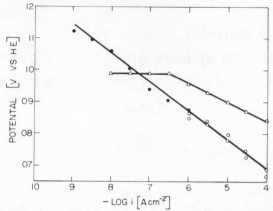

Fig. 1. V-log i plots for oxygen reduction on Pt electrodes in acid solution. •, points obtained starting from the reversible potential; ○, points obtained by reversing anodic current and fast measurements; △, points obtained on a prereduced (oxide-free) electrode. [From Damjanovic and Bockris.[1]]

2. Catalysis of Oxygen Reduction at Oxide-Free Electrodes

2.1. Effect of *d* Electrons on the Mechanism of Reaction

The kinetics of oxygen reduction has been determined on various oxide-free noble metals and alloys.[12-19] Potential log i relationships are shown in Figure 2 for a few Pt-Au and Pd-Au alloys, including the end members. At gold and gold-rich alloys the Tafel slopes are close to $-2(2.3\ RT/F)$, whereas at platinum, palladium, and palladium-rich alloys, the slopes are close to $-2.3\ RT/F$.[13] These slopes are plotted vs. alloy composition in Figure 3. It is evident that an abrupt change in the slopes occurs at a particular alloy composition.[13,20] The change of slope occurs at about 40-50 atomic percent of Au (in Pd), i.e., at the composition at which one would expect that the *d* orbital vacancies of Pd would be filled with the *s* electrons of Au, if the electronic structure in Pd-Au alloys changes with composition in the same way as in Cu-Ni alloys. The number of unpaired *d* electrons in the Pd-Au alloys decreases with alloying with Au and becomes zero at 60 at% Au. This change in the electronic structure is also illustrated in Figure 3.

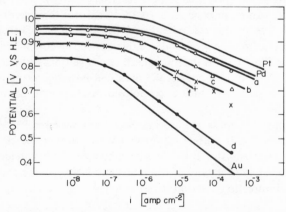

Fig. 2. Potential–current density relationship at palladium and Pd-Au alloys with data for platinum, gold, and a Pt-Au alloy included; (a) 10 at.% gold in palladium; (b) 25 at.% gold in palladium; (c) 50 at.% gold in palladium; (d) 75 at.% gold in palladium [From Damjanovic and Brusic.[12]]

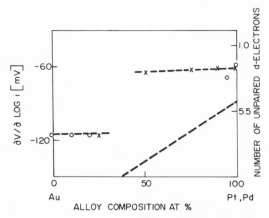

Fig. 3. Tafel slopes against alloy composition: o, platinum, Pt–Au alloys, and gold; X, palladium and Pd–Au alloys. Dashed line indicates a number of unpaired d electrons [From Damjanovic and Brusic.[13]]

Table 1. Oxygen Adsorption at Different Noble Metals[21]

Metal	Observed coverage $(\mu C/cm^2)$	Coverage for a monolayer $(\mu C/cm^2)^a$	Fraction of surface covered	No. of unpaired d electrons
Palladium	110	510	0.22	0.55
Platinum	135	500	0.27	0.55–0.6
Rhodium	480	530	0.90	1.7[b]
Iridium	440	525	0.84	1.7[b]
Ruthenium	500	530	0.95	2.2[c]
Gold	<15		<0.03	0

[a] Assuming 0 as adsorbed specie.
[b] Expected for the metals in the same group as cobalt.
[c] Expected for the metal in the same group as iron.

The abrupt change in the Tafel slopes implies a change in the mechanism of oxygen reduction with the presence of unpaired d electrons. Here, we have clear evidence that unpaired d electrons profoundly affect the mechanism of an electrode reaction.

The explanation for the change of mechanism at the composition at which the density of the unpaired d electrons approaches zero goes along the following line.[10,13,20] At platinum, palladium, and gold alloys which are rich in platinum or palladium, the unpaired d electrons participate directly in adsorption of oxygen (or OH radicals). In fact, the extent of adsorption* is linearly related to the number of unpaired d electrons per atom.[7,21,22] This has been well illustrated in the Bockris laboratory, and confirmed elsewhere,[23,24] for platinum, palladium, rhodium, iridium, and gold as well as for Pt–Rh alloy electrodes (Table 1 and Figure 4). From Figure 4 it is evident that the coverage at Pt–Rh alloys is linearly related to the atomic composition of the alloy and to the expected number of unpaired d electrons per atom (for atoms in the bulk).

It follows that for oxygen reduction at Pt, Pd, and Pt- and Pd-rich alloy electrodes, the mechanism involves relatively strong bonding and intermediate coverages (<0.5, cf.

*Saturation coverage observed at oxygen pressure of 1.0 atm.

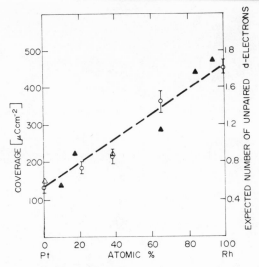

Fig. 4. Results from two independent sources of maximum coverage by oxygen-containing species on Pt–Rh alloys as a function of alloy composition. ○, Damjanovic and Rao;[22] ▲, Hoare;[23] and △, Thacker and Hoare.[24]

Table 2. Kinetic Parameters for Oxygen Reduction at Oxide-Free Platinum[17]

	$\left(\dfrac{\partial V}{\partial \log i}\right)_{pH}$	$\left(\dfrac{\partial V}{\partial \, pH}\right)_{i,pH}$	$\left(\dfrac{\partial V}{\partial \log p}\right)_{i,pH}$
	(mV)	(mV)	(mV)
Perchloric acid	$-60^{(12,17)}$ $-120^{(25)a}$	-100 to $-110^{(17)}$ $-100^{(12)}$	$-60^{(12)}$
Sulfuric acid	-53 to $-55^{(19)}$ $-65^{(15)}$ $-60^{(9,10,16,18)}$ $-120^{(25)a}$	$-90^{(10)}$	$-55^{(19)}$
Phosphoric acid	$-60^{(14)}$		$-60^{(14)}$

^{a}Obtained at high currents or in transient measurements.

Table 1) with O or reaction intermediates, whereas at Au and Au-rich alloys the reaction occurs at a coverage which is basically zero. It is then expected that the mechanism of oxygen reduction at the former electrodes is controlled by Temkin, whereas at the latter it proceeds under Langmuir, conditions of adsorption.

2.2. Mechanism of Oxygen Reduction at Oxide-Free Platinum and Palladium

The essential kinetic parameters for the oxygen reduction at platinum are summarized for convenience in Table 2. The rate equation derived from these parameters can be written formally as[12]

$$i_{cath} = kp_{O_2}^n \, [H^+]^m \exp \left[-\alpha_{cath} FV/RT\right] \qquad (1)$$

where n and α_{cath} for the reaction in acid solution are both equal to 1, and m is close to 1.5.

It has been shown in the Bockris laboratory that it is not possible (mostly because of fractional reaction order for H^+) to interpret the experimental data on the basis of any mechanism involving Langmuir adsorption.[12] This is not surprising considering that in the potential range from about 0.8 to 1.0 V (vs HE) the oxygen coverage is intermediate.

The analysis of the kinetic data lead to the suggestion that the first electrochemical step, such as

$$O_2 + H^+ + e^- \xrightarrow{\text{RD}} \text{products} \tag{2}$$

is rate determining under Temkin conditions of adsorption.[12] For such a step, the rate equation can be written in a simple form:

$$i = k p_{O_2} \, c_{H^+} \exp\left(\frac{-\Delta G^{\ddagger} + \alpha r\theta + \beta FV}{RT}\right) \tag{3}$$

Here, $G^{\ddagger}$ is the chemical part of the activation energy at low coverages, $r\theta$ represents the change of energy of adsorption of reaction intermediates with (total) coverage, and other symbols have their usual significance. It has been shown[12] experimentally that coverage, θ, changes nearly linearly with potential, irrespective of whether the electrode is in an oxygen- or nitrogen-saturated solution (Figure 5). Moreover, the coverage depends on pH

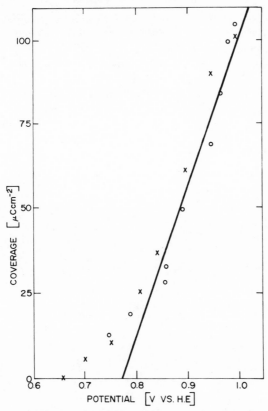

Fig. 5. Coverage by oxygen-containing species on platinum as a function of potential. ○, open circuit conditions at different oxygen partial pressures; X, electrodes potentiostatted in nitrogen-saturated solution. [From Wroblowa *et al.*[26]]

according to

$$r\theta = FV - RT \ln c_{H^+} + \text{const.} \qquad (4)$$

With equations (3) and (4), and with $\alpha = \beta = \frac{1}{2}$, it follows that

$$i = k p_{O_2}\, c_{H^+}{}^{3/2} \exp\left(-\frac{FV}{RT}\right) \qquad (5)$$

This equation is in full accord with the observed parameters and equation (1). Basically, the model for the reduction is then as follows: At each potential total coverage with adsorbed oxygen-containing species is established. The heat of adsorption of reaction products in the rate-determining step decreases with increasing coverage and, hence, the activation energy increases as the total coverage and electrode potential increase. If the first electrochemical step is rate determining, this then leads to the rate equation (5).

Since the coverage decreases with decreasing electrode potential and falls below 0.05 already at about 0.80 V (Figure 5), Temkin conditions are not expected to hold below about 0.80 V. Once Temkin conditions cease to operate, the kinetic equation for oxygen reduction should reduce to

$$i = k p_{O_2}\, c_{H^+} \exp\left(-\frac{\beta FV}{RT}\right) \qquad (6)$$

providing, of course, that the first electrochemical step is still rate determining. Consequently, with $\beta = \frac{1}{2}$, $dV/d \log i$ should change from -60 at low currents (high potentials) to -120 mV at high currents (low potentials) when the coverage becomes virtually zero. This was indeed observed at potentials cathodic to 0.80 V when, with the rotating disk electrode, the limiting, diffusion-controlled current was pushed towards higher values.[25]

Paucirova et al.[17] in the Bockris laboratory have provided additional evidence in support of the mechanism using a transient technique. If a low oxygen coverage can be maintained for a short time when the potential of an electrode is suddenly increased from a value below 0.8 V to a higher value (e.g., 0.9 V), then, according to the proposed mechanism of oxygen reduction, the initial current in the $i_t - t$ transient should obey rate

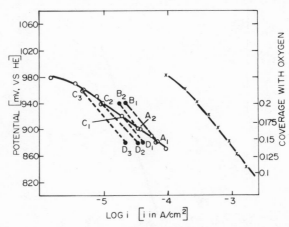

Fig. 6. Potential–log i relations: X, initial, extrapolated currents obtained in transients when starting potential was 0.50 V; $\circ$, steady-state data; $\bullet$, initial, extrapolated current after sudden increase (B_1 and B_2) or sudden decrease (D_1, D_2, and D_3) of electrode potential from the indicated starting steady state points. [From Paucirova, et al.[17]]

equation (6) rather than (1) or (5). Consequently, the increment in the current density for a given change of potential should be such that $dV/d \log i_t$ is -120 mV. This is observed[17] when potential is increased from below 0.8 above 0.84 V as illustrated in Figure 6. Subsequently, the current reaches a steady-state value (in about 50–100 sec) with $dV/d \log i = -60$ mV, as expected when coverage ultimately reaches the value corresponding to the higher potential.

The same workers have also analyzed reaction kinetics at constant coverage using this transient technique. When the initial electrode potential is suddenly increased, or decreased, from a value at which the coverage is θ, the initial current at the new value of the potential, after charging the double layer, is expected to be controlled, for a short time, by the as yet unchanged coverage inherited from the previous potential. Consequently, the initial increment in the current should again be governed by equation (6) and not by (5). The slope of a line in the V–$\log i$ diagram joining the starting point to the point which corresponds to the initial current at the new value of potential is then expected to be -120 mV, as demonstrated in Figure 6.

2.3. Change of the Heat of Adsorption with Coverage

Paucirova et al.[17] have calculated the coefficient r in equation (3) from the initial currents, i_t, in transients observed when an electrode is brought to higher potentials from a potential where the coverage is virtually zero ($V = 0.5$ V), and from the steady-state currents, i_θ. For a constant potential, pH, and oxygen partial pressure:

$$i_\theta = i_t \exp\left(\frac{-\alpha r\theta}{RT}\right) \tag{7}$$

and

$$r = \frac{RT}{\alpha\theta} \ln \frac{i_t}{i_\theta} \tag{8}$$

With fractional coverage θ taken from Figure 5, the r factors calculated for four potentials are given in Table 3. As expected, the factor r is invariant with potential. It is about 20 kcal/equiv, in full agreement with the coulometric data ($23^{[12]}$ and $18^{[26]}$kcal/equiv).

We may therefore conclude that the mechanism of oxygen reduction at oxide-free noble metals and alloys with d electrons available for adsorption involves the first discharge step as rate determining, and the process proceeds under Temkin conditions of adsorption. Once the coverage decreases below a certain value (<0.05), due to the decrease of electrode potential, Langmuir conditions prevail and the Tafel slope changes from -60 to -120 mV. At gold and gold-rich alloys where coverage is very limited, the first charge-transfer step is still rate controlling but under Langmuir conditions. Here,

Table 3. Calculated Values of r[17]

Potential V vs. HE	θ^a	i_t (A/cm^2)	i_θ (A/cm^2)	r (kcal/equiv)
0.857	0.125	1.4×10^{-3}	1.6×10^{-4}	20.8
0.885	0.15	8.5×10^{-4}	6.6×10^{-5}	20.4
0.910	0.175	5.3×10^{-4}	2.8×10^{-5}	20.2
0.935	0.2	3.2×10^{-4}	1.2×10^{-5}	19.5

^{a}Fractional coverage.

therefore, is an example of how the electronic structure of an electrocatalyst indirectly affects the kinetics of a reaction.

2.4. Effect of Bond Strength on Catalysis

The observed change in the catalytic activity of Pd–Au alloys with alloy composition (Figure 2) has been traced to the change in the activation energy which, in turn, depends on the energy of adsorption and the bond strength of adsorbed species on the metal surface. In general, as the bond energy increases, the activity increases. In Figure 7, cur-

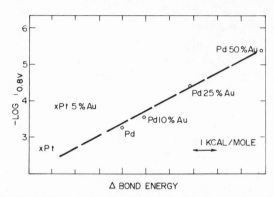

Fig. 7. Dependence of log i at 0.8 V on the bond energy of adsorbed intermediates. [From Damjanovic and Brusic.[13]]

rent density at 0.8 V changes linearly with the bond energy calculated using the bond energy expression given by Pauling. This linear relationship is, in fact, expected for the mechanism of oxygen reduction discussed above (cf. equation 3).

3. Anodic Oxide Films at Platinum above 1.0 V

3.1. Nature of the Anodic Film

Recent ellipsometric work of Bockris and co-workers,[27,28] particularly the determination of optical constants ($n = 2.8$ and $k = 1.7$) and thickness of the anodic films from ellipsometric data alone,[28,29] has shown that potentiostatic polarization of platinum electrodes produces above 1.1 V vs. HE a film that must be described as a definite oxide phase.[28] The optical constants of the films are consistent with the reported value[30] for bulk $PtO \cdot H_2O$, thus confirming unequivocally the phase nature of the anodic oxide film formed* above about 1.1V. For a fixed time of polarization, the film thickness increases with the polarization potential (Figure 8). Low-energy electron diffraction[31] and field ion spectroscopy[32] studies of the anodic films formed at high potentials point also to the phase oxide nature of the anodic films. At other noble metals, too, it is expected that phase oxide films form at high anodic potential. For instance, at gold electrodes, an oxide film forms[33] above about 1.2 V.

From these studies it follows that the oxygen evolution reaction at platinum—as well

---

*Early work of Reddy et al.[27] indicates that the oxide phase starts to grow already at 1.0 V. The difference in the initiation potential is probably due to the method of surface preparation.[28] Reddy et al. used mechanically polished, whereas Kim et al.[28] used sputtered, platinum as electrodes.

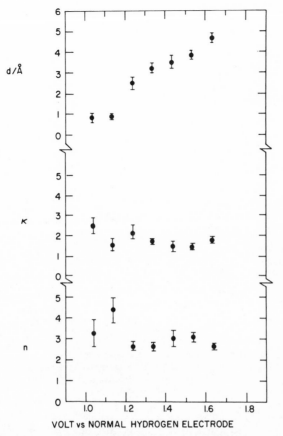

Fig. 8. The calculated values of the thickness (d), the real and imaginary parts of the refractive index of the film (n and κ) at various potentials. [From Kim *et al.*[28]]

as at other noble (and nonnoble) metals and alloys, including Au—occurs only at electrodes covered with a relatively thin film of an oxide phase. The catalysis of the reaction is therefore expected to be controlled by the nature of the oxide films (e.g., by their semiconductor characteristics) and/or by their thickness (e.g., in case of electron tunneling through the thin insulating films) rather than by the properties of the metal, as was the case in oxygen reduction at oxide-free metal and alloy electrodes.

The formation of the anodic oxide films at platinum at about 1.0 V restricts oxygen reduction at oxide-free electrodes to potentials below 1.0 V (cf. Figures 1 and 2). Oxygen reduction at oxide-covered electrodes may, however, occur both at potentials above 1.0 V—all the way up to the reversible potential for the reaction—and at potentials below 1.0 V. The latter is due to the hysteresis in the reduction of the oxide films. The hysteresis is best illustrated in potentiodynamic cycling traces. These show that the reduction of the oxide films is quite irreversible and occurs at relatively low potentials (cf. reference 34).

Before we discuss the effect of the anodic oxide films on the oxygen evolution (and reduction) reaction, it is necessary to point to some recent developments concerning the *kinetics of growth* of the oxide films. It will be shown that oxide films continue to grow, although slowly, both under galvanostatic and potentiostatic polarization and during oxygen evolution.[35–38] This, in turn, affects the kinetics of the oxygen reaction.

3.2. Initial Growth of Anodic Films

When a constant anodic current is applied to a prereduced, oxide-free platinum electrode in oxygen-saturated acid solution, a linear or nearly linear V-t relationship has frequently been observed,[36,38-41] starting from about 1.0 V and with coverage close to 2.10^{-4} C cm^{-2} (cf. reference 37). The initial linear increase in potential is then followed by a much slower rise of potential with time (Figure 9). The linear relationship represents

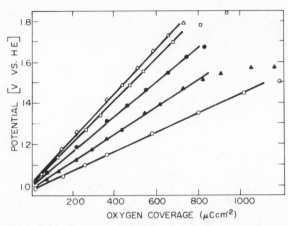

Fig. 9. Linear change of potential with oxygen coverage at initial stages of oxide growth at different constant currents of polarization. △, 3.6×10^{-2}; □, 1.8×10^{-2}; ●, 5.5×10^{-4}; ▲, 1.8×10^{-5}; and ○, 7.3×10^{-7} A/cm^2. Solution 2 N H$_2$SO$_4$. [From Damjanovic *et al.*[38]]

the buildup of an oxide phase without a side or parallel reaction, e.g., oxygen evolution. The latter has been confirmed with a rotating disk-ring electrode.[37] When a constant current is applied to the disk electrode, no current is detected at the ring electrode for as long as the potential of the disk remains in the linear region. Close to the knee in the V-t curves (Figure 9), oxygen starts to evolve at the disk and is detected at the ring electrode. With chronoellipsometry, too, a linear change in thickness of a phase oxide with time, or potential, is observed[38] during the initial stages of polarization (Figure 10).

The dependence of potential on integrated charge $q(= i_{og}t)$ is shown in Figure 9 for various constant current densities. The $(\partial V/\partial q)_i$ values increase linearly with logarithm of applied current density (Figure 11) for several decades of current density, i.e.,

$$\left(\frac{\partial V}{\partial q}\right)_i = a \ln i_{og} + b \tag{9}$$

with a equal to 57 V cm^2 C^{-1}, and b to 1240 V cm^2 C^{-1} when i_{og} is in A cm^{-2}.

The kinetic data of Figures 9 and 11 lead to the following equation for the growth of oxide films[38]

$$i_{og} = i_{og,o} \exp\left(\frac{V - V_0}{aq}\right) = i_{og,o} \exp\left(\frac{V_{of}}{ard}\right) \tag{10}$$

Here, V_0 is the potential when thickness $d(= q/r)$ is zero, i.e., V_0 is the potential at which the oxide film starts to grow, and V_{of} is the potential difference across the oxide film. Factor r transforms the thickness of the oxide films to coulombs. Constant $i_{og,o}$ [$\sim \exp(-b/a)$] is the exchange current density for oxide growth. It is equal to 1.8×10^{-10} A cm^{-2} (cf. references 35 and 36). V_0 is nearly independent of applied cur-

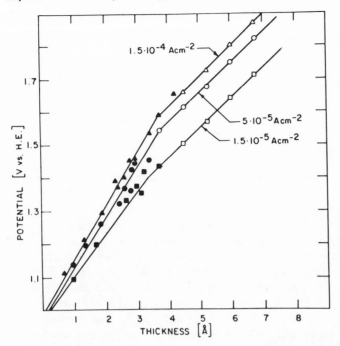

Fig. 10. The dependence of potential on ellipsometric thickness in the initial linear (solid symbols) and subsequent logarithmic growth (open symbols) of oxide films. ■ and □, 1.5×10^{-5}, ● and ○, 5×10^{-5}; and ▲ and △, 1.5×10^{-4} A/cm^{-2}. [From Damjanovic *et al.*[38]]

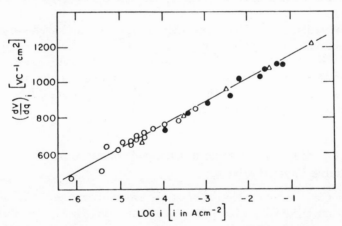

Fig. 11. The dependence of slopes in V-q diagrams on current density. ●, oscilloscope recording; ○, recording with a mechanical recorder, △, data from Vetter and Schultze.[36] [From Damjanovic *et al.*[38]]

rent density. It is readily obtained by extrapolation of the linear V-t curves to zero time or charge. From Figure 9, $V_0 \approx 1.00$ V vs. HE.

The form of rate equation (10) for the growth of oxide films under galvanostatic conditions is the same as that for the growth of insulating films under high field conditions according to the Cabrera–Mott mechanism, i.e.,

$$\frac{d\,d(t)}{dt} = K \exp\left(\frac{ze\lambda \, V_{of}}{kTd}\right) \tag{11}$$

where K is a constant. Parameter λ is the half-jump distance of an ion of charge ze migrating in an electric field V_{of}/d, where d is the film thickness. Other symbols have their usual significance. Equation (11) reduces to

$$\frac{d\,d(t)}{dt} = \text{const.} \exp\left(\frac{\alpha V_{of}}{d}\right) \tag{12}$$

with transfer coefficient $\alpha = z\lambda F/RT$. From equations (10), (11), and (12) it follows that

$$\alpha = (ar)^{-1} \qquad [= 39.6z\lambda] \tag{13}$$

If α is determined from coulometric measurements and the conversion factor r is determined from estimates of the density and stoichiometry of the oxide phase, then λ can be calculated and compared with structural parameters. It turns out[38] that the assumption of PtO for the stoichiometry of the anodic phase (i.e., $z = 2$) gives the only reasonable solution for the half-jump distance ($\lambda \sim 2.0$ Å) using the measured transfer coefficient ($\alpha = 158$ Å V^{-1}) and the appropriate value of the conversion factor ($r = 1.1 \times 10^4$ C cm^{-3} for PtO). The analysis of the transfer coefficient therefore leads to the same conclusion as ellipsometry: The anodic oxide film at platinum above 1.0 V vs. HE contains divalent Pt^{2+} ions, i.e., the oxide phase is PtO. The significant feature of this result is that it implies the existence of a high electric field across the anodically growing film. This in turn implies that the film itself must be electronically insulating in order to sustain the fields of 10^6–10^7 V/cm encountered during ionic growth. The insulating nature of the anodic oxide film has rarely been taken into account in kinetic analyses of reactions occurring at electrodes covered by such films. Certainly, such films will be barriers to charge transfer in electrochemical reactions, as will be discussed further in the next section.

The above results point also to the kinetic nature of anodic oxide films. They grow in thickness with time, and the thickness at the same potential depends on applied current density. Even when oxygen starts to evolve and potential changes only slowly with time, the thickness continues to increase, although very slowly and according to a different law of growth, as will be shown later. This extended growth affects the catalysis of reactions that occur at such oxide-covered surfaces.

4. Anodic Oxide Films as Barrier to Charge Transfer in Series with Double-Layer Barrier

In previous analyses of the kinetics of oxygen evolution reaction[42–44], it was tacitly assumed that the anodic oxide films were conducting electronically, i.e., that the potential difference across the oxide film was virtually zero. From the preceding discussion of the mechanisms of growth of anodic oxide films at platinum, it appears that these oxide films are rather poor conductors, e.g., they can easily sustain relatively high electric fields during the growth. Furthermore, since the catalytic activity for oxygen evolution decreases with thickness of the oxide film (cf. Section 6 and reference 45), the film itself must be a barrier to charge transfer and a potential difference should exist across it. The potential difference, ΔV, with respect to a suitable reference, e.g., reversible oxygen potential, is then the sum of the potential difference across the oxide film, ΔV_{of}, and that across the double layer,[45] ΔV_{dl}

$$\Delta V = \Delta V_{of} + \Delta V_{dl} \tag{14}$$

In the steady state, when the current for oxide growth is negligible and the thickness of the oxide film remains nearly constant, ΔV_{of} and ΔV_{dl} for the same ΔV adjust themselves in such a way that

$$i_{obs.} = i_{dl} = i_{of} = k a_{H^+}^{-p} \exp \left(\frac{\alpha F \Delta V}{RT} \right) \tag{15}$$

Here, i_{dl} and i_{of} are current densities across the double layer and oxide film, respectively, a_{H^+} is the activity of H^+ ions, α is the overall transfer coefficient ($= \frac{1}{2}$), and p is the negative of the reaction order for H^+ ($= \frac{1}{2}$).

The concept of the dual-barrier control in a series type of arrangement has been suggested by Meyer[46] in an analysis of cathodic processes at passive Zr electrodes. MacDonald and Conway[47] used the same concept to explain some unusually high Tafel slopes in oxygen evolution at Au and Pd electrodes (see also references 48 and 49). This concept has recently been applied in the analysis of the kinetics of oxygen evolution at platinum with the aim of accounting for the unusual pH dependence of the reaction.[45] It has now been confirmed[45] with solution-exchange experiments that when care is taken the thickness of the oxide film remains the same in solutions of different pH, the $(\partial V / \partial pH)_i$ is -60 mV, i.e., the reaction order for H^+ is negative and *fractional*. This reaction order cannot be accounted for by any simple mechanism either under Langmuir or Temkin conditions of adsorption.*

In a simple way, current across the double layer is given as[45]

$$i_{dl} = Y a_{H^+}^{n} \exp \left[\frac{\alpha_{dl} F \Delta V_{dl}}{RT} \right] \tag{16}$$

It follows then that i_{of} also is of the form

$$i_{of} = X a_{H^+}^{m} \exp \left[\frac{\alpha_{of} F \Delta V_{of}}{RT} \right] \tag{17}$$

Here, α_{dl} and α_{of} are transfer coefficients for the individual reactions, and m and n are the orders of the reactions with respect to H^+. X and Y are constants for a given thickness of the oxide film. The kinetic equation based on the dual-barrier model is then obtained from equations (16) and (17) in a general form

$$i = \left[X^{\alpha_{dl}} Y^{\alpha_{of}} a_{H^+}^{(\alpha_{dl}m + \alpha_{of}n)} \right] \frac{1}{\alpha_{dl} + \alpha_{of}} \exp \left[\frac{\alpha_{of} \alpha_{dl} F \Delta V}{(\alpha_{of} + \alpha_{dl})RT} \right] \tag{18}$$

Comparison of the corresponding terms in this rate equation with those in the experimental rate equation (cf. equation 15) yields

$$k = \left[X^{\alpha_{dl}} Y^{\alpha_{of}} \right] \frac{1}{\alpha_{dl} + \alpha_{of}} \tag{19}$$

$$\alpha = \frac{\alpha_{dl} \alpha_{of}}{\alpha_{dl} + \alpha_{of}} \qquad [= \tfrac{1}{2}] \tag{20}$$

$$P = - \frac{n\alpha_{of} + m\alpha_{dl}}{\alpha_{of} + \alpha_{dl}} \qquad [= \tfrac{1}{2}] \tag{21}$$

---

*It was previously reported[43] that this pH dependence was close to -30 mV, and it was ignored. This then led to the conclusion that the first electrochemical discharge of H_2O is the rate-determining step.

from which α_{dl} and α_{of} for various values of m and n can be evaluated. Since the reaction across the oxide/solution interface is more likely to depend on pH than the rate across the oxide film, m is taken to be zero. Then, the only simple solution is the one with $n = -1$ which gives $\alpha_{of} = \alpha_{dl} = 1$. The rate equation for the reaction across the double layer is then given by

$$i_{dl} = Ya_{H^+}^{-1} \exp\left(\frac{F\Delta V_{dl}}{RT}\right) \tag{22}$$

i.e., there is *no fractional reaction order* with respect to H^+. The overall rate is given by

$$i = (XY)^{1/2} a_{H^+}^{-1/2} \exp\left[\frac{F\Delta V}{2RT}\right] \tag{23}$$

and the fractional reaction order appears in the rate expression. This reaction order now becomes fully accounted for in terms of the dual-barrier control of the reaction. Factor X reflects the dependence of the reaction rate on the thickness of the oxide films. It is constant for a given thickness but sharply decreases as the film thickness increases, as will be shown in following sections. The rate across the oxide film is given by equation (17) when $\alpha_{of} = 1$ and $m = 0$. The significance of this rate equation is discussed in a subsequent section.

5. Reaction Path and Rate-Determining Step for the Reaction Across the Double Layer

Equation (22) for the rate across the double layer, i_{dl}, is compatible with a chemical step as the rate-controlling reaction step where this step follows a charge transfer step in which a water molecule is the reactant, e.g.,

$$H_2O \rightleftharpoons H^+ + OH + e^- \tag{24}$$

$$OH + M \longrightarrow [M \cdots OH] \text{ or } [M \cdots O^- \text{ or } H^+] \tag{25}$$

Here, M is a site on the surface of the oxide film.

An attractive reaction path has been recently suggested.[45] According to this path, an oxygen atom in the surface of the oxide film participates directly in the oxygen evolution reaction. For instance, after step (24) the reaction proceeds according to:

With the second step as rate determining.* The oxygen vacancy in the surface of the oxide film is then replenished in a *parallel* reaction, e.g., according to:

$$
\begin{array}{c}
\underset{|}{-}O\underset{}{\underset{\cdot\cdot}{-\!-\!-}}\overset{|}{\underset{|}{Pt}} \\[4pt]
\underset{|}{-}\overset{|}{Pt} \quad\quad + H_2O \longrightarrow \\[4pt]
\underset{|}{-}O\underset{}{-\!-\!-}\overset{|}{\underset{|}{Pt}}
\end{array}
\qquad
\begin{array}{c}
\overset{|}{\underset{|}{-O-Pt}} \\[4pt]
\overset{|}{\underset{|}{-Pt-O}} \quad + 2H^+ + 2e^- \\[4pt]
\overset{|}{\underset{|}{-O-Pt}}
\end{array}
\qquad (27)
$$

Evidence that an oxygen atom in the surface of the oxide film may participate in the oxygen evolution reaction is provided by an early work of Rosenthal and Veselovski.[50] These workers have found, using a tracer technique with ^{18}O, that if the platinum-surface oxide film is initially enriched with ^{18}O by anodic polarization in ^{18}O-enriched solutions, then the first gas anodically evolved at platinum in ^{18}O unenriched solutions is enriched with ^{18}O. At silver electrodes, too, direct participation of oxygen in surface oxides has been detected in oxygen evolution in alkaline solution.[51]

6. Dependence of the Rate of the Oxygen Evolution Reaction on the Thickness of the Surface Oxide Film

Ellipsometry and standard electrochemical techniques have been combined to determine how the thickness of an anodic oxide film affects the catalysis of the oxygen evolution reaction at platinum. A prereduced electrode is first polarized in sulfuric acid solution with a constant current density, i_p, for a given time and then the thickness, d, of the anodic film and the electrode potential, V, are determined for various current densities $i \leqslant i_p$.

It is found[52] that both the thickness and electrode potential change linearly with the logarithm of the current density, i_p, with which a prereduced electrode was anodically prepolarized for a given time. For all times of polarization, the $dV/d(\log i_p)$ slopes are close to 170 mV/decade of current density, and the $dd/d(\log i_p)$ slopes are close to 1.0 Å/ decade of current density (Figures 12 and 13).

After an electrode has been polarized for 750 sec with a constant current density, i_p, and the current density, i, is decreased in steps, no change in thickness, d, is observed (from A to B in Figure 14). In contrast to this, the potential, V, decreases as the current density, i, is decreased (Figure 13). The $dV/d(\log i)$ slope is now close to the familiar and frequently reported value of 115 mV/decade or to $2.3(2RT/F)$. The same behavior is observed after the electrode has been prepolarized for 125 or 1800 sec, as shown in Figure 13.

An arrest in the growth of the anodic oxide film occurs when, after the initial polarization of a prereduced electrode with a constant applied current density, the current density is decreased. Such "hysteresis" has been reported previously in coulometric studies in the potential region lower than that of this ellipsometric study.[53,54]

The linear V-log i relationship observed for the oxygen evolution reaction for over several decades of current density with slope close to $\partial V/\partial \ln i = 2RT/F$ is therefore obtained only under the condition of nearly constant thickness of the oxide film. Under

---

*"Bonds" and "positions" of platinum and oxygen atoms only illustrate the model.

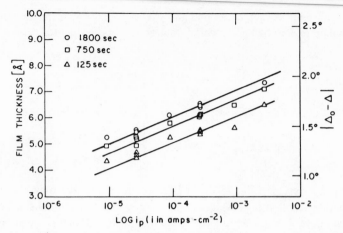

Fig. 12. Plot of anodic film thickness after 125, 750, and 1800 sec vs. log current density of polarization. [From Damjanovic *et al.*[(52)]]

this condition

$$\left(\frac{\partial V}{\partial \ln i}\right)_d = \frac{2RT}{F} \quad \text{and} \quad \left(\frac{\partial V}{\partial \ln i}\right)_d \neq f(V, d, i) \tag{28}$$

The catalytic activity for the oxygen evolution reaction decreases with increasing thickness of the surface oxide film. This is evident from the change of the electrode potential at the same current density with the thickness of the film as illustrated in Figure 15. In this figure, the potentials from Figures 12 and 13 are plotted vs. thickness for two current densities. As the thickness increases, the potential at the same current density increases linearly, i.e., the catalytic activity decreases. The ratio of the potential to the

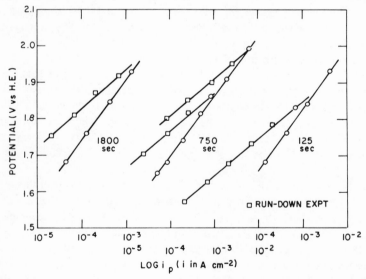

Fig. 13. Plot of electrode potential (HE) after 125, 750, and 1800 sec (circles) vs. log current density of polarization. Squares represent subsequent potential–current density relationships for decreasing current densities. [From Damjanovic *et al.*[(52)]]

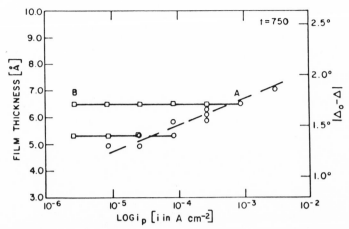

Fig. 14. Plot of anodic film thickness after 750 sec (circles) vs. log current density of polarization. Squares represent subsequent film thickness–current density relationships for decreasing current densities. [From Damjanovic *et al.*[52]]

thickness appears to be independent of current density, i.e.,

$$\left(\frac{\partial V}{\partial d}\right) = 0.085 \text{ V/Å} \qquad \text{and} \qquad \left(\frac{\partial V}{\partial d}\right) \neq f(V, i, d) \tag{29}$$

An analysis of equations (28) and (29) shows that, to a good approximation, $\log i$ is linearly related to both V and d, i.e., the function

$$f(V, d, \log i) = \text{constant} \tag{30}$$

represents a plane surface in three dimensions. This follows from the fact that, as illustrated in Figure 16, the projections of the straight line AB on two perpendicular planes are parallel to the corresponding projections of the straight line A′B′ while intersecting the projection of the third straight line AD. All these lines lie in the plane $f(V, \log i, d) = $ constant.

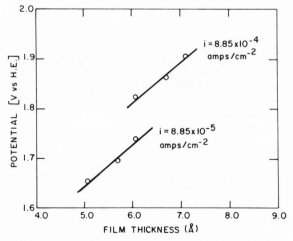

Fig. 15. Plot of electrode potential (HE) vs. film thickness after polarization at constant current density for various times. [From Damjanovic *et al.*[52]]

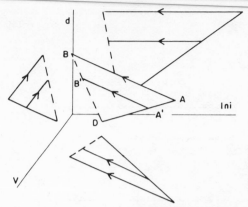

Fig. 16. Representation of the plane $f(V, \ln i, d)$ = constant in three dimensions. [From Damjanovic et al.[52]]

The analytical solution to $f(V, \log i, d)$ = constant is

$$\epsilon V - \left(\frac{2RT\epsilon}{F}\right) \ln i + \frac{2RT}{F} \left(1 - \frac{\kappa F}{2RT}\right) d = \text{constant} \qquad (31)$$

where

$$\epsilon = \left[\frac{\partial d}{\partial (\ln i_p)}\right]_t \qquad (32)$$

is equal to 1.0 Å/decade of current density. Constant κ is given by

$$\kappa = \left(\frac{\partial V}{\partial \ln i_p}\right)_t \qquad (33)$$

and is equal to 0.170 V/decade. Now with

$$\frac{2RT - \kappa F}{\epsilon RT} = -\delta \qquad (34)$$

equation (31) transforms to

$$i = C \exp \left(\frac{-\delta d}{2}\right) \exp \left(\frac{FV}{2RT}\right) \qquad (35)$$

This equation gives the dependence of the rate of the oxygen evolution reaction both on the electrode potential and on the thickness of the surface oxide film.* For a constant thickness, the equation reduces to the well-known rate equation for the oxygen evolution reaction in an acid solution of a given pH:

$$i = i_0 \exp \left(\frac{FV}{2RT}\right) \qquad (36)$$

The catalytic activity, expressed as i_0, decreases exponentially with the thickness of the oxide film:

$$i_0 = C \exp \left(-\frac{\delta d}{2}\right) \qquad (37)$$

---

*Vetter and Schultze[55] have arrived at the same relationship from different arguments.

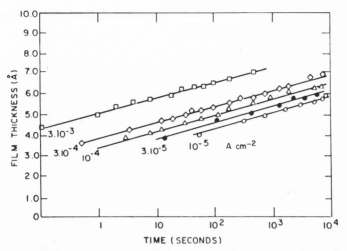

Fig. 17. Change of ellipsometric thickness with time of polarization of a platinum electrode at various constant current densities. The major reaction is O_2 evolution. [From Damjanovic and Ward.[37]]

Constant C contains, among other terms, the concentration dependence. Differences in catalytic activities can now be accounted for in terms of the differences in thicknesses of surface oxide films in different experiments (cf. reference 45).

Finally, from the dual-barrier model and equation (23), it follows that

$$X = X_0 \exp{(-\delta d)} \tag{38}$$

and the current across the oxide film is given by

$$i_{of} = i_{obs.} = X_0 \exp{(-\delta d)} \exp{\left(\frac{F\Delta V_{of}}{RT}\right)} \tag{39}$$

The link between the catalytic activity for oxygen reaction and the thickness of the oxide film is now evident. The significance of this rate equation is discussed below.

7. Extended Growth of Anodic Oxide Films at Noble-Metal Anodes

Following the initial linear growth,* which can be described by the Cabrera–Mott model of oxide film formation, once the oxygen evolution reaction becomes the major reaction (Section 3.2), *in situ* ellipsometry shows[37] that at a constant current of polarization, i_p, the thickness of anodic oxide films at platinum electrodes increases *logarithmically* with time. This is observed for all constant current densities in the range 10^{-5}–10^{-2} A/cm². It may be noted that only a small part of the current is used for oxide formation. For instance, at 10^{-3} A/cm² during the first 10^3 sec only about 0.3% of the total charge goes for oxide formation. Hence, $i_p \approx i_{0_2}$. In Figure 17, the dependence of the ellipsometric thickness on time is shown for five constant current densities of polarization. It is observed that[37]†

$$\left(\frac{\partial d(t)}{\partial \ln t}\right)_{i_{0_2}} = \psi = \text{const} \neq f(i_{0_2}) \tag{40}$$

---

*Linear growth occurs only at a constant current density. At a constant potential, the growth is described by the "inverse logarithmic law."

†A similar logarithmic dependence of $d(t)$ on time at a single constant current was reported by Makrides[56] for gold anodes.

Further, from the separation of the parallel $d(t)$–$\ln t$ lines

$$\left(\frac{\partial d(t)}{\partial \ln i_{0_2}}\right)_t = \epsilon = \text{const} \neq f(i_{0_2}) \tag{41}$$

At a constant potential, too, the surface coverage with oxygen species, as determined by coulometry,[35,37,57] or the ellipsometric thickness of the oxide film[37] increases logarithmically with time as shown in Figure 18. It follows that*

$$\left(\frac{\partial d(t)}{\partial \ln t}\right)_V = \text{const} = \omega \neq f(V) \tag{42}$$

Here too the slopes are independent of other variables. Also, $\omega \neq \psi$. Further, the d–$\ln t$ lines at different potentials are spaced in such a way that

$$\left(\frac{\partial d(t)}{\partial V}\right)_t = \text{const} = \sigma \neq f(V) \tag{43}$$

Integration of equations (40) and (41) gives the dependence of thickness on time during this extended growth:

$$d(t) = \epsilon \ln i_{0_2} + \psi \ln\left(\frac{t}{t_0}\right) + C' \tag{44}$$

for constant current, or

$$d(t) = \sigma V_p + \omega \ln\left(\frac{t}{t_0}\right) + C'' \tag{45}$$

for constant potential mode of polarization. Equations (44) and (45) describe the "direct logarithmic law" of growth.

The current density for oxide growth, i_{og}, is given, for instance, from equation (45), by

$$i_{og} = r\frac{dd(t)}{dt} = \sigma\frac{dV}{dt} + \frac{\omega}{t_0}\exp\left[-\frac{d(t)}{\omega} + \frac{\sigma V}{\omega}\right] \tag{46}$$

The rate equation reduces to a familiar form when, instead of potential difference across the whole interface, ΔV, the potential difference across the oxide film only, ΔV_{of}, is introduced into the rate equation. According to the dual-barrier model and the analysis of the kinetics and mechanism of the oxygen evolution reaction discussed above, this potential difference is related to the polarization current by equation (39). With equation (39) from equation (46) the following rate equation for oxide growth is obtained:

$$i_{og} = i_{og,0}\exp\left[-\frac{1 + \epsilon\sigma}{\psi}d(t) + \frac{\epsilon F\Delta V_{of}}{\psi RT}\right] \tag{47}$$

This equation describes the extended growth of anodic oxide films both under constant potential and under constant current conditions. An equation identical in form to equation (47) is frequently observed for the growth of anodic oxide films at other metals too, e.g., at iron electrodes. In contrast to platinum electrodes, at which the major

---

*Feldberg et al.[57] have found $dq/d \log t$ slopes independent of applied electrode potential. According to Vetter and Schultze,[35] however, the slopes increase with increasing potential. Nonetheless, in the potential range where oxygen evolution is the major reaction, e.g., from 1.6 to 1.9 V, the q–$\log t$ lines appear to be nearly parallel (cf. Figure 5 in reference 35).

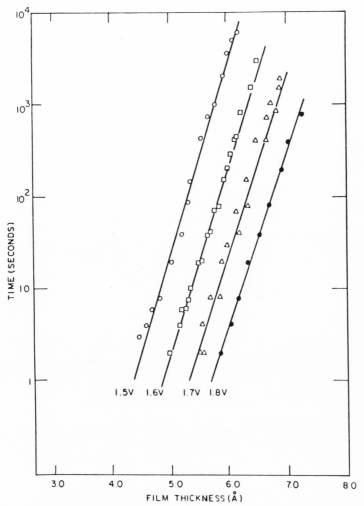

Fig. 18. Change of ellipsometric thickness with time of polarization of a platinum electrode at various constant potentials. [From Damjanovic and Ward.[37]]

reaction is the oxygen evolution reaction, the major reaction at iron electrodes is the growth of the oxide film itself. The change of potential with time at the iron electrodes therefore represents the change arising only from the increase in thickness of the oxide film.[58-60] At platinum anodes, the change of potential with time may represent the change due to the major ongoing reaction, the oxygen evolution reaction, and is only indirectly related to the thickness of the oxide film.

The change of growth mechanism from the initial growth that proceeds according to equation (10), to the extended growth, that proceeds according to equation (47), occurs only after O_2 starts to evolve or becomes the major reaction. It is not clear why the mechanism changes at this point, or whether the change is related to the onset of an electronic current through the oxide film. Further study here is desirable.

The extended, slow growth of anodic oxide films is expected to occur also at other noble metals, noble-metal alloys, and at some transition-metal electrodes. If the rate of an electrochemical reaction is controlled by a relationship similar to equation (35), then a consequence of the extended growth would be a continuous decrease of the catalytic activity of the electrode for that reaction. This is shown to be the case in the next section.

8. Decrease of the Catalytic Activity of Platinum Electrodes with Time of Polarization

Bockris and Huq in 1956 demonstrated for the first time in a clear way that the catalytic activity for the oxygen evolution reaction at platinum decreases significantly with time of polarization,[43] e.g., the electrode potential at a given constant current density increases logarithmically with time. A similar effect was observed at gold electrodes.[56] Recently, Bockris and Huq's data were confirmed. The kinetic analysis was expanded to include a series of constant current densities ranging from 3.10^{-5} to 10^{-2} A cm^{-2}. For any constant current density of polarization, i_p, it was found that (Figure 19)

$$\left(\frac{\partial V}{\partial \ln t}\right)_{i_p} = \text{const} = \alpha \neq f(i_p, t) \tag{48}$$

Further, the V-log i lines at different i_p's are parallel, and, for an equal increment in log i_p, are shifted away one from another by the same increment ΔV, i.e.,

$$\left(\frac{\partial V}{\partial \ln i_p}\right)_{t} = \text{const} = -\frac{\alpha}{\beta} \neq f(i_p, t) \tag{49}$$

with $\alpha = 33$ mV and $-\alpha/\beta = 76$ mV.

Similarly, at a constant potential, V_p, catalytic activity for the oxygen evolution reaction decreases with time according to (Figure 20)[37, 61]

$$\left(\frac{\partial \ln i_{0_2}}{\partial \ln t}\right)_{V_p} = \text{const} = \beta \neq f(V_p, t) \tag{50}$$

Again, for the same increment ΔV, log i_{0_2}–log t lines are shifted with respect to each

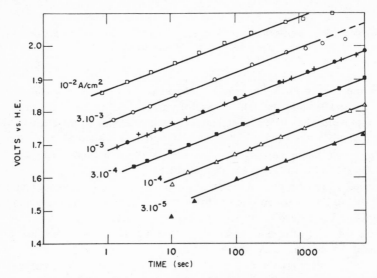

Fig. 19. Change of potential at platinum electrodes with time at various constant current densities of polarization, i_p. [From Damjanovic and Ward.[37]]

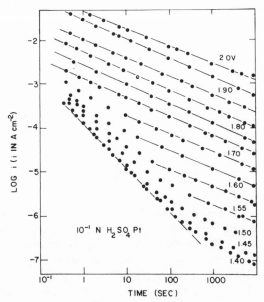

Fig. 20. Change of current density with time at various constant potentials, platinum anodes. [From Damjanovic and Ward.[37]]

other by the same increment $\Delta \log i_{0_2}$, i.e.,

$$\left(\frac{\partial \ln i_{0_2}}{\partial V_p}\right)_t = \text{const} = -\frac{\beta}{\alpha} \neq f(V_p, t) \tag{51}$$

It may be noted that α and β have the same numerical values as for galvanostatic polarization. The partial derivative (with respect to $\log i_p$) of equation (48) has a significantly higher value (~ 175 mV) than that usually observed for the Tafel slopes for O_2 evolution at platinum anodes (~ 120 mV). This is because the former derivative is obtained under the condition of extended continuous growth of the anodic oxide films, whereas the Tafel slopes of 120 mV refer to a constant thickness of the oxide films, as discussed in a preceding section (Figure 13).

It follows from the above partial derivatives that both the potential at a constant i_p and the current density at a constant V_p change with time according to

$$V = -\frac{\alpha}{\beta} \ln \frac{i}{i_0} + \alpha \ln \frac{t}{t_0} + C \tag{52}$$

This empirical equation describes the dependence of the catalytic activity for oxygen evolution with time.

The observed decrease with time of the catalytic activity for oxygen reaction at platinum electrodes* is explained along the lines suggested in Section 7. With equation (45) into equation (35), the rate equation (52) results, i.e., as the thickness of the anodic oxide film increases with time of polarization, so does the barrier for charge transfer according to equation (35). In other words, the decrease in the catalytic activity with time is a consequence of the continuous increase in the thickness of the anodic oxide film.

---

*Probably also at gold electrodes.

9. Quantum Mechanical Tunneling in Thin Anodic Films

The exponential dependence of the exchange current density, or of the activity, for a given reaction on thickness of the oxide film has led to the conclusion that charge transport in such films occurs via a quantum mechanical tunneling process. The tunneling transition occurs from an occupied electronic level in (adsorbed) reactants to the unoccupied electronic level of the same energy in the metal beneath the oxide film.[62,63] For this model of charge transport across the oxide film, the anodic current, according to Schultze and Vetter,[62] is proportional to (cf. reference 64)

$$i \sim \int_{-\infty}^{+\infty} D_m(E) \cdot f(E_F - E) \cdot D_e(E) \cdot f(E - E_F) \cdot \nu(E) \cdot dE \tag{53}$$

where $D_m(E)$ and $D_e(E)$ are the distribution functions of electron states in the metal, m, and electrolyte, e, and $\nu(E)$ is the frequency factor. The number of the unoccupied electron states in the metal is given by $D_m(E) \cdot f(E_F - E)$ where f is the Fermi function. The corresponding product $D_e \cdot f(E - E_f)$ gives the number of occupied electron states due to species adsorbed at the oxide surface. The dependence of the current on potential arises from the shift of the Fermi level E_F with respect to the Fermi level in the electrolyte. The frequency factor $\nu(E)$ is given by[62]

$$\nu(E) \sim \exp\left[- \frac{4\pi d}{h} \sqrt{2 m_e \Delta E_t}\right] \tag{54}$$

where E_t is the mean height of the potential barrier and m_e the electron mass. Factor ν depends on the square root of E_t. Such dependence is not evident from the experimental results on oxygen evolution discussed above. It is likely, as pointed out by Schultze and Vetter,[62] that the relative changes in potential are small and do not greatly affect the frequency factor.

It is questionable whether Fermi statistics, rather than Boltzman statistics, should be applied to the (adsorbed) reactants (cf. reference 65). Further, for very thin anodic films with a high energy gap, the idealized barriers, which are trapezoidal under potential bias, may significantly differ from the practical barriers when the effect of the image forces rounds off the corners of the idealized barrier.[66] This may then lead to a different expression for the tunnel probability. It is evident that a large "gap" exists between the experimental achievements and theoretical interpretations. This is particularly true in the case of thin, nonconducting films as electrodes when electron tunneling is expected to control the rate of the reaction. One may expect a significant effort in the future to interpret the mechanisms of reactions and analyze the catalysis at such electrodes in terms of quantum mechanics.

A few examples are given here to illustrate further how anodic oxide films affect the rate of a given reaction. The rate of the Ce^{4+}/Ce^{3+} redox reaction at platinum has been found[63] to depend exponentially on the thickness of anodic films. Again, tunneling is assumed to control the rate. Oxidation of ethylene and some other hydrocarbons[3,4] at noble-metal and alloy electrodes suddenly decreases as the electrode potentials become sufficiently anodic for surface oxides to form. Similarly, the (limiting) rate of hydrogen oxidation at platinum electrodes rapidly decreases[67] at potentials at which oxide films are expected to form at the electrode surfaces (above 1.0 V for platinum). These "passivations" of electrode reactions are caused by the formation of the anodic oxide films and the change in the mechanisms which now probably include electron tunneling.

Not only kinetics of reactions but also mechanisms are expected to be affected by the presence of anodic films. An illustration of this is oxygen reduction at platinum (cf. Figure 1). At oxide-free electrodes[12] the rate-determining step is the first discharge under Temkin conditions of adsorption of reaction intermediates with the Tafel slope of -60 mV. At oxide-covered electrodes, the Tafel slope is close to -120 mV and electrons tunnel through the oxide film.

These examples illustrate the importance of the *thin* anodic surface oxide films in determining the mechanisms of various reactions and in studies of electrocatalysis. They also point to the need to determine the nature of the surface of any catalyst. The texture (e.g., porosity) of the oxide films, their composition, crystallinity, thickness, and conducting properties should be taken into account when analyses are made of mechanisms of various reactions and when catalysis of a given reaction on various substrates is studied. New techniques, e.g., Auger spectroscopy, LEED, and RHEED, are expected to be useful for such determinations and studies and should supplement well-established electrochemical techniques, including ellipsometry. Evidently, one can find much room here for further research and studies.

The role of *thick*, semiconducting oxide films or of bulk oxides, e.g., oxide bronzes, in controlling the mechanism of an electrochemical reaction and electrocatalysis is far better understood compared to the role of thin anodic films. Thick oxide films may equally well prove useful as electrocatalysts for O_2 reduction. These are not considered in the present paper as we have restricted discussion to electrocatalysis of the oxygen reaction at bare electrodes and at electrodes covered by a *thin* anodic oxide film.

‚10. Conclusions and Predictions

It has been illustrated here that at bare electrodes the catalysis of a reaction is greatly affected by the electronic structures of electrodes. The exchange current density for oxygen reduction, for instance, increases as the number of unpaired d electrons increases in a series of Au–Pd alloys (cf. Figure 2). How far this trend will continue beyond 0.6 unpaired d electrons for Pd, e.g., with Pd–Rh alloy electrodes, is not clear. Rhodium is less active than palladium, and so a volcano-type relationship, similar to that observed by Bockris[68] for ethylene oxidation, may be expected. If indeed the rate at rhodium is less than that at palladium electrodes, then one should examine whether the volcano behavior in oxygen reduction is due to the change in the mechanism of reaction going from palladium to rhodium or whether the effect is due to a gradual change of catalytic activity. One should be aware that rhodium oxidizes more readily than palladium and if an oxide film exists at the rhodium surface, then the rate of oxygen reduction may be expected to decrease due to "resistance" of the oxide film, as shown for platinum electrodes. When the activities at various metal and alloy electrodes are plotted vs. bond strength of adsorbed species (cf. Figure 7), instead of against the number of unpaired d electrons (cf. reference 13) a linear relationship is again obtained. This illustrates the interrelationship of various physical parameters of metals and alloys and points to the need to separate, if at all possible, the prime effects from the secondary effects.

For fuel-cell applications a small-grain platinum black Teflon-bonded catalyst is a candidate. The performance of such a cathode deteriorates with time, particularly at elevated temperatures. The cause for the decay of activity has been shown to be the dissolution of platinum and subsequent redeposition and possible crystallization.[69] Little is known about the factors affecting the activity of "small-grain catalysts." Empirical approaches[69] have shown that these activity losses can be minimized by alloying, e.g., Au with Pt or Ni, without greatly affecting the initial catalytic activity. The oxidation state of the finely divided catalyst surface clearly deserves more rigorous attention. Finely divided nickel is well known to be pyrophoric on account of its very high rate of aerial oxidation. How does the air-formed oxide differ from that produced anodically? Can a similarly active surface be simulated under the highly controlled conditions of anodic growth? Is the activity simply related to the surface area, or do other factors, such as crystallinity, grain size, grain boundary, surface defects, and surface energy determine the activity? In view of the importance of dispersed small-grain catalysts, the catalysis of such systems should be studied as a function of the parameters maintained above, both from a practical and theoretical point of view. Above all, the improvements in long-term stability resulting from alloying[69] should be placed on a sound theoretical basis.

Many metals and alloys that can be considered as potential catalysts for various reactions at high potentials, e.g., for oxygen evolution or reduction, are far less "noble" than platinum and will readily oxidize and, in time, form anodic films of variable thicknesses over their surfaces. If these films are poor electronic conductors, the catalytic activities for these reactions will decrease with time as shown in this paper for the oxygen reaction at platinum.

An obvious way to try to enhance the catalysis at oxide-film electrodes, one would think, is to decrease the thickness of the films. This may not be an easy task, particularly when "insulating" films are already too thin. At least two requirements should be met by these films. First, they should not extensively grow with time when used as electrocatalysts. Second, they should be compact enough to prevent metal dissolution. The thinner an insulating oxide film is, the faster it will grow if conditions for growth are favorable.* Both the tendency for extended growth and protection against dissolution ultimately depend on the structure of the oxide phase and the related rate of ion diffusion through the oxide phase. This points to the possibility, as yet unexplored, of tailoring surface oxide films with desirable properties. For instance, thin crystalline films may be grown from a gas phase at a suitable temperature for a short time, or a poorly crystalline film, e.g., anodically formed, may be crystallized or recrystallized. Doping with an equivalent cation may also alter structural and diffusional properties of an oxide phase without affecting electrical properties appreciably. What one would wish to achieve with these procedures, and perhaps with others, it to *stabilize* the oxide-film catalyst. Technological, development-type research is needed here and is expected to be rewarding.

Considering how frequently electrode reactions, even those at relatively low potentials, e.g., the hydrogen reaction, occur at thin and relatively insulating oxide films, extensive research and study can be expected in the future with the aim of elucidating basic mechanisms of the reactions at such interfaces.

An alternative approach to the reduction in thickness of the insulating film is to alter conduction properties of the thin oxide films, e.g., by doping to produce a p- or n-type semiconductor surface that conducts well. In this respect, it may be recalled that at platinum anodes during oxidation with a high current density (e.g., 100 mA cm^{-2}) the potential–time plot shows a characteristic maximum although the thickness of the film continues to increase.[70] Further, during a potentiostatic oxidation,[71] current–time plots show a minimum. In both cases, therefore, the catalytic activity increases with increasing thickness of the oxide film. The evidence indicates[70] the presence of two oxide phases, perhaps PtO and PtO$_2$. Apparently, the new phase conducts well, and both the mechanism of reaction and catalytic activity change.

---

*For example, if the electric field within the oxide phase is high enough to assist diffusion.

References

1. A. Damjanovic and J. O'M. Bockris, *Electrochim. Acta 11*, 376 (1966).
2. A. Damjanovic, A. Dey, and J. O'M. Bockris, *J. Electrochem. Soc. 113*, 739 (1966).
3. J. O'M. Bockris and H. Wroblowa, *J. Electroanal. Chem. 7*, 428, (1964).
4. H. Wroblowa, B. J. Piersma, and J. O'M. Bockris, *J. Electroanal. Chem. 6*, 401 (1963).
5. S. Srinivasan, H. Wroblowa, and J. O'M. Bockris, *Adv. Catal. 17*, 351 (1967).
6. A. Damjanovic, D. Sepa, and J. O'M. Bockris, *J. Res. Inst. Catal. (Hokkaido Univ.) 16*, 1 (1968).
7. J. O'M. Bockris, A. Damjanovic, and J. McHardy, *Proc. Third Int. Symp. Fuel Cells*, Brussels, Vol. VI, 16 (1969).
8. A. J. Appleby, *Catal. Rev. 4*, 221 (1970).
9. J. O'M. Bockris and J. McHardy, *J. Electrochem. Soc. 120*, 53, 61 (1973).
10. A. Damjanovic, in: *Modern Aspects of Electrochemistry* (J. O'M. Bockris and B. E. Conway, eds.), Vol. 5, p. 369, Plenum Press, New York (1969).
11. T. P. Hoare, *The Electrochemistry of Oxygen*, Wiley-Interscience, New York (1968).
12. A. Damjanovic and V. Brusic, *Electrochim. Acta 12*, 615 (1967).
13. A. Damjanovic and V. Brusic, *Electrochim. Acta 12*, 1171 (1967).
14. A. J. Appleby, *J. Electrochem. Soc. 117*, 328 (1970).
15. D. S. Gnanamuthu and J. V. Petrocelli, *J. Electrochem. Soc. 114*, 1036 (1967).
16. A. Damjanovic, M. A. Genshaw, and J. O'M. Bockris, *J. Electrochem. Soc. 114*, 466 (1967).
17. M. Paucirova, D. M. Drazic, and A. Damjanovic, *Electrochim. Acta 18*, 945 (1973).
18. M. R. Tarasevich and V. A. Bogdanovskaya, *Elektrokhimiya 7*, 1072 (1971).
19. Y. Ya. Shepelev, M. R. Tarasevich, and R. L. Burshtein, *Elektrokhimiya 7*, 999 (1971).
20. A. Damjanovic, V. Brusic, and J. O'M. Bockris, *J. Phys. Chem. 71*, 2741 (1967).
21. M. L. B. Rao, A. Damjanovic, and J. O'M. Bockris, *J. Phys. Chem. 67*, 2508 (1963).
22. A. Damjanovic and M. L. B. Rao, in: *18th Annual Power Sources Conference*, p. 3, PSC Publications Committee (1964).
23. J. P. Hoare, *Res. Inst. Catal. (Hokkaido Univ.) 16*, 19 (1968).
24. R. Thacker and J. P. Hoare, paper presented at spring meeting, Electrochemical Society, New York (1969).
25. A. Damjanovic and M. A. Genshaw, *Electrochim. Acta 15*, 1281 (1970).
26. H. Wroblowa, M. L. B. Rao, A. Damjanovic, and J. O'M. Bockris, *J. Electroanal. Chem. 15*, 139 (1967).
27. A. K. N. Reddy, M. A. Genshaw, and J. O'M. Bockris, *J. Chem. Phys. 148*, 671 (1968).
28. S. H. Kim, W. Paik, and J. O'M. Bockris, *Surface Sci. 33*, 617 (1972).
29. W. Paik and J. O'M. Bockris, *Surface Sci. 28*, 61 (1971).
30. W. Visscher, *Optic 26*, 407 (1967).
31. S. Shibata and P. M. Sumino, *Electrochim. Acta 17*, 2215 (1972).
32. C. C. Schubert, C. L. Page, and B. Ralph, *Electochim. Acta 18*, 33 (1973).
33. W. E. Reid and J. Kruger, *Nature 203*, 402 (1964).
34. H. Angerstein-Kozlowska, B. E. Conway, and W. B. A. Sharp, *J. Electroanal. Chem. 43*, 9 (1973).
35. K. J. Vetter and J. W. Schultze, *J. Electroanal. Chem. 34*, 131 (1972).
36. K. J. Vetter and J. W. Schultze, *J. Electroanal. Chem. 34*, 141 (1972).
37. A. Damjanovic and A. T. Ward, in: *International Review of Science*. Physical Chemistry Series Two, Volume 6, Electrochemistry. (J. O'M. Bockris, ed.), Butterworths, London, 1976, p. 103.
38. A. Damjanovic, A. T. Ward, B. Ulrick, and M. O'Jea, *J. Electrochem. Soc. 122*, 471 (1975).
39. J. L. Ord and F. C. Ho, *J. Electrochem. Soc. 118*, 46 (1971).
40. R. Thacker and J. P. Hoare, *J. Electroanal. Chem. 30*, 1 (1971).
41. M. Rosen, D. R. Flinn, and S. Schuldiner, *J. Electrochem. Soc. 116*, 1112 (1969).
42. T. P. Hoare, *Proc. 8th Meeting CITCE*, p. 439, Butterworths, London (1956).
43. J. O'M. Bockris and A. K. M. S. Huq, *Proc. R. Soc. A237*, 271 (1956).
44. A. Damjanovic, A. Dey, and J. O'M. Bockris, *Electrochim. Acta 11*, 791 (1966).
45. A. Damjanovic and B. Jovanovic, *J. Electrochem. Soc. 123*, 374 (1976).
46. R. E. Meyer, *J. Electrochem. Soc. 107*, 847 (1960).
47. J. J. MacDonald and B. E. Conway, *Proc. Roy Soc. 269*, 419 (1962).
48. H. Göhr and E. Lange, *Z. Elektrochem. 63*, 673 (1958).
49. V. I. Veselovskii, Proceedings of the 4th Conference on Electrochemistry, Moscow (1956).
50. K. I. Rozenthal and V. I. Veselovskii, *Dokl. Akad. Nank USSR III*, 647 (1956).
51. K. I. Rozenthal and V. I. Veselovskii, *Zh. Fiz. Khim. 35*, 2481 (1961).

52. A. Damjanovic, A. T. Ward, and M. O'Jea, *J. Electrochem. Soc. 121*, 1186 (1974).
53. H. A. Laitimen and C. G. Enke, *J. Electrochem. Soc. 107*, 773 (1960).
54. S. Gilman, *Electroanal. Chem. 2*, 111 (1967).
55. K. J. Vetter and J. W. Schultze, *Ber. Bunsenges. Phys. Chem. 77*, 945 (1973).
56. A. C. Makrides, *J. Electrochem. Soc. 113*, 1158 (1966).
57. S. W. Feldberg, C. G. Enke, and C. E. Bricker, *J. Electrochem. Soc. 110*, 826 (1963).
58. N. Sato and M. Cohen, *J. Electrochem. Soc. 111*, 512 (1964).
59. J. O'M. Bockris, M. Genshaw, V. Brusic, and H. Wroblowa, *Electrochim. Acta 16*, 1859 (1971).
60. H. Wroblowa, V. Brusic, and J. O'M. Bockris, *J. Phys. Chem. 75*, 2823 (1971).
61. J. W. Schultze, *Z. Phys. Chem. 73*, 29 (1970).
62. J. W. Schultze and K. J. Vetter, *Electrochim. Acta 18*, 889 (1973).
63. P. Kohl and J. W. Schultze, *Ber. Bunsenges. Phys. Chem. 77*, 953 (1973).
64. H. Gerischer, *Z. Phys. Chem. NF 26*, 223, 325 (1960).
65. H. Gerischer, in: *Physical Chemistry, An Advanced Treatise*, Vol. IX, p. 463, (H. Eyring, ed.), Academic Press, New York (1968).
66. J. G. Simmons, *J. Appl. Phys. 34*, 1793 (1963).
67. S. Schuldiner, *J. Electrochem. Soc. 115*, 362 (1968).
68. A. T. Kuhn, H. Wroblowa, and J. O'M. Bockris, *Trans. Faraday Soc. 63*, 1458 (1967).
69. M. S. Freed and R. J. Lawrance, paper presented at 147th Electrochemical Society Meeting, Toronto, Canada (1975).
70. W. Visscher and M. Blijlevens, *Electrochim. Acta 19*, 387 (1974).
71. S. Shibata and P. M. Sumino, *Electrochim. Acta 16*, 1089 (1971).

7

Electrocatalysis by *d* and *sp* Metals

Hideaki Kita

1. Introduction

The hydrogen electrode reaction, $2H^+ + 2e^- = H_2$, is a heterogeneous catalysis where an electrode material acts as catalyst. The electrochemical approach to heterogeneous catalyses provides many advantages in comparison with the usual chemical one. Much emphasis should be given especially to the fact that we can easily apply a wide range of potential to an electrode, so that one can study kinetics even on a very inactive catalyst such as mercury. The hydrogen electrode reaction is the only case in which the kinetics have been observed on most of metals except for alkali and alkali earth metals. Hence, it is a good example in which to see the catalytic action of metals.

The purpose of the present article is to survey the catalytic action of electrode metals for the hydrogen electrode reaction as well as for the electrolytic reduction of organic compounds, and to deduce a common nature in the catalytic action. There will also be discussion of which property of a metal is the most important in connection with the hydrogen electrode reaction, what kind of regularities are found in the electrolytic reduction of organic compounds, and what kind of electronic distribution is expected at the metal surface. Such a common nature and regularities will provide useful predictions in discussing electrocatalysis.

2. The Hydrogen Electrode Reaction

2.1. Current Density–Overvoltage Relation

In 1905, Tafel[1] reported a linear relation between the cathodic current density, i (A/cm²), and the overvoltage, η (volt), of the hydrogen evolution reaction, i.e.,

$$\eta = a + b \log i \tag{1}$$

where a and b are constant. The term η is defined as

$$\eta \equiv [\bar{\mu}(e^-) - \bar{\mu}(e^-)_{rev}]/F \tag{2}$$

in terms of the electrochemical potentials of electron at the working and the reversible hydrogen electrodes, $\bar{\mu}(e^-)$ and $\bar{\mu}(e^-)_{rev}$, and the farad, F.

Since Tafel, the above relation (1) has been extensively examined by many workers and nowadays the data on most electrode metals are available under various experimental conditions. Equation (1) has been confirmed in most cases and has been named the Tafel line. The constant b is called the Tafel slope.

Careful examination of experimental data, however, shows that in spite of the wide applicability of the Tafel line, many electrodes reveal a rather complicated behavior when

---

Hideaki Kita • Department of Chemistry, Faculty of Science, Hokkaido University, Sapporo, Japan.

the cathodic current density is further extended to higher or lower values. For instance, the mercury electrode shows so-called depolarization in acidic solution at relatively low current densities. Some electrodes such as Ni and Cu reveal a saturation current at large negative polarization. We will first describe the general patterns observed for the current density–overvoltage relations on various electrodes. The present author limits himself to treating the experimental data obtained in acidic solutions which do not contain surface-active organic substances. The results in alkaline solutions accompany the possibility that the alkali cation deposits on the electrode surface and thus may disturb the hydrogen electrode reaction.

The relations between the current density and overvoltage in acidic solution are classified into three types* as outlined below.

2.1.1. Type I

Current density–overvoltage relations in cathodic and anodic polarizations are represented by the respective Tafel lines, both of which intersect at a potential far more negative than the reversible potential of the hydrogen electrode, as shown in Figure 1. The potential at the intersection, i.e., the rest potential is a corrosion potential at which the evolution of hydrogen and the dissolution of metal take place with an equal rate. The anodic Tafel line represents the kinetics of the electrochemical dissolution of electrode metal unless the passivation takes place. The type I relation is observed on the transition metals of the first long period (except for Mn and Zn) and on Zr and Nb of the second long period, as indicated in Figure 1. In the second and third long periods, the transition

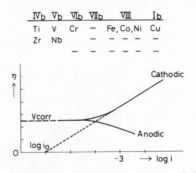

Fig. 1. The first type of current density i vs. overvoltage η relation in acidic solution.

metals positioned after Tc (VIIb) and W (VIb), respectively, do not show the type I relation. Mo (VIb) and Hf (IVb) in the second and third long periods are expected to show the type I relation owing to their corrosive property in acidic solution. The slope of the cathodic Tafel line, i.e., b in equation (1), is close to 120 mV at room temperature except for a few cases where a higher value is reported, e.g., 145–204 mV on V.[3]

2.1.2. Type II

The characteristic of the second type is that the rest potential reproduces the value of the reversible hydrogen electrode. Figure 2 schematically illustrates the current

---

*Bockris and Potter[2] mentioned three other types which show a hysteresis or complicated shapes owing to the presence of surface impurities or oxides. They are not discussed here.

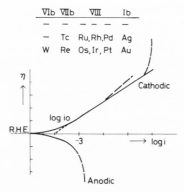

Fig. 2. The second type of the current density *i* vs. overvoltage η relation in acidic solution. R.H.E. stands for the reversible hydrogen electrode potential.

density–overvoltage relations in the cathodic and anodic polarizations. The anodic reaction is in most cases the ionization of the hydrogen molecule dissolved in solution. The low solubility of hydrogen gives rise to a limiting diffusion current in the anodic polarization as observed on the platinum-group metals. Thus, the well-defined anodic Tafel line is hardly obtained. The type II relation is observed on Tc, Ru, Rh, Pd, and Ag of the second long period and on W, Re, Os, Ir, Pt, and Au of the third long period, respectively.

A break in the cathodic Tafel line is reported in a few cases observed on the electrodes of W[4] and Ag[5] as shown by the broken line in the figure. At high current densities, the overvoltage often deviates upwards from the Tafel line, showing the presence of a limiting current on electrodes such as Ni,[6] Cu,[7] Pt,[8] and Au.[9]

The slope of the cathodic Tafel line is reported as approx. 120 mV in most cases. A platinum electrode, however, shows an exceptionally small value of 30 mV in acidic solution. A few data on Rh,[4] Pd,[10] and Re[11] also show 30 mV but others almost exclusively show approx. 120 mV. An intermediate value of approx. 60 mV was observed when there is a break in the Tafel line, i.e., in the lower current-density region on Ag[5] and Au,[12] but in the higher current-density region, the slope was approx. 120 mV in all cases.

2.1.3. Type III

The current density–overvoltage relation in cathodic polarization is characterized by its S-shape as illustrated in Figure 3. At lower current densities polarization from the rest

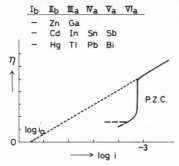

Fig. 3. The third type of the current density *i* vs. overvoltage η relation in acidic solution. P.Z.C. stands for the potential of zero charge.

potential is either very small or in some cases there exists a short linear relation, as shown by the broken lines. With the increase of current density, the electrode potential jumps to a high value of overvoltage at a certain current density and then the well-defined Tafel line follows with a slope of approx. 120 mV. The type III relation is observed in Zn, Ga, and Mn of the first long period, Cd, In, Sn, and Sb of the second long period, and Hg, Tl, Pb, and Bi of the third long period. The rest potential deviates widely from the value of the reversible hydrogen electrode, i.e., in the negative direction when the electrode is corrosive or to a value determined by the presence of a certain surface compound of low solubility such as Hg_2SO_4.[13] All metals mentioned above except for Hg are thermodynamically corrosive in 1 N H_2SO_4 solution.[14]

2.2. Catalytic Activity of Electrode Metals

Since the value of b is found to be about 120 mV on most of metals, the constant a of equation (1) may reflect how active the electrode metal is. The larger a is, the less active the electrode. In the present article, we will take the exchange current density, i_0, as a measure [15] of the activity. Exchange current density represents the unidirectional rate of the reaction at equilibrium and is estimated by the extrapolation of the Tafel line to $\eta = 0$ (dotted lines in Figures 1, 2, and 3), i.e.,

$$\log i_0 = -(a/b) \qquad (3)$$

Purity of the system under examination is one of the most important conditions to produce reliable data. Thus, the purification of electrode and electrolyte solution has received much attention for the last 20 years. The exchange current density of each metal was estimated from the available results mainly reported since 1950. When there is a break in the Tafel line (Figure 2), a lower current-density region is extrapolated by assuming that the region satisfies the condition of $\eta \gg RT/F$. In the case of the type III relation (Figure 3), on the other hand, the Tafel line at higher current densities was extrapolated, since at lower current densities the condition of |polarization| $\gg RT/F$ is not satisfied in many cases and no reliable Tafel line seems to be observed over a sufficient range of the current density. In general, it can be stated that the exchange current density was estimated by extrapolating the Tafel line around 1 mA/cm^2 as shown in Figures 1, 2, and 3, except for a few metals of Pt, Pd (larger than 1 mA/cm^2), and W (smaller).

Values of log i_0 on various metals are plotted against the atomic number of the electrode metals in Figure 4. Only one value was taken from each publication for the respective metals except for the single crystal. In the latter case, values of each lattice plane were chosen. When the kinds of acids, their concentration, and temperature were varied, the values chosen were those of H_2SO_4 and HCl and closest to 1 N and 25°C. Data obtained in heavy water were not included for the sake of simplicity. As seen from Figure 4, values of log i_0 fluctuate considerably on each metal, i.e., 10^2–10^3 times in terms of i_0. Mercury is supposed to be the metal on which it is easiest to reproduce a clean, flat surface, but its i_0 value fluctuates by approx. 10^2 times. Such a large fluctuation may be partly due to the long extrapolation of the Tafel line. In Figure 4, the symbol X represents the mean value of log i_0 and the X's are connected by lines to each other. Dotted lines indicate a trend with some uncertainty because of the scanty data on log i_0 or the possibility of the molecular metal hydride formation instead of the hydrogen evolution reaction. Table 1 reproduces the mean value of log i_0 on the respective metals. The exchange current density varies by 10^{10} times between the most and the least active metals. It must

Table 1. Mean Value of Log i_0 (A/cm^2) in Acidic Solution[a]

Element	Log i_0	Element	Log i_0
Ti	−6.9	Ag	−6.4
V	−6.2	Cd	−12.0
Cr	−6.4	In	−10.9
Mn	−	Sn	−9.2
Fe	−5.8	Sb	(−8.7)
Co	−4.9	Te	(−7.5)
Ni	−5.2	Hf	−
Cu	−7.4	Ta	−7.8
Zn	−10.5	W	−6.4
Ga	−9.8	Re	−5.1
Ge	(−8.7)	Os	−4.0
As	(−7.3)	Ir	−3.3
Zr	−6.7	Pt	−3.3
Nb	−7.3	Au	−5.7
Mo	−6.5	Hg	−11.9
Tc	−4.1	Tl	−11.5
Ru	−3.3	Pb	−12.6
Rh	−2.5	Bi	−10.2
Pd	−2.4	Po	−

[a]Values in parentheses are uncertain because of the possibility of hydride molecule formation.

be emphasized that only an electrochemical study can tell about such an anomalously large change of the catalytic activity by metals.

An seen from Figure 4, the log i_0 values of the metals are taken to be basically a periodic function of the atomic number in spite of the fluctuation of experimental data. Such a periodic variation of log i_0 or of the constant a of the Tafel line has been mentioned by several authors.[16−19] According to this periodicity, it is generally stated that the log i_0 value increases first with the atomic number, reaches a maximum at the metals of group VIII, decreases quite sharply with a minimum at the metals of Zn, Cd, Hg, or Pb, and then increases again with further increase of the atomic number. Periodic variation in Figure 4 proceeds in a similar way in the second and third long periods, revealing a precise periodicity. However, an abnormal behavior is now noticed in the first long period. The log i_0 value on Mn is expected to be exceptionally small.* It is of interest to note that a similar "cut-in" at Mn is observed for some physical properties, e.g., the melting point of metals. The melting point of Mn is exceptionally low. Tc in the second long period shows also a less sharp cut-in with respect to the melting point, but its log i_0 value does not show the cut-in (Figure 4), although only one value is available.

Log i_0 of the rare earth metals is of great interest in order to examine the effect of the f-orbital electrons. The reported results[22] are inserted in Figure 4, but the comparison with other data seems, at present, to be difficult because of the less reliable value of the Tafel slope which changes irregularly from 100 to 333 mV by the rare earth metals.

*Hurlen and Valand[20] studied the cathodic and anodic polarizations on Mn in 0.02 M HCl + 0.98 M KCl solution. Mn is very corrosive, and the diffusion of H$^+$ controls the corrosion potential (−1.232 V). The observed cathodic Tafel line gives −12.8 for log i_0 which is extremely small. This value, however, was not included in Figure 4 since these authors concluded there was discharge of water molecule in acidic solution. Kuhn *et al.*[21] quoted a value of −7.8 for log i_0 on Mn.

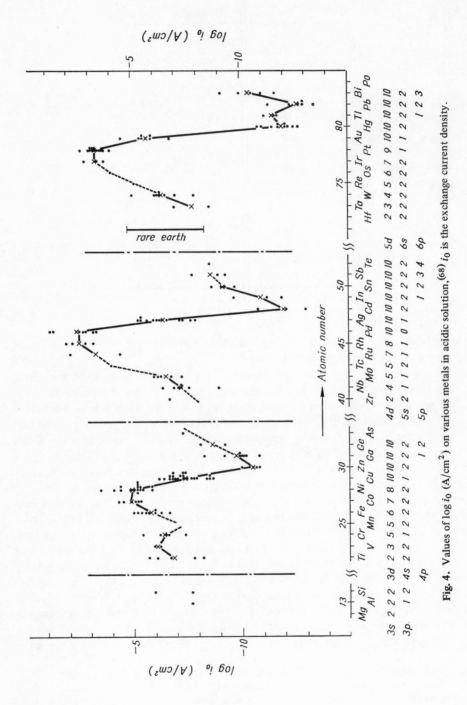

Fig. 4. Values of log i_0 (A/cm^2) on various metals in acidic solution,[68] i_0 is the exchange current density.

2.3. Effects of Various Conditions on Log i_0

Experimental results collected in Figure 4 are discussed below from various aspects.

2.3.1. Pretreatment of Electrode

Nickel will be a good example to see the effect of the pretreatment of electrode, since it has been subjected to a variety of pretreatments, i.e., reduction in H_2, chromic/sulfuric acid treatment, precathodization, quenching of the melt in He, anodic and cathodic polarizations, evaporation of Ni on glass, electrodeposition, etc. Log i_0 values on polycrystalline electrodes subjected to these pretreatments are shown in Table 2. They show an excellent agreement with each other. Table 2 also includes the values observed in alkaline solutions and in acidified $NiSO_4$ solution for comparison. Values agree within an order of magnitude. Hence, the surfaces, carefully prepared, seem to give a catalytic activity of the same order of magnitude, independent of pretreatment.

2.3.2. Kinds of Acid and Solvent

Table 2 also indicates that the log i_0 value appears independent of the kind of acid, i.e., -5.4 and -5.22 in 1 N HCl[23] and in 0.5 N H_2SO_4,[4] on the electrode subjected to the same pretreatment of H_2 reduction. When an acid shows some interaction with the electrode, however, a specific effect will be expected. An example is the depolarization on Hg, which depends sensitively on the kind of hydrogen halogenide acids. Such an effect has been discussed in detail by Frumkin[32] in connection with the specific adsorption of anion. Another example, the hydrofluoric acid easily attacks Ti electrodes and the resulting current density-overvoltage relation is totally different from that observed in H_2SO_4 solution.[33]

Table 3 shows the log i_0 values obtained in acidic solution of different solvents on Hg[34] and on Fe.[35] Log i_0 values fluctuate 10–10^2 times with the kind of solvent, but no systematic variations seem to be deducible.

Table 2. Log i_0 (A/cm^2) on Polycrystalline Ni Electrode

Treatment	Log i_0	Electrolyte	Concentration and reference
Heated in H_2	-5.4 ± 0.1	HCl	1 N[23]
Heated in H_2	-5.87 ± 0.15	HCl	0.1 N[5]
Heated in H_2	-5.24	HCl	0.1 N[24]
Heated in H_2	-5.22 ± 0.53	H_2SO_4	0.5 N[4]
Chromic/sulfuric acid	-5.3	$HClO_4$	1 M[25]
Cathodic polarization	-5.25	H_2SO_4	1 N[26]
Treated in boiling 10% KOH, cathodized	-5.2	H_2SO_4	0.1 N[27]
Heated in H_2	-6.4 ± 0.2	NaOH	0.1 N[23]
Anodic, cathodic polarization	-5.03	NaOH	0.1 N[28]
Evaporated *in vacuo*	-5.1	NaOH (pH = 13.68)[29]	
Melt in He	-5.32	NaOH	0.95 N[30]
Heated in H_2	-6.1 ± 0.1	NaOH	0.5 N[4]
Electrodeposition	-6.1	$NiSO_4$	0.5 M[31]

Table 3. Effect of Solvents on Log i_0 (A/cm^2)

Electrode	Solvent	Log i_0	b (mV)	a (V)
Hg[34]	H_2O (HCl, 0.1 M)	−11.8	117	
	CH_3OH (HCl, 0.1 M)	−11.5	107	
	C_2H_5OH (HCl, 0.1 M)	−10.8	104	
	nC_3H_7OH (HCl, 0.1 M)	−10.8	102	
Fe[35]	H_2O (HCl, 1 N)	−6.9	120	0.69
	CH_3OH (HCl, 1 N)	−7.3	120	0.51
	C_2H_5OH (HCl, 1 N)	−5.9	110	0.62
	Ethyleneglycol (HCl, 1 N)	−5.1	100	0.55
	Glycerin (HCl, 1 N)	−6.8	100	0.58

2.3.3. Affinity of Metal toward Hydrogen

Different metals have a different affinity towards hydrogen. Some metals form an alloy with hydrogen, some metals form a volatile hydride molecule, and others are rather inert, only allowing an adsorption or not at all.

2.3.3.1. Group IIIb to Vb Metals. Transition metals of groups IIIb to Vb are known to form the nonstoichiometric, interstitial hydrogen alloy. The alloys of $M(Vb)H_{<1}$ and $M(IVb)H_{<2}$ are stable but $M(IIIb)H_{<3}$ reacts with water giving hydrogen molecule. An example of the effect of the hydrogen absorption on overvoltage is shown in Figure 5[36] where curves 1 and 2 represent the changes in the electrode potential of Nb and its lattice constant during the cathodization with a constant current of 5 mA/cm^2 in 1 N H_2SO_4. The overvoltage decreases with the adsorption of hydrogen for the first 30 hr and then reaches a constant value with the saturation of hydrogen, i.e., with the formation of a β phase which has a lattice constant of 3.44 kX. The above decrease in the overvoltage corresponds to an increase in the catalytic activity by an order of magnitude. On vanadium, Kudriashov and Falin[3] also reported the decrease of overvoltage with cathodic polarization. In the case of Ti, precathodization of 1 hr with a current density of 5 mA/cm^2 gives almost the same value of log i_0 as that obtained by chemical activation of the electrode by dissolving the surface with 1 N HF for 1 min. The average value of log i_0 for both treatments is estimated as −6.99 and −6.75 by using the reported data.[33,37-40] On Zr, the current density–overvoltage relation at a steady state shows a marked hysteresis when observed with increasing and decreasing current density.[41] Artemova[39] reported the increase of the kinetic parameters of a and b from the initial values of 0.92 V and 120 mV to 1.23 V and 185 mV from repeated observations. Such

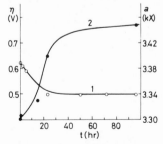

Fig. 5. Changes in overvoltage (curve 1) and in lattice parameter (curve 2) during the cathodic polarization of Nb with a constant current of 5 mA/cm^2 in 1 N H_2SO_4.[36]

an increase in overvoltage is just the reverse of what happens with Nb. These different effects of precathodization were briefly discussed by Krishtalik.[42] Nevertheless, the effect of the hydrogen adsorption on the catalytic activity will not be so large as to disturb the periodic change of log i_0.

2.3.3.2. Group IVa to VIa Metals. In contrast to the above transition metals, metals of groups IVa to VIa form the respective volatile hydride molecules of the type MH_n. As seen from the standard potentials[14] for the reaction $MH_n \rightleftharpoons M + nH^+ + ne^-$, the possibility arises at high cathodic polarization that hydride molecule formation takes place. In fact, SnH_4 for example is observed to form at a polarization more negative than -1.03 V compared to the standard hydrogen electrode.[43,44] Thus, the current density–overvoltage relation in such a potential range does not necessarily reflect the kinetics of the hydrogen evolution reaction itself. Figure 6 shows the current density–overvoltage relation observed on a Te electrode in 1 N HCl.[45] It was concluded that the relation in

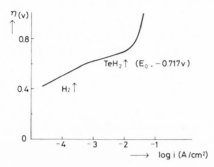

Fig. 6. The current density–overvoltage relation on Te in 1 N HCl.[45]

the lower current-density region represents the Tafel line for the hydrogen evolution reaction and then for the evolution of TeH_2 in the middle current-density region. The Tafel slope is different between them. In the higher current-density region, the relation bends upward where the dissolution of Te occurs.

Hydride molecules will be unstable on account of the low values of their standard potentials and easily decompose to metal and hydrogen molecules.

2.3.4. Single Crystals and Crystal Modification

Horiuti and Kita[46] carried out a theoretical estimation of the rates of the recombination of adsorbed hydrogen atoms, $2H(a) \rightarrow H_2$, on (111), (100), and (110) lattice planes of Ni and showed that a large difference in the rates will be expected in the following order: (111) $\gg$ (100) $\gg$ (110). Each difference amounts to two or more orders of magnitude in terms of the current density at $\eta > 0.6$ V. It is thus of interest to examine experimental results reported on the respective lattice planes. There is now a growing amount of published work on single-crystal electrodes. Table 4 shows the values of log i_0 on the single crystals of V in 0.2 N H_2SO_4,[3] Ni in 0.1–0.15 N $HClO_4$[47] and 1 N H_2SO_4,[48] Cu in 1 N H_2SO_4,[49,50] Mo in 1 N H_2SO_4,[51] W in 1 N H_2SO_4,[52] and Pt in 1 N H_2SO_4.[53] Any distinct difference in the crystal lattice plane, however, seems not to have been observed. For the sake of comparison, the mean values on the corresponding polycrystalline electrodes are quoted from Table 1. Results on both single and polycrystalline electrodes are almost the same. Such similarity could be due to the fact that, although the orientation of the crystal plane is properly arranged on a macroscopic scale,

Table 4. Log i_0 (A/cm^2) on Single-Crystal Electrodes in Acid Solution

Plane	V[3]	Ni[47]	Ni[48]	Cu[49]	Cu[50]	Mo[51]	W[52]	Pt[53]
(110)[a]	−6.22	−5.2	−4.4	−7.43	−8.65		−6.32	
(100)	−6.00	−4.85	−4.4	−7.26	−8.55	−5.96	−6.32	−4.0
(111)		−5.2	−4.1	−7.20	−8.45	−6.58	−7.00	−3.1
[Poly]	−6.2	−5.2		−7.4		−6.5	−6.4	−3.4

[a] A set of data on the respective lattice planes are obtained by the same author.

the real atomic array on the surface will be disturbed to a considerable extent. Even though assumed to be the ideal one in a microscopic scale, the surface will be cultivated during contact with the solution or during the measurements. Hence the characteristics of each lattice plane will be more or less smeared out.

Crystal modification, such as α (b.c.c.)-, β (hexagonal)-Cr and α (h.c.p.)-, β (f.c.c.)-Co, is the other factor to be examined. Values of log i_0 for the respective cases[54,55] show about a 10-fold difference in the exchange current density by crystal modification. However, it seems difficult to deduce a systematic variation of log i_0.

Another fact to be mentioned is that in Figure 4 the value of log i_0 of liquid mercury conforms, in order of magnitude, to the periodicity obtained at solid electrodes of the first or second long period. In addition, the log i_0 on Ga is much the same whether the electrode is in the liquid or solid state, as observed at temperatures slightly above or below the melting point of 29.78°C.[56] Such minor differences in the catalytic activity between the solid and liquid states provide evidence for the crystal plane model for the catalyst and against the so-called active center theory.

2.4. Catalytic Activity and Properties of the Metal

The periodic variation of the catalytic activity should be ultimately understood with the electronic structure of the metal. It will be worthwhile, however, to examine the relation with the physical properties of the metal which are also known to show a periodic gradation within the long periods.

Various kinds of physical properties are conventionally classified into three groups. The first group consists of metallic radius, density, compressibility, etc., most of which show an extreme value at the subgroup metals near VIII. The second group consists of heat of fusion, heat of vaporization, melting point, boiling point, etc., which show an extreme value at the subgroup metals of VIb. These groupings were proposed by Bond[57] in connection with heterogeneous catalysis in the gas phase. The third one is the work function and related properties such as the potential of zero charge. General discussions on the effects of these properties are given by Heifets et al.,[58] Kuhn et al.,[21] Trasatti,[59] and Kita.[60]

2.4.1. Properties of Group I

With respect to the metallic radius, it was mentioned[61,62] that the overvoltage becomes a minimum at an atomic distance of the metal of 2.76Å which is equivalent to the distance between the water molecules in the solid state. Thus, the importance of water molecule orientation on the metal surface was discussed. When the present values of log i_0 are plotted against the metallic radius, a maximum appears indeed at a value of

1.37Å with respect to b.c.c. and f.c.c. metals. However, no distinct relation is observed with h.c.p. metals. In addition, the metals Pt, Pd, Rh, W, and Mo, which have almost the same metallic radius (1.34–1.38Å),[57] show exchange current densities differing by approx. 10^5 times.[60] Hence, the metallic radius cannot be taken as the main factor to be related in the determination of the catalytic activity. Generally speaking, $\log i_0$ has a broad tendency to decrease with the increase of the metallic radius. Lorentz[63,64] reported a linear relation between the constant a and the reciprocal of the square root of the compressibility. Goralnik[65] further mentioned a similar linear relation with the square root of Young's modulus and the shear modulus. These results lead to the following general trend:

> A plot of $\log i_0$ with metallic radius, density, compressibility, or hardness gives a monotonously decreasing relation. (i)

2.4.2. Properties of Group II

A linear relation between the constant a and the heat of fusion or of vaporization was reported by Kobozev.[66,67] He discussed the relation by assuming the equilibrium of the discharge step of a proton. However, the additional data which are now available on other metals do not support his linear relation. The plot appears rather valley-like. When $\log i_0$ values are plotted against the heat of atomization[19] or the boiling point, one obtains the volcano-type relation which is also known in heterogeneous catalysis in gas phase. Similar relation holds with the melting point. Thus:

> A plot of $\log i_0$ with melting point, boiling point, heat of fusion, heat of sublimation, or heat of vaporization gives a volcano-type relation. (ii)

2.4.3. Properties of Group III

Since a metal electron takes part in the reaction giving a hydrogen intermediate, the ease of releasing an electron from the metal, and the affinity of the metal towards hydrogen may affect the catalytic activity. The effect of the work function has been examined by many workers.[16,18,21,59,66,68–70] Kobozev[66] reported a linear relation between the constant a and the work function by using data for 12 metals, but later more data on other metals became available. Conway and Bockris[69] showed the presence of a linear relation between $\log i_0$ and the work function on most of the metals except for Hg, Pb, and Tl of which $\log i_0$ values deviate far from the linear relation. Later, Petrenko[18] showed that the low- and high-overvoltage metals reveal different dependences of the constant a on the work function and that Ib metals Cu and Ag belong to neither of the two groups. In contrast, Kita[19] included Ib metals in the low-overvoltage metals.

The present mean values of $\log i_0$ in Table 1 are plotted against the work function in Figure 7, where the values of the work function are those used by Trasatti.[59] The family of metals is divided into two groups; one is the transition metals from IIIb to Ib and the other is the metals after IIb in the periodic table. Two dotted lines in the figure represent the relations reported by Trasatti.[59] The quantitative relation between $\log i_0$ and the work function, however, seems to depend largely on the work function data. When the values of Eastman,[71] observed by the photoelectic method, on the polycrystalline films of the transition metals and those of Heine and Hodges'[72] for the nontransition metals are used, different slopes are obtained.[68] The plot of $\log i_0$ on the nontransition metals against the work function of Michaelson[73] or Bond[57] gives a relation almost independent of the work function.[19]

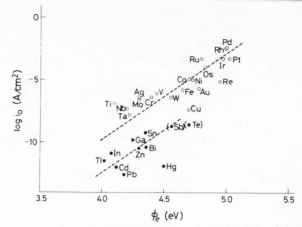

Fig. 7. Plot of the mean log i_0 values (Table 1) against the work function ϕ_e.[59] The two dotted lines represent the relations reported by Trasatti.[59]

Heifets *et al.*[58] examined the constant a in connection with the potential of zero charge and reported the presence of two linear relations with the same slope which hold for both high- and low-overvoltage metals. The Ib metals Cu, Ag, and Au are included in the high-overvoltage metals, whereas Ga is in the other group. Since the potential of zero charge of many metals is discussed as a linear function of the work function, one can state as follows:

A plot of log i_0 with work function divides the family of metals into two groups, each showing different linear dependencies.   (iii)

If the heat of adsorption of hydrogen is calculated by Eley–Stevenson's equation,[74] the same grouping of metals is obtained.[19]

2.5. Experimental Results Bearing on the Relations

The question now arises as to which property among the three groups described above is most important in understanding the catalysis of electrode metals and the reaction mechanism of the hydrogen electrode reaction. Let us inspect other experimental results which may serve the purpose of solving the question.

Horiuti and Okamoto[75] found that the separation factor of deuterium differs on the two groups of metals studied, i.e., approx. 7 on Ni, Au, Cu, Pt, (at high overvoltage), and Pb (in alkaline solution), whereas it is approx. 3 on Hg, Sn, and Pb (in acidic solution). Kita[19] extended their grouping as follows: The separation factors for deuterium and tritium are relatively large on the transition metals, including Cu and Ag, and low on the metals after IIb in the periodic table.

Effect of surface-active substances on the overvoltage at a constant current density is also different between the two groups of metals. Petrenko[18] concluded that any type of surface-active substance increases the overvoltage of metals which adsorb hydrogen well, while on the metals which do not adsorb hydrogen well, cationic and neutral types of surface-active substances, increase and the anionic one decreases the hydrogen overvoltage. Table 5 summarizes the available results for the effects of I^- and $(C_4H_9)_4N^+$ surface-active substances on the hydrogen overvoltage. I^- ions increase the hydrogen overvoltage on the transition metals including Ag, while decreasing it on the other metals studied, although a few exceptional cases are reported. On Cr[84] the hydrogen over-

Table 5. Effect of I^- and $(C_4H_9)_4N^+$ on the Hydrogen Overvoltage

Adsorbate	Fe	Co	Ni	Cu	Zr	Ag	Zn	Ga[a]	Hg	Pb
I^-	+[76]	+[78]	+[26]	+[79]	+[41]	+[80]	_[81]	(-)[82]	_[83]	_[80a]
$(C_4H_9)_4N^+$	+[77]						+[81]	+[82]	+[83]	

[a]Effect on Ga is uncertain (see the text).

voltage is reported to decrease in the presence of I^-, although Cr is a transition metal. On solid Ga, Morozov *et al.*[85a] and Bagotskaya and Khalturina[85b] reported the increase of hydrogen overvoltage in contrast to the other data in Table 5. The cation, $(C_4H_9)_4N^+$, on the other hand, increases the overvoltage on all electrodes studied. Generally speaking, the effect of surface-active substances seems to be different between the two groups of metals.

Figure 8 illustrates the effect of solution pH on the hydrogen overvoltage at 1 mA/cm^2 on various metals.[86] It is seen from this figure that the hydrogen overvoltage on the transition and Ib metals is higher in alkaline than in acidic solution, while that on metals after IIb is lower in alkaline than in acidic solution. Effect of solution pH is thus different between these two groups of metals.

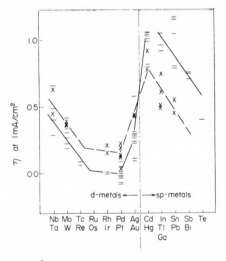

Fig. 8. Overvoltage η at 1 mA/cm^2 on various metals in acidic (---) and in alkaline (X) solutions.[86]

Comparison of the above results with relations (i), (ii), and (iii) indicates that the respective results are in harmony with relation (iii) with respect to the grouping of metals. Kita[19] denoted the transition metals, including Ib, as "*d* metals" and the metals following them in the periodic table as "*sp* metals." These terms arise from the characteristic difference in the electronic configuration of the outer shells of the respective elements. Dowden[87] first used the terms, *d* and *sp* types of metals, although the grouping is somewhat vague, in connection with chemisorption phenomena of gases.

2.6. Reaction Mechanism

Grouping the metals into two groups indicates that the reaction mechanism of the hydrogen electrode reaction differs between them. Horiuti and Okamoto[75] proposed the catalytic mechanism,

$$H^+ + e^- \rightleftharpoons H(a); \qquad 2H(a) \xrightarrow{\;\Lambda\;} H_2 \tag{4}$$

for the metals of high separation factor and the electrochemical mechanism,

$$2H^+ + e^- \rightleftharpoons H_2^+(a); \qquad H_2^+(a) + e^- \xrightarrow{\;\Lambda\;} H_2 \tag{5}$$

for the metals of low value, where $H(a)$ and $H_2^+(a)$ denote the adsorbed hydrogen atom and hydrogen molecule ion, and $\xrightarrow{\Lambda}$ the rate-determining step. The present author explained qualitatively his results for log i_0 vs. the work function plots,[19] assuming the catalytic and electrochemical mechanisms for the d and sp metals, respectively.

The point which should be stressed here is that the reaction intermediate on the d metals is the neutral adsorbed hydrogen atom, $H(a)$ and that on the sp metals it is the charged hydrogen molecule ion, $H_2^+(a)$. Presence of $H_2^+(a)$ has not been confirmed experimentally yet. Even if we hold onto the so-called "slow-discharge mechanism," the rate-determining step includes the charged species. Hence, so far as a charge is concerned, one could characterize the rate-determining step of the d and sp metals as being noncharged and charged, respectively.

3. Electrolytic Reduction of Organic Compounds

Electrolytic reduction of organic compounds in aqueous solution occurs via the electrolytic hydrogenation reaction, where the source of hydrogen is a proton in the solution and an electron in the metal. Hence, a close relation might be expected between the electrolytic reduction of organic compounds and the hydrogen electrode reaction.

Previous work reporting on the electrolytic reduction has been mainly concerned with synthesis of organic compounds, and less attention has been paid to its kinetics. Thus, it is difficult to compare the catalytic action of each metal in terms of the activity, as in the case of the hydrogen electrode reaction. Since many organic compounds have a variety of functional groups to be reduced and generally have many steps of partial reduction, it will be of interest to see what group and to what degree each metal electrode reduces organic compounds. Antropov has already pointed out[88] that the metals which adsorb hydrogen, behave in a different way from those which do not adsorb hydrogen in the electrolytic reduction of organic compounds. The reported results are summarized below.

3.1. Electrolytic Reduction of Carbonyl Group

Table 6 lists which metals are good or poor catalysts for the electrolytic reduction of ketone and aldehyde.[88–94] It is seen from the table that each organic compound of the

Table 6. Electrolytic Reduction of RR'CO on Various Metals[88–94] ($\Phi = C_6H_5$)

R	R'	Good	Poor	Ref.
CH_3-	$-H, -CH_3, -C_2H_5,$			
CH_3-	$-C_3H_7, -C_3H_6OH$	Hg, Pb, Cd, Zn, Sn, Al	Fe, Ni, Cu, Ag	88, 89
Φ	$-\Phi, -CH_3, -C_2H_5$	Hg, Pb, Cd, Zn, Sn, Bi	Cu, Ni, Pt	88, 90
$(CH_3)_2CH-$	$-CH(C_3H)_2$	Hg, Pb, Cd, Zn		91
$(CH_3)_2CHCH_2-$	$-CH_2CH(CH_3)_2$	Hg, Pb, Cd		89
$H-$	$-CH(OH)CH_2OH$	Pb, Cd, Zn, Sn	Fe, Ni, Cu, Ag, Pt	92
⬡=O		Hg, Pb, Cd, Zn, Sn	Cu, Ag, Ti	93
Streptomycin	$-CHO$	Zn, Cd, Sn, Pb	Ni, Cu stainless	94

Table 7. Electrolytic Reduction of Acetone[96-98] a

		Hg	Pb	Cd	Zn	Cu	Ni
Acidic[96]	P^b	73 (2)	86 (25)	90 (100)	86 (97)	3 (100)	1 (100)
	A	73 (95)	86 (68)		86 (3)		
	D	73 (3)	86 (7)				
Neutral[97]	P	95 [0]	76 [0]	70 [0]			
	A	95 [7.00]	76 [0.82]	70 [0.66]			
	D	95 [3.37]	76 [3.08]	70 [3.46]			
Alkaline[98]	P						
	A						
	D	{25}	{17}	{23}	{30}	{3}	{7}

^{a}Numbers before or in parentheses indicate the current efficiency or the distribution of products in (%), [gr] and {yield}, respectively.
^{b}P: $CH_3CH_2CH_3$, A: $(CH_3)_2CHOH$, D: $(CH_3)_2(OH)C—C(OH)(CH_3)_2$.

type RR'CO is well reduced on *sp* metals and hardly reduced on the *d* metals. The carbonyl group of carboxylic acid and esters is also reduced on metals such as Hg and Pb [e.g., $(COOH)_2$,[88] $X—C_6H_4—COOH$[95]].

Electrolytic reduction of acetone might be a typical example for studying current efficiency and product distribution. Reported results[96-98] are summarized in Table 7. Current efficiency is very high (73-90%) on the *sp* metals such as Hg, Pb, Cd, and Zn and very low (1-3%) on the *d* metals such as Cu and Ni. In the latter case, most of the electricity is exhausted in the production of hydrogen molecule.

An interesting example from the point of view of stereochemistry is the electrolytic reduction of 2,4-dimethylcyclohexanone to alcohol.[99] Copper as electrode of the *d* metals produces *cis*-2,4-dimethylcyclohexanol, while Hg and Pb electrodes of the *sp* metals produce the *trans* isomer. Such a difference clearly suggests that the reaction mechanism is different between these two groups of metals.

3.2. Electrolytic Reduction of Unsaturated Bond

According to Antropov,[88] the triple or double bond in organic compounds such as $HC≡CR$ (where R is H—, $CH_2=CH—$), R—COOH [R, $CH_3CH=CH—$, $CH_3CH=CHCH=CH—$, $CH_3(CH_2)_7CH=CH(CH_2)_7—$], and phenol is reduced on metals which adsorb hydrogen well and is hardly reduced on metals which do not adsorb hydrogen well. Partial reduction of alkynes to olefins takes place on Ni, Cu, and Ag. Cathodic reduction of dialkyl acetylenes (e.g., $CH_3CH_2CH_2·C≡C·CH_2CH_2CH_3$) and diphenylacetylene ($Φ·C≡C·Φ$) on a spongy Ni cathode in alcoholic H_2SO_4 solution gives exclusively *cis*-olefin, but no reduction was observed at Cd, Pb, or amalgamated Pb electrodes.[100] From these facts, a general conclusion can be drawn that the *d* metals are good catalysts and that the *sp* metals are poor catalysts in the electrolytic reduction of an unsaturated bond. This tendency is the reverse of the case of the electrolytic reduction of the carbonyl group.

Reduction of the triple to double bond provides the possibility of *cis* or *trans* addition of hydrogen. The above examples of dialkyl- and diphenylacetylenes show the formation of *cis* isomer on Ni. In solutions which contain alkali metal ions, dialkylacetylenes are reported to be reduced to olefin on Hg.[101] The products in this case, however, are the *trans*-olefin (>92% in all cases studied). Similar selectivity is found in the electrolytic reduction of 2-butyne-1,4-diol in KOH solution.[102] *Cis* addition selectively takes place on the *d* metals studied, i.e., Ag, Cu, Ni, Co, Pt, Fe, and Pd, and *trans* addition, although its rate is low, takes place on the *sp* metals of Pb, Sn, Cd, Hg (Cu amalgam), and Zn. An interesting example is the electrolytic reduction of dimethylmaleic and -fumaric acids on Hg in HCl–KCl buffer solution (pH = 1.9).[103] The product from maleic acid is the racemic form of α,α′-dimethylsuccinic acid and that from fumaric acid is the *meso* form. These results directly prove the

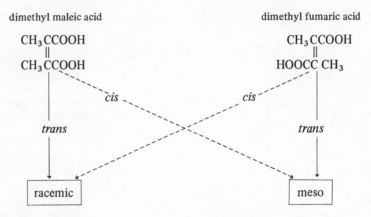

dimethyl maleic acid   dimethyl fumaric acid

trans addition of hydrogen on Hg.

Thus, the formation of *cis* and *trans* isomers seems to be characteristic for the *d* and *sp* metals, respectively. *Trans* isomer formation on the *sp* metals may include the possibility that the electrolytic reduction does not take place via direct reduction at the electrode but instead by alkali metal formed *in situ*.

3.3. Selectivity in the Electrolytic Reduction of Organic Compounds

From the results described in Sections 3.1 and 3.2, we extend the conclusion of Antropov[88] and Petrenko[18] as follows:

Electrodes of the *sp* metals reduce selectively polar bonds, and electrodes of the *d* metals reduce selectively nonpolar bonds.   (iv)

The above conclusions are confirmed by data from the electrolytic reduction of a molecule which contains both polar and nonpolar groups to be reduced. An example is the electrolytic reduction of CH_2=CHC≡N. Iron of the *d* metals (0.7 N NaOH) yields CH_3CH_2C≡N with a current efficiency of 61.8%,[104] whereas Pb of the *sp* metals (H_2SO_4 acidic methanolic solution) yields CH_2=$CHCH_2NH_2$ with a current efficiency of 55% and the by-product $CH_3CH_2CH_2NH_2$ with a current efficiency of 13%.[105] Recently, it was shown that electrolytic reduction of β,γ-unsaturated ketone, 2,2,5,5-tetramethyl-3-methylene cyclohexanone, gives the corresponding unsaturated alcohol on Hg.[106]

$$+ 2H^+ + 2e^- \longrightarrow$$

Electrolytic reduction of 2-cyclohexene-1-one in 1 N H_2SO_4 aqueous solution gives cyclohexane-1-one on Pt but not on Hg.[107]

4. Discussion

4.1. Common Nature in the Catalytic Action for Cathodic Reactions

We see that the behavior of each group of metals is distinctly different both in the hydrogen electrode reaction and in the electrolytic reduction of organic compounds as stated in Sections 2 and 3. The intermediate in the hydrogen electrode reaction is suggested to be the neutral $H(a)$ on the d metals and the charged H_2^+ (a) on the sp metals. In the electrolytic reduction of organic compounds, electrodes of the d metals reduce selectively nonpolar unsaturated bonds and those of the sp metals reduce selectively polar groups such as $>C=O$ or $-C\equiv N$. Hence, so far as electricity is concerned, there is a similarity between these two reactions.

From the analogy that a polar solvent dissolves a polar solute and a nonpolar solvent dissolves a nonpolar solute, the above facts lead to the following suggestion on the nature of the electrode surface:

> Electronic distribution at the metal surface may be rather homogeneous for the d metal and may be rather heterogeneous or localized for the sp metal, since the nonpolar processes take place on the d metal and the polar processes take place on the sp metal.   (v)

This suggestion has been discussed in detail by Johnson[108] from the point of view of his theory of "an interstitial electron model for metals."[109]

Table 8 indicates the deviation in the lattice from the ideal; each figure gives the deviation of the ratio of c_0/a_0 from unity in the cases of f.c.c., b.c.c., rhombic, and b.c.tet., the deviation of $(c_0/a_0)/1.63$ from unity in the case of h.c.p., and the deviation of angle/90 from unity in the case of rhombohedral, respectively. The terms c_0 and a_0 are lattice constants. The sp metals show a large deviation in contrast to the d metals, except for Al and Pb. Thus, it may be expected that the ion core of these sp metals deforms from the spherical one and hence that the electronic distribution will not be isotropic. According to Johnson's theory, the sp metals are characterized by a weak positive field perpendicular

Table 8. Deviation in the Lattice from the Ideal

V_b	VI_b	VII_b	VIII			I_b	II_b	III_a	IV_a	V_a
								Al f.c.c. 0	Si	
V b.c.c. 0	Cr b.c.c. 0	Mn	Fe b.c.c. 0	Co h.c.p. 0	Ni f.c.c. 0	Cu f.c.c. 0	Zn h.c.p. 0.14	Ga rhombic 0.04	Ge	As
Nb b.c.c. 0	Mo b.c.c. 0	Tc	Ru h.c.p. −0.03	Rh f.c.c. 0	Pd f.c.c. 0	Ag f.c.c. 0	Cd h.c.p. 0.16	In b.c.tet. 0.08	Sn b.c.tet. −0.46	Sb * −0.37
Ta b.c.c. 0	W b.c.c. 0	Re h.c.p. −0.01	Os h.c.p. −0.03	Ir f.c.c. 0	Pt f.c.c. 0	Au f.c.c. 0	Hg * −0.17	Tl h.c.p. −0.02	Pb f.c.c. 0	Bi * −0.37

*Rhombohedral.

to the surface but having a strong positive field in the lateral direction.[108] On the other hand, most of the d metals may have spherical ion cores which extend a symmetrical positive field. The electronic distribution on the surface is much smoothed and rather homogeneous[108] at the d metals.

In Section 3 a general tendency in electrolytic reduction of organic compounds was indicated. To know details of the electrolytic reduction, it is necessary to examine many other factors such as effects of solution composition, electrolysis condition, the neighboring group next to the one to be reduced, etc. An attempt has been reported to find the relation between the rate constant and the polarization or hindrance factor of a neighboring alkyl group in the electrolytic reduction of $RR'CO$.[89] The electrolytic reduction of nitrobenzene has been studied since the beginning of this century. Products vary widely depending on the electrode metal, solution composition, and electrolysis conditions. Establishment of the effect of secondary factors on the catalysis will further make it possible to predict the selectivity on a more detailed level.

4.2. Future Aspects of the Electrocatalysis

The word "electrocatalysis" was first proposed by Kobozev in 1936[110]. However, little attention seems to have been paid to it for the following 30 years. An electrochemical approach to catalysis has many advantages. Electrode processes themselves are the processes which take place at room temperature and atmospheric pressure. In addition, it will be noted that (1) a wide range of potential can be applied to the electrode, (2) a precise control of potential can easily be attained by modern electronic techniques, (3) an electron acts as a reducing or oxidizing reagent, and (4) the reagents which act as catalysts can be regenerated on the electrode. These advantages are briefly discussed below.

One can easily apply a voltage of, say, 1 V to an electrode. If this amount of voltage is applied to the reversible hydrogen electrode in the negative direction, the hydrogen evolution accompanies the free-energy change of -2 eV/mol H_2. This change is equal to that produced when the hydrogen pressure is decreased by 10^{33}. Attainment of such an enormous change in pressure is practically impossible. Thus, the electrochemical method is most convenient for conducting various reactions over a wide range of the conditions, from the extremely reductive atmosphere to the extremely oxidative one.

An organic compound can, in principle, be reduced or oxidized to various kinds of compounds. Each reduction or oxidation reaction may occur in a different potential range, which will be determined by an equilibrium potential of the reaction and the electrode material to be used. Therefore, precise control of the electrode potential is required for having a certain electrochemical reaction in a selective manner. For instance, the potential at which the Kolbe reaction starts is commonly referred to as the critical potential. Critical potentials for a number of different carboxylates have a value between 2.1 and 2.8 V (NHE).[111] Electrode potential can be easily controlled within a few mV.

In usual chemical processes, metals such as Zn, Mg, Al, Na, and K are often used as the reducing reagent and metal oxides such as $KMnO_4$ and MnO_2 are used as the oxidizing reagent. After the reaction, these reagents remain in the products as oxides, hydroxides, or salts. Thus, the products must be isolated from them. However, in electrochemical processes the electron acts as the reducing or oxidizing reagent without giving such metal compounds. Electrochemical processes can take place under much simplified conditions and give products of high purity.

Another aspect to be mentioned is the electrochemical regeneration of inorganic reagents which are capable of oxidizing or reducing an organic species in solution. Small amounts of cerium or vanadium salts lead to high yields of anthraquinone in the anodic oxidation of anthracene in 40% H_2SO_4.[112] Electrogenerated manganic salts oxidize substituted toluenes to benzaldehydes in aqueous H_2SO_4.[113] Many other polyvalent metal ions can be used as the catalyst.

As seen from the above characteristic aspects, electrocatalysis is undoubtedly one of the most promising fields in the future. Science and technology have been making progress with accelerating speed, and it is difficult to predict the situation after the next two or three decades. Professor Bockris's very interesting article on electrochemistry discussed many predicted developments likely to occur within the next 15-30 years.[114] However, much more emphasis should be placed on electrocatalysis as the primary tool in the future chemical industry.

References

1. J. Tafel, *Z. Physik. Chem. 50*, 641 (1905).
2. J. O'M. Bockris and E. C. Potter, *J. Electrochem. Soc. 99*, 169 (1952).
3. I. V. Kudriashov and L. A. Falin, *Elektrokhimiya 8*, 1029 (1972); A. Bélanger and A. K. Vijh, *J. Electrochem. Soc. 121*, 225 (1974).
4. J. O'M. Bockris and S. Srinivasan, *Electrochim. Acta 9*, 31 (1964).
5. B. E. Conway, *Proc. R. Soc. A256*, 128 (1960).
6. J. O'M. Bockris and A. M. Azzam, *Trans. Faraday Soc. 48*, 145 (1952).
7. O. Nomura and H. Kita, *J. Res. Inst. Catal. (Hokkaido Univ.) 15*, 35 (1967).
8. M. Hammerli, J. P. Mislan, and W. J. Olmstead, *J. Electrochem. Soc. 116*, 779 (1969).
9. G. M. Schmid, *Electrochim. Acta 12*, 449 (1967).
10. M. C. Barton and F. A. Lewis, *Z. Phys. Chem. N. F. 33*, 99 (1962).
11. M. J. Joncich, L. S. Stewart, and F. A. Posey, *J. Electrochem. Soc. 112*, 717 (1965).
12. N. Pentland, J. O'M. Bockris, and E. Sheldon, *J. Electrochem. Soc. 104*, 182 (1957).
13. H. Kita and T. Kurisu, *J. Res. Inst. Catal. (Hokkaido Univ.) 18*, 1 (1970).
14. M. Pourbaix, ed., *Atlas of Electrochemical Equilibria in Aqueous Solutions*, Pergamon Press, New York (1966).
15. A. N. Frumkin, *Elektrokhimiya 1*, 394 (1964).
16. J. O'M. Bockris, *Trans. Faraday Soc. 43*, 417 (1947).
17. N. E. Khomutov, *Zh. Fiz. Khim. 39*, 532 (1965).
18. A. T. Petrenko, *Zh. Fiz. Khim. 39*, 2097 (1965).
19. H. Kita, *J. Electrochem. Soc. 113*, 1095 (1966).
20. T. Hurlen and T. Valand, *Electrochim. Acta 9*, 1077 (1964).
21. A. T. Kuhn, C. J. Mortimer, G. C. Bond, and J. Lindley, *J. Electroanal. Chem. 34*, 1 (1972).
22. S. L. Morse and N. D. Greene, *Electrochim. Acta 12*, 179 (1967).
23. J. O'M. Bockris and E. C. Potter, *J. Chem. Phys. 20*, 614 (1952).
24. B. E. Conway, E. M. Beatty, and P. A. DeMaine, *Electrochim. Acta 7*, 39 (1962).
25. A. K. M. S. Huq and A. J. Rosenberg, *J. Electrochem. Soc. 111*, 270 (1964).
26. E. I. Mikhailova and Z. A. Jofa, *Elektrokhimiya 1*, 107 (1965).
27. L. V. Tamm, Yu. K. Tamm, and V. E. Past, *Elektrokhimiya 9*, 1382 (1973).
28. A. C. Makrides, *J. Electrochem. Soc. 109*, 977 (1962).
29. A. Matsuda and T. Ohmori, *J. Res. Inst. Catal. (Hokkaido Univ.) 10*, 203 (1962).
30. H. Kita and T. Yamazaki, *J. Res. Inst. Catal. (Hokkaido Univ.) 11*, 10 (1963).
31. J. Yeager, J. P. Cels, E. Yeager, and F. Hovorka, *J. Electrochem. Soc. 106*, 328 (1959).
32. A. N. Frumkin, in: *Advances in Electrochemistry and Electrochemical Engineering* (P. Delahay, ed.), Vol. 1, p. 65. Interscience, New York (1961).
33. M. E. Straumanis, S. T. Shih, and A. W. Schlechten, *J. Phys. Chem. 59*, 317 (1955).
34. S. Minc and J. Sobkowski, *Bull. Akad. Polon. Sci. 8*, 29 (1959).
35. V. I. Vigdorovich, L. E. Tsygankova, and T. V. Abamoma, *Elektrokhimiya 9*, 1151 (1973).
36. A. L. Rotinyan and N. M. Kozhevnikova, *Elektrokhimiya 1*, 664 (1965).
37. Ya. M. Kolotyrkin and P. S. Petrov, *Zh. Fiz. Khim. 31*, 659 (1957).
38. A. T. Petrenko, *Zh. Fiz. Khim. 36*, 1527 (1962).
39. V. M. Artemova, *Elektrokhimiya 3*, 1219 (1967).

40. N. T. Thomas and K. Nobe, *J. Electrochem. Soc. 117*, 622 (1970).
41. V. M. Artemova, B. G. Govaroucha, and V. S. Derkulskaya, *Elektrokhimiya 7*, 1227 (1971).
42. L. I. Krishtalik, *Elektrokhimiya 2*, 616 (1966).
43. N. de Zoubov and E. Deltombe, Rapport Technique RT. 26 Cebelcor (1955).
44. E. Deltombe, N. de Zoubov, and M. Pourbaix, *Proc. 7th Meeting of CITCE*, p. 216, Butterworths, London, (1957).
45. V. A. Fishman, V. I. Lainer, and I. G. Erusalimchik, *Elektrokhimiya 5*, 530 (1969).
46. J. Horiuti and H. Kita, *J. Res. Inst. Catal. (Hokkaido Univ.) 12*, 122 (1964).
47. R. Piontelli, L. P. Bicelli, and A. La Vecchia, *Accad. Nazionale Dei Lincei VIII 27*, 312 (1959).
48. I. Mieluch, *Bull. Acad. Polon. Sci. Ser. Sci. Chim. 17*, 43 (1969).
49. I. V. Kudriashov and L. A. Falin, *Elektrochimiya 7*, 1770 (1971).
50. V. V. Batrakov, Yu. Dittrikh, and A. N. Popov, *Elektrokhimiya 8*, 640 (1970).
51. I. V. Kudriashov and S. D. Kamyshchenko, *Elektrokhimiya 7*, 1284 (1971).
52. I. V. Kudriashov, S. D. Kamyshchenko, and E. M. Makaryan, *Elektrokhimiya 9*, 478 (1973).
53. I. I. Physhnograeva, A. M. Skundin, Yu. B. Vasiliev, and V. S. Bagotsky, *Elektrokhimiya 6*, 142 (1970).
54. St. G. Christov and N. A. Pangarov, *Z. Elektrochem. 61*, 113 (1957).
55. Z. A. Iofa and Vei Bao-Min, *Elektrokhimiya 2*, 755 (1966).
56. K. Sabo and I. A. Bagotskaya, *Dokl. Akad. Nauk SSSR, 149*, 139 (1963).
57. G. C. Bond, *Catalysis by Metals*, p. 475, Academic Press, London, New York (1962).
58. V. L. Heifets, B. S. Krasikov, and A. L. Rotinyan, *Elektrokhimiya 6*, 916 (1970).
59. S. Trasatti, *J. Electroanal. Chem. 39*, 163 (1972).
60. H. Kita and M. Honda, *Denki Kagaku (Electrochemistry) 38*, 17, 93 (1970).
61. H. Leidheiser, *J. Am. Chem. Soc. 71*, 3634 (1949).
62. N. E. Khomutov, *Zh. Fiz. Khim. 24*, 1201 (1950).
63. A. K. Lorentz, *Zh. Fiz. Khim. 24*, 853 (1950).
64. A. K. Lorentz, *Zh. Fiz. Khim. 27*, 317 (1953).
65. A. S. Goralnik, *Elektrokhimiya 2*, 1193 (1966).
66. N. Kobozev, *Zh. Fiz. Khim. 26*, 112 (1952).
67. N. Kobozev, *Zh. Fiz. Khim. 26*, 438 (1952).
68. H. Kita and T. Kurisu, *J. Res. Inst. Catal. (Hokkaido Univ.) 21*, 200 (1973).
69. B. E. Conway and J. O'M. Bockris, *J. Chem. Phys. 26*, 532 (1957).
70. D. B. Mattews, Thesis, University of Pennsylvania, Philadelphia (1965).
71. D. E. Eastman, *Phys. Rev., B 2*, 1 (1970).
72. V. Heine and C. H. Hodges, *J. Phys., C 5*, 225 (1972).
73. H. B. Michaelson, *J. Appl. Phys. 21*, 536 (1950).
74. D. D. Eley, *Disc. Faraday Soc. 8*, 34 (1950).
75. H. Horiuti and G. Okamoto, *Sci. Papers Inst. Phys. Chem. Res., Tokyo, 28*, 231 (1936).
76. Z. A. Iofa and L. A. Medvedeva, *Dokl. Akad. Nauk 69*, 213 (1949).
77. Z. A. Iofa and E. I. Liakhovetskaya, *Dokl. Akad. Nauk 86*, 577 (1952).
78. Z. A. Iofa and Vei Bao-Min, *Zh. Fiz. Khim. 37*, 2300 (1963).
79. Z. A. Iofa and V. A. Makarova, *Elektrokhimiya 1*, 231 (1965).
80. (a) Ya. M. Kolotyrkin, *Trans. Faraday Soc. 55*, 455 (1959); (b) Ya. M. Kolotyrkin and L. A. Medvedeva, *Dokl. Akad. Nauk 140*, 168 (1961).
81. Tza Chuan-Sin and Z. A. Iofa, *Dokl. Akad. Nauk 131*, 137 (1960).
82. K. Sabo and I. A. Bagotskaya, *Dokl. Akad. Nauk 150*, 128 (1963).
83. Z. A. Iofa and B. Kabanov i dr., *Zh. Fiz. Khim. 14*, 1105 (1939).
84. Ho Ngok Ba and Nguen Dyk Vi, *Elektrokhimiya 4*, 990 (1968).
85. (a) M. Morozov, I. A. Bagotskaya, and E. A. Preis, *Elektrokhimiya 5*, 40 (1969); (b) I. A. Bagotskaya and T. I. Khalturina, *Elektrokhimiya 6*, 1013 (1970).
86. H. Kita and T. Kurisu, *J. Res. Inst. Catal. (Hokkaido Univ.) 18*, 167 (1970).
87. D. A. Dowden, *Chemisorption* W. E. Gardner, ed., p. 3, Butterworths, London, (1957).
88. L. I. Antropov, *Zh. Fiz. Khim. 24*, 1428 (1950); *26*, 1688 (1952).
89. A. Yamura and T. Sekine, *Denki Kagaku 34*, 115 (1966).
90. I. A. Avrutskaya, M. Ya. Fioshin, S. D. Kabanova, and L. E. Gerasimova, *Elektrokhimiya 5*, 951 (1969).
91. S. Swann, Jr., D. K. Eads, and L. H. Krone, Jr., *J. Electrochem. Soc. 113*, 274 (1966).
92. A. P. Tomilov, A. A. Sergo, and S. L. Varshavsky, *Elektrokhimiya 1*, 1126 (1965).
93. T. Arai, *Denki Kagaku 30*, 31 (1962).

94. N. E. Khomutov, V. V. Skornyakov, and T. P. Fadeeva, *Zh. Fiz. Khim. 38*, 102 (1964).
95. C. Mettler, *Berichte 38*, 1745 (1905); *39*, 2933, 2940 (1906).
96. T. Sekine, A. Yamura, and K. Sugino, *J. Electrochem. Soc. 112*, 439 (1965).
97. A. Yamura, T. Sekine, and K. Sugino, *Denki Kagaku 34*, 110 (1966).
98. M. S. Sminova, V. A. Sminov, and L. I. Antropov, *Chem. Abstr. 54*, 24011 (1960).
99. J. Mizuguchi and T. Satoh. *Denki Kagaku 38*, 148 (1970).
100. K. N. Campbell and E. E. Young, *J. Am. Chem. Soc. 65*, 965 (1943).
101. R. A. Benkeser and C. A. Tincher, *J. Org. Chem. 33*, 2727 (1968).
102. M. Sakuma, *Denki Kagaku 28*, 164 (1960).
103. I. Rosenthal, J. R. Hayes, A. J. Martin, and P. J. Elving, *J. Am. Chem. Soc. 80*, 3050 (1958).
104. I. L. Knunyants, S. L. Varshavskii, A. P. Tomlilov, and L. V. Kabaky, *Chem Abstr 64*, 1660 (1966).
105. T. Nonaka and K. Sugino, *J. Electrochem. Soc. 114*, 1255 (1967).
106. E. Kariv and J. Eermolin, *Electrochim. Acta 18*, 417 (1973).
107. H. Kita and M. Nakamura (to be published).
108. O. Johnson, *J. Res. Inst. Catal. (Hokkaido Univ.) 19*, 152 (1972).
109. O. Johnson, *Bull. Chem. Soc. Jpn. 45*, 1599, 1607 (1972).
110. N. Kobozev and W. Monblanova, *Zh. Fiz. Khim. 7*, 645 (1936); *Acta Physicochim., U.S.S.R. 4*, 395 (1936).
111. L. Eberson, *Chemistry of the Carboxyl Group*, S. Patai, ed., Chapter 2, Wiley-Interscience, London (1969).
112. K. Shirai and K. Sugino, *Denki Kagaku 25*, 284 (1957).
113. R. Ramaswamy, M. S. V. Pathy, and V. K. Udupa, *J. Electrochem. Soc. 110*, 202 (1963).
114. J. O'M. Bockris, *J. Electroanal. Chem. 9*, 408 (1965).

8

Surveying Electrocatalysis

A. T. Kuhn

1. An Overview*

Although not all might find the fact palatable, the truth remains that electrocatalysis is *the* most important subject in the field of electrochemistry. The rate at which a reaction proceeds and the course it takes are demonstrably more important than the structure of the double layer or the question of adsorbed species. Granted such matters underlie the whole question of electrocatalysis, but by sheer empirical methods, the practitioners of the subject can and have brought to it an ordered structure which was previously largely absent.

Looking at the present—before considering both past and future—one must face up to the fact that problems of electrocatalysis are the only ones that lie between us and the successful and widespread application of fuel cells in the vision that John Bockris and his colleagues and coworkers have seen. Thus the able engineers at Pratt & Whitney have demonstrated not only the reliability and advantages of fuel cells in space flight, but also that such units can be engineered at a price that would make them attractive in domestic and industrial use as well. But what has deprived them of success in this so far has been the very limited life of the electrodes in such cells. Simply stated, the catalytic activity has fallen off over a space of two years or so, when—for the device to be economic—it should have lasted 10 years or more.

At this very point in time we face a crucial watershed. There are those on the one hand who believe that the very same factors in an electrode which promote high activity also promise a rapid decay of that activity, perhaps due to sintering of the metal, perhaps due to poisoning. These pessimists, or realists, as they would prefer to be called, see only a mechanical-type answer to the problem or an answer which avoids the direct solution. Thus we now have much discussion of the redox battery advocated by Beccu[1] and others, in which the use of concentrated solutions with dissolved "fuels" permits the use of a stable, long-lived planar electrode. It is too early yet to appraise the idea, although of course redox fuel cells in a slightly different form were studied extensively in the 1960s and the verdict on the many United States agency-funded researches has been well summarized by Austin.[2]

So we see that electrocatalysis is the vital link to success in fuel cell technology. What of the other main area of applied electrochemistry, industrial processing? Here, one must be honest and state that while for many applications an improved electrode material would be most welcome, it does not stand between success and failure as is the case with the fuel cell. Thus the chlorine manufacturers welcomed the advent of the ruthenium oxide electrode to replace the graphite previously used and are even now searching for a

*With E. W. Brooman and J. S. Clarke.

A. T. Kuhn • Department of Chemistry and Applied Chemistry, University of Salford, England.

better and cheaper one still. But it is probably true to say that these anodes did not, in the slightest degree, affect the total amount of chlorine produced around the world. Looking at possible candidates for economic industrial electrochemical processes, it is hard to think of one whose future success depends on the creation of an electrode material unknown today. Having said this, however, one should add that—in financial terms—the subject is of immense importance, and the ruthenium oxide anodes referred to have proved and are proving very rewarding, not only to their inventor, Henri Beer, to whom too little credit is given, but also the firm of De Nora who backed Beer at a time when others lacked the courage and foresight. Having set the scene for the subject, let us then consider its past and future.

1.1. Electrocatalysis in the Past

Even in the most distant and dark days of electrochemistry, it was recognized that the material of which the electrode was made was of crucial importance. From the middle of the 19th century to the 1920s, the main emphasis lay in electro-organic synthesis, and Haber and many others, most of whose work has been recorded by Fichter,[3] knew that the rate of an electro-organic reaction, as well as its course, depended on the choice of the appropriate metal. Of course such knowledge was reached largely on purely preparative grounds, in terms of yield, etc. But the message was there.

In the 1920s, largely under the influence of Bowden and Rideal, renewed interest was expressed in the more physicochemical aspects of electrochemistry, in the kinetics and mechanistic aspects which had been earlier examined by Haber himself. In this time began an era of electrocatalysis which, although declining, is still with us today. We shall refer to this as "single-phase" electrocatalysis.

1.2. Single-Phase Electrocatalysis

We are considering here a situation in which the electrode was a single-phase metal or alloy, or possibly situations in which it was thought to be, hoped to be, or indeed one in which no thought had been given to the matter at all. The variation of the rate of reaction with electrode material has been exhaustively studied, especially for hydrogen evolution. Early on, attempts were made to bring together results of different metals and to correlate them with some physical or physicochemical parameter. Kuhn et al.[4] surveyed and reviewed these attempts. One of their conclusions was that many earlier correlations were tainted with subjectivity. The authors concerned, confronted with a spread of values, selected some and discarded others in what was, although they might have denied it, an arbitrary fashion. Kuhn et al. therefore decided to break away from this by laying down criteria for selection. They chose only those workers who had used pre-electrolysis in their studies. By so doing, they undoubtedly excluded much good work. However, the procedure also served to exclude the many papers written by those under the impression that acceptable measurements are made by filling a cell with electrolyte and commencing the measurements.

A subsequent paper by Trasatti[5] was praiseworthy in many ways, but returned to the old habit of arbitrarily selecting data values. In addition, a whole series of electronic work functions were used, all derived from a normalization procedure which many criticized. Most of the work in this period was devoted to the hydrogen-evolution reaction, and with hindsight we may one day find that this was a mistake, as we shall see subsequently. Brooman and Kuhn[6] extended the earlier work to an examination of

Table 1. Reported Tafel Data for the Anodic Ionization of Hydrogen in Acid Solutions

Metal	Reference	Electrolyte[a]	Anodic		Cathodic[b]	
			b_a (mV/decade)	$-\log i_0$	$-b_c$ (mV/decade)	$-\log i_0$
Rh	7	1 N H$_2$SO$_4$(1,2)	38	3.4	32 or 62	3.2–3.8
Pd	7	1 N H$_2$SO$_4$(1)	100	2.89	100	2.89–3.2
Ir	7	1 N H$_2$SO$_4$(2)	50	3.4	–	2.7
Pt	7	1 N H$_2$SO$_4$(1)	32	3.0	30	2.6–3.1
Pt	7	1 N H$_2$SO$_4$(2)	30	3.52	30	2.6–3.1
Pt	9	1 N H$_2$SO$_4$(1)	28	–	25–32	2.6–3.1
Pt	8	2 N H$_2$SO$_4$(1)	25	~4.1	25	2.6–3.1
Au	7	1 N H$_2$SO$_4$(1)	105	5.15	105–116[c]	4.4–5.8

[a](1) = electrochemical electrode pretreatment; (2) = thermal electrode pretreatment.
[b]Data for i_0 (mA/cm^2) taken from reference 1 for $-b_c$ as follows: RH = 7; Pd = 7; Ir = 7; Pt = 7, 8, 10; Au = 7, 11.
[c]Two slopes are reported in reference 12: $b = 50$ ($i_0 = 4.75$) and $b = 105$ ($i_0 = 3.92$).

hydrogen-evolution kinetics on alloys and found the situation to conform with the generally accepted picture.

To provide a wider base for the correlation of the rate of electrochemical reactions and the properties of metals, the published data for the ionization of hydrogen in acid and alkaline solutions was compared with similar data for the evolution of hydrogen.

Tables 1 and 2 summarize all known kinetic data for these reactions at 25°C and 1 atm pressure, unless otherwise stated. Any data calculated or extrapolated from published polarization curves are shown in these tables as approximate values. The exchange current density, i_0, is taken as a measure of the activity of the electrode material.

These tables demonstrate both the lack of data reported and its limitation essentially to the noble metals. In acid solution, dissolution of the nonnoble metals precludes

Table 2. Reported Tafel Data for the Anodic Ionization of Hydrogen in Alkaline Solutions

Metal	Reference	Electrolyte[a]	Anodic		Cathodic[b]	
			b_a (mV/decade)	$-\log i_0$	$-b_c$ (mV/decade)	$-\log i_0$
Fe	12	0.2 N NaOH(2)	60	~5	120	5.82
Ni	13	0.1 – 1 N NaOH (1,2) 35°C	~25	~6	89–130	5.03–6.4
Rh	7	0.1 N NaOH(1)	120	4.0	118	4.0–4.85
Pd	7	0.1 N NaOH(1)	105	3.60	105	3.60
Pt	7	0.1 N NaOH(1)	105	3.4	105	3.4–4.17
Pt	9	1 N NaOH (1)	70	3.63	114–117	3.4–4.17
Au	7	0.1 N NaOH(1)	125	5.89	120	5.4

[a](1) = electrochemical electrode pretreatment; (2) = chemical electrode pretreatment.
[b]Data for i_0 (mA/cm^2) taken from reference 1 for $-b_c$ as follows: Fe = 7; Ni = 7, 14; Rh = 7; Pd = 7; Pt = 7, 10; Au = 7.

accurate measurements, and in alkaline solutions, the presence of passivating films adds further complexities to the investigations.

Before the data can be reviewed critically a number of factors arising both from the experimental techniques used and the properties of the electrode and its environment require examination.

1.2.1. Diffusion Control of the Reaction

Since hydrogen is so insoluble in aqueous media, the reaction becomes mass-transport limited very close to the reversible potential. In the case of platinum useful comparisons can be made of the evaluated parameters and the methods used to eliminate or account for the diffusion overpotential. For example Schuldiner[8] observed limiting current conditions during rapid H_2 purging of the electrolyte and recorded steady-state measurements while Harrison and Khan,[9] during rotating disk measurements in acid solution, observed the reaction to be completely diffusion controlled over the rotation speeds examined. In alkaline solution,[9] the reaction was not totally diffusion controlled, and by extrapolation from the diffusion region a Tafel slope of 70 mV was recorded.

Most of the published data is based on the work of Mannan,[7] and in light of the poor agreement between her results on platinum and other studies based on the rotating disk electrode (r.d.e.), as well as the paucity of experimental details, doubt is thrown on the remaining values, although for the less active metals with low i_0 values, some of the results were quite possibly correct, since diffusion overpotential is less important here.

The role of adsorbed hydrogen is also an important factor in determining the kinetic data for the ionization hydrogen. For the cathodic reaction, the variation of coverage of adsorbed hydrogen, Θ_H, with potential has been found not to change significantly with overvoltage and therefore to contribute little to the interpretation of the data.[15] On the anodic side of the reaction, work on the platinum metals has shown[16-18] that Θ_H varied from unity at the reversible potentials, down to zero at anodic potentials less than 250 mV. Thus the change of surface coverage may influence significantly the parameters derived from the Tafel equations.

As Vetter[15] indicated, and as we have also found by "normalizing" the data in Table 1 to constant (i.e., unity) hydrogen coverage using Breiter's data on hydrogen coverage, changes in Tafel slope ($dE/d \log i$) of approximately 0.02 V can be found, with corresponding changes in i_0, by either assuming a Volmer model or first-order dependence. However, it is unfortunately true that in many cases the diagnostic parameters reported for this reaction make these particular assumptions invalid.

A number of factors characteristic of both the properties of the metal and the electrolyte determine the chemisorption of hydrogen and the part it plays in the reaction. These can be identified briefly.

1.2.2. Position of Metal in the Periodic Table

For the transition metals (the so-called d metals) hydrogen is easily chemisorbed. This is true particularly for the platinum metals where the calculated values of $\Delta G_{H_{ads}}$ vary in the following order:[11] Pt > Pd > Rh > Ir > Ru > Os in acid solution. For the nonnoble metals the problems associated with corrosion or passivating films make determinations more difficult, as shown for iron[19] and nickel.[20]

The sp metals fall into the second category and, as Trasatti[5] pointed out, no chemisorption is expected because the energy for dissociation of hydrogen molecules is about twice that of the hydrogen metal bonds. Thus the hydrogen adsorption coverages observed for gold,[21] copper, and silver[22] are all less than 5% (mostly <1%).

1.2.3. Type of Adsorbed Hydrogen

The manner in which hydrogen is sorbed onto platinum metal electrodes has been shown to vary both with the metal and alternative surface sites. Thus Breiter[16] postulated two types of surface site for platinum and iridium in acid solution (as also mentioned by Gokhshtein[18]) with only one type for rhodium in acid or alkaline solutions. The two types of site are arbitrarily called "weakly bonded" and "strongly bonded" hydrogen sites and various attempts[16-18, 23-28] have been made to correlate the separate maxima on the i–E polarization curves to adsorption isotherms with varying bonding energies for each site. Similar phenomena have been observed at nickel electrodes.[20]

The question then arises as to whether the totally adsorbed hydrogen, at sites of differing bond energies, is available to take part in the electron-transfer processes or whether it is restricted to the more weakly bonded sites of a transitory nature. This issue has been discussed by Schuldiner.[29] The recent work with optical techniques may provide some resolution of this problem, but the question is still unresolved as a recent publication of Angerstein-Kozlowska et al.[210] shows.

1.2.4. Influence of Electrolyte Medium

An additional factor in all Θ_H data is that of the medium, namely the interfering effect of specific anion adsorption[8, 16] and the dependence of coverage on pH. In the latter case, for platinum and similar metals $d\theta/d_{pH} = 0$, while on gold it may be inferred[33] that Θ_H changes with pH. The influence of halide ion adsorption at rotating platinum electrodes was clearly demonstrated by Frumkin and Aikasjan[30] and by Makowski[31] as a reduction in the limiting current.

In theory,[32] kinetic data obtained during the ionization of hydrogen should correspond with that obtained during the evolution of hydrogen, provided that the conditions for the h.e.r. are as similar as possible (especially with respect to the conditions of the electrode surface). Figure 1 shows the extent of the agreement between published anodic and cathodic hydrogen evolution reaction (h.e.r.) data in terms of $-\log i_0$ values. To a first approximation the agreement is satisfactory in acid and alkaline solutions, so that the hypothesis stated above about the reversibility of the h.e.r. may be said to be correct.

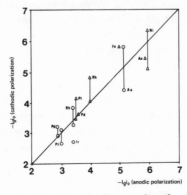

Fig. 1. Correspondence between anodic and cathodic rate data for the ionization and evolution of hydrogen. Data taken from tables 1 and 2: O = acid solutions; Δ = alkaline solutions.

As more accurate data becomes available and more metals are investigated, it would be interesting to observe whether the agreement between anodic and cathodic kinetic data improves.

The other two classes of inorganic reaction mechanism which we must mention are oxygen evolution and the redox reactions. Of the first, correlations were attempted, but these[34] were open to a double criticism. In the first place, the authors plotted over-voltage at constant current density, thereby submerging any problems—nonetheless very real and important—of different Tafel slopes. In other words, their results were a function of the (arbitrary) potential at which their comparison was made. In the second place, it was gradually appreciated that the existence of oxide layers on the anode surface, some-times of substantial thickness, resulted in a potential difference (p.d.) within this layer and the oxygen evolution reaction being driven by a smaller–metal solution p.d. than was actually measured. The problem has not been examined since, although techniques and understanding of the system now approach the state where it could be fruitful.

As regards the redox systems, formally at least, these are the simplest of all electrode reactions requiring a pure act of charge transfer. Their rate should depend only on the electronic work function of the metal and nothing else, and in a complete circuit (see below) even this factor drops out. However, when the data are assembled, as done by Galizziole and Trasatti[35] for example, the picture is far from simple. This poses con-siderable problems. However, Peter[36] has suggested that a simple explanation might be found. He is presently making an experimental study of this type of reaction and believes that much of the data in the literature may be in error as a result of unclean or passivated electrodes. If he is correct, the electrocatalytic situation regarding the redox electrodes may not be the anomalous one we are led to believe.

Electro-organic reactions, too, have been examined in the same light. Kuhn et al.[37] studied the anodic oxidation of ethylene and found the rate to vary in volcano-like fashion with the properties of the metal. Of course some of the anode materials used were, at the potential of comparison, oxide covered and therefore in a sense not strictly comparable to those which were not. On the other hand it was precisely these physical properties which were responsible for that condition, i.e., metal–metal bond strength etc. Byrne and Kuhn[38] found a similar volcano in the kinetics of the cathodic reduction of ethylene, and here, of course, the metals were all in their active states. Acetylene too[39] was reduced in the same way as ethylene and a volcano-like curve was constructed for this reaction. The reduction of oxygen can take place on a filmed or unfilmed metal surface. In the former case, the objections raised in the case of the o.e.r. remain. In the latter case, a correlation is possible and Damjanovic and coworkers[75] have published excellent data correlating rates with various properties of the metals and their alloys.

What has so far been demonstrated is that correlations can be shown to exist between the rates of many simple electrochemical reactions and a range of physical or physico-chemical properties. In the main, disagreement about the validity of one point or another does not and should not affect the overall picture, although as Kuhn et al.[4] showed, there can sometimes be such gross distortion, either in the choice of points or the manner in which they are graphically presented, that severe disagreements can in fact arise. The main area of discord in the experimental data relates to the use of the metals Au and Pd in such correlations. While Dahms and Bockris[40] and later Johnson et al.[41] have pub-lished numerous correlations, mainly organic oxidation reactions, in which these metals or their alloys are important, a school of thought which includes Goodridge,[42] Blake et al.,[43] and Cwiklinski and Perichon[44] has provided much evidence that these metals, especially in the presence of organics, dissolve anodically to form palladous ions for example. These react homogeneously with the substrate, in a manner analogous to the Hoechst–Wacker process. This being true, it is wholly inappropriate to include them with those metals which are physically and chemically stable, and it also means that the carefully constructed theories explaining why these metals (and to all intents only these metals) given partial oxidation products are ill-founded.

The work cited above enables us to predict, with some confidence, the rate at which a reaction will proceed on a given type of metal. The data relating to the almost infinitely large field of electro-organic chemistry is more limited, but here too, a picture is beginning to emerge. It has been stated[45] that the rate and course of electroreductions can be affected by the purity of the cathode metal, small amounts causing major changes in both parameters. More careful examination of the facts on which this statement is made raise some doubts. It is the author's belief that where single-phase materials are concerned, and in many cases of multiphase materials, small additions of material X cause only small changes in electrocatalytic property, where these changes are related only to the mole fraction of X and its intrinsic catalytic activity relative to the matrix in which it is inserted. While we shall return to this below, there seem to be no grounds for believing that addition of 1% of a metal to another, as a single phase, could cause any significant change. Too many data in the literature describe so-called electrocatalytic effects (variations of 2 or 3) without consideration of the surface roughness and the variable electrode area. Such effects, for example, are the probable explanation of the differences in the rate of electrofluorination on Ni anodes with various pretreatments.[46]

We should not dismiss the theoretical side of the subject. There was, for example, the controversy as to the conditions under which rates should be compared, with one school opting for constant (metal–solution) potential, the other for comparison at constant (rational) potential, relative to the potential of zero charge (p.z.c.). Quite apart from the rights and wrongs of the matter—for it took a surprisingly long time before it was realized that the differences in electronic work function (which must be important) are cancelled out by the contact p.d. where the electrode metal joins the rest of the circuit and that summed round a complete circuit the term disappears—the second school had to contend with the fact that the value of the p.z.c. for solid metals was often in doubt by 100 mV or more. The building-in of this error can hardly have helped in their analyses. On a more positive side, the volcano-like shape of the electrocatalytic plots was theoretically predicted by Parsons and by other workers.[47] In recent years, with the exception of one persistent Canadian worker, we have seen fewer correlations of the type described here.

The explanation is at least in part a philosophical and conceptual one. It was mainly Kita who drew attention to the correlation between the rate of reaction and the simple periodicity of the metal used as electrode. At this point, many realized that, insofar as almost any one physical or chemical property can depend on another, the argument becomes circular, and the search for a magic parameter which will uniquely and successfully correlate activity with some other quantity is seen to be a mirage. Indeed, Bond, in his book,[48] depicts graphically this very circularity. At this point then, when it was realized that almost everything depends on something else, the emphasis on correlation dropped off, to be preserved only by a few practitioners of the "correlation game" as it then became. In its least creative form, this was reduced to an exercise in which one set of data (electrochemical) was correlated with another set (physicochemical) using any concept, any suggested equation, that might conveniently be found and drawing a line or a curve on such plots with blatant disregard to any statistical significance or lack of it.

The picture that we have attempted to draw then, is one in which the situation is basically understood, at least where single-phase systems are concerned. Where does future progress lie? We can put forward two alternative scenarios. First, we can envisage the development of a new "wonder catalyst"—some material which is a good conductor, resists anodic dissolution, and at the same time catalyzes one or more electrode reactions. Such materials were the subject of many exhaustive searches in the course of the United States fuel-cell program in the early 1960s. The result, as we know, was unsuccessful. Second, we can seek—less demandingly—a material which will resist anodic dissolution and conduct electrons, although possibly lacking catalytic ability. This last we can create by "doping" the catalytically inert matrix with some other material, quite possibly a noble metal.

One of the major signposts (and at the same time an interesting story) was the publication by Sepa et al.,[49] in which a high degree of electrocatalytic activity was claimed for sodium tungsten bronze as an oxygen reduction electrode. Later Bockris[50] realized that the activity resulted from minute traces of Pt which derived from the Pt crucible used to fuse the bronze. He clarified the situation[50] by showing how catalytic activity varied as a function of Pt concentration. When all was said and done, however, the fact remained that *very small amounts* of Pt induced activity comparable to that of pure Pt metal. This clear result has provided all workers in the field with much food for thought and experimentation.

For this reason, we felt it would be useful to summarize—for this has not hitherto been done—the presently known range of nonmetallic conducting materials which are or have been candidates for novel electrode materials. As is clear, we now should re-examine this, not as catalysts *per se*, but simply as possible vehicles for subsequent activation by doping with some active species.

2. Nonmetals as Electrode Materials*

2.1. Nonmetallic Electrode Materials and Their Behavior in Aqueous Media

As we have seen, the electrocatalytic behavior of the single-phase metals and alloys can now be said to be broadly understood, and this fact, together with the emergence of a host of new materials derived from electronics work, marks the end of the "platinum and mercury" era. While we have, therefore, not only this enormous range of conducting nonmetallic materials, and also a considerable theoretical background to help us in our understanding of the conduction processes, there is much less information available concerning their behavior in solution. To review the extent of our present knowledge, and to make what interpretative comments we can, is the purpose of this section.

Nonmetallic electrodes are not, of course, completely novel in electrochemistry. Apart from carbons of various forms and graphites, silicon and germanium have both been widely used and are reviewed by Boddy.[51] However, their range of use is limited by anodic dissolution on the one side and their cathodic reduction (to form silane or germane) on the other. Oxides of tin and zinc are likewise well known and have found application wherever techniques involving simultaneous optical observations are called for.[52,53] As we shall see, tin oxide has recently acquired a new and commercially far more important application.

Many simple oxides other than those mentioned above are metallically conducting and have been widely used as electrodes. These include PbO_2, NiO, MnO_2, and RuO_2,[54] while magnetite has been known and used for over half a century in certain applications.

*With R. Smailes.

Many of these have been or are now of real commercial value, and their use is described elsewhere.[55]

But leaving all these materials aside, we attempt here to review, for the first time as far as we are aware and with the exception of a more limited treatment of Von Sturm,[56] the available knowledge regarding the stability, under aqueous anodic conditions, of those materials which do not fall into the simple oxide category above. In order to limit our scope somewhat, we have excluded from our considerations the organic electrocatalysts such as the phthalocyanines and anthracene. Then too, we have limited ourselves to studies in aqueous media, since the entire behavior and stability ranking of many of the compounds listed here is totally different in molten salts.

Finally, we have decided to exclude (for they have been discussed elsewhere) the class of compounds known as the tungsten bronzes (either hydrogen or sodium, for example) which form the subject of many papers recently available,[57-94] as well as alloys and intermetallics for which, in any case, little data exist.

Apart from these fairly specific exclusions we believe the table which forms the center of this review (Table 3) is as complete as is reasonably possible. The first and chief purpose of the table is to enable the reader to check all known work on a particular compound or class of compounds. The table then summarizes, in the limited available space, the type of study and the nature of the results so obtained. It will readily be appreciated that the scope and ambition of the work here referred to varied enormously. Some workers carried out simple open-circuit corrosion tests which, for electrochemical applications, have, frankly, a minimal value. Other workers were assessing the catalytic activity of the materials for possible use in fuel cells and, of these, some separated the corrosion currents from whatever faradaic reaction process was being observed—not all these attempts were successful, however!

Table 3, then, lists all known electrochemical studies within the limits set above. Nonelectrochemical studies at open circuit have been omitted, except in rare instances where little or no other electrochemical data are available, since results obtained in these laboratories and elsewhere indicate that such "floating" corrosion tests provide no reliable guide to corrosion rates under an applied potential.

Of course it will be appreciated that the range of electrically conducting nonmetallic compounds is many orders of magnitude greater than that listed here. The properties of these have been cataloged in multivolume compendia such as references 95 and 96, although the majority can be seen to be, or have been found to be, unstable in an aqueous environment and hence of no further interest in the present exercise.

From Table 3 it will be seen that a very large body of compounds has at least a limited application as electrode materials or conductors in an aqueous environment. A number of points are not adequately brought out in Table 3 and must be discussed further. First, because of the nonstoichiometric nature of many of these compounds (and the great problems in analysis of their composition), much apparently conflicting data will be found in the literature. Thus TiC is stated by Bianchi *et al.*[124] not to corrode until relatively anodic potentials (e.g., 1.8 V) while reference 99 describes corrosion at 0.8 V. In other words one must be cautious before accepting the given formula or structure at face value. A detail which might also be covered here is the loose usage of the phrase "semiconductor electrode materials" which is too prevalent. Many of the compounds listed here (notably the tungsten bronzes) are categorically not semiconducting as defined in terms of their absolute (and very high) electrical conductance, their mode of conduction, and the behavior associated with this.

Table 3.

Compound	Conditions[a]	Reaction studied[b]	Fabrication[c]	Corrosion information[d]	References
		A. Borides			
CoB_2	2NS at RT	H_2An	S	CQL	97
Co_2B	2NS at RT	CO An	TB	ND 0–1.2 V (DHE)	98
Cr_5B_3	2NK at 75°C	O_2 Cat	TB	CQT 100 mV (max 1.0 V)	99
	2NK at 75°C	O_2 Cat	I	CQL	100
Cr_4B	2NK at 75°C	O_2 Cat	I	CQL (low i)	100
Cr_3B	2NK at 75°C	O_2 Cat	I	CQL (low i)	100
Cr_2B	2NK at 75°C	O_2 Cat	Disk	CQT (800 mV)	101
	0.05NS	Redox	—	ND	102
	2NK at 75°C	O_2 Cat	I	CQL	100
CrB	0.05NS	Redox	—	ND	102
	2NK at 75°C	O_2 Cat	TB	CQT (300 mV)	99
	2NK at 75°C	O_2 Cat	I	CQL	100
CrB_2	2NK at 75°C	O_2 Cat	TB	CQT (200 mV)	99
	2NK at 75°C	O_2 Cat	I	CQL	100
	0.05NS	Redox	—	ND	102
	20% Na	Decomp of Na Amalgam	PrS	CQL	103
	1MS at RT	H_2An	TB	CQL	104
LaB_6	1MS at RT	H_2An	HPr	CQL	104
	0.1M KCL	Redox	ROD	ND	105
	0.1MCl at 25°C	Corr	HPr	ND (i ltd)	106
MoB	2NK at 25°C	O_2 Cat	TB	CQT (200 mV)	99
	0.05NS	Redox	—	ND	102
	2NK at 25°C	O_2 Cat	I	CQT (200 mV)	100
MoB_2	1MS at RT	O_2 Cat	TB	CQL	97
	0.05NS	Redox	—	ND	102
	3NS at 80°C 6NCL at 80°C 5NK at 80°C 18NP at 80°C	Corr	—	CQL, OCV	107
	2NK at 25°C	O_2 Cat	TB	CQT (200 mV)	99
	2NK at 25°C	O_2 Cat	I	CQT (200 mV)	100
	0.05NS	Redox	—	ND	102
Mo_2B_5	0.05NS	Redox	—	ND	102
Mo_2B_4	0.05NS	Redox	—	ND	102
NbB	2NK at 75°C	O_2 Cat	TB	CQT (0 V)	99
	2NK at 75°C		I	CQL	100
NbB_2	1MS at 75°C	H_2An	ST	CQT (0.06 mA at 200 mV)	104
	2MK at 75°C·	O_2 Cat	I	CQL	100
	2MK at 75°C	O_2 Cat	TB	CQT (0 V)	99
NiB	6NK at 80°C	H_2 CH_3OH An O_2 Cat	Disks	CQT	108

[a] Electrolyte media: 6NS = 6 N H_2SO_4; 6NK = 6 N KOH; 6NCl = 6 N HCl; 6NP = 6 N H_3PO_4; 6NNa = 6N NaOH.

[b] Test reaction used: H_2 An = Anodic H_2 oxidation; Redox = Fe^{2+}/Fe^{3+} or similar; O_2 Cat = cathodic O_2 Reduction; Corr = corrosion study in N_2 purged and other solutions; o.e.r. = oxygen-evolution reaction; h.e.r. = hydrogen-evolution reaction.

[c] Electrode fabrication details: S = sintered; TB = Teflon bonded; G = glass; Cer = ceramic; HPr = hot pressed; Pr = pressed; I = ingot; ST = sintered with Teflon.

[d] Data cited in quoted reference: CQT (300 mV) = quantitative corrosion information (measure corrosion observed at 0.3V); CQL = qualitative corrosion information; ND (0 → 1.2V) = corrosion data not determined (range of experiment = 0 – 1.2V); (s) = imprecise stoichiometry; (P) = protective film formed (NP) 1.8V = non protective film formed at potentials > 1.8V; OCV = tested on open circuit.

Table 3 (*Continued*).

Compound	Conditions[a]	Reaction studied[b]	Fabrication[c]	Corrosion information[d]	References
NiB$_2$	1MS	H$_2$ An	TB	CQL	97
Ni$_2$B	2MK at 75°C	O$_2$ Cat	I	ND	100
	45% K	H$_2$ An	TB	CQT (720 mV)	109
	6NK at 90°C	H$_2$ An	TB	CQL	110
	30% K at 90°C	H$_2$, N$_2$H$_4$ An	TB on Ni screen	CQT	111
	2MK at 25°C	H$_2$ An	S	CQT (225 mV)	112
	6NK at 85°C	H$_2$, CH$_3$OH An O$_2$ Cat	Disk	CQT	108
	8NK at 80°C and 150°C	H$_2$ C$_2$H$_4$ An	On Ni screen	ND	113
	6NK at 90°C	N$_2$H$_4$ An	TB	ND	110
	30% K at 90°C	H$_2$ An	TB	ND	110
	1N HClO at RT + 90°C	H$_2$ An	TB	CQT (350 mV) CQL	110
	2NK at 75°C	O$_2$ Cat	Disk	CQL	101
	6NK at 90°C	KBH$_4$ An	TB	CQL	110
	Ac/Ac pH3	CO An	TB	ND (0–1.2 V)	98
Ni$_3$B	2MK at 75°C	O$_2$ Cat	TB	CQT (1.2 V)	99
	2MK at 75°C	O$_2$ Cat	I	CQL	100
	7MK	H$_2$ An	S rods	CQT	114
	6NK at 80°C	H$_2$ An CH$_3$OH O$_2$ Cat	Disk	CQT	108
	6NK at 90°C	H$_2$ An	TB	CQL	110
	1M HCLO$_4$ at RT + 90°C	H$_2$ An	TB	CQT (350 mV at RT) + CQL	110
Ni$_4$B$_3$	6NK at 80°C	H$_2$ An CH$_3$OH O$_2$ Cat	Disk	CQT	108
TaB	2NK at 75°C	O$_2$ Cat	TB	CQT (0 V)	99
	2NK at 75°C	O$_2$ Cat	I	CQL	100
TaB$_2$	2NK at 75°C	O$_2$ Cat	TB	CQT (0 V)	99
	2NK at 75°C	O$_2$ Cat	I	CQL	100
	1MS at 25°C	H$_2$ An	TB	CQL	97
	1MS at 50°C	H$_2$ An	TB	CQL	97
TiB$_2$	1MS at 25°C	H$_2$ An	TB	CQL	104
	20% Na	Decomp of Na amalgam	SPr	CQL	103
	0.05NS	Redox	—	ND	102
	2NK at 75°C	O$_2$ Cat	TB	CQ (200 mV)	99
	2NK at 75°C	O$_2$ Cat	I	CQL	100
	3NS, 6NCl, 18NP	Corr	—	CQL, OCV	107
VB$_2$	2NK at 75°C	O$_2$ Cat	TB	CQT (0 V)	99
	2NK at 75°C	O$_2$ Cat	I	CQL	100
	3NS 6NCl 18N, H$_3$PO$_4$ 5NK	Corr	—	CQL, OCV	107
WB	1MS at RT	H$_2$ An	TB	CQL	104
	2NK at 75°C	O$_2$ Cat	TB	CQT (100 mV)	99
	2NK at 75°C	O$_2$ Cat	I	CQL	100

Table 3 (*Continued*).

Compound	Conditions[a]	Reaction studied[b]	Fabrication[c]	Corrosion information[d]	References
W_2B	2NK at 75°C	O_2 Cat	TB	CQT (200 mV)	99
	2NK at 75°C	O_2 Cat	I	CQT (100 mV)	100
W_2B_5	2NK at 75°C	O_2 Cat	TB	CQT (100 mV)	99
	2NK at 75°C	O_2 Cat	I	CQT (100 mV)	100
ZrB_2	1MS at RT	H_2An	TB	CQL	104
	2MK at 75°C	O_2 Cat	TB	CQT (0 V)	99
	2MK at 75°C	O_2 Cat	I	CQL	100
	20% Na	Decomp of Na amalgam	SPr	CQL	103

B. Carbides

Compound	Conditions[a]	Reaction studied[b]	Fabrication[c]	Corrosion information[d]	References
B_4C	2MK at 75°C	H_2An	I	CQL	100
	2NK at 75°C	O_2 Cat	Disk	CQL	101
	85% P at 150°C	Corr	Powder	CQL, OCV	{ 115 116
	1MS 0.1M per Cl	Redox	Rods	CQL	117 118
	MS, MCl	Redox	—	CQL	119
	0.05NS	Redox	—	ND	102
	1NS	h.e.r. + redox	Rods	ND	71
	—	Corr		CQT	120
	$0.1MK_2SO_4$	O_2 Cat	Rods	ND $0 \to 5$ V (SCE)	121
Cr_3C_2	2MK at 75°C	O_2 Cat	TB	CQT (300 mV)	99
	2MK at 75°C	O_2 Cat	I	CQT (600 mV)	100
	2NS at RT	H_2An	TB	ND	97
	0.05NS	Redox	—	ND	102
	20% Na	Decomp of Na amalgam	PrS	CQL	103
		Corr.	HPr	CQT	122
$Cr_{23}C_6$ (Cr; Fe)$_{23}$ (C$_6$(10-33% Fe)	0.01 3NS at 25, 50 and 85°C				
Cr_7C_3	0.05NS	Redox	—	ND	102
Fe_2C	2NK at 75°C	O_2 Cat	Rods	CQL	101
	2NK at 75°C	O_2 Cat	I	CQT (770 mV)	100
Fe_3C	0.05NS	Redox	—	ND	102
HfC	2NK at 75°C	O_2 Cat	I	CQT (500 mV)	100
	2NK at 75°C	O_2 Cat	Rods	CQL	101
MoC	2NS at RT	H_2An	TB	CQL	97
	0.05NS	Redox	—	ND	102
Mo_2C	2NK at 75°C	O_2 Cat	I	CQL	100
	2NK at 75°C	O_2 Cat	Rods	CQT (200 mV)	101
	0.1NS 3NS	O_2 Cat	Rods	CQL	92
	6MCl 18NP at 80°C 5NK	Corr	—	CQL, OCV	107
	20% Na	Decomp of Na amalgam	PrS	CQL	103
	0.05NS	Redox	—	ND	102
NbC	2NK at 75°C	O_2 Cat	Disk	CQT (600 mV)	101
	2NK at 75°C	O_2 Cat	I	CQT (600 mV)	100

Table 3 (*Continued*).

Compound	Conditions[a]	Reaction studied[b]	Fabrication[c]	Corrosion information[d]	References
	1MS at RT	H_2 An	TB	CQL	104
	0.5MS	o.e.r. + redox	Formed on metal foil	CQT (1600 mV) NP > 1.6 V	124
	0.05NS	Redox	—	ND	102
NiC	0.1MS	O_2 Cat	Rods	ND	92
Ni_3C	Ac/Ac	CO An	TB	ND (0 → 1.2 V)	98
	Alkali	O_2 Cat	TB	CQT (stable 700 mV), 30 mA/cm^2)	125
SiB_4 C	2NS at 50°C	H_2 An	TB	ND	97
SiC	2NS at 50°C	H_2 An	TB	ND	97
	2NK at 75°C	O_2 Cat	Ingots	CQT	100
	85% P at 150°C	Corr	Powder	CQL, OCV	115
	10^{-3}MFe$(CN)_6^{4-}$	Redox	Rod	ND	126
	0.05NS	Redox	—	ND	102
TaC	1MS	H_2 An	TB	CQL	104
	10^{-3}MFe$(CN)_6^{4-}$	Redox	Rod	ND pH 10.0 → 1.0	126
	2NK at 75°C	O_2 Cat	I	CQT (630 mV)	100
	2NK at 75°C	O_2 Cat	Disk	CQT (700 mV)	101
	0.5MS	o.e.r. + redox	Formed on metal foil	CQT (1800 mV) NP > 1.8 V	124
	5NS at 30°C	H_2 An	Formed on on gold-plated Pt screen	ND	72
	0.5MS	Redox	Formed on metal foil	CQL	127
	85% P	Corr	Powder	CQL, OCV	115
	0.05NS	Redox	—	ND	102
	INS	h.e.r. + redox	Rods	ND	71
TaC/NbC	1MS	H_2 An	TB	ND	104
TiC	2NK at 75°C	O_2 Cat	I	CQT (800 mV)	100
	2NK at 75°C	O_2 Cat	TB	CQT (800 mV)	99
	14.6MP at 145°C	H_2 An CO An	Pr	ND (0–700 mV)	128
	2NS at RT	H_2 An	TB	ND	104
	3NS at 95°C 85%P at 140°C	Fuel cell	TB	ND	129
	0.5MS	o.e.r. + redox	Formed on metal foil	Not determinable (P)	124
	1NS	h.e.r. + redox	Rods	ND	71
	20% Na	Decomp of Na amalgam	PrS	CQL	103
	0.05NS	Redox	—	ND	102
	S	Corr	Disk	CQT (800 mV)	130
	1N Na	H_2 An	—	CQL	131
	3NS 3NCl } 20°C 18NP	Corr	—	CQL, OCV	107
TiC/WC	2NS at RT	H_2 An	TB	ND	104
VC	0.1MS	O_2 Cat	Rods	CQT	92

Table 3 (*Continued*).

Compound	Conditions[a]	Reaction studied[b]	Fabrication[c]	Corrosion information[d]	References
	0.05NS	Redox	—	ND	102
	2NS at RT	H_2An	TB	CQL	97
	2NK at 75°C	O_2 Cat	I	CQT (880 mV)	100
	2NK at 75°C	O_2 Cat	Disk	CQT (850 mV)	101
	3NS ⎫				
	6NCl ⎬ 20°C	Corr	—	CQL, OCV	107
	18NP ⎪				
	5NM ⎭				
WC	0.5MS	Redox	Formed on metal foil	CQT (>O_2 Evol. pot)	127
	0.1NNa 0.5MNa$_2$SO$_4$	Redox	Formed on metal foil	CQL	127
	2NS 2MP	Corr	Slug	CQT (450 mV An)	132
	0.2N Cerric sulfate + 2NS	Corr	Slug	CQL, OCV	132
	2NS	H_2An	S	ND	133
	NS	H_2 Ads	Powder	ND	134
	85%P at 150°C	Corr	Powder	CQL, OCV	115
	10^{-3}MFe (CN)$_6^{4-}$	Redox	Rod	ND	126
	2NK at 75°C	O_2 Cat	I	CQL	100
	2NS at RT	H_2An	TB	ND	104
	6NS at 80°C	Corr	—	CQL, OCV	107
	4NS at 70°C	H_2An	TB	CQL	135
	4NS at 70°C	H_2An	TB	CQL	136
	2NS at 75°C	CO and HCOOH: An aldehydes	—	ND	137
	3NS at 95°C 85%P at 140°C	Fuel cell	TB	ND	129
	2NS at 70°C	Corr	TB	CQT (P)	138
	2NS at 60°C	H_2An CH$_3$OH	TB, HPr	CQT (500 mV)	139
	0.5MS 0.1MNa + 0.5M Na$_2$SO$_4$	o.e.r. + redox	Formed on metal foil	CQT (1100 mV, 900 mV) (P)	124
	1NS ⎫ P ⎬ at 20°C Cl ⎭	H_2An	Powder	CQT (350 mV)	140
	3NS at 70°C	HCOH An	—	ND	141
	5NS at 30°C	H_2An	Powder on Au/Pt screen	ND	72
	2NS at 75°C	H_2An N$_2$H$_4$ An	ST	CQT (300 mV)	142
	3NS at 75°C	CH$_3$OH HCOOH An	S	CQT	143
	S	H_2An	S	ND	144
	14.7M P at 145°C		ST	CQT (600 mV)	145

Table 3 (*Continued*).

Compound	Conditions[a]	Reaction studied[b]	Fabrication[c]	Corrosion information[d]	References
	2NS at 70°C	H_2, NH_4 An	ST	CWT (300 mV)	146
	K	HCHO An	ST	CQT (500 mV)	146
	P at 90°C	CH_3CHO, CO, HCOOH An	ST	CQL	146
	0.05NS				
	0.05NS	Redox	—	ND	102
	0.05NS	Redox	—	ND	102
W_2C	2NS at RT	H_2An	TB	ND	104
	0.05NS	Redox	—	ND	102
80WC–20Co	2NK at 75°C	O_2 Cat	TB	CQT (100 mV)	99
	2NK at 75°C	O_2 Cat	I	CQL	100
$WTiC_2$	0.05NS	Redox	—	ND	102
ZrC	2NS at RT	H_2An	TB	ND	97
	3NS 6NCl at 20°C	Corr	—	CQL, OCV	107
	0.1NS	O_2 Cat	Rods	CQL	92
	2NK at 75°C	O_2 Cat	I	CQT (400 mV)	100
	2NK at 75°C	O_2 Cat	Disk	ND (0 → 400 mV)	101
	20% Na	Decomp of Na amalgam	PrS	CQL	103

C. Nitrides

Compound	Conditions[a]	Reaction studied[b]	Fabrication[c]	Corrosion information[d]	References
AlN	10^{-3}MFe $(CN)_6^{4-}$	Redox	Rods	ND	126
CrN	0.05NS	Redox	—	ND	102
Cr_2N	2NK at 75°C	O_2 Cat	I	CQT (400 mV)	100
	2NK at 75°C	O_2 Cat	TB	CQT (900 mV)	99
HfN	2NK at 75°C	O_2 Cat	I	CQT (500 mV)	100
	2NK at 75°C	O_2 Cat	TB	CQT	99
MoN	2NS at RT	H_2An	TB	CQL	97
NbN	2NK at 75°C	O_2 Cat	I	CQT (270 mV)	100
	2NK at 75°C	O_2 Cat	TB	CQT (300 mV)	99
Si_3N_4	2NS at 50°C	H_2An	TB	ND	97
	10^{-3}MFe $(CN)_6^{4-}$	Redox	Rods	ND	126
TaN	2NK at 75°C	O_2 Cat	TB	CQT (0 V)	99
	2NK at 75°C	O_2 Cat	I	CQT (100 mV)	100
	2NS at RT	H_2 An	TB	ND	104
TiN	2NK at 75°C	O_2 Cat	I	CQT (770 mV)	100
	2NK at 75°C	O_2 Cat	TB	CQL	99
	10^{-3}MFe $(CN)_6^{4-}$	Redox	Rods	ND	126
	2NS	H_2 overvoltage	S.		147
	NS	Redox	Formed on metal foil	CQL	127
	0.5MS	o.e.r.	Formed on metal foil	CQT (1500 mV) (P) 0.7–1.5 V	124
	0.05NS	Redox	—	ND	102
	20% Na	Decomp of Na amalgan	SPr	CQL	103
Ti_2BN	0.05NS	Redox	—	ND	102
VN	2NK at 75°C	O_2 Cat	TB	CQT (100 mV)	99
	2NK at 75°C	O_2 Cat	I	CQT (180 mV)	100
	0.05NS	Redox	—	ND	102

Table 3 (*Continued*).

Compound	Conditions[a]	Reaction studied[b]	Fabrication[c]	Corrosion information[d]	References
ZrN	2NK at 75°C	O_2 Cat	TB	CQT (200 mV)	99
	0.05NS	Redox	—	ND	102
	20% Na	Decomp of Na amalgam	SPr	CQL	103

D. Silicides

Compound	Conditions[a]	Reaction studied[b]	Fabrication[c]	Corrosion information[d]	References
$CoSi_2$	2NK at 50°C	H_2 An	TB	ND	97
	2NK at 75°C	O_2 Cat	TB	CQT (0 V)	99
	2MK at 75°C	O_2 Cat	I	CQL	100
	2NS at 50°C	H_2 An	TB	ND	97
$CoSi_2$ + + 1%RhSi	10^{-3}MFE $(CN)_6^{4-}$	Redox	Rods	ND pH1–10	126
$CrSi_2$	2MK at 75°C	O_2 Cat	T.B.	C.Qt (500 mV)	99
	2MK at 75°C	O_2 Cat	I	C.QL	100
	2NS at 50°C	H_2 An	T.B.	N.D.	97
	10^{-3}MFe $(CN)_6^{4-}$	Redox	Rods	ND pH1–pH10	126
	85%P at 150°C	Corr	Powder	CQL	115
$CrSi_2$ + 1%MnSi	10^{-3}MFe $(CN)_6^{4-}$	Redox	Rods	ND	126
Cr_3Si	2MK at 75°C	O_2 Cat	TB	CQT (90 mV)	99
	2MK at 75°C	O_2 Cat	I	CQL	100
	0.05NS	Redox	—	ND	102
$FeSi_2$	2NS at RT	H_2 An	TB	ND	104
Mn_3Si	0.05NS	Redox	—	ND	102
$MnSi_2$	10^{-3}MFe $(CN)_6^{4-}$	Redox	Rods	ND	126
	2MK at 75°C	O_2 Cat	TB	CQT (0 V)	99
	2MK at 75°C	O_2 Cat	I	CQL	100
$MoSi_2$	0.05NS	Redox	—	ND	102
	2NS at RT	H_2 An	TB	ND	104
	2NS at RT	H_2 An	S	ND	97
	2MK at 75°C	O_2 Cat	TB	CQT (0 V)	99
	2MK at 75°C	O_2 Cat	I	CQT (200 mV)	100
	85%P at 150°C	Corr	Powder	CQL, OCV	115
	0.1NS	O_2 Cat	Rods	CQT (0.3 V)	92
	10^{-3}MFe $(CN)_6^{4-}$	Redox	Rods	ND	126
$NbSi_2$	2NK at 75°C	O_2 Cat	TB	CQT (0 V)	99
	3NS 5MK } at 80°C	Corr	—	CQL, OCV	107
	2NK at 75°C	O_2 Cat	I	CQL	100
	2NS	H_2 An	TB	ND	104
NiSi	M. per Cl pH 0.04 + 0.1MNaClO$_4$ pH10.8	Corr (An) O_2 Cat	Rods	CQL	148
	0.1M per Cl at pH0.04	h.e.r.	Rods	CQL (H_2 Tafel Slopes)	149
	0.1M HClO$_4$ 0.1M Na + 1M NaClO$_4$ 6K + P buff pH 1	h.e.r. + o.e.r. O_2 Cat H_2 An	TB	CQL	150

Table 3 (*Continued*).

Compound	Conditions[a]	Reaction studied[b]	Fabrication[c]	Corrosion information[d]	References
TaSi$_2$	2NS at RT	H$_2$ An	TB	ND	104
	2NS at RT	H$_2$ An	S	ND	97
	2NK at 75°C	O$_2$ Cat	I	CQL	100
	2NK at 75°C	O$_2$ Cat	TB	CQT (0 V)	99
	85%P at 150 150°C	Corr	Powder	CQL, OCV	115
Ta$_5$Si$_3$	2NK at 75°C	O$_2$ Cat	I	CQL	100
		O$_2$ Cat	TB	CQT (0 V)	99
TiSi	0.05NS	Redox	—	ND	102
TiSi$_2$	2NK at 75°C	O$_2$ Cat	TB	CQT (0 V)	99
	2NK at 75°C	O$_2$ Cat	I	CQL	100
	85%P at 150°C	Corr	P	CQL, OCV	115
	1MS at 50°C	H$_2$ An CH$_3$OH	STB	CQT (Sl at 600 mV)	151
	2NS at RT	H$_2$ An	S	ND	97
	2NK at RT	H$_2$ An	TB	ND	104
Ti$_5$Si$_3$	2NK at 75°C	O$_2$ Cat	TB	CQT (0 V)	99
	2NK at 75°C	O$_2$ Cat	I	CQL	100
VSi$_2$	2NS at RT	H$_2$ An	S	ND	97
	2NS at RT	H$_2$ An	TB	CQL	104
	0.1NS	O$_2$ Cat	Rods	CQL	92
	10^{-3}MFe (CN)$_6^{4-}$	Redox	Rods	ND (pH 1–10)	126
	2NK at 75°C	O$_2$ Cat	I	CQL	100
	2NK at 75°C	O$_2$ Cat	TB	CQT (0 V)	99
WSi$_2$	S at 25°C	H$_2$ An CH$_3$OH	STB	CQT	151
	2NK at 75°C	O$_2$ Cat	I	CQT (0 V)	100
	2NK at 75°C	O$_2$ Cat	TB	CQT (0 V)	99
	2NS at RT	H$_2$ An	TB	ND	104
	2NS at RT	H$_2$ An	S	ND	97
	0.1NS	O$_2$ Cat	Rods	CQL	92
ZrSi$_2$	2NK at 75°C	O$_2$ Cat	I	CQL	100
	2NK at 75°C	O$_2$ Cat	TB	CQT (0 V)	99
	2NK at RT	H$_2$ An	TB	ND	104
	85%P at 150°C	Corr	Pr	CQL	115
E. Sulfides					
CdS	0.5MK$_2$SO$_4$	Det of charging curves	Crystal	ND	152
	0.5MK$_2$SO$_4$ + 0.1MK$_4$Fe (CN)$_6$	Redox	Crystal	ND	153
	1M KCl	Anodic behavior in light and dark		ND	154
CuS, Cu$_2$S, PbS, Ag$_2$S,	NKNO$_3$	emf	—	ND	155
Ni$_3$S$_2$	6NK at 75°C	CO An	TB	ND	112
Sb$_2$S$_3$	0.1N		—	ND	156
Cu$_2$S, PbS, Bi$_2$S$_3$, CoS MnS, Zns, FeS$_2$, Ag$_2$S, CuS, FeS,	NKCl	emf			

Table 3 (*Continued*).

Compound	Conditions[a]	Reaction studied[b]	Fabrication[c]	Corrosion information[d]	References
MoS	2NK at 60°C	CO, H_2 An	HPr + Thermoplastic + carbon	CQL	157, 158
MoS_2	0.05NS	Redox	—	ND	102
WS	2NS at 60°C	CO_2, H_2 An	HPr + Thermoplastic + carbon	CQL	157
ZnS	5N NaCl + NH_4Cl	Cat		CQT	159

F. Oxides

Compound	Conditions[a]	Reaction studied[b]	Fabrication[c]	Corrosion information[d]	References
Co-Fe oxide	5NK 25°C + H_2O_2	O_2 Cat	ST	CQT (500 mV)	160
$CoMoO_4$	5NS	CO An	TB	ND (s)	161
	KiNa, LiOH, S, P, Cl, N etc. up to 500°C, 1–50 atm	H_2 An	—	ND	162
$LaCoO_3$	5NK at 25°C + acid	O_2 Cat	Cer disk	CQL	163
	75% KAT 230°C	O_2 Cat	Freeze-dried	CQT	164
$LaCrO_3$	5NK at 25°C	O_2 Cat	Cer disk	ND	163
MoO_2	5NS	CO An	TB	ND (s)	161
Mixed perovskites		H_2 An		CQL	165
AB′ × B″ $(1 - x)O_3$, Ca 25 samples	Acid	O_2 Cat			
$PrCoO_3$	5NK at 25°C	O_2 Cat	Cer disk	ND	
Spinels, e.g. CoOFeO-Al_4O_6	5MK, 5MCl, 5MS	O_2 Cat	TB	CQL	166
Ta oxide	85% P at 150°C	Corr	Powder	CQL, OCV	115
K tantalates	Neutral	O_2 An	Single crystal	ND	167
"Tantalite" (MnFe) $(TaNO)_2O_6$	Oleic acid, pH 1–14			ND	168
Tellurides	Acid/alkali	Fuel cell	—	ND	169
Cd telluride	NNa		Single crystal	CQL	170
TiO_2 + oxides of Mo, Zn W or V	Sat NaCl/Cl_2 gas	Cl_2 evol	—	CQT(1.4 V) (TiO_2/V_2O_5)	171
W oxide	85% P at 150°C	Corr	Powder	CQL, OCV	115
Ferrites Cobalites Manganites Vanadites Chromates Molybdates	3NK at 60°C	H_2 An O_2 Cat	SNi	CQL	172

Table 3 (*Continued*).

Compound	Conditions[a]	Reaction studied[b]	Fabrication[c]	Corrosion information[d]	References
		G. Miscellaneous			
Lithium borosilicate glass	UNIV Buffer RANGE of pH	Strength of H bond in glass		CQL	173
		Coordination of B		CQL	174
CdF_2	IMLiClO$_4$				
ZnF_2	at RT			(P)	

2.1.1. Factors Affecting the Rate of Corrosion

The rate of corrosion of the materials in question is governed by two main factors: (1) thermodynamic stability of the compound and corrosion product, and (2) protective properties of subsequently formed corrosion layers. These will now be considered in turn.

2.1.1.1. Thermodynamic Stability of Compound and Corrosion Product. Several authors have attempted to demonstrate the link between such thermodynamic data as ΔG (formation) of the compound and the rate of corrosion or, better still, the actual ΔG value for the corrosion reaction itself. This approach is difficult to follow due to a lack of thermodynamic data and occasionally not knowing the reaction mechanism.

However, the work of Juza[176] has shown a link between ΔH of the first-row transition-metal nitrides and their chemical as well as their thermal resistance. Vijh[22] has related corrosion resistance of metals in anhydrous HF to the stability of the subsequently formed fluorides.

Juza[176] has shown that for the nitrides of the first-transition metals increasing atomic number was accompanied by decreasing corrosion resistance, whereas an interpretation of the data presented in this paper is as shown in Table 4. This pattern can be linked with the ΔH_f for these compounds as shown in Table 5. This approach neglects the variable ΔH_f of the corrosion product, and this is only a guide to the ΔG (corrosion) which is the true driving force.

2.1.1.2. Protective Properties of the Corrosion Layers. While for sample metals or alloys the protective phenomena known as "passivation" have been the subject of some study, few data relating to the properties of the corrosion layer itself have been assembled. There is a suggestion that the inclusion in a compound of small amounts of easily dis-

Table 4. Electrochemical Corrosion of the Nitride Series[a]

IIIA	IVA	VA	VIIA		VIII			IB	
Sc	Ti^{VG}	V^F	Cr^G	Mn	Fe	Co	Ni	Cu	Zn
Y	Zr^F	Nb^F	Mo^P	Tc	Ru	Rh	Pd	Ag	Cd
	Hf^F	Ta^F	W	Re	Os	Ir	Pt	Au	Hg

[a]VG = very good; G = good; F = fair; P = poor.

Table 5. Selected ΔH_f Values for Nitrides

Compound	ΔG	ΔH_f (kcal/mole)
ScN	–	–67.5
TcN	–72.52	–79.67
VN	–45.70	–51.88
γ-CrN	–	–29.01
Cr_2N	–24.2	–30.8
$Mn_{(2.5)}V$	–	–24.1
ϵ-Mn_4N		–30.3
γ-Fe_2N		–3.0
Ni_3W		–0.2
Cu_3W		+17.84

solvable metals which might "dope" a surface layer does accelerate corrosion, while several authors mention oxides (possibly hydrated) as being formed when carbides, for example, corrode.

Unlike many authors cited, we feel too much importance has been attached to the catalytic activity of the compounds here described. It is our belief that, whereas the activity of a substance such as $NaWO_3$ can be drastically enhanced by the addition of small amounts (1000 ppm of noble metal[78]), the corrosion behavior is far less dependent on such additions. In other words we suggest that the approach should be one of finding corrosion-resistant, electrically conducting substances and then setting about rendering them catalytically active by some form of "doping." If this view is accepted, the problem will be seen to be an analytical and preparative one. We have to explore methods for making these materials and determining their composition. X-ray methods are of course invaluable here. But it is important to be able to study the surface composition and the bond structure that exists there, not only in the "as-formed" state but also after, or preferably during, their exposure to the electrochemical environment. In this context, Raman spectroscopy offers a tool hitherto unavailable and this, it is felt, should be strongly pursued.

The fear is sometimes expressed that the fabrication of these materials may prove a barrier to their successful use. The drawbacks of magnetite electrodes as being bulky and subject to thermal shock have often been referred to. However, it has been known for some time that this compound can be formed on Ti surfaces,[177-179] and there is no reason why the same should not be true of other compounds. New techniques such as flame-spraying, plasma arc deposition, electron plating as well as modifications of nitriding and analogous processes must all be considered.

All these and many other reasons exist for a much more thorough examination of these compounds and others. At present we have no criterion for selecting the "best performers," and while concepts such as "strong lattice" are attractive, they too break down as in the series Pt, Ir, Ru, Os, where the latter compound appears totally useless in an electrochemical sense. Likewise the use of free energies of compound formation are far from being infallible. On the practical side, the electrical properties of amorphous compounds (which almost all of these are) vary enormously with conditions of fabrication such as temperature, atmosphere when pressed, powder size, etc., and the corrosion behavior too might be expected to follow along similar lines. These ideas must figure in future experimental work.

Other trends are also discernible. The use of glassy coatings on electrode surfaces is incompletely understood. Mortimer[180] has shown that comparatively thick layers of TiO_2 can be coated onto RuO_2 electrodes without greatly impeding their electrochemical activity, and Tseung[181] and others are pursuing similar ideas. Such coatings may serve to stabilize surfaces which would otherwise corrode or passivate. Indeed as Tseung[181] and others previously have pointed out, the key to success in this field lies in an achievement of a stable surface, whose composition will quite likely differ from that of the bulk material. For this reason we should be cautious in placing too much emphasis on bulk properties even though there are times when they can be convincingly correlated with catalytic activity.[181]

With fuel-cell studies quiescent, two fields—or should one say two needs—dominate electrocatalytic studies. These relate to chlorine manufacture and the electrowinning of metals. In the first of these, we have seen how the RuO_2 surfaces were modified to include larger and larger amounts of TiO_2 without suffering loss of performance. More recently, a patent[123] has appeared specifying further additions of tin oxide and reducing still more the amount of precious metal oxide needed. The studies are doubtless proceeding at a feverish pace in various industrial laboratories around the world, and the goal of course is the total or near-complete elimination of the noble metal. The electrowinning process demands an acid-resistant anode (the RuO_2 type is not completely stable in this medium), and again the emergence of various patents indicates that some degree of success is already being met here.

It is interesting to note that, after a decade of academic publications and theorizing in the field of electrocatalysis, we now have a clear and very exciting lead to follow from those working in an industrial environment. Apart from the compounds cited in the patents and this paper, one should make the general point that many classes of compounds, such as the carbides, nitrides, and hydrides or silicides have a surprisingly high anodic stability, and could be possibly used as anodes when few would have predicted this. In particular these remarks apply to the valve metals such as Ti. The past, albeit somewhat limited, successful application of silicon-rich metal phases (call them silicides if one wishes) seems to us to be another pointer to the future. Although a recent suggestion that Ta_2Si might be useful was somewhat trivial (in view of the stability of the metal itself, in its passive state), we feel that if conductivity could be induced in the siliceous phase which (as far as is known) protects these materials, a useful anode material could be created. Such conductivity might be achieved by various additions of precious metals or their oxides, titanium oxides, or even introduction of species such as argon, which have been found to promote conductivity in this class of material. A rich field awaits investigators in industry and academia alike.

3. Electrocatalysis by Dispersed Metals: Poisoning and Other Morphological Phenomena*

In Section 1 we have seen how, when the electrode material is a simple one such as a single-phase metal, the electrocatalytic properties can be predicted.[4-6] In Section 2 we have seen how a range of nonmetallic electrically conducting materials can be used as electrodes, although we have pointed out that they are mainly lacking in catalytic activity. In Section 3 we shall consider catalysis not from a bulk properties point of view as in Section 1, but on the atomic level. This immediately raises the following questions, many of which have too often been side-stepped in the past.

---

*With E. W. Brooman and M. Hayes.

The question of whether the electrode is homogeneous is often coupled with questions arising from the method of preparation of the metal, stress, preferred orientations arising from rolling, etc. On another level, the relative abundance of edge, corner, and face atoms is unknown, as are the number of defects. The catalytic effects of such as 100, 110, or 111 planes may or may not be equal. More simply expressed, we do not know whether all the exposed atoms of Pt, for example, on a Pt sheet, contribute equally to the overall rate of reaction, or whether a small percentage of "active sites" contributes overwhelmingly to the total observed rate. Protagonists can be found both for "geometric" and "electronic" factors in catalysis and the latter have to consider problems arising from changes in the electronic character of an electrocatalyst resulting from absorption during the experiment, either of reactants or products. This in turn leads to questions concerning the effect of poisoning—whether small amounts (in terms of surface fraction covered) of poisons exert an effect arithmetically proportional to their coverage or not. If such active sites exist, can they be characterized as edge, corner, or defect sites?

In recent years, electrochemists have begun to follow the example of the gas-phase chemists in preparing and evaluating electrodes which consist of an inert (although electrically conducting) matrix on which "islands" of catalytically active material such as Pt are dispersed. The question then arises as to the optimum size and shape of such islands and the effect that variation of these parameters has on the reaction rate.

Before considering what the published electrochemical data suggest, one would be wise to be reminded of the conclusions in the same field reached by the gas-phase catalysis workers. Their main finding is that the effectiveness of a particular dispersed-phase catalyst, such as supported Pt, varies according to the test reaction. They have grouped the more commonly used test reactions into two classes—"facile" and "demanding" reactions.[182] Facile reactions are insensitive to the method of preparation of the disperse phase, i.e., to the size and shape of the metal crystallites. Demanding reactions depend critically on these factors. Reviews of this have been carried out by Anderson[183] and Boudart *et al.*[182] The electrocatalytic studies we shall discuss below refer mainly to hydrogen evolution, methanol oxidation, and hydrogen oxidation; the latter reaction tends to be suspect insofar as the very low solubility of hydrogen leads to mass-transport control, whereby the catalytic effects disappear and are replaced instead by structural effects (in the case of porous electrodes) such as pore size, distribution, wettability, etc. In view of the very strong parallels which exist between electrocatalysis and gas-phase catalysis, therefore, one is left suspecting that one or more of the commonly used electrochemical "test reactions" may well be "demanding" but that we do not know which these might be. We shall, however, cite evidence to suggest that hydrogen evolution is a "facile" reaction.

3.1. The Characterization of Dispersed Electrocatalysts

An electrocatalyst requires preparation, electrochemical evaluation (using one or more test reactions), and characterization with respect to the surface area, the crystallographic data and, in the case of dispersed structures, the number, size, and size distribution of the crystallites of catalytically active material. The latter can be performed simultaneously with the electrochemical evaluation, by using duplicate samples, or following the electrochemical work using the same sample. This will be further considered below. The methods of characterization are, in the main, straightforward applications of techniques long used and accepted in gas-phase catalysis. In addition, the electrochemical nature of the system allows further purely electrochemical characterization to be performed.

Surface areas of samples are usually too small (10–100 cm^2) to allow the classical gas-phase techniques such as B.E.T. and its variations to be used, although Mortimer[184] has so characterized a comparatively large and rough ruthenium oxide electrode. Furthermore, based on physical adsorption, these methods give the total surface area rather than the area of catalytically active substance. Almost all the work cited here relies, therefore, on an electrochemical method for surface-area measurement, usually the amount of electrodeposited chemisorbed hydrogen or the amount anodically removable. From surface-area measurements, crystallite sizes can be calculated assuming each crystallite to be of the same size and (regular) geometric shape, e.g., a cube. The total number of exposed Pt atoms is derived from the coulombic charge of hydrogen deposited or removed.[185, 186] The above assumptions are supported by direct measurement using transmission electron microscopy.[185, 187]

Determination of particle-size distributions requires physical methods of which electron microscopy appears to be the most sensitive,[186] being able to resolve to 10 Å. Unfortunately the procedure has to be repeated many times for different parts of the electrode to obtain a statistically significant value for the size distribution.

3.2. Choice of Test Reactions

It is surprisingly difficult to select a satisfactory test reaction. Anodic reactions are dogged by problems arising from catalyst loss as a result of corrosion, while oxide growth is a process which continues (at exponentially decreasing rate) for many hours and can cause changes in surface area as well as possible obscuration of catalyst particles. Such growth additionally changes the electronic properties of the electrode, leading to loss of metallic character. This will have greater effect on small crystallites than on large ones.[188] Oxygen reduction suffers from mass-transport limitation except at small current densities or high stirring rates.

Grave doubts exist concerning the stability of electrodes with time. Thermal sintering of platinum on carbon was found by Bett et al.[187] to have surface diffusion of the platinum crystallites as the rate-determining step. Stonehart and Zucks[190] and Malachesky et al.[191] found that platinum-black electrodes in an electrochemical context sintered much more rapidly, and this was related to the anodic potential. Adsorbed CO or other species inhibited the process. However, supported catalysts may be more resistant to sintering than unsupported ones.[192] These findings further detract from anodic test reactions, especially at higher temperatures.

Hydrogen oxidation is such a fast reaction that diffusion control rapidly sets in and structural factors (porosity, etc.) obscure the catalytic ones.

Cathodic processes such as hydrogen evolution are the easiest and most commonly used. However, the appreciable solubility of hydrogen in many metals leads to doubts as to whether the sample under investigation is that originally prepared or a partial hydride with different electronic properties. A more serious problem is that the mechanism and, to some extent, the rate of the hydrogen evolution reaction depend critically on extreme cleanliness of solution. The accepted rate-determining step for this reaction on Pt displays a 30-mV Tafel slope and is mass-transport controlled in respect to the rate of escape of products. It is all too easy to find instead a 120-mV slope, indicating that the reaction has been to some degree "poisoned," although such data tend still to give a very similar value of i_0 to that found under ideal conditions. There are redox reactions, such as Ce^{3+} oxidations, which suffer less from this disadvantage. Finally there is a large range of possible organic test reactions, including methanol or (a simpler compound) formic acid oxidation. These reactions are characterized by formation over a period of hours[189] of a "poisoning" chemisorbate, and the resulting activity tends to reflect the ability of the electrode to free itself from these poisons as much as to promote the main faradaic reaction. As a result, interpretation is difficult.

3.3. The Effect of Impurities

The deliberate introduction of poisons to a system to gain information on reaction mechanisms has been practiced for a long time. It has become apparent that in many cases exceedingly small amounts of a substance can have a substantial effect.

Bockris and Conway[195] have shown that as little as 10^{-10} mole/liter of impurity could affect the hydrogen evolution reaction on nickel. More recently Taylor et al.[196] have reported the reduction in rate of oxidation of formic acid caused by chloride ion concentrations as low as 10^{-7} M. They found that 0.1 of a monolayer of chloride reduced the oxidation rate by 75%.

The usual method of purification of solutions is by pre-electrolysis between large-area noble-metal electrodes, usually platinum black. Unfortunately, it has been discovered that this can result in the dissolution of the anode metal with subsequent deposition on to the working electrode.[197-199] Aylward and Smith[198] found that 4×10^{-6} monolayers of platinum on smooth gold could be detected by the effect on the hydrogen evolution current at an overpotential of 10 mV.

Purification of solutions can be achieved by holding the pre-electrolysis electrode either at open circuit or at a low anodic potential.[197,206]

According to recent reports[201-205] carbon is a major impurity of platinum surfaces and is so difficult to remove that it may always be present. This happens even when the platinum has a high catalytic activity, according to Huff,[205] who also found that platinum surfaces are very quickly contaminated by trace amounts of hydrocarbons. His attempts to clean platinum electrodes electrochemically by potential cycling succeeded in removing from the surface silicon and calcium completely but only a small amount of the carbon. However, he did not quote the potentials used, so the use of more anodic potentials may be successful.

The use of activated charcoal to purify solutions has been practiced for a number of years.[206,207] Jenkins and Weedon[207] found large quantities of carbon on their lead electrodes after use in perchloric acid that had been treated with activated charcoal. However, doubts have been expressed as to the origins of the carbon (i.e., pick-up from the atmosphere, pump oil, etc.), and we regard the case "not proven" either way.

Since Will[26] reported the differences in the heights of the anodic hydrogen peaks at different platinum crystal faces, many workers have used this method to explain the change in shape of their potentiodynamic current–potential curves with time.[218-220] However, a certain amount of caution is necessary here because the shape of the curves is very sensitive to the presence of minute amounts of impurity.[208] It is affected by well-known poisons such as carbon monoxide and "reduced carbon dioxide"[209] and adsorbed metal atoms that are smaller than platinum, e.g., copper.[210] The adsorption of anions from solution[210-214] is known to have an effect, as are several of the possible methods of electrode pretreatment. Breiter[215] has described how hot concentrated nitric acid produced long-lasting changes in the shape of the hydrogen adsorption-desorption curves; many hours of potential cycling were needed to reach a stable curve. Mechanical abrasion also had long-lasting effects. Heating in a yellow natural-gas flame produced strong poisoning, presumably by deposits of carbon or hydrocarbons, that could only be removed by mechanical abrasion.

More recently Biegler[216] has confirmed the deleterious effect of nitric acid pre-treatment on the shape and stability of hydrogen adsorption–desorption curves. This is particularly important because many of the reported results have been based on electrodes pretreated with nitric acid.[217–218b] In fact Bett *et al.*[217] found that the curves for platinum-on-carbon electrodes were different from those for platinum-black electrodes, and they attributed this to a particle size effect. However, because the carbon-supported electrodes had first been immersed in concentrated nitric acid to facilitate wetting, whereas the platinum-black electrodes had not, the difference could be due to poisoning of the electrode. On the other hand, Kinoshita *et al.*[218a] found that boiling concentrated nitric acid produced well-defined peaks and went on to interpret, using Will's results, the changes in the crystallographic orientation of the platinum-catalyst surface.

3.4. Electrochemical Studies Relating to Dispersed Electrocatalysts

3.4.1. Effect of Particle Size

The atoms on a catalyst surface can be placed in one of three categories: edge atoms, corner atoms, and face atoms, the difference being one of coordination number and interatomic spacing. The relative proportion of each depends on the crystallite size and crystal structure.[190, 193, 219, 220]

According to Poltorak and Boronin,[219] crystallites of <7 Å are made up of corner atoms. As the crystallite size increases, the number of edge and face atoms increases, the edge predominating up to ~14 Å and then giving way to the face atoms which rapidly become the major type of surface atom at ~50 Å where they constitute ~90% of the surface.

Thus reactions on crystallites smaller than 20 Å should give information about the activity of atoms with low coordination numbers, whereas those on crystallites in the 20–50 Å range should make it possible to distinguish between the activity of edge and face atoms. These conclusions, of course, assume that the crystallites are ideally crystalline, which is by no means certain, and that there is a narrow crystallite size distribution. This latter point is particularly important, since it is possible for two dispersed catalysts to have the same average crystallite size but different specific activities.[194] Furthermore, as the particle size decreases, the amount of bulk metal diminishes and so will the bulk electronic properties, i.e., the particle will lose its metallic character.

3.4.2. Effect of Crystal Morphology

Single crystals of platinum have been studied by several workers with differing results. Will[26] was the first to report the results of hydrogen adsorption/desorption during potentiodynamic sweeps in 8 N H_2SO_4. He observed two peaks whose potentials did not change with crystallographic orientation, although the ratio of their heights did. He attributed one peak to the (100) plane and the other to the (110) plane. A small third peak, occurring between the other two and visible only on the anodic sweep, was thought to be due to desorption from the (111) plane. This of course meant that several planes were exposed on each sample and that they were in different proportions for each sample, as Will himself pointed out.

However, it took more than 200 excursions to +1.5 V before reproducible results were obtained. This may have been caused by the electrode pretreatment, which included polishing with alumina.[215]

Schuldiner *et al.*[221] investigated the effect of crystallographic orientation on hydrogen overvoltage at low values of overpotential: They found no dependence on orientation. However, their electrodes were treated with concentrated nitric acid, and the potential had to be cycled for several days before changes in area ceased. This was thought to be due to the removal of pieces of polythene that had inadvertently covered the surface while molten during the electrode preparation. It is also possible that the initial electropolishing at 700°C could have rearranged the surface: the X-ray check they made would only indicate the bulk orientation.

Bagotskii *et al.*[213] were able to obtain reproducible results in 1 N sulfuric acid without the need for extensive cycling. They cycled the potential between 0 and 0.4 V and found that the curves for hydrogen adsorption were very similar for platinum (100) and (111) single crystals and polycrystalline platinum. The weak adsorption peak for the (100) sample was slightly lower and the strong peak slightly higher than those of the polycrystalline sample and (111) sample.

Angerstein-Kozlowska *et al.*[210] found that the shape of the hydrogen adsorption-desorption curves in 0.01–1 M sulfuric acid was dependent on the crystallographic orientation of the electrode. However, a given peak is not itself characteristic of a particular plane. They found four peaks in the anodic I–V curve, each corresponding to a different state of hydrogen chemisorption. These multiple states of bonding of hydrogen therefore exist at each crystallographic plane and are in proportions determined by the geometry of the plane. Change in electrolyte concentration causes shifts in the relative heights of the peaks but not their potentials, i.e., the effect was not a change in the states but only the distribution of hydrogen in them.

The authors conclude that the answer is most probably to be found in the density-of-states function, which describes the number of electron states of a particular energy at the crystal surface. It is reasonable to expect the hydrogen to adsorb on the highest energy states first and for them to be affected by the number available at that energy. When these are filled, either by hydrogen or some other adsorbed species, states of the next available energy would have to be used, thereby producing a redistribution of the adsorbing atoms.

One of the crystals used by these workers was also characterized by LEED and found to have suffered only minor faceting after 6525 cycles between the potentials +0.06 and +1.35 V.

The Pt (100) plane was found by Bagotskii *et al.*[213] to be up to one order of magnitude more active than the other faces for the evolution of hydrogen, oxygen, and the oxidation of methanol. The apparent activation energy of each reaction on all faces was the same as that on a polycrystalline sample. However, these results are not supported by those of Biegler,[216] who was able to alter the hydrogen adsorptive properties of platinum, but not the specific activity for methanol oxidation by varying the electrode pretreatment. Because the I–V profile in the "hydrogen region" is so sensitive to impurities and poisons, it is possible that Biegler was in fact poisoning his electrodes, in the manner described earlier.

Different crystal faces gave different surface energies and so a change in crystal morphology or change in the distribution of faces could affect the overall surface energy of an electrode. This would affect the sorptive properties of an electrode and result in an observable electrocatalytic effect. However, no large changes in crystal face distribution with changing crystallite size have so far been found as determined by changes in the hydrogen adsorption/desorption peaks.[218, 220] It has been reported by Somorjai[201] that impurities adsorbed on the surface can alter the surface free energies and hence stabilize apparently less-stable faces. Although these LEED studies relate to gas-phase measurements, the consequences in electrochemical situations would be expected to be similar.

Not only can the surface structure be rearranged by the movement of surface atoms,[187,219,222-224] the crystallites themselves can undergo drastic changes.

It has been known for some time[225] that small particles in contact with a conducting solution are unstable with respect to larger ones and dissolve preferentially. Connolly et al.[226] found that supported Pt crystallites grew rapidly when in contact with a conducting solution, causing a loss of area of the Pt catalyst. This loss, which was far greater than that for thermal sintering, was explained by assuming that the small crystallites became anodic because of their higher surface energies and the resulting complex ions diffused to and recrystallized at the larger cathodic crystallities; no Pt was lost to the solution.

Stonehart and Zucks[190] and Kinoshita et al.[227] studied the loss of surface area of platinum-black electrodes in hot concentrated H_3PO_4 and found that the rate of sintering was dependent on the potential and that adsorbed carbon monoxide or other surface active impurities from the H_3PO_4 retarded the sintering rate. Thermal sintering of platinum on carbon was found by Bett[187] to have surface diffusion of the crystallites as the rate-determining step.

Potential cycling is known to produce dissolution of platinum[228] and changes in the surface area of platinum electrodes. Kinoshita et al.[227] found that for Pt-sheet electrodes slow triangular wave and square wave potential cycling produced different effects. In the former case, the Pt surface area increased as did the number of (111) planes. In the latter mode the Pt dissolved more slowly, the surface area did not increase, and the morphology changed to exclude the (111) orientation. The lower dissolution rate of Pt found for the high surface Pt electrodes is a result of the porous nature of the electrodes, the soluble Pt complex being held within the structure until the low potential region of the cycle causes it to deposit on the existing crystallites. This produces crystal growth and loss of surface area, with the largest crystals growing at the expense of the smallest. The morphology of the surface was not changed.

Potential cycling also produces an activated electrode.[219] Whether this is due to surface area increase, surface rearrangement, or impurity removal is still open to debate.[230,231]

Shibata[232] has reported that repeated anodization and reduction of the oxide of Pt leads to a platinized electrode. He found[233,234] that after activation of an electrode the life of the activated state depended on the extent of oxidation and suggested that a thin layer of unstable atoms was formed by the oxidation. These atoms were moved into new, less stable positions, and the decay process was their moving to more stable, less active positions. The time taken for this to happen would depend on the extent of displacement and the strength of the Pt–O bond in the film. The proponents of the poisoning school would doubtless argue that longer oxidation at higher anodic potentials produced a higher surface area electrode, capable of accepting a larger amount of diffusion controlled "poisons" from the solution before becoming inactive. However, Biegler[216] has attributed the activation of cold-worked Pt electrodes to the anodic dissolution of a structurally disturbed layer of surface Pt atoms of lower activity than a crystalline surface. Adsorbed oxygen has been observed to move Ni and W atoms[234] and LEED studies[223] have suggested that both small-scale (displacement) and large-scale (reconstruction) movements of surface atoms are possible on adsorption from the gas phase. Schmidt and

Luss[224] have reported that the crystal structure of Pt/Rh catalysts changed to mainly (100) planes on the adsorption of H_2S. Sulfur also has a similar effect on Ni and Ag, changing the (111) planes to (100).

3.4.3. Surface Defects

The catalytic and electrochemical properties of crystalline materials have for a long time been explained on the basis of active sites or centers[235] that are the result of surface defects in the crystal lattices.[236] Bagotskii et al.[213] have investigated the effect of surface defects on Pt electrodes and concluded that for bulk electrode materials crystal-structure defects did not act as especially active centers. One of their arguments is that the maximum number of defects to be expected after heavy cold working of a material[237] is only 10^{11}-10^{12} defects/cm^2. In annealed polycrystalline metals there are only about 10^7-10^8/cm^2. When compared with the density of surface atoms (approx 10^{15} cm^{-2}) the defect density is small, and so if they were the active sites, they would have to affect the rate of a reaction by many orders of magnitude to permit an observable change in electrocatalytic activity.

However, the LEED studies by Lang et al.[204] of chemisorbed gases on Pt indicate stepped sites to have a greater reactivity, with (111) planes being more reactive than (100). They concluded that this was the result of the differing atomic structures at the steps, there being more nearest neighbors than on a flat surface and hence more metal orbitals for an adsorbed molecule to interact with. They also concluded that the atomic structure of highly dispersed platinum is likely to resemble that of stepped surfaces.

The effect of crystal defects on electrochemical behavior of dispersed metals is not known. Bett et al.[217] produced Pt dispersions by various means, some of which were expected to give rise to crystallites containing many defects. However, they were unable to detect changes in the specific activity for oxygen reduction in 1 M sulfuric acid at 70°C.

The electrochemical sintering of fine particles is expected to occur at a significant rate[217] and, even at temperatures not far above ambient, surface and volume diffusion will be rapid enough to quickly reduce the number of surface defects to a very small value. Also, for dispersions of small crystallite size each crystallite has a higher probability of being a single crystal, and surface disruptions due to grain boundary emergence, for example, will also be reduced.

The foregoing data suggest that surface defects do not cause large changes in electrocatalytic activity.

3.4.4. Reactions

Relatively few electrochemical studies on dispersed catalysts have been reported. They concern the adsorption, oxidation, and/or reduction of hydrogen[198,213,238-253]; oxygen reduction[217,240,253-258]; oxide formation and reduction[218]; Fe^{2+}-Fe^{3+} exchange[259]; and for organic reactions the oxidation of methanol,[216,243,244,247,260] formic acid,[243,260] acetate ion,[262] formaldehyde,[263] and various hydrocarbons.[264]

When platinum was electrodeposited on to gold or pyrocarbon, Kanevskii et al. found[245,246] the rate of oxidation and reduction of hydrogen at a fixed overvoltage to increase sharply at low platinum coverages (and hence small crystallite size—less than 50 Å). However, even large crystallites (150 Å) had an activity greater than that of massive platinum. Because the behavior of gold–platinum alloys is directly dependent on their composition and is between that of the pure metals,[6,212] alloy formation can be ruled out.

Based on the fact that enhancement was seen going from massive platinum to 150 Å crystallite sizes as well as from 150 down to 50 Å, the authors concluded that low coordination atoms were certainly not the only source of activity enhancement. They later reported[247] that such crystallites were energetically equivalent to bulk platinum and so the enhanced activity must be a result of an increase in the number of active sites. Bagotskii et al.[243] attributed this effect to the platinum crystallites activating the adjacent substrate areas by changing their electronic properties. Bagotskii and coworkers have also found that a platinum electrode deposited on to oxidized refractory metals[242] exhibited the same dependence of the specific rate of hydrogen evolution on S_{rel} (where S_{rel} is the ratio of true platinum area to the geometric area of the substrate). Similar results with such differing substrates have now led them to consider that activation of the platinum by the substrate is more probable. They also observed that the adsorption of hydrogen depended on the voltage sweep rate, as it did for pyrographite supports.[245] This was attributed to diffusion of hydrogen on to the support. Evidence of the migration of adsorbed species has been reported by other authors.[211,251,265,266]

Platinum crystallites on smooth ruthenium were found by Tuseeva et al.[244] to behave like bulk platinum even at very low coverages.

Fleischmann and Grenness[248] have observed that during the electrodeposition of ruthenium on to vitreous carbon the ruthenium crystallites had to exceed 200 Å before hydrogen evolution was measurable. The authors themselves concluded that this figure was "surprisingly large" and that a background current due to reduction of solution impurities interferred.

Fleischmann et al.[250] have also found that during the electrodeposition of ruthenium on to mercury, hydrogen evolution took place only on the edges of the growth center; the surface of the fully formed film had low catalytic activity.

Giles et al.[249] later reported that the hydrogen evolution reaction on electrodeposits of platinum and ruthenium on carbon and mercury took place via a different mechanism for the two substrates. Platinum on mercury behaved differently from both pure platinum and pure mercury.

Hobbs and Tseung[251] investigated the oxidation of hydrogen on porous WO_3 electrodes that had been platinized to varying degrees in the range 15–250 monolayers. When the ratio of Pt:WO_3 particles was greater than unity an enhancement in the reaction rate was observed and they proposed a spillover mechanism to explain it.

Tungsten oxide electrodes have also been studied by Grenness et al.,[252] who added their platinum by ion implantation at 200 keV. They found an enhancement in the hydrogen evolution reaction only after several cathodic polarization sweeps. The activation was thought to be due either to a hydrogen–bronze formation or to reduction of the oxide, thereby exposing more platinum.

Voinov et al.[253] have implanted platinum into pyrocarbon but at much lower energies, viz., 1 keV. The hydrogen oxidation reaction rate in 1 M perchloric acid was found to increase as the platinum loading was increased from 1.8 to 4.1 monolayers. However, neither hydrogen adsorption/desorption peaks, nor any indication of platinum oxide formation could be obtained. They concluded that the platinum atoms, although catalytically active, behaved electrochemically in a manner different from the usual platinum surface, at least as far as adsorption and desorption of hydrogen and oxygen were concerned. Heating to 700°C completely removed the catalytic activity, presumably by loss of the platinum into the substrate.

Aylward and Smith[198] found that in 1 M sulfuric acid the electrodeposition of platinum on to the surface of a smooth gold electrode changed the potentiodynamic current–potential curves only slightly until the equivalent of 0.04 monolayers of platinum had been deposited. No adsorption of hydrogen could be seen until at least 4 monolayers of platinum had been deposited, although the effect of the presence of as little as 4×10^{-6} monolayers of platinum could be seen from the increase in the hydrogen evolution current at 10 mV and the increase in the peak resulting from the subsequent oxidation of this hydrogen later in the potential cycle. However, the adsorption peaks were still not quite the same as for bulk platinum even at 40 monolayers.

This difference in the behavior of hydrogen adsorption has also been noted by Bagotskii et al.[243] who found that the fraction of strongly bound hydrogen decreased as crystallite size decreased. Stonehart and Lundquist[238] have also reported that the specific rate constant for oxidation–deposition of hydrogen adsorbed on graphite-supported Pt was lowered as the crystallite size decreased. They attributed it to the loss of metallic properties with decreasing size.

Urisson et al.[240] found that the specific activity of Pt–C electrodes in 8.5 N KOH increased as the Pt loading decreased and that, although the substrates themselves were inactive towards hydrogen oxidation, different carbons exerted effects of varying magnitudes on the specific activity of the Pt crystallites. Similar effects have also been reported by Hillenbrand and Lacksonen,[267] who were able to minimize them by prior treatment of the carbon at 1000°C first in N_2 to clean the surface and then in CO_2 to reoxidize it.

The electroreduction of oxygen has been reported by several authors.[217, 240, 254, 255, 257] Bett et al.[217] varied the mean crystallite diameter of Pt supported on carbon between 30 and 400 Å, thereby changing surface densities of corner and edge atoms 200-fold and 30-fold, respectively. No difference in activity (when expressed in terms of real area of Pt) was observed, indicating that no type of atom is much more reactive than any other.

The dispersions of Pt on C with crystallite sizes less than 15 Å, that were prepared by Blurton et al.,[254] had a much lower activity for oxygen reduction than did Pt black in H_2SO_4. This was thought to be due to interaction between the crystallite and the substrate or loss of metallic character as discussed earlier.

However, it is also quite likely that the method of preparation (pyrolysis at 700–900°C of an ion-exchange resin after addition of the Pt) would have introduced carbon impurities on to the crystallites. Zeliger[255] found that the rate of oxygen reduction in 5 N H_2SO_4 was indifferent to Pt crystallite size in the range 50–100 Å. Bagotskii et al.[256] found that the specific activity increased as the Pt loading decreased, both on C[240] and on Ta,[256] but no crystallite sizes were given. Lundquist and Stonehart[218] have studied the effect of Pt crystallite size (carbon supported) on oxide formation and reduction in acid solutions. A definite relationship between the activity for reduction of oxide and the fraction of atoms at the crystallite surfaces was observed. They attribute the 7-fold change in coverages over the crystallite size range 30–200 Å to changes in the surface free energies of the crystallites as the ratio of surface atoms to bulk atoms changed.

Pyrocarbon was found by Voinov et al.[253] to be active in the reduction of oxygen in 1 M perchloric acid after implantation of silver or platinum. As the platinum loading was varied from 1.8 to 4.1 monolayers, the reaction rate increased.

Bockris and McHardy[257] have reported the effect of small, known amounts of platinum in sodium tungsten bronze on the reduction and evolution of oxygen. Their platinum was also present in the bulk of the sample and always had aluminum present with it. Addition of 400 ppm of platinum increased the rate of reduction by four orders of magnitude and approached that of bulk platinum. They concluded that a spillover of adsorbed intermediates must take place. Bagotskii et al.[243, 260] oxidized methanol on platinum electrodeposited on gold and pyrographite. The results were the same for both substrates, with a minimum in specific activity (mA/true cm^2) at a fixed potential at an S_{rel} figure of 0.3–0.5, where S_{rel} is the ratio of true area of Pt to the apparent area of the electrode. There was a sharp rise at lower S_{rel}, but crystallite sizes were not given. Since methanol oxidation involves three Pt atoms per methanol molecule, a decrease in methanol adsorption and oxidation rate is expected as the Pt crystallite size decreases. However, the sharp increase in rate constant at lower coverages implies that small amounts of Pt exert some sort of long-range activating action on their supports. Pt crystallites on $Ru^{(244)}$ had an activity which increased with decreasing S_{rel} but did not show a minimum.

Biegler[216] has predicted that the oxidation of methanol should have maximum activity on a well-ordered surface structure. He found[261] that a Teflon-bonded Pt-black electrode (crystallite size 150 Å) had the same specific activity as smooth Pt. Freshly platinized electrodes, however, were less active, but it was not known whether this was a result of surface disorder or a particle-size effect.

Bagotskii and co-workers[243–246] found that the effect of crystallite size on the specific activity for formic acid oxidation was similar to that of hydrogen, with a rapid increase at low Pt coverages on both pyrographite and gold. Here again Pt–support interactions were suggested as the explanation.

The specific activity for electro-oxidation of acetate[262] on Pt was found by Woods to be independent of the mode of preparation of the electrode over a 10^3 range of surface roughness. Similarly, invariant specific activities for the electro-oxidation of formaldehyde[263] and hydrocarbons[264] on platinized electrodes have been obtained.

The rate of Fe^{2+}-Fe^{3+} exchange at Pt on gold was found by Bagotskii et al.[259] to vary with S_{rel} in exactly the same way as, for example, hydrogen evolution or formic acid oxidation. Alloy formation would have produced intermediate properties and so was ruled out.

4. Conclusions

Dispersed metal catalysts are in general more active than smooth (bulk) metals when compared on a weight basis, as is expected. When a comparison is made based on true catalyst area, the activity is seen to depend on the reaction being considered. Chemisorbed reactants such as *atoms* of hydrogen and oxygen on platinum are sensitive to the crystallite size,[268] i.e., they are "demanding." *Molecules* of these elements are weakly adsorbed and the electrode reactions are not affected by crystallite size, i.e., they are "facile." The sensitive reactions are controlled by the fraction of catalyst atoms in the crystallite that are at the surface.

Since adsorption of a substance changes the electronic character of the crystallite, it is reasonable to expect the adsorption of the crystallite onto the substrate to have some effect, both on the adsorbate and on the substrate. This effect should be reduced as the ratio of "bulk" atoms in the crystallite to the number of metal atoms bonded to the substrate increases, i.e., as the crystallite size increases. The results of Bagotskii and Aylward and Smith indicate a beneficial interaction between the gold and pyrocarbon supports and platinum, although why such apparently widely differing supports should have the same effect is not clear. There also appears to be a critical size below which the activation of the catalyst increases rapidly. Unfortunately the sizes of the crystallites were not reported.

Other examples of interaction between the substrate and submonolayer coverages have been reported[218, 269-270] and also for coverages greatly in excess of a monolayer.[271, 272]

It is therefore clear that electronic factors are of importance in electrocatalysis but cannot be completely isolated from geometric factors.

It would be interesting to investigate the effect of a dispersion of single-catalyst atoms on the properties of a supporting material. Crystallites, spread over the surface of an electrode in small clusters, could only affect the support over restricted local areas, whereas a dispersion of atoms uniformly distributed across the surface of a suitable support could modify the properties on a more global basis and hence yield a more efficient electrocatalyst.

Ion implantation is a technique ideally suited to give a controlled dispersion of an electrocatalyst on an atomic basis. Work proceeding at this university is beginning to reveal the workings of this system and will be published in due course, showing the extent to which our simple model is valid.

It has been our intention in this paper to indicate, if only in outline, the beginnings of the subject, the ways in which it has moved forward, and some of the areas now of immediate interest. We believe, just as electrochemistry is generally conceded to be a complex subject, that electrocatalysis is more difficult yet. Using simple mercury electrodes, a consensus of opinion could be found, as to the "right" way of making a particular electrochemical measurement. We do not believe the same is true where studies on inhomogeneous electrode materials are concerned. We have tried to show some of the pitfalls into which many investigators (and we do not except ourselves) have fallen. All the old difficulties of solution cleanliness are still here, along with all the additional problems of surface and bulk composition, surface films, etc. We have strenuously tried to avoid any suggestion that this or that is the proper way to investigate the subject. But the difficulties are not insuperable, and we feel confident that scientists will increase their understanding of electrocatalysis and that inventors, both among the ranks of the academics and the industrialists, will unveil new and interesting electrode materials.

Who knows what future awaits us in the area of organic semiconducting electrodes, or what we can learn from the physiological processes in living animals, many of which are electrochemical or analogous to it? Every day, quite literally, we learn of new discoveries in this field or of a more successful return to older ideas. Tseung has shown that the performance, even of gas-evolving electrodes, can be improved by addition of hydrophobic substances such as polytetrafluoroethylene, while recent patents by De Nora refocus our attention on the silicides.

We shall make no predictions for the future. But at the same time, we should state this: The metal of choice, as shown again and again, is Ti, the cheapest of the "valve" metals. The interesting electrodes are mainly deposits on this metal. Those deposits, beginning with precious metals such as Pt, have drifted, as we have said, to oxides of Ru. From the 100 elements in the periodic table, a mere handful crop up again and again as means of modifying and perhaps ultimately replacing the precious metals and their oxides. They are Mn, Pb, Co, and Sn. This is the first thread of our forward look. Then we see the little understood role of N in the surface or subsurface layer. Early on, Henri Beer patented TiN as an electrode. Somehow, we feel—perhaps combining the electrical properties of this compound with those of titanium oxide (which in at least one form is electrically conducting)—we shall attain our goal of an acid-resisting precious-metal-free anode material. Intuitively, one feels close to success, much more so than those in search of fuel-cell anode and cathode materials, although these may benefit from the studies now in progress.

For all of us in this field it is an exciting time in what has always been one of chemistry's most exciting fields. The next five years will certainly show enormous progress.

Acknowledgments

The authors express their gratitude to the many friends and colleagues who have helped and advised them in this. In particular, they would mention Dr. M. Andrew (Shell Research), Drs. Sunderland and Roscoe (Electricity Council Laboratories, Capenhurst), Peter Hayfield (Imperial Metal Industries), Dr. A. Tseung (City University), Dr. Zirngiebel (B.A.S.F.), and Drs. Hampshire and Wakeman here at the University of Salford. Finally, they would like to set on record the crucial role of their university library in compiling this review. Without the support of the librarian (A. C. Bubb) in making it a superb electrochemical library, or the aid of the reader's advisers there, and none more so than Brenda Marsden, this would not have been started, never mind completed.

References

1. K. Beccu, *Chem. Ing. Tech.* (February 1974).
2. L. G. Austin, NASA Report SP120.
3. F. R. Fichter, *Electro-organische Chemie*, Steinkopf Verlag (Reprinted Salford University Bookshop).
4. A. T. Kuhn, C. J. Mortimer, G. C. Bond, and J. Lindley, *J. Electroanal. Chem. 34*, 1 (1972).
5. S. Trasatti, *J. Electroanal. Chem. 39*, 163 (1972).
6. E. W. Brooman and A. T. Kuhn, *J. Electroanal. Chem. 49*, 325 (1974).
7. R. J. Mannan, Ph.D. thesis, University of Pennsylvania (1967).
8. S. Schuldiner, *J. Electrochem. Soc. 115*, 362 (1968).
9. J. A. Harrison and Z. A. Khan, *J. Electroanal. Chem. 30*, 327 (1971).
10. J. O'M. Bockris and S. Srinivasan, *Electrochim. Acta 9*, 31 (1964).
11. V. L. Kheifets, B. S. Krasikov, and A. L. Rotinyan, *Elektrokhimya 6*, 916 (1970).
12. S. Schuldiner and C. M. Shepherd, *J. Electrochem. Soc. 115*, 916 (1968).
13. R. N. O'Brien and K. V. N. Rao, *J. Electrochem. Soc. 112*, 1245 (1965).
14. H. Kita, *J. Electrochem. Soc. 113*, 1095 (1966).
15. K. J. Vetter, *Electrochemical Kinetics*, Academic Press, New York (1967).
16. M. W. Breiter, *Electrochemical Processes in Fuel Cells*, Springer-Verlag, New York (1969).
17. F. G. Will and C. A. Knorr, *Z. Elektrochem. 64*, 258 (1960).
18. A. Ya. Gokhshtein, *Elektrokhimya 9*, 285 (1973).
19. I. Epelboin, P. Movel, and H. Takemati, *J. Electrochem. Soc. 118*, 1282 (1971).
20. M. Bonnemay and H. Kagan, *Electrochim. Acta 9*, 337 (1964).
21. G. M. Schmid, *Electrochim. Acta 12*, 449 (1967).
22. A. K. Vijh, *J. Electrochem. Soc. 118*, 1963 (1971).
23. V. Breger and E. Gileadi, *Electrochim. Acta 16*, 177 (1971).
24. A. A. Sutyagina, I. N. Golyanitskaya, and G. D. Vouchenko, *Electrokhimya 8*, 908 (1972).
25. Yu. F. Balybin and L. F. Kozin, *Vestn. Akad. Nauk Kaz. SSR 29*, 70 (1973).
26. F. G. Will, *J. Electrochem. Soc. 112*, 451 (1965).
27. T. C. Franklin and S. L. Cooke, Jr, *J. Electrochem. Soc. 107*, 556 (1960).
28. O. A. Petrii and T. Ya. Kolotyrkina, *Elektrokhimya 9*, 254 (1973).
29. S. Schuldiner, *J. Electrochem. Soc. 107*, 452 (1960).
30. A. N. Frumkin and E. A. Aikasjan, *Dokl. Akad. Nauk USSR 100*, 315 (1955).
31. M. P. Makowski, Ph.D. thesis, Western Reserve University, Cleveland, Ohio, (1964).
32. J. O'M. Bockris and S. Srinivasan, *Fuel Cells: Their Electrochemistry*, pp. 131–133, McGraw-Hill, New York, (1969).
33. A. T. Kuhn and M. Byrne, *Electrochim. Acta 16*, 391 (1971).
34. P. Ruetschi and P. Delahay, *J. Chem. Phys. 26*, 532 (1957); *23*, 556 (1955); P. Ruetschi and B. D. Cahan, *J. Electrochem. Soc. 104*, 406 (1957).
35. D. Galizziole and S. Trasatti, *J. Electroanal. Chem. 44*, 367 (1973).
36. L. Peter, private communication.

37. A. T. Kuhn, H. Wroblowa, and J. O'M. Bockris, *Trans. Faraday Soc. 63*, 1458 (1967).
38. M. Byrne, A. T. Kuhn, *J. Chem. Soc. Faraday Trans. 1 68*, 1898 (1972).
39. M. Byrne, A. T. Kuhn, and V. J. Whittle, *J. Chem. Soc. Faraday Trans. 1 69*, 787 (1973).
40. H. Dahms and J. O'M. Bockris, *J. Electrochem. Soc. 111*, 728 (1964).
41. J. W. Johnson, S. C. Lai, and W. J. James, *Electrochim. Acta 16*, 1763 (1971).
42. See discussion in J. O'M. Bockris and A. K. N. Reddy, *Modern Electrochemistry* Chapter 75 Plenum Press, New York (1970).
43. A. R. Blake, A. T. Kuhn, and G. Sunderland, *J. Chem. Soc. A* 3015 (1969).
44. C. Cwiklinksi and J. Perichon, *C. R. Acad. Sci. (Paris) Ser. C 272*, 1930 (1971).
45. H. D. Popp and F. P. Schultz, *Chem. Rev. 62*, 19 (1962).
46. A. T. Kuhn, in: *Electrochemistry of the Elements*, Vol. 4 (J. Bard Ed.), Marcel Dekker, New York (1975).
47. R. Parsons, *Surface Sci. 1*, 418 (1969); J. O'M. Bockris and A. K. N. Reddy, *Modern Electrochemistry*, p. 1164ff, Plenum Press, New York (1970).
48. G. C. Bond, *Catalysis by Metals*, Academic Press, New York (1954).
49. D. B. Sepa, A. Damjamovic, and J. O'M. Bockris, *Electrochim. Acta 12*, 746 (1967).
50. J. O'M. Bockris, *Proc. 3rd. Int. Fuel Cell Symp.*, Brussels (1969).
51. P. J. Boddy, *J. Electroanal. Chem. Interfacial Electrochem. 10*, 199 (1965).
52. T. Kuwana *et al.*, *Anal. Chem. 36*, 2023 (1964).
53. J. W. Stojek, *J. Am. Chem. Soc. 90*, 1353 (1968).
54. D. Galizzioli *et al.*, *J. Appl. Electrochem. 4* (1), 57 (1974).
55. A. T. Kuhn, *Industrial Electrochemical Processes*, Elsevier, Amsterdam (1971).
56. F. von Sturm, I.U.P.A.C. Additional Publication 24th Int. Cong., Vol. 5 (1973).
57. E. O. Brim *et al.*, *J. Am. Chem. Soc. 73*, 5427 (1951).
58. D. B. Sěpa *et al.*, *J. Electrochem. Soc. 119*, 1285 (1972).
59. D. B. Sěpa *et al.*, *J. Electroanal. Chem. Interfacial Electrochem.* (in press).
60. M. A Wechter *et al.*, *Anal. Chem. 44*, 850 (1972).
61. A. Weser and E. Pungnor, *Acta Chim. Acad. Sci. Hung 59*, 319 (1967).
62. M. V. Vojnovic *et al.*, *Croat. Chim. Acta 44*, 89 (1972).
63. M. V. Vojnovic and D. B. Sepa, *J. Chem. Phys. 51*, 5344 (1969).
64. B. Broyde, *J. Catal. 10*, 13 (1968).
65. B. Broyde, *Inorg. Chem. 8*, 1588 (1967).
66. L. E. Conroy and T. Yokokawa, *Inorg. Chem. 4*, 994 (1965).
67. Esso. Ecom., 03743-F. AD 823-886, *USGRDR*, 1969, Part 19, p. 84.
68. R. A. Bernoff and L. E. Conroy, *J. Am. Chem. Soc. 82*, 6261 (1960).
69. M. J. Sienko, *J. Am. Chem. Soc. 81*, 5556 (1959).
70. A. C. C. Tseung and B. S. Hobbs, *Platinum Met. Rev. 13*, 46 (1969).
71. R. J. Mannan, Ph.D. thesis. University of Pennsylvania (1967).
72. A. C. C. Tseung and B. S. Hobbs, *Platinum Met. Rev. 13*, 146 (1969).
73. G. H. Bouchard *et al.*, *Inorg. Chem. 6*, 1682 (1967).
74. C. V. Collins and W. O. Sterag, *J. Am. Chem. Soc. 88*, 3171 (1966).
75. A. Damjanovic *et al.*, *J. Res. Inst. Catal. (Hokkaido Univ.) 16* (1), 1 (1968).
76. P. G. Dickens and M. S. Whittingham, private communication (1975).
77. P. G. Dickens and M. S. Whittingham, *Trans. Faraday Soc. 61*, 1226 (1965).
78. B. S. Hobbs and A. C. C. Tseung, *J. Electrochem. Soc. 119*, 580 (1972).
79. B. S. Hobbs and A. C. C. Tseung, *J. Electrochem. Soc. 120*, 766 (1973).
80. B. S. Hobbs *et al.*, Nature *222*, 556 (1969).
81. J. McHardy and J. O'M. Bockris, *From Electrocatalysis to Fuel Cells* (G. Sandstede, ed.), p. 109, University of Washington Press (1972), Seattle and London.
82. A. R. Mackintosh, *J. Chem. Phys. 38*, 1991 (1963).
83. W. McNeill and L. E. Conroy, *J. Chem. Phys. 36*, 87 (1962).
84. L. W. Niedrach and H. I. Zelinger, *J. Electrochem. Soc. 116*, 152 (1969).
85. W. Ostertag, *Inorg. Chem. 5*, 758 (1966).
86. M. J. Sienko, *J. Am. Chem. Soc. 90*, 6568 (1968).
87. M. Voinov and H. Tannenberger, *From Electrocatalysis to Fuel Cells* (G. Sandstede, ed.), p. 621, University of Washington Press (1972), Seattle and London.
88. D. B. Sěpa *et al.*, presented at XVIII meeting of Serbian Chemical Society (January 1973).
89. Army Electronic Comm. AD 816-318, *USGRDR*, 1968, Part 16, p. 93.
90. D. B. Sěpa *et al.*, *Electrochim. Acta 12*, 746 (1967).

91. L. D. Ellerbech *et al.*, *J. Chem Phys. 31*, 298 (1961).
92. J. O'M. Bockris *et al.*, AD 693-256, *USGRDR*, 1969, Part 21, p. 86.
93. J. O'M. Bockris *et al.*, *J. Electroanal. Chem. Interfacial Chem. 18*, 349 (1968).
94. J. H. Fishman *et al.*, *Electrochim. Acta 14*, 1314 (1969).
95. Inspec. Class. B. J. Field. (Ed), Inst. Elect. Eng. (1973), London.
96. Electronic Properties of Materials D. L. Grigsby, D. H. Johnson, M. Nevberger, and S. J. Wells eds., 3 Vols., Plenum Press, N.Y. 1967.
97. J. R. Aylarard *et al.* AD 656-468, *USGRDR*, 1967, Part 19, p. 84.
98. U. S. Bureau of Mines Report PB 198-108, *USGRDR*, 1976, Part 10, p. 85.
99. Tyco Labs. 3rd Quart. Report N66-26759 (March 1966). *USGRDR*, 1966, Part 21, p. 81.
100. Tyco Labs. Final Report N69-10585 (1968), *USGRDR*, 1969, Part 03, p. 103.
101. Tyco Labs. 2nd Quart. Report N66-24550, *USGRDR*, 1966, Part 21, p. 80.
102. E. Pungor and A. Weser, *Acta Chim. Acad. Sci. Hung. 61*, 241 (1969).
103. V. P. Pancheshnaya *et al.*, *Poroshkovaya. Metall.* (*Sov. Powder Metall. Met. Org.*) 7, 21 (1967).
104. R. D. Armstrong *et al.*, *J. Electrochem. Soc. 118*, 568 (1971).
105. D. J. Curran and K. J. Fletcher, *Anal. Chem. 42*, 1663 (1970).
106. D. J. Curran and K. J. Fletcher, *Anal. Chem. 40*, 78 (1968).
107. H. Böhm and F. A. Pohl, *Wiss. Ber. AEG-Telefunken 41*, 46 (1968).
108. H. Jahnke, *Rev. Eng. Prim. 2*, 33 (1966).
109. S. G. Meibuhr, *Electrochim. Acta 12*, 1059 (1967).
110. AFCRL - 66 - 134 - Vol. 2, AD 638-363, *USGRDR*, 1966, Part 20, p. 118, R. Jasinski, *18th Ann. Proc. Power Sources Conf.*, Atlantic City (1964); R. Jasinski, *J. Electrochem. Tech. 3*, 40, 127 (1965).
111. R. Jasinski *et al.*, *Rev. Eng. Prim. 2*, 38 (1966).
112. AFCRL - 68 - 0669, AD 686-470, *USGRDR*, 1969, Part 12, p. 72.
113. R. Thacker, *Nature 206*, 186 (1965).
114. I. Lindholm, *Rev. Eng. Prim. 2*, 26 (1966).
115. Englehard 3rd Quart. Rep. AD 455-564 (1964).
116. T. R. Mueller and R. N. Adams, *Anal. Chim. Acta 25*, 482 (1961).
117. A. M. Hartley and D. Axelrod, *J. Electroanal. Chem. Interfacial Electrochem. 18*, 115 (1968).
118. E. Jackson and D. A. Pantony, *J. Appl. Electrochem. 2*, 353 (1972).
119. T. R. Mueller and R. N. Adams, *Anal. Chim. Acta 23*, 467 (1960).
120. T. R. Mueller, *Diss. Abs. 24*, 4952 (1964). Ph.D. thesis, University of Kansas (1963).
121. M. T. Sawyer and E. T. Seo, *J. Electroanal. Chem. Interfac. Electrochem. 3*, 410 (1962).
122. V. A. Suprunov *et al.*, *Sov. Powder Metall. Met Cer. 13*, 307 (1973).
123. O'Leary. U.S. Pats. 3855092, 3776834.
124. G. Bianchi *et al.*, *Z. Phys. Chim. 226*, 40 (1964).
125. Tyco Quart. Rep. N68-25891. *USGRDR*, 1968, Part 17, p. 96.
126. J. F. Alder and B. Fleet, *J. Electroanal. Chem. Interfacial Electrochem. 30*, 427 (1971).
127. F. Mazza and S. Trassatti, *J. Electrochem. Soc. 110*, 847 (1963).
128. W. T. Grubb and J. M. King (G.E.C.) AD 640-521.
129. J. D. Voorhies *et al.*, in: *Hydrocarbon Fuel Cell Technology* (B. S. Baker, ed.), p. 455 Academic Press, New York (1965).
130. R. D. Cowling and H. E. Hinterman, *J. Electrochem. Soc. 117*, 1447 (1970).
131. V. P. Pancheshnaya *et al.*, *Elektron Str. Fiz. Svoistva. Tverd. Teca. 1972* (2), 98–105 (Russ.) (1972).
132. J. D. Voorhies, *J. Electrochem. Soc. 119*, 219 (1972).
133. H. Böhm, *Electrochim. Acta 15*, 1273 (1970).
134. V. Palanker and D. V. Sokolskii, *Sov. Electrochem. 7*, 1193 (1971).
135. Fr. Pat. 2005-567 Battelle
136. Fr. Pat. 2005-568 Battelle
137. H. Binder *et al.*, *Angew. Chem. Int. Ed. 8*, 757 (1969).
138. H. Binder *et al.*, *Energy. Convers. 10*, 25 (1970).
139. Fr. Pat. 1,598,442.
140. D. V. Sokolskii *et al.*, *Sov. Electrochem. 8*, 1702 (1973).
141. Ger. Pat. 1903521 H. Binder.
142. H. Binder *et al.*, *Nature 224*, 1299 (1969).
143. D. Baresel *et al.*, *Angew. Chem. 83*, 213 (1971); *Angew. Chem. Int. Ed. 10*, 194 (1971).
144. H. Bohm and F. A. Pohl, 3rd Int. Symp. on Fuel Cells, p. 180, Brussels (1969).

145. W. T. Grubb and J. H. King G.E.C. Rep. No. 10 AD 649-895.
146. K. Van Benda *et al.*, in: *From Electrocatalysis to Fuel Cells* (G. Sanstede, ed.), University of Washington Press, p. 85 (1973), Seattle and London.
147. J. I. Vasilenko *et al.*, *Sov. Electrochem. 7*, 1566 (1971).
148. A. K. Huq *et al.*, AD 628-071, *USGRDR*, 1966, Part 7, p. 60; *J. Electrochem. Soc. 111*, 278 (1964).
149. A. K. Huq *et al.*, *J. Electrochem. Soc. 111*, 270 (1964).
150. Tyco Final Tech. Rep. AD 634-241 (1962), *USGRDR*, 1966, Part 15, p. 68.
151. Pratt and Whitney Inc. AD 651-194, *USGRDR*, 1967, Part 12, p. 78.
152. V. A. Tyagai, *Sov. Electrochem. 1*, 328 (1965).
153. V. A. Tyagai, *Sov. Electrochem. 1*, 335 (1965).
154. R. Al. Van Dem. Berghe *et al.*, *Ber. Bunsen ges. Phys. Chem. 77*, 289 (1973).
155. W. Noddack and P. Wraetz. *Z. Electrochem. 59*, 76 (1955).
156. W. Noddack and P. Wraetz. *Z. Electrochem. 59*, 752 (1955).
157. H. Bohm, *Nature 227*, 483 (1970).
158. H. Bohm *et al.*, *Energy Convers. 10*, 119 (1970).
159. P. Ganesan *et al.*, *Trans. Indian Inst. Met. 25*, 20 (1973).
160. J. R. Goldstein and A. C. C. Tseung, *J. Phys. Chem. 76*, 3646 (1972).
161. L. W. Niedrach and I. B. Weinstock, *Electrochem. Tech. 3*, 270 (1965).
162. C. E. Thompson, U.S. Pat. 3116169 (ESSO).
163. D. B. Meadowcroft, *Nature 226*, 847 (1970).
164. A. C. C. Tseung and H. L. Bevan, *J. Electroanal. Chem. Interface Electrochem. 45*, 429 (1973).
165. Esso Semi Ann. Rp. AD472-900, *USGRDR*, 1968, Part 15, p. 90.
166. Econ. 3354, AD715-707, *USGRDR*, 1971, Part 03, p. 90.
167. P. J. Boody *et al.*, *Electrochim. Acta 13*, 1311 (1968).
168. S. I. Polkin *et al.*, *Russ. Metall. Min. 1963* (2), 113 (1963).
169. Ger. Pat. 213 6394.
170. P. P. Kononov *et al. Sov. Electrochem. 4*, 198 (1968).
171. Ger. Pat. 2044 260. J. K. O'Leary.
172. G. Feuillade, *Rev. Eng. Prim. 2*, 49 (1966).
173. V. S. Bobrov *et al.*, *Sov. Electrochem. 2*, 269 (1966).
174. V. S. Bobrov *et al.*, *Sov. Electrochem. 2*, 389 (1966).
175. A.F.C.R.L. 72-0605, AD757-883, *USGRDR*, 1973, Part 10, p. 106.
176. R. Juza, *Adv. Inorg. Chem. Radiochem. 9*, 98 (1966).
177. U.S. Pat. 3,491,014 (1970).
178. Ger. Pat. 1,964,999 (1971).
179. Fr. Pat. 2,009,337 (1970).
180. C. J. Mortimer, Private communication (1975).
181. A. C. C. Tseung, Private Communication (1975).
182. M. Boudart, A. Aldag, J. E. Berson, N. A. Dougharty, and C. G. Hopkins, *J. Catal. 6*, 92 (1966).
183. J. R. Anderson, Adv. Catal. *23*, 1 (1973).
184. C. J. Mortimer, Ph.D. thesis, Salford University (1973).
185. General Electric Co., Direct Energy Operation, Tech. Summ. Rep. 6, 1964 AD 612 766, p. 123 (1964).
186. A. D. O. Cinneide and J. K. A. Clarke, *Catal. Rev. 7*, 213 (1972).
187. J. A. Bett, K. Kinoshita, and P. Stonehart, *J. Catal. 35*, 307 (1974).
188. O. M. Poltorak and V. S. Boromin, *Russ. J. Phys. Chem. 40*, 1436 (1966).
189. S. Schuldiner and B. J. Piersma, *J. Phys. Chem. 74*, 2823 (1970).
190. P. Stonehart and P. A. Zucks, *Electrochim. Acta 17*, 2333 (1972).
191. P. A. Malachesky, L. Leung, and H. Feng, Tyco Labs. Inc. Report C154, AD 742 263, p. 27 (1972).
192. J. H. Sinfelt, *Ann. Rev. Mater. Sci. 2*, 641 (1972).
193. R. Van Hardeveld and F. Hartog, *Adv. Catal. 22*, 75 (1972).
194. D. Luss, *J. Catal. 23*, 119 (1971).
195. J. O'M. Bockris and B. E. Conway, *Trans. Faraday Soc. 45*, 989 (1949).
196. A. H. Taylor, R. D. Pearce, and S. B. Brummer, *Trans. Faraday Soc. 66*, 2076 (1970).
197. P. Malachesky, R. Jasinskii, and B. Burrows, *J. Electrochem. Soc. 114*, 1104 (1967).
198. J. R. Aylward and S. W. Smith, U.S. Army Electronics Command Tech. Report ECOM 02205-F, AD 656 468 p. 33 (1967).

199. S. B. Brummer, *J. Electrochem. Soc. 112*, 633 (1965).
200. T. Biegler, D. A. J. Rand, and R. Woods, *J. Electroanal. Chem. 29*, 269 (1971).
201. G. A. Somorjai, *Catal. Rev. 7*, 87 (1973).
202. R. M. Lambert, W. H. Weinberg, C. M. Connie, and J. W. Linnet, *Surface Sci. 27*, 653 (1971).
203. W. H. Winberg, R. M. Lambert *et al.*, 5th Int. Congress on Catalysis.
204. B. Lang, R. W. Joyner, and G. A. Somorjai, *Surface Sci. 30*, 454 (1972).
205. J. R. Huff, in: *Proc. of Symposium on Electrocatalysis* (M. W. Breiter, ed.), publ. by Electrochemical Society, 305 (1974).
206. R. Parsons and F. G. R. Zobel, *Trans. Faraday Soc. 62*, 3511 (1966).
207. D. A. Jenkins and C. J. Weedon, *J. Electroanal. Chem. 31*, App. 13 (1971).
208. B. E. Conway, H. Angerstein-Kozlowska, W. B. A. Sharp, and E. E. Criddle, *Anal. Chem. 45*, 369 (1972).
209. P. Stonehart and G. Kohlmayer, *Electrochim. Acta 17*, 369 (1972).
210. H. Angerstein-Kozlowska, W. B. A. Sharp, and B. E. Conway, in: *Proc. of Symposium of Electrocatalysis* (M. W. Breiter, ed.), Electrochemical Society, 94 (1974).
211. P. Stonehart, *Electrochim. Acta 15*, 1853 (1970).
212. M. W. Breiter, *Electrochim. Acta 8*, 925, 973 (1963).
213. V. S. Bagotskii, Yu. B. Vassiljev, and I. I. Pysknograeva, *Electrochim. Acta 16*, 2141 (1971).
214. V. S. Bagotskii, Yu. B. Vassiljev, J. Weber, and J. N. Pirtskholava, *J. Electroanal. Chem. 27*, 31 (1970).
215. M. W. Breiter, *J. Electroanal. Chem. 8*, 230 (1964).
216. T. Biegler, *Aust. J. Chem. 26*, 2571 (1973).
217. J. Bett, J. Lundquist, E. Washington, and P. Stonehart, *Electrochim. Acta 18*, 343 (1973).
218. J. Lundquist and P. Stonehart, *Electrochim. Acta 18*, 349 (1973).
218a. K. Kinoshita, J. Lundquist, and P. Stonehart, *J. Electroanal. Chem. 48*, 157 (1973).
218b. K. Kinoshita, J. Lundquist, and P. Stonehart, *J. Catal. 31*, 325 (1973).
219. O. M. Poltorak and V. S. Boronin, *Russ. J. Phys. Chem. 40*, 1436 (1966).
220. G. C. Bond, *Proc. 4th Int. Cong. Catalysis* (II), Moscow (1968).
221. S. Schuldiner, M. Rosen, and D. R. Flinn, *J. Electrochem. Soc. 117*, 1251 (1970).
222. A. Masson, J. F. Melois, and R. Kern, *Surface Sci. 27*, 463 (1971).
223. J. W. May, *Adv. Catal. 21*, 151 (1970).
224a. L. D. Schmidt and D. Luss, *J. Catal. 22*, 269 (1971).
224b. R. Bouwman, G. F. M. Lippits, and W. M. H. Sachtler, *J. Catal. 25*, 350 (1972).
225. G. W. Greenwood, *Acta Metall. 4*, 243 (1956).
226. J. F. Connolly, R. J. Flannery, and B. L. Meyers, *J. Electrochem. Soc. 114*, 241 (1967).
227. K. Kinoshita, K. Routsis, J. A. Bett, and C. S. Brooks, *Electrochem. Acta 18*, 953 (1973).
228. D. A. J. Rand and R. Woods, *J. Electroanal. Chem. 35*, 209 (1972).
229. T. Biegler, *J. Electrochem. Soc. 114*, 1261 (1967).
230. G. E. Barker, *J. Electrochem. Soc. 113*, 1024 (1966).
231. T. Biegler, *J. Electrochem. Soc. 116*, 1131 (1969).
232. S. Shibata, *Electrochim. Acta 17*, 395 (1972).
233. S. Shibata, *Bull. Chem. Soc. Jpn. 36*, 525 (1963).
234. S. Shibata and M. P. Sumino, *Electrochim. Acta 16*, 1511 (1971).
235. H. S. Taylor, *Proc. R. Soc. London A108*, 105 (1925).
236. J. M. Thomas, *Adv. Catal. 19*, 293 (1969).
237. D. McClean, *Mechanical Properties of Metals*, John Wiley, New York (1962).
238. P. Stonehart and J. Lundquist, *Electrochim. Acta 18*, 907 (1973).
239. K. Kinoshita and P. Stonehart, *Electrochim. Acta 20*, 101 (1975).
240. N. A. Urisson, G. V. Shteinberg, and V. S. Bagotskii, *Sov. Electrochim. 9*, 1107 (1973).
241. W. Vogel, J. Lundquist, P. Ross, and P. Stonehart, *Electrochim. Acta 20*, 79 (1975).
242. T. V. Balshava, V. S. Bagotskii, A. M. Skundin, and L. G. Gindin, *Sov. Electrochem. 8*, 1558 (1972).
243. V. S. Bagotskii, L. S. Kanevskii, and V. Sh. Palanker, *Electrochim. Acta 18*, 473 (1973).
244. E. K. Tusseeva, A. M. Skundin, and V. S. Bagotskii, *Sov. Electrochem. 10*, 974 (1974).
245. L. S. Kanevskii, V. Sh. Palanker, and V. S. Bagotskii, *Sov. Electrochem. 6*, 262 (1970).
246. L. S. Kanevskii, V. Sh. Palanker, and V. S. Bagotskii, *Sov. Electrochem. 6*, 1799 (1970).
247. L. S. Kanevskii, V. V. Emel'yanenko, and V. S. Bagotskii, *Sov. Electrochem. 8*, 221 (1972).
248. M. Fleischmann and M. Grenness, *J. Chem. Soc., Faraday Trans. 1 68*, 2305 (1972).
249. R. D. Giles, J. A. Harrison, and H. R. Thirsk, *J. Electroanal. Chem. 20*, 47 (1969).

250. M. Fleischmann, J. Koryta, and H. R. Thirsk, *Trans. Faraday Soc. 63*, 1261 (1967).
251. B. S. Hobbs and A. C. Tseung, *J. Electrochem. Soc. 120*, 766 (1973).
252. M. Grenness, M. W. Thompson, and R. W. Cahn, *J. Appl. Electrochem. 4*, 211 (1974).
253. M. Voinov, D. Buhler, and H. Tannenberger, *Transactions of the Symposium on Electrocatalysis*, (M. W. Breiter, ed.), Electrochemical Society, 268 (1974).
254. K. F. Blurton, D. Greenberg, H. G. Oswin, and D. R. Rutt, *J. Electrochem Soc. 19*, 559 (1972).
255. H. I. Zeliger, *J. Electrochem. Soc. 114*, 144 (1967).
256. V. S. Bagotskii, T. V. Baloshova, L. G. Gindin, L. S. Kanevskii, and A. M. Skundin, *Sov. Electrochem. 10*, 429 (1974).
257. J. O'M. Bockris and J. F. McHardy, *J. Electrochem. Soc. 120*, 61 (1973).
258. A. C. C. Tseung and K. Dhara, *Electrochim. Acta 19*, 845 (1974).
259. V. V. Emel'yanenko, A. M. Skundin, V. S. Bagotskii, and E. N. Baibatyiov, *Sov. Electrochem. 8*, 728 (1972).
260. L. S. Kanevskii, I. I. Astakhov, V. Sh. Palanker, and V. S. Bagotskii, *Sov. Electrochem. 7*, 1284 (1971).
261. J. Biegler, *Aust. J. Chem. 26*, 2587 (1973).
262. R. Woods, *Electrochim. Acta 13*, 1967 (1968).
263. D. F. A. Koch, Electrochemical Society Spring Meeting, Dallas, Texas 1967, Extended Abstracts, Vol. 5 (1967).
264. R. Thacker, in: *Hydrocarbon Fuel Cell Technology* (S. Baker, ed.), Academic Press, New York and London (1965).
265. K. M. Sancier, *J. Catal. 20*, 106 (1971).
266. R. Lewis and R. Gomer, *Surface Sci. 17*, 333 (1969).
267. L. J. Hillenbrand and J. W. Lacksonen, *J. Electrochem. Soc. 112*, 245, 249 (1965).
268. P. Stonehart, K. Kinoshita, and J. A. Bett, *Proc. of Symposium on Electrocatalysis*, (M. W. Breiter, ed.), Electrochemical Society, 275 (1974).
269. G. M. Sabe, *Disc. Faraday Soc. 41*, 252 (1966).
270. W. F. Taylor, D. F. C. Yates, and J. H. Sinfelt, *J. Phys. Chem. 68*, 2962 (1964).
271. N. V. Korovin, A. G. Kicheev, and I. G. Schmachbova, *Proc. of Symposium on Electrocatalysis* (M. W. Breiter, ed.), Electrochemical Society, 156 (1974).
272. R. C. De Geiso and L. B. Rogers, *J. Electrochem. Soc. 106*, 433 (1959).

9

Looking Back on Electrode Kinetics

B. E. Conway

1. Introduction

Since 1945, the year to which we look back in this symposium, the growth of the field of electrode kinetics has been so extensive that a review of its progress in a short paper would be both presumptuous and inadequate. Instead, a personal view of some highlights will be presented here in relation to the parallel and connected growth of a number of areas of physical chemistry and chemical physics. Indeed, an important aspect of the contributions of Bockris has been the emphasis of this connection—the understanding of kinetics and mechanisms of electrode processes in relation to their physicochemical significance and to the consequences of various models examined from a chemical physics viewpoint.

A number of areas in which these relations have been developed in a profitable way for the growth of the subject of electrode kinetics are shown in Table 1. Developments of the subject in an historical perspective are illustrated in Table 2.

2. Solvation Effects in Electrode Kinetics

The relation between kinetics of ion discharge and the solvation behavior of ions was implicit in the theories of Gurney,[1] Horiuti and Polanyi,[2] and especially Butler,[3] who considered the activation energy for proton discharge in terms of a potential energy diagram for proton desolvation and one for H adsorption at the electrode metal (Figure 1). Figure 1 serves to illustrate the three principal energy constraints that can influence the activation energy, and hence the kinetics, of an electrode reaction: (1) electrode potential; (2) solvation energy of the reacting ion; and (3) adsorption energy of a discharged intermediate. High pressure is a further possible constraint.

In some of the earliest work by Bockris, a series of studies was made[4,5] of the effects of changing solvent (from water to other solvents, and in aqueous–nonaqueous mixtures) on the kinetics of the hydrogen-evolution reaction (h.e.r.). This was continued in work by Bockris and Ignatowicz[6] and by Bockris and Parsons,[7] who also considered[8] the basis of potential energy curve representation of the electrochemical energy barrier for proton discharge. Parallel work with Rosenberg[9] investigated the selective solvation of the proton in aqueous–nonaqueous solvent mixtures through conductance measurements which led[10] to an important advance in the understanding of the state of the proton in hydroxylic solvents and the mechanism of its migration (cf. reference 11).

Solvation effects enter into electrode kinetics in three distinguishable ways:

1. In determining, in part, the energy barrier for the discharge step of the electrochemical reaction;

B. E. Conway • Chemistry Department, University of Ottawa, Ottawa, Canada.

Table 1. Some of the Principal Areas in Electrode Kinetics Where Main Advances Have Been Made in the Past 30 Years

1. Relation of electrode kinetics to ionic solvation behavior.
2. Relation of kinetic parameters and behavior to electrode surface properties.
3. Relation of electrode kinetic behavior to properties and electric potential profile of the double layer.
4. Developments of models and theories of electron transfer in relation to structure and properties of solvated ions and their interactions with electrode surfaces.
5. Relation of electrode kinetic behavior to adsorption isotherms for reactants and intermediates.
6. Quantitative study of adsorbed intermediates on electrode surfaces in relation to kinetics and mechanisms of electrode reactions, especially in electrocatalysis.
7. Progress in electronic instrumentation.
8. Development of pulse and relaxation methods.
9. Solution of diffusion kinetic equations for complex reaction schemes under non-steady-state conditions.
10. Application and development of *in situ* optical methods for following the course of electrode processes (fast scanning spectroscopy, ATR spectroscopy, automatic ellipsometry, electric modulated reflectance spectroscopy) and hence their kinetics from a surface-chemistry viewpoint.
11. Applications of non-steady-state techniques and electrode-kinetic theory to problems of electrochemical phase growth processes.
12. Application of kinetic, ohmic, and diffusion polarization theory to processes in porous electrodes.

2. In determining, in a way related to the properties of the energy barrier, the molecular dynamics of electron transfer,[12-15] e.g., in a regular redox reaction such as $Fe_{aq}^{3+} + e \rightarrow Fe_{aq}^{2+}$ or $Fe(CN)_6^{3-} + e \rightarrow Fe(CN)_6^{4-}$, or in a process involving atom as well as electron transfer, (e.g., reference 12) as in the gas evolution reactions or in a phase deposition;

3. In determining the state of ions in the double layer, especially the state of specifically adsorbed ions where changes of hydration energy, e.g., are associated with chemisorption bonding and ion polarization effects at the electrode interface. Such effects must occur with, e.g., halide and Tl^+ adsorption at Hg and SO_4^{2-} adsorption[16] at Pt and Au, where distortion of the solvation envelope accompanies chemisorptive and electrostatic (e.g., image potential) binding, as discussed, for example, by Anderson and Bockris.[17] Also, the solvent determines the dipole surface-potential.

The relation between the exchange current, i_0, for the h.e.r. and the energy of solvation of the proton raised interesting problems in the mechanism of electron and proton transfer, which were later discussed in theoretical detail by Dogonadze *et al*.[12] in terms of a double adiabatic mechanism with distinguishable quantum effects associated with proton and electron transfer.

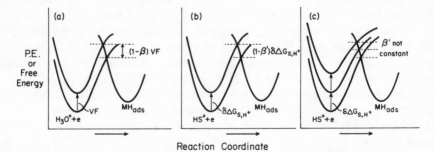

Fig. 1. Representation of Brønsted coefficients for electrochemical proton transfer as a function of (a) changing Fermi level; (b) changing free energy of solvation of the proton with constant effective force constant; and (c) changing solvation of the proton with solvent-dependent effective force constant.

**Table 2. Some Principal Advances in Electrode-Kinetic Work in Relation
to Those in Ionic Solvation**

Suhrmann and Breyer, Spectroscopic studies on hydration		1930 Erdey–Gruz and Volmer theory
		1931 Volmer, phase growth theories
		1931 Gurney–Fowler theory
Bernal and Fowler, Ionic hydration, state of proton	1933	
		1933 Frumkin and Slygin, charging curves
		1934 Bawn and Ogden, proton tunneling
		1935 Horiuti and Polanyi, p.e. curves and H adsorption
		1936 Butler, Theory of electron transfer and adsorption
		1937 Proskurnin and Frumkin, Double-layer a.c. studies
Eley and Evans, Ionic hydration	1938	
		1940 Hickling's potentiostat, a.c. modulation and reaction kinetics (Dolin and Ershler)
Frank and Evans, Solvent structure effects in ionic hydration	1945	
		1946 Solvation effects in h.e.r., Bockris, Faraday
		-1947 Soc. Discussion on electrode processes
		1947 A.c. impedance theory (Randles, Ershler)
		1947 Grahame's review on the double layer
Bockris, Primary hydration numbers	1949	
		1949 Bockris and Parsons, Bockris and Conway,
		-1956 PE curve treatment of various reactions
		1952 Delahay's treatments of diffusion-controlled fast processes
		1952 Bockris and Potter, Solvent orientation in kinetics
Samoilov, Dynamical theory of ionic solvation	1954	
		1955 Potential step treatments, Gerischer
		1956 Bockris, General treatment of consecutive reactions
		1956 Marcus–Levich theories of electron transfer
Faraday discussion on ionic interactions	1957	
		1958 Christov, Conway, Theory of proton
		-1959 tunneling at electrodes
		1955 Fleischmann and Thirsk, Treatment for
		-1959 growing phases
		1960 Will and Knorr, Cyclic voltammetry of Pt surface processes
		1960 Mott and Watts-Tobin solvent orientation model of electrode interfaces
		1961 Complex-plane analysis, Sluyters
		1963 BDM isotherm from solvent orientation
Walrafan, IR and Raman studies on ionic hydration	1966	
		1966 Experimental work on low T, H^+ tunneling, mixed tunneling + classical H^+ transfer, (Bockris and Matthews; Conway and Salomon)
		1966 Kinetic theory of sweep method -1974
		1966 Kinetic studies on pure surface processes -1974 H, OH, O, metal atoms
Bockris and Saluja, hydration numbers and solvation energy	1974	
		1966 Kinetic optical studies on electrodes, -1974 Bockris and Genshaw, Devanathan, Reddy, McIntyre, Bewick, Kolb.

On the theoretical side of electrode kinetics undoubtedly the most interesting but controversial topic has been the treatment of adiabatic electron transfer and reorganization energy in the various papers of Marcus,[13, 15] and Levich and his co-workers.[12, 14] Bockris and several co-workers[8, 51] had developed the Butler–Volmer theory in some detail in the early 1950s in attempts to evaluate relative energies of activation for charge- and atom-transfer processes using detailed models of reorganization of the ionic solvation shell. The introduction of Marcus's treatment for adiabetic electron transfer was based on an oversimplified view (Born model) of ion–solvent interaction and led to the view that long-range solvent polarization fluctuations were associated with the activation process. This view tended to direct attention away from specific ion–solvent interactions characteristic of the hydration shell. A consequence of this type of treatment (arising from harmonicity assumptions) was the prediction that ln i would be a *quadratic* function* of potential. While some indications that this might experimentally be the case have appeared, in those cases where the current–potential curve can be obtained[20] over a wide range of ln i, e.g., for the h.e.r., a *linear* relation is found. Nonlinear relations can also arise for quite different reasons connected with potential dependence of coverage by intermediates. It appears that models based on more detailed descriptions of the solvent/ solvate shell reorganization process, required to allow adiabatic electron transfer to occur, are preferable.

In this period, also, quantum mechanical *proton* tunneling processes were also considered, e.g., in proton conductance[10] and the h.e.r.[21, 23]

Generally, a dependence of ln i_0 on the free energy of solvation of H^+ will be expected[18] according to a Brønsted relation, analogous to that obtaining for homogeneous proton transfer; log i_0 will be expected to decrease with increasing (negative) free energy of solvation with a Brønsted coefficient β' similar to the β factor (symmetry coefficient) determining the dependence of a proton transfer rate constant on electrode potential. However, β' will not necessarily be constant for a series of solvents from which H^+ is discharged (Figure 1). The shape of the PE surface can change significantly with different solvents, S, because the PE profile for desolvation of H^+ is a complex one, depending on (1) the nature of the primary "onium" ion species and hence, the force constant for the H^+-S interaction and (2) the nature of the secondary solvation of the HS^+ ion (H_3O^+ in the case of water, $CH_3OH_2^+$ for methanol, etc.) and the dependence of its energy with state of deformation of the HS^+ ion in the double layer in the process of activation. For changing electric potential V, β is expected (cf. reference 2) to remain constant (as it does for the h.e.r. over a very wide range of potential; cf. reference 20) since changing electrode potential changes primarily the level of the Fermi surface ($\phi = \phi_0 \pm eV$) but does not significantly affect the force constant or configuration of the H^+/H_2O or H^+/S solvate complex.[21] Indirect effects of potential may, of course, arise due to the field dependence of solvent molecule orientation[22] in the double layer, which can affect the solvation of H^+ in the double layer. The experimental results,[20] however, indicate that such effects must be small. Recent calculations[12] indicate that perturbation effects must be allowed for in the sense that the form of the potential barrier itself changes in the activation and electron-transfer process. Also, delocalization of the proton over an $H_9O_4^+$ complex in the double layer (Figure 2) in a time scale of $> \sim 10^{-13}$ sec (cf. reference 10) has to be included in any realistic treatment of the event of electrochemical proton transfer and neutralization.

---

*It is interesting that quadratic effects in field in the double layer could also arise from electrostriction effects, since electrostrictive tension P is proportional to E^2.

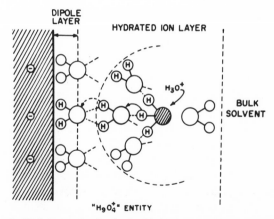

Fig. 2. $H_9O_4^+$ complex in the double layer at an electrode surface.

Contrary to the predictions of the Marcus–Levich theories of adiabatic electron transfer,[12–15] β is remarkably constant[20,23] with changing potential for the h.e.r. on metals where proton discharge is believed to be the rate-controlling step. In special cases, however, where the PE curves for ion discharge and chemisorption intersect near the zero-point level, β can become potential-dependent.[24]* This would occur for very high or very low activation energies;[24] in the ultimate limiting cases, these situations correspond to the conditions of so-called "barrierless" (i.e., when $\Delta H^{\ddagger}$ = the endothermic ΔH° of the reaction and $\beta = 1$ over a short potential range) or "activationless" ($\Delta H^{\ddagger} = 0$; limiting maximum current) discharge distinguished by Krishtalik.[25]

Solvation effects in electrode kinetics must also be examined in terms of the indirect effects which arise from solvent dependence of specific adsorption of ions and the relative hydration of various ions. During the past 30 years, much very detailed information on specific adsorption of ions and its related effects in electrode kinetics has accumulated, following the main confirmations of the Frumkin theory of double-layer effects in electrode kinetics by Frumkin's school itself,[30] and by Gierst[26] and others. Various isotherms for specifically adsorbed ions have been tested,[27] and model calculations for configurations of hydration of ions in the double layer have been quantitatively examined by Anderson and Bockris.[17] The interactions between adsorbed hydrated anions have been examined by Bockris *et al.*[22] as a basis for explaining the hump in the capacitance behavior of Hg which is difficult to attribute to reorientation of solvent molecules since, in aqueous medium at Hg, this reorientation appears to arise at approx. -2 to -3 $\mu C/cm^2$ rather than positive to the potential of zero charge as indicated by studies of the charge for maximum adsorption of the rigid (overall) nonpolar molecule pyrazine[28] and the charge for maximum entropy of the interphase.[29]

The order of strength of specific adsorption of anions at Hg and other metals follows approximately the inverse order of their free energies of hydration; it is also related to availability of lone-pair electrons and electronic polarizability. It is thus useful to think of specific adsorption as a donor–acceptor ligand interaction with the metal influenced by ionic solvation and solvent adsorption.

---

*In reactions with a complex sequence, the transfer coefficient α can become apparently potential-dependent when the isotherm for adsorption of an electroactive intermediate in the reaction sequence has a large valued term for the coverage dependence of the energy of adsorption of the intermediate. This effect is to be distinguished from that arising from any direct potential dependence of β.

In parallel with studies on ion-specific adsorption in electrode kinetics, much work has been published on the nature of ionic hydration in the past 30 years that has helped to provide a better understanding of the role of hydration in specific adsorption and in the molecular mechanics of ion discharge. Prior to 1945, the last important paper on ion hydration was that of Eley and Evans[31] [cf. Bernal and Fowler[11]], who made detailed molecular-model and statistical mechanical calculations of the heats and entropies of ion hydration; Everett and Coulson[32] made related calculations of ionic heat capacities. These treatments gave a good account of the structural features of ion hydration, at least insofar as the "primary hydration shell" was concerned (a useful term introduced by Bockris[33] in 1949). In 1945, Frank and Evans[34] [cf. Everett and Wynne-Jones[35]] published their now famous paper providing proof of the subtle ion-specific structural effects that arise in ionic hydration, corresponding to the so-called "structure-breaking" and "structure-making" effects in water. Comparative aspects are illustrated in Table 2.

These ideas were somewhat long in being appreciated by workers in electrode kinetics, but they are undoubtedly important in giving rise (at least in part) to ion specificity in double-layer effects in electrode kinetics and in accounting for the entropy properties of the double-layer which are determined by

1. The excess entropy associated with ion adsorption Γ_i
2. The entropy change associated with a changed state of hydration of ions in the double-layer in comparison with that in the bulk
3. Electrode-charge dependent orientation of water molecules at the electrode interface[29]
4. Changes of vibrational frequencies of oriented water molecules with charge on the electrode and with changing presence of other ions in the double layer
5. Possible changes of entropy of metal atoms in the metal surface, especially when specific adsorption corresponds to ligand interaction with surface metal atoms.
6. Gurney co-sphere overlap in the double-layer.[38]

Again, in recent years, the possibility of treating adsorption at metals in terms of ligand interactions in relation to the electronic configuration of the metal and the "ligand" adsorbate has been recognized.[36]

Two other aspects of recent evaluations of ionic hydration behavior in bulk water[37] which are relevant to the specific effects of adsorbed ions on kinetics of electrode processes are: (1) the role of hydration cosphere* interaction effects in the double layer [recently treated by Conway and Dhar[38]] and (2) the role of "hydrophobic interactions"[39] in the case of "hydrophobic" ions (R_4N^+) and molecules. The hydrophobic properties of large R_4N^+ ions have, for example, recently been used advantageously by Baizer[40] in developing the electrochemical hydrodimerization process for adiponitrile synthesis from acrylonitrile.

In the case of lateral interactions in the double layer which give rise to a falling value of $-\Delta G^\circ_{ads}$ with coverage, the hydration–cosphere overlap effect is a major one,[38] comparable with the direct coulombic repulsions, since lateral interionic distances in the double layer are comparable with those in 2-3 M bulk aqueous ionic solutions where the cosphere overlap effect is an important one in the thermodynamics of aqueous electrolytes.[37]

The state of solvation of specifically adsorbed ions and their binding to the surface

---

*The term "cosphere" for the sphere of significant influence of an ion on solvent properties was originally employed by Gurney. Cosphere interaction effects are now known to be a major factor in ionic interactions at moderate and high concentrations. They correspond to effects represented in Hückel's empirical linear concentration-dependent term in the extended Debye–Hückel theory.

affects the kinetics of their discharge. Generally, it will be more facile, since (1) some partial desolvation has already occurred on account of the specific adsorption; and (2) some partial electron transfer has arisen if the specific adsorption (contrast that of hydrophobic ions, R_4N^+) involves ligand–metal interactions as it probably does with halide (Cl^-, Br^-, I^-) and pseudohalide (CNS^-, CN^-) ion adsorption. In fact, the spectrochemical series for ligand field splitting effects, viz.:[41] $I^- < Br^- < Cl^- < F^- < OH^- \simeq C_2O_4^{2-} < H_2O < -NCS^- <$ pyridine $< NH_3 <$ En $<$ Dipy $< o$-Phen $< NO_2^-$, parallels quite closely but not exactly, the inverse order of effectiveness of anions in modifying the kinetics of cathodic reactions and the order of adsorbability (standard free energies of adsorption where known) of these anions at Hg. Also, interaction effects in the transition state due to adsorption of anions can modify the activation energy for an electrode process according to the treatment of Parsons.[42]

Specific adsorption almost certainly modifies the surface electronic properties of metals, especially those with partially vacant d bonds. This has two effects: (1) The work function is diminished. (2) The d hole character of the metal (when it is a transition metal) is changed so that the properties of the electrode as an adsorbent for stabilization of intermediates of the electrode process can be altered. Factor (1), however, will not normally change the rate constant since a parallel change of potential of zero charge arises and the effect of changed work function cancels out in the treatment of the kinetics.

3. Kinetics and the Nature of the Electrode Metal

Complementary to solvent effects in electrode kinetics is the dependence of the kinetics of a given process, e.g., the H_2 or O_2 evolution reactions, on the properties of the electrode metal. This is a central aspect of electrocatalysis. Butler,[3] in 1936, showed theoretically why the activation energy of the h.e.r. should be lower on Ni than on Hg, due to the higher adsorption energy of H on Ni than Hg. This provided the basis for a number of examinations of the relation, e.g., of i_0 values to metal properties such as the electronic work function ϕ.

Conway and Bockris[43] showed (cf. reference 44) that no direct relation between $\log i_0$ for the h.e.r. and ϕ values would be expected since ϕ terms cancel out in the measurement of electrode process rates at a given experimental electrode potential, e.g., the reversible potential, when i_0 is considered. The apparent dependence of $\log i_0$ on ϕ originates indirectly because of a linear dependence of initial heat of adsorption of H, $\Delta H_{ads,H}$ on ϕ for a number of metals. The dependence of $\log i_0$ on metal really originates in a primary way from the dependence of $\log i_0$ on $\Delta H_{ads,H}$. Again, for a given rate-determining mechanism, a Brønsted-type relation[18] will be expected between $\log i_0$ and $\Delta H_{ads,H}$, as with changing solvation energy of the proton.

The relations between $\log i_0$ and $\Delta H_{ads,H}$ are complicated because of the several possible rate-controlling mechanisms in the h.e.r.: one involving *deposition* of adsorbed H, the other two its *desorption*. For the first type of mechanism, $\log i_0$ will tend to increase with larger (negative) energy of adsorption, while for the second two, $\log i_0$ will tend to decrease with increasing energy of adsorption. The effects will be further complicated if coverage dependence of heat of adsorption of intermediates is taken into account. The general case was treated by Gerischer[46] and by Parsons.[45]

In the case of H chemisorbed on metals, the correct relation between work function and heat of adsorption is obtained when allowance for polarity in the M–H bond is made[43] using the square of the electronegativity difference $\chi_M - \chi_H$, according to Pauling's relation. Similar effects of χ values determine underpotential deposition of atomic metal monolayers.

More detailed examination of $\log i_0$, ϕ, and $\Delta H_{ads,H}$ relations by Trasatti[47] has indicated that several families of relations exist, depending on the specific chemisorption and hence orientation of water–solvent dipoles, and thus the surface potential of adsorbed water.

More recently, the energies for states of submonolayer deposition of atoms of various metals, e.g., Cu, Pb, Tl, Bi, on noble metals at potentials positive to the bulk reversible potential have been measured. Multiple states of chemisorption occur, as with H deposition. The energies are found[48] to depend on the work-function difference and hence electronegativity difference between the adsorbate and substrate metals. The situation with base metals deposited in monatomic layers on noble metals is thus quite analogous to that with H,[43] and similar polar bonding can arise.

The possibility that deposition of submonolayer quantities of atoms involves only partial transfer of the stoichiometric electron charge, ze per ion, has been the subject of recent controversy[49] in relation to partial-charge transfer in anion-specific adsorption. The half-widths of peaks resolved in the cyclic voltammetry i–V profile for underpotential deposition of metallic monolayers show [cf. the analysis of Conway et al.[50]] that any residual charge on metal atoms faradaically deposited on Pt or Au must be very small, otherwise large negative g factors in the adsorption isotherm would arise, giving[50] broad pseudocapacitance peaks that could not exhibit the experimentally observed resolutions. The question of the charge on electrodeposited adatoms or adions was treated by Conway and Bockris[51] in their theory of the elementary steps in metal deposition.

4. Electrode Kinetics and Surface Science

4.1. Adsorbed Intermediates and Kinetic Relations

The period on which our symposium is focused began at a time when the role of chemisorption of intermediates in electrode reactions and other heterogeneous catalytic reactions was becoming recognized as a major factor in electrode kinetics. Prior experimental work by Frumkin and Slygin[52] and others had directly indicated chemisorption of H at Pt by charging curves, and similar results were found for O species on Pt. Butler's theoretical demonstration of the importance of considering the energy of chemisorption of intermediates (H), produced in an electrochemical discharge reaction in treatments of the activation energy, took Gurney's theory to a point where it could more realistically be applied to many electrode processes involving multiple steps with chemisorbed intermediates.

Using Christiansen's method, Bockris[53] made a useful general analysis of the complex kinetic pathways in the anodic O_2 evolution reaction where OH and O chemisorbed intermediates are involved. This steady-state treatment provided a series of diagnostic criteria based on Tafel slopes and forms of the kinetic equations for various rate-limiting steps and consequent adsorption conditions. Despic and Jovanović[53] extended this type of approach to other complex reaction sequences. From these treatments, three conclusions emerge:

1. Limiting cases in the steady-state treatment must normally be considered if useful diagnostic criteria are to be deduced for a given complex reaction sequence.

2. When this is done, the conditions correspond to one step being clearly rate controlling and, as a consequence, the result is the same as that which can be deduced by the method which treats all steps prior to the rate-limiting step as at quasi-equilibrium; then the kinetic expression can be more easily written in terms of quasi-equilibrium constants for steps prior to the rate-determining one.

3. When a sequence of steps in a complex reaction sequence is examined in this way, the following limiting cases emerge: (a) Discharge steps of the type $X^- + M \rightarrow MX_{ads} + e$, $MX + X^- \rightarrow M + X_2$, etc., have Tafel slopes of the form $b = RT/(n + \beta)F$ where n is the number of the (rate-controlling) electron-transfer step beyond the first ($n = 0$) step and $\beta \doteq 0.5$. (b) Chemical recombination steps, e.g., $2\,OH_{ads} \rightarrow O_{ads} + H_2O$, $2\,O_{ads} \rightarrow O_2$, have slopes of the form $b = RT/2nF$ where the recombination step is the nth step following the formation of the chemisorbed intermediate in the initial discharge process.

Consideration of this type of problem occupied an important section of the subject of electrode kinetics for some years for it provided a basis for evaluation of electrode processes both in a fundamental and practical way. Thus, in practice, it is not just the exchange current i_0 that determines the reversibility of an electrode process but also the rate at which the activation polarization η increases with log i; low i_0 behavior can thus be traded against a low Tafel slope which would arise if a step "well down" a reaction sequence were rate controlling instead of the initial discharge step. However, if such a step *is* rate controlling, it is implicit that its standard rate constant is *less* than that for the initial step which would have the highest Tafel slope $RT/(n + \beta)F$ with $n = 0$.

With the developing interest in fuel cells that took place in the period, application of electrode kinetic principles to electrocatalytic reactions was made. The kinetic problems in this field are more complex due to (1) chemisorption of the fuel molecule; (2) kinetics of its dissociation on the catalyst; (3) kinetics of the electro-oxidation of the chemisorbed fragments in relation to steady-state coverage by the fragment; (4) complexity of interaction effects in the ad-layer and their effect on the kinetics of the various steps; and (5) codeposition of OH and O species on most anodes at which electrocatalytic oxidations are proceeding at elevated rates (potential region > 0.75 V E_H on Pt, for example).

In contrast to the behavior of most cathodic reactions and of the anodic gas-evolution reactions, most electrocatalytic oxidations can exhibit limiting currents due to the finite rate of the initial dissociative chemisorption and/or inhibition effects (passivation) due to potential-dependent buildup of coadsorbed O species or of other organic species (e.g., as in HCOOH or MeOH oxidation).

Also, intermediates can be formed which are appreciably adsorbed but are not the kinetically significant intermediates in the overall reaction, i.e., those which are involved in the steady-state sequence of steps in the main reaction path. This is a general problem in catalysis as well as electrocatalysis.

4.2. Adsorption Isotherms and Kinetic Behavior of Intermediates

A general problem of great interest in heterogeneous kinetics (and thus in electrode kinetics) of electrocatalytic processes is how the rate constants of various steps depend on coverage by chemisorbed intermediates. In early representations of the kinetics of electrode processes this factor was not taken into account.

In the quasi-equilibrium treatment of complex reaction sequences, a relation between coverage Θ_X of an intermediate (X) and electrode potential (V) and concentration (c) is automatically generated; it is equivalent to an electrochemical isotherm for the adsorption of X as a function of V and c. Its equilibrium constant K will not be dependent on Θ_X unless a coverage dependence of the rate constants for formation and removal of X has been introduced. For most chemisorption processes, however, K is Θ-dependent due to interactions and/or surface heterogeneity, and for electrode processes this type of dependence was treated by Temkin who gave a limiting logarithmic isotherm for an heterogeneous surface. A more generally applicable type of isotherm is that similar to Frumkin's which may be written

$$\Theta_X/(1 - \Theta_X) = \bar{K} \exp [-g\Theta_X] \cdot c \tag{1}$$

where g measures interaction effects or extent of dependence of free energy of adsorption on coverage in an assumed linear way. For intermediate values of Θ_X and for large g values (e.g., >10), equation (1) becomes equivalent to Temkin's isotherm (Θ_X proportional to $\ln c$). $\bar{K} \exp [-g\Theta_X]$ is seen to be a Θ-dependent equilibrium constant for chemisorption of X.

Normally, for electrochemical adsorption, where X is a species formed in a charge-transfer step,

$$\bar{K} = K \exp zVF/RT \tag{2}$$

so that

$$\Theta_X/(1 - \Theta_X) \cdot \exp [g\Theta_X] = K \exp zVF/RT \cdot c \tag{3}$$

When an electrochemical isotherm of type (3) applies, several interesting consequences arise* which are important in the electrode kinetics of the reaction in which X arises:

1. Θ_X varies with V over a range of V dependent on the magnitude and sign of g (normally g is positive for adsorption on a heterogeneous surface or for repulsive interactions). Correspondingly, any passivation effects due to X extend over a potential range determined by g.

2. $C_\phi = q_{\Theta_X=1}(d\Theta_X/dV)$ is a pseudocapacitance which exhibits a maximum in its dependence on V or Θ_X. Plotted with respect to V, its halfwidth is a measure of g [50]; $q_{\Theta_X=1}$ is the charge for a monolayer of X (220 μC/cm for H, 440 μC/cm for O on Pt). C_ϕ is given as $f(\Theta_C)$ by

$$C_\phi = \frac{q_{\Theta_X=1}F}{RT} \frac{\Theta_X (1 - \Theta_X)}{1 + g \Theta_X (1 - \Theta_X)} \tag{4}$$

3. If the free energy of adsorption does not vary in a linear way with Θ_X (e.g., it could depend on $\Theta_X^{1/2}$ or $\Theta_X^{3/2}$, etc.), C_ϕ is not symmetrical in Θ_X or V, so the *form* of the dependence on Θ_X or V can in principle (limitations are discussed below) be used to distinguish between a $\Theta^{1/2}$, Θ^1, or $\Theta^{3/2}$ dependence.

4. When g is appreciable, η–log i behavior for steps other than simple discharge shows less well-defined (cf. reference 53) limiting cases. There is a more gradual change of slope for a process such as $H^+ + e + M \rightleftharpoons MH_{ads}$ or $MH_{ads} + H^+ + e \rightarrow H_2$ with increasing V or Θ_H, and the limiting slopes of $RT/\beta F$ and $RT/(1 + \beta)F$ with $\Theta_H \sim 1$ or $\Theta_H \ll 1$, respectively, are only observed at extremes of potential or coverage depending on the value of g. Corresponding complexity in the e.m.f. decay behavior arises.

Experimental information on the behavior of absorbed intermediates must be obtained largely by means of *non-steady-state* electrochemical techniques and, more recently, from the application of optical techniques such as electrically modulated reflectance and ellipsometry[88] and from measurements of changes of surface conductance, a technique developed by Suhrmann in the case of gas/solid adsorption studies.

5. Kinetics of Electrode Processes in the Non-Steady State

The ability to modify the rate of an electrode reaction by change of potential provides one of the most convenient cases in the study of chemical kinetics for application of

*The occurrence of "Temkin" kinetics in many electrode processes involving chemisorbed species has been demonstrated in a number of papers by Russian workers, especially Bagotskii, Vasil'ev, and Frumkin.

relaxation techniques. In fact, the development of the so-called relaxation method, employed by Eigen in T-jump and P-jump techniques, originates in electrochemistry with the work of Dolin and Ershler[56] using a.c. modulation of the reaction rate. The latter technique was further worked out in papers by Ershler[57] and Randles and Somerton.[58] Recent developments by Rehback and Sluyters,[59] Armstrong,[60] and Epelboin[61] and coworkers have brought this method to a high degree of sophistication using the complex-plane method (based on Cole–Cole type of plots originating in the representation of the relaxation behavior of dielectrics).

Three principal types of modulation of potential with time, $V(t)$, have been employed:

1. Linear voltage sweep (or repetitively, in the method of cyclic voltammetry). This method originates in polarography, especially from the work of Sevčik.
2. Potential step (cf. Eigen's T- and P-jump methods)—e.g., see ref. 67.
3. Sinusoidal.

Corresponding periodic current modulation may be made and the conjugate potential changes observed (cf. the galvanostatic charging method).

In the non-steady-state methods, diffusion processes must often be taken into account, and much of the complex mathematical representation of non-steady-state electrode kinetics is associated with the solution of Fick's equations coupled with electrode kinetic rate relations for various boundary conditions and various forms of $V(t)$. For the sweep method, Nicholson and Shain,[62] and Savéant and Vianello,[63] for example, have worked out the behavior of various complex electrochemical + chemical (e.g., e,c; e,c,e; e,c,e,c; e,c,c,e; etc.) sequential steps in terms of the characteristic forms of $i-V(t)$ profiles which result under cyclic conditions. In much of this kind of work, solution-soluble species only (reactants and intermediates) were considered, adsorption being unfortunately neglected. In more recent work, e.g., by Laviron,[64] the role of adsorption (cf. reference 55) has been more properly recognized and of course has a major influence on the electrode kinetic results. Conway et al.[65] have examined the response of various surface processes to linear potential sweep modulation. (Here diffusion complications are absent, but the behavior depends in important ways on the isotherm for the electrodeposited species). For the study of faradaic surface processes involving adsorption up to a monolayer, steady-state techniques cannot be used and the non-steady-state methods, especially (1) and (3), have led to a great deal of detailed information on the kinetic and adsorption behavior of species on electrode surfaces (see Section 6).

The potential step method is especially useful as the relaxation kinetic equations which are developed are simpler than those for conditions (1) or (3). The method was treated in detail by Gerischer and co-workers.[67] It has the advantage of analogy to T- and P-jump procedures and, apart from the initial change of potential, the kinetics of the "relaxing" electrode process itself can be treated at *constant* potential. The method has also been applied to surface processes by Gilroy et al.[68] and by Urback.[69] Interesting applications have been made in the repetitive mode to the study of relaxation effects in organic electrode processes[70,74] (Kolbe, Crum–Brown–Walker, and Hofer–Moest reactions) and in H/D separation, especially with regard to the dependence of product-yield patterns on pulse height and repetition rate. Most of the effects observed in the Kolbe reaction arise, however, from relaxation effects in the formation of Pt surface oxide rather than from direct relaxation effects in the sequence of steps in the Kolbe reaction itself. This is demonstrated by the absence of such effects when the Kolbe reaction is performed in strictly anhydrous solutions where no anodically formed oxide film can be generated.

6. Relation to Experimental Behavior of Adsorbed Intermediates at Electrodes

The development of a new technique often brings qualitatively, as well as quantitatively, new information into a field. When fast electronic potentiostats were developed in the 1950s (Hickling had developed an instrument earlier, but its potentialities were not appreciated at that time) and applied by Will and Knorr[72] in 1959–1960 to the study of the Pt electrode surface with a cyclic regime of linear time-dependent potential change, the detailed form of C_ϕ [cf. equation (4)] as a function of V was clearly revealed for the electrochemisorption of H and O species at Pt. In later work, the adsorption of H and O species was studied at other metals, especially Rh, Ir, Au, and Pd.

The important general feature that is demonstrated by the application of cyclic voltammetry to surface processes is that *multiple* states of chemisorption arise *below a monolayer*[72] of electrochemisorbed species, even on well-prepared and adequately characterized single-crystal surfaces.[73] For such widely different chemical species as OH and O on Pt and Au; H on Pt, Rh, and Ir; Cu (as an underpotential deposited film) on Pt or Au; and Pb on Pt and Au, a number of states of chemisorption are well resolved at coverages below the monolayer limit. Not only can the states be resolved but the kinetics of their formation in electrodeposition and the kinetics of their electrochemical desorption can be followed accurately by measuring the peak potentials of C_ϕ for the states as a function of sweep rate. For a surface process, for which the peak current i_p is proportional to $C_{\phi,p}$ and s the sweep rate ($i_p = C_{\phi,p} \, dV/dt = C_{\phi,p} \, s$), a plot of the peak potential E_p vs. log s is equivalent to a Tafel line, so that the kinetic parameters of the surface process can be observed. Figures 3 and 4 show such plots for H desorption on Pt and Cu desorption from Pt. Examples of multiple states of chemisorbed H and Pb are shown in Figure 5.

The presence of multiple states of chemisorption at electrodes has interesting implications for the validity of many treatments of adsorption effects at electrodes. It is clear

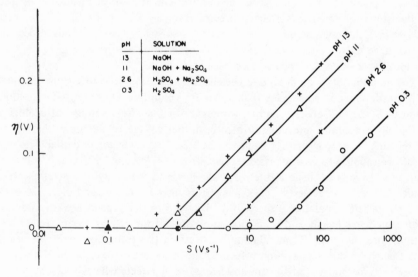

Fig. 3. Kinetic plots (peak potential vs. log [sweep rate]) for electrochemical deposition of H on Pt below the monolayer coverage limit.

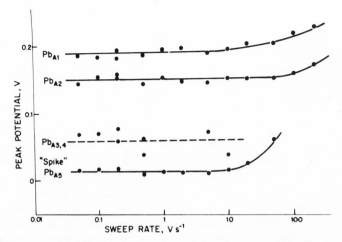

Fig. 4. As in Figure 3 but for electrochemical desorption of Pb from Au.

that a Langmuir or Temkin isotherm for electroactive chemisorbed intermediates is a seriously oversimplified representation.

Probably the multiple states arise by induced heterogeneity effects associated with successive formation of lattices[74] of chemisorbed atoms on the host lattice surface, occupying progressively more of the surface of the electrode. This was the basis of Kozlowska, Conway, and Sharp's[74] model for accounting for the multiple states of O chemisorption on Pt electrodes below the monolayer. Three states can be resolved below the monolayer (1 OH per Pt). The changing energy of the states arises from electronic interactions[75] and adsorbed dipole repulsions; the discrete states arise from the discrete geometries of successively filled lattice arrangements. Similar lattices are developed in adsorption from the gas phase and have been observed by means of LEED. Distinguishable states of chemisorption in various adsorbate/adsorbent systems have been observed[76,77] also by means of temperature-programmed desorption and by field-emission

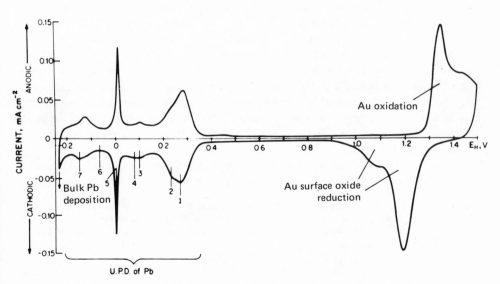

Fig. 5. Multiple states of chemisorption of Pb on Au.

energy distribution[77] (FEED) measurements. In the case of H on Pt electrodes in aqueous solution, and on dispersed Pt in contact with the gas phase, a range of energies of adsorption of 7.5 kcal/mole over three or four states has been observed. Theoretical explanations for the electronic effects must probably be sought in the fine structure of the electron density-of-states function for the metal surface.

At electrodes, the energies of the states of adsorbed H, and the relative coverages $\delta\Theta$ in the states ($\Sigma\delta\Theta = \Theta = 1$ at full coverage), vary with the presence of different anions, but in sufficiently dilute acid solutions (0.02 M H_2SO_4, $HClO_4$) the distribution of states of chemisorbed H is independent of the anions and of concentration. In very pure water solutions [pyrodistilled[78]], four* states of H chemisorption are always resolved at energies which are independent of pH; however, the relative occupancies of the states are different above and below pH 7. This is connected with anion adsorption and the potential of zero charge on the pH-dependent H_2/H^+ electrode scale (i.e., in alkaline solutions, the p.z.c. of Pt becomes negative to the reversible H_2 potential while in acid medium it is about +0.2 V E_H. Hence in the potential region for H adsorption in alkaline solutions, anion specific adsorption is small or absent).

While the states of adsorption of H and other species at Pt can be well resolved by cyclic voltammetry[72] or by electronic differentiation[79] of galvanostatic charging curves, the interesting question for electrode kinetics arises whether these states are those involved as intermediates in overall reactions, e.g., H_2 evolution at appropriate electrodes. The answer is probably in the negative; strongly bound chemisorbed species are the detectable species, but these are not necessarily those kinetically involved in the electrode reaction.

In the case of H_2, several pieces of evidence lead to this view:

1. At active Pt, a Tafel slope of $RT/2F$ is observed over about two decades of current density. Classical electrode kinetic treatments predict this slope if Θ_H is potential-dependent and $\ll 1$. This is not the case since $\Theta_H \to 1$ as the reversible H potential is reached from the positive side. Hence the H coverage, which varies with overpotential and is at much less than saturation coverage, cannot be the H measured in cyclic voltammetry or by charging curves.

2. In cyclic voltammetry at Pt in aqueous solution in the presence of dissolved or electrogenerated H_2, the i–V profile for H chemisorption and desorption is completely independent of the overall faradaic i–V profile for H_2 oxidation. It is difficult to see how the H observed in cyclic voltammetry experiments can have any connection with the H arising by dissociation of H_2 in the electrocatalytic oxidation of H_2.

3. In electrogeneration of H_2 at Pt, the faradaic H_2 current–potential profile (like that for H_2 oxidation) adds to the H deposition–desorption profile in a way dependent only on the partial pressure of H_2. The two processes appear to be independent, i.e., the H species which are intermediates in the cathodic H_2-evolution reaction may chemisorb on top of a layer of strongly chemisorbed H; it is the latter which is probably measured in cyclic voltammetry.

The same conclusions seem to apply to the strongly bound chemisorbed OH or O species that are electrodeposited on Pt some 0.6 V negative to the O_2-evolution potential.

---

*In early work, only *two* states (weak and strongly bound) of chemisorbed H were referred to. It is clear, however, that there are at least four distinguishable states below a monolayer and six or seven in the case of electrochemisorption of Pb on Au and four or five of Cu on Pt and Au. Models in which, e.g., H, is pictured with a positive and a negative surface dipole, depending on its location in the surface lattice of Pt, are therefore oversimplified. The multiple states of H chemisorption probably arise because of electronic interactions involving progressive occupancy of d-orbitals rather than successive 2-dimensional lattices as in adsorption of OH and O species.

7. Electrochemical Kinetics of Growth of Phases

Growth of new phases is an important type of process in electrochemistry. It arises in metal deposition, in corrosion of metals to form oxide layers, and in a number of battery processes where growth of insoluble films of oxide, sulfate, or halides commonly occurs in the charging or discharging of secondary batteries. Understanding of the kinetics of such processes has formed an important area of electrode kinetics.

While empirical knowledge on electrolytic growth of metals and oxides existed for many years, the basic treatment of phase growth in electrode processes commenced with the work of Volmer[80] on metal deposition in which the role of the development of nuclei of critical size was recognized. A minimum quasi-thermodynamic overpotential η_c is required to generate nuclei of an initial size since they have a more positive free energy than the bulk phase because of their large surface/volume ratio. Once formed, the phase growth process can continue at a less impeded rate through ion discharge and surface diffusion.

In the case of metal deposition, the role of nucleation and growth was demonstrated in the elegant experiments of Kaishew and Budewski,[81] who succeeded in observing the growth of Ag from *single* nuclei on a single crystal plane of Ag in a capillary.

The electrochemical processes of crystal growth are closely related to those from the gas phase and from ordinary supersaturated solutions where the extra chemical potential of supersaturation $\Delta\mu_s$ plays the same role as η_c in electrocrystallization. The elementary processes in nonelectrochemical crystal growth were considered by Burton and Cabrera.[82] Conway and Bockris[51] showed how similar elementary steps must be considered for the successive stages of metal electrodeposition onto surface planes, edges, and kink sites. The role of successive stages of desolvation at various types of sites in ion discharge and crystal building was emphasized. Preferred pathways in the crystal-building process were deduced by calculating relative values of activation energies. Bockris and Despić extended this type of treatment by quantitatively calculating the surface diffusion behavior of adions between growing steps.

Phase growth (of oxide films) in corrosion of metals was elegantly treated by U. R. Evans[83] in a classical paper in which the kinetics of randomly nucleated growing circles of oxide was treated. The mathematical development of this type of model was given also by Avrami.[84] Fleischmann and Thirsk[85] and other co-workers made elegant experimental and theoretical studies on the phase growth of calomel, based on the growing circle (or circular cylinder) and growing hemisphere models. Diagnostic kinetic equations were worked out for various growth geometries and the initial growth of the phase in successive layers could be detected by the periodicity in the transient current arising from response to a potential step. Similar periodicities were beautifully demonstrated in the experiments of Kaischew and Budewski[81] on Ag deposition at single-crystal, defect-free surfaces.

Barton and Bockris[86] made a quantitative electrode and crystal-growth kinetic analysis of the growth of dendrites. This area is of great interest mechanistically in electrochemical crystal growth and also practically in metal electrodeposition and charging of special secondary batteries involving, e.g., Zn or Li. The role of diffusion was emphasized and the theory elegantly treated[86] growth of the advancing dendrite tip in terms of activation overvoltage and spherical diffusion. Reddy[87] made experimental studies on dendrite growth from melts and also concluded that diffusion-controlled processes were critical. He also emphasized the role of nucleation centers along twin planes.

The possibilities of *in situ* observation of the kinetics of phase growth by chronoellipsometry were initiated by Devanathan *et al.*[88] In the case of calomel formation, a dissolution/precipitation mechanism was indicated from the induction period in the change of optical parameters of the surface in response to an electric current pulse.

References

1. R. W. Gurney, *Proc. R. Soc. London A134*, 137 (1931).
2. J. Horiuti and M. Polanyi, *Acta Physicochim. U.R.S.S. 2*, 505 (1935).
3. J. A. V. Butler, *Proc. R. Soc. London A157*, 423 (1936).
4. J. O'M. Bockris, *Trans. Faraday Soc. 44*, 519, 860 (1948).
5. J. O'M. Bockris, *Disc. Faraday Soc. 1*, 95, 229 (1947).
6. J. O'M. Bockris and S. Ignatowicz, *Trans. Faraday Soc. 44*, 519 (1948).
7. J. O'M. Bockris and R. Parsons, *Trans. Faraday Soc. 44*, 860 (1948).
8. R. Parsons and J. O'M. Bockris, *Trans. Faraday Soc. 47*, 916 (1951).
9. H. Rosenberg, Ph. D. thesis, London (1950).
10. B. E. Conway, J. O'M. Bockris, and H. Linton (née Rosenberg, cf. ref. 9), *J. Chem. Phys. 24*, 835 (1956).
11. J. D. Bernal and R. H. Fowler, *J. Chem. Phys. 1*, 515 (1933).
12. R. R. Dogonadze, A. M. Kuznetzov, and V. G. Levich, *Electrochim. Acta 13*, 1025 (1968); *Elektrokhimiya 3*, 280 (1967).
13. R. A. Marcus, *J. Chem. Phys. 24*, 966 (1956); *41*, 2624 (1964); *43*, 679 (1965); *37*, 1835 (1962).
14. V. G. Levich, *Adv. Electrochem. Electrochem. Eng. 4*, 249–371 (1965).
15. R. A. Marcus, *Annu. Rev. Phys. Chem. 15*, 155 (1964).
16. N. A. Balashova and V. E. Kazarinov, *Elektrokhimiya 1*, 512 (1965); I. I. Labkovskaya, V. T. Lukyanskaya, and V. S. Bagotskii, *Elektrokhimiya 5*, 580 (1969).
17. T. N. Anderson and J. O'M. Bockris, *Electrochim. Acta 9*, 347 (1964).
18. B. E. Conway and M. Salomon, *J. Chem. Educ. 44*, 554 (1967).
19. A. N. Frumkin, *Z. Phys. Chem. 160*, 116 (1934).
20. H. W. Nürnberg, *Fortschr. Chem. Forsch. 8*, 241 (1967).
21. M. Salomon and B. E. Conway, *Disc. Faraday Soc. 39*, 223, (footnote on p. 236) (1965).
22. J. O'M. Bockris, M. A. V. Devanathan, and K. Müller, *Proc. R. Soc. London, A274*, 55 (1963); R. Watts-Tobin, *Phil. Mag. 6*, 133 (1961); *8*, 333 (1963).
23. B. E. Conway, D. J. MacKinnon, and B. V. Tilak, *Trans. Faraday Soc. 66*, 1203 (1970).
24. A. R. Despič and J. O'M. Bockris, *J. Chem. Phys. 32*, 389 (1960).
25. L. K. Krishtalik, e.g., in *Elektrokhimiya 2*, 624 (1966).
26. L. Gierst, in: *Transactions of the Symposium on Electrode Processes*, Philadelphia, *1958*, E. Yeager, ed., The Electrochemical Society N.Y., p. 136. (1960).
27. R. Parsons, *Trans. Faraday Soc. 51*, 1518 (1955).
28. B. E. Conway and H. P. Dhar, *Croat. Chem. Acta 45*, 109 (1973).
29. B. E. Conway and L. G. M. Gordon, *J. Phys. Chem. 73*, 3609 (1969).
30. A. N. Frumkin, in: *Advances in Electrochemistry and Electrochemical Engineering* (P. Delahay and C. Tobias, eds.), Vols. 2, 3, Wiley-Interscience, New York (1962, 1963).
31. D. D. Eley and M. G. Evans, *Trans. Faraday Soc. 34*, 1093 (1938).
32. D. H. Everett and C. A. Coulson, *Trans. Faraday Soc. 36*, 633 (1940).
33. J. O'M. Bockris, *Q. Rev. Chem. Soc. London 3*, 173 (1949).
34. H. S. Frank and M. Evans, *J. Chem. Phys. 13*, 507 (1945); H. S. Frank and W. Y. Wen, *Disc. Faraday Soc. 24*, 133 (1957).
35. D. H. Everett and W. F. K. Wynne-Jones, *Proc. R. Soc. London A177*, 499 (1940); *Trans. Faraday Soc. 37*, 373 (1941).
36. A. Clark, *Theory of Adsorption and Catalysis*, Academic Press, New York, (1970).
37. P. S. Ramanathan, C. V. Krishnan, and H. L. Friedman, *J. Solution Chem. 1*, 237 (1972).
38. B. E. Conway and H. P. Dhar, *J. Colloid Interface Sci. 48*, 73 (1974).
39. G. Nemethy and H. A. Scheraga, *J. Chem. Phys. 36*, 3401 (1962); *J. Phys. Chem. 66*, 1773 (1962).
40. M. M. Baizer, *J. Electrochem. Soc. 111*, 215 (1964).
41. S. Petrucci, in: *Ionic Interactions* (S. Petrucci, ed.), p. 161, Academic Press, New York (1971).
42. R. Parsons, *J. Electroanal. Chem. 21*, 35 (1969).
43. B. E. Conway and J. O'M. Bockris, *J. Chem. Phys. 26*, 532 (1957).

44. P. Delahay and P. Ruetschi, *J. Chem. Phys. 23*, 195 (1955).
45. R. Parsons, *Trans. Faraday Soc. 54*, 1053 (1958).
46. H. Gerischer, *Z. Phys. Chem. N.F. 8*, 137 (1956).
47. S. Trasatti, *J. Electroanal. Chem. 28*, 257 (1970); *33*, 351 (1971); *39*, 163 (1972); *44*, 367 (1973).
48. H. Gerischer, D. M. Kolb, and M. Przanyski, *Surface Sci. 43*, 662 (1974); see also *J. Electroanal. Chem. 54*, 25 (1974).
49. J. W. Schultze, *Ber. Bunsenges. 74*, 705 (1970); see also K. J. Vetter and J. W. Schultze, *Ber. Bunsenges. 76*, 920, 927 (1972); *J. Electroanal. Chem. 44*, 63 (1973); *53*, 67 (1974).
50. B. E. Conway, E. Gileadi, and M. Dzieciuch, *Electrochim. Acta 8*, 143 (1963).
51. B. E. Conway and J. O'M. Bockris, *Electrochim. Acta 3*, 340 (1961).
52. A. N. Frumkin and M. Slygin, *Acta Physicochim., U.R.S.S. 3*, 791 (1935); *4*, 991 (1936).
53. J. O'M. Bockris, *J. Chem. Phys. 24*, 817 (1956); see also A. R. Despič and S. Jovanovič, *Doc. Chem. Yugoslav. 25*, 427 (1960).
54. J. O'M. Bockris and J. Kita, *J. Electrochem. Soc. 108*, 676 (1961).
55. B. E. Conway and E. Gileadi, *Trans. Faraday Soc. 58*, 2493 (1962).
56. P. Dolin and B. V. Ershler, *Acta Physicochim., U.R.S.S. 13*, 747 (1940).
57. B. V. Ershler, *Disc. Faraday Soc. 1*, 11 (1947).
58. J. E. B. Randles and R. Somerton, *Trans. Faraday Soc. 48*, 951 (1952).
59. M. Rehback and J. H. Sluyters, *Rec. Trav. Chim. 80*, 469 (1961); *81*, 301 (1962); *82*, 525, 535 (1963).
60. R. G. Armstrong, e.g., in *Disc. Faraday Div. Chem. Soc. 56*, 244 (1974).
61. I. E. Epelboin, *Disc. Faraday Div. Chem. Soc. 56*, 264 (1973).
62. R. S. Nicholson and P. Shain, *Anal. Chem. 36*, 706 (1964); *37*, 121 (1965).
63. J. M. Savéant and E. Vianello, *Electrochim. Acta 8*, 905 (1963); *10*, 905 (1965); *12*, 1545 (1967); see also E. Laviron, *J. Electroanal. Chem. 39*, 1 (1972).
64. E. Laviron, *J. Electroanal. Chem. 52*, 355 (1974); E. Laviron and A. Vallet, *J. Electroanal. Chem. 46*, 421 (1973).
65. P. Stonehart, H. A. Kozlowska, and B. E. Conway, *Proc. R. Soc. London A310*, 541 (1969).
66. S. Srinivasan and E. Gileadi, *Electrochim. Acta 11*, 321 (1966).
67. H. Gerischer and W. Mehl, *Z. Elektrochem. 59*, 1049 (1955).
68. D. Gilroy, B. E. Conway, and M. A. Sattar, *Electrochim. Acta 14*, 677, 711 (1969).
69. H. Urback, *J. Electrochem. Soc. 117*, 1500 (1970).
70. W. J. Lippincott and C. L. Wilson, *J. Am. Chem. Soc. 78*, 4290 (1956); *J. Electrochem. Soc. 103*, 672 (1956).
71. M. Fleischmann, J. R. Mansfield, and W. F. K. Wynne-Jones, *J. Electroanal. Chem. 10*, 522 (1965); *Electrochim. Acta 12*, 967 (1967); *Disc. Faraday Soc. 45*, 254 (1968).
72. F. G. Will and C. A. Knorr, *Z. Elektrochim. 64*, 258 (1960).
73. F. G. Will, *J. Electrochem. Soc. 112*, 451 (1966).
74. H. A. Kozlowska, W. B. A. Sharp, and B. E. Conway, *J. Electroanal. Chem. 43*, 9 (1973).
75. D. M. Newns, *Phys. Rev. 178*, 1123 (1969); *J. Chem. Phys. 50*, 4572 (1969); see also T. B. Grimley, *Adv. Catal. 12*, 1 (1960); *Proc. Phys. Soc. 90*, 751 (1967); *92*, 776 (1967).
76. L. D. Schmidt, *Catal. Rev., Sci. Eng. 9*(1), 115 (1974).
77. J. W. Gadzuk and E. W. Plummer, *Rev. Mod. Phys. 3*, 487 (1973).
78. B. E. Conway, H. A. Kozlowska, W. B. A. Sharp, and E. Criddle, *Anal. Chem. 45*, 1331 (1973).
79. H. A. Kozlowska and B. E. Conway, *J. Electroanal. Chem. 7*, 109 (1964).
80. M. Volmer, *Kinetik der Phasenbildung*, Akad. Verlag. Dresden (1939); *Z. Phys. Chem. A157*, 165 (1931).
81. R. Kaishew and E. Budewski, *Contemp. Phys. 8*, 89 (1967); *Electrochim. Acta 9*, 477 (1964).
82. W. K. Burton and N. Cabrera, *Disc. Faraday Soc. 5*, 33, 48 (1949).
83. U. R. Evans, *Trans. Faraday Soc. 41*, 365 (1945).
84. M. Avrami, *J. Chem. Phys. 7*, 1103 (1939).
85. M. Fleischmann and H. R. Thirsk, *J. Electrochem. Soc. 110*, 688 (1963); *Trans. Faraday Soc. 58*, 2200 (1962).
86. J. Barton and J. O'M. Bockris, *Proc. R. Soc. London A268*, 485 (1962).
87. T. B. Reddy, *J. Electrochem. Soc. 113*, 117 (1966).
88. M. A. V. Devanathan, J. O'M. Bockris, and A. K. Reddy, *Proc. R. Soc. London A279*, 327 (1964).

10

Electrode Processes—The Future

A. K. N. Reddy

1. Introduction

Technological forecasting[1] is attracting considerable attention these days, particularly since the smooth development of technology was so rudely interrupted by the energy crisis. One of the obvious techniques of forecasting is that of extrapolating current trends and hoping that unexpected events (like the energy crisis) will not radically change the course of development. Another interesting approach is the Delphi technique,[2] analogous to the Oracle of Delphi, which involves the gathering of "expert" opinions on future developments in a particular area of technology. Unfortunately, the whole exercise of technological forecasting is tormented by a basic problem—technology is only one ingredient in social change, and if the other crucial ingredients, politics and economics, have effects that are far more decisive, then predictions about technology must be subordinated to forecasting developments in politics and economics.

In forecasting for basic science, as distinct from technology, one is dealing with a relatively more autonomous process. If basic science is at all subject to social pressures, the medium is through technology. Thus, the science of electrode processes will respond to those needs of electrochemical technology in the areas of electrosynthesis, energy conversion, and corrosion, which are being spelled out by the other participants at this symposium. It is obvious, however, that inasmuch as energy costs have escalated dramatically, a major rethinking will take place with regard to energy-intensive electrolytic methods of production. It remains to be seen whether the new reckoning will not tilt the balance in favor of chemical methods. In contrast, all energy sources, such as the sun, wind, and tides, which display diurnal, seasonal, or random time variations, must be backed up with energy-storage devices; this need will generate a powerful demand for more efficient electrodes and better electrocatalysts. In turn, this demand will call for a deeper understanding of electrode processes. Likewise, metals, which will have to be looked upon as repositories of energy invested during the extraction processes, must be defended against corrosive attacks from the environment. These techniques of defense, elaborated elsewhere at this symposium, must include the use of surface-active substances; hence the mechanistic origins of adsorption will receive constant attention.

The technological determinants shaping the course of basic electrode process research are perhaps easier to visualize than the factors leading to the autonomous growth of the subject. Here, one has to adopt an approach analogous to the morphological analysis[3] technique of technological forecasting, i.e., one must delve into the structure of the subject, spotlight current dilemmas, and envisage advances which will ensure enhanced functional capability. In other words, one must extrapolate from current limitations in theory

A. K. N. Reddy • Department of Inorganic and Physical Chemistry, Indian Institute of Science, Bangalore 560012, India.

to future unifications, even though one cannot imagine the precise nature of the leaps of imagination required to achieve such syntheses.

However, the whole endeavor is fraught with the risk that the discovery of totally new phenomena will shatter the predictions. One must beware of the fate of Lord Kelvin who confidently proclaimed that all that remained for physics was measurement to the next decimal place. And then, in rapid succession came the discoveries of electrons, X-rays, and radioactivity and the formulation of the quantum and relativity theories. Forecasters have no choice, therefore, except to proceed with humility, and it is with this spirit that the following picture of the future of electrode processes is sketched.

2. Systems Approach

An electrochemical cell with electronic conductors in contact with ionic conductors must be viewed as a system[8]. Through the further device of a nonpolarizable electrode, change can be suppressed at one electrode/electrolyte interface, so that virtually all changes in the cell can be attributed to those at the other interface. Then, a single electrode/ electrolyte interface becomes the system of interest. Just as this system is a part of a larger system (viz., the cell), the interface itself consists of components, parts, or subsystems. To the extent that all systems in nature are in ceaseless change, one must view systems as processes. This implies that the activity of an electrode/electrolyte interface is made up of component activities or processes. Now, subsystems interact and component processes interlock—that is how systems survive and persist as systems. Thus, in order to understand the behavior of a system in response to external stimuli, one must identify the *components* which are assembled to form the system and acquire a knowledge of the *interaction* of these components. It is in this way that function and structure become correlated.

What the components of an interfacial process are depend on the level of analysis. For instance, on the phenomenological level, one may consider the following components of the interfacial system:

1. Charge separation, involving a relationship between the double-layer charging current, $i_{nf} = dq_m/dt$, and the changes in the potential E and the surface excesses, Γ_i

2. Adsorption–desorption which relates the Γ's to the activities c_i and the potential ΔE through isotherms

3. Charge transfer at the electrode which provides the connection between the faradaic current, i_f, the potential, ΔE, and the interfacial activities of species involved in the charge transfer reaction

4. Mass transfer with or without bulk sources/sinks, which links surface concentration and fluxes

5. Homogeneous chemical reaction (complexing, dissociation, etc.)

6. Space charge formation

7. Phase formation on electrode (anodic films, O_2 layer)

8. Diffusion in the electrode

9. Heterogeneous chemical reaction (recombination, disproportionation)

10. Surface-transport phenomena (surface diffusion, etc.)

These components of an interfacial process have been known for a long time, and most of them have received mathematical treatment as specific and special processes. They come, however, "not as single spies, but in battalions," for the components of an interface are invariably complex; one interface differs from another in the number of components which are involved in the coupling, i.e., in the complexity of the interactions.

The traditional path through this complexity has been to treat simple classes of coupling between the components and, for such simple cases, to unravel the interfacial response to perturbing stimuli. As examples, one may cite the Randles–Ershler expression[4,5] for the CD type of coupling, i.e., charge transfer/diffusion, or the Frumkin and Melik-Gaikazyan result[6] for the charge separation/adsorption/diffusion ABD coupling (equilibrium adsorption), or the Lorenz treatment[7] for the same type of coupling when the adsorption–desorption is a nonequilibrium process.

From the point of view of the development of the subject, these contributions must be seen as discrete bursts of ingenuity, but directed towards very specific types of coupling. It is also appropriate to stress that, with increasing complexity of coupling between the phenomenological components, the resulting expressions began to get more and more messy. And, what was worse, not only the final results, but also the methodology became increasingly painful. To avoid this painful approach, one had to become selective in choosing the interfaces for analysis. Clearly, what was required was an embracing unification which would reveal all preceding treatments as various special cases.

It is this grand synthesis which has been recently achieved by Rangarajan. In a series of papers which may well prove path-breaking, Rangarajan discovered the simple mathematical structure underlying the solutions to the special cases.[8-11] Essential for this accomplishment was the development of a new formalism and a treatment in terms of operators and matrices.

To illustrate Rangarajan's approach, consider that the electrode/electrolyte interface involves four phenomenological components: (1) charge separation; (2) adsorption–desorption; (3) charge transfer at the electrode; and (4) mass transfer with or without volume sources/sinks. In the case of linear systems, the first three of these components can be described quantitatively by the following equations:

$$i_{nf} = \frac{dq_m}{dt} = \left(\frac{\partial q_m}{\partial E}\right)_{\Gamma_j} \frac{dE}{dt} + \sum \left(\frac{\partial q_m}{\partial \Gamma_j}\right)_{\Gamma_{i \neq j}, E} \frac{d\Gamma_j}{dt} \tag{1}$$

$$\frac{d\Gamma_j}{dt} = \sum \left(\frac{\partial \Gamma_j}{\partial c_k}\right)_E \frac{dc_k}{dt} + \left(\frac{\partial \Gamma_j}{\partial E}\right)_c \frac{dE}{dt} \tag{2}$$

$$\Delta E = \sum r_k \Delta c_k + \left(\frac{\partial E}{\partial i_f}\right)_c i_f \tag{3}$$

Conceptually, the important step is to write these linear equations in the operator form so that the various coupling coefficients linking the variables become operator coefficients (denoted by a tilde "~" over the symbol):

$$i_{nf} = \tilde{C}_E \Delta E + \sum \tilde{C}_j (\Delta \Gamma_j) \tag{4}$$

$$\Delta \Gamma_j = \sum \gamma_{jk} (\Delta c_k) + \tilde{\gamma}_{jE} (\Delta E) \tag{5}$$

$$\Delta E = \sum r_k (\Delta c_k) + r_t i_f \tag{6}$$

The mass-transport problem can be considerably simplified by avoiding actual expressions for the interfacial concentrations, c_i, and seeking only the representations for the (integral) operators, m, linking the concentration perturbations, Δc_j, with the fluxes, J_k:

$$\Delta c_j = \sum \tilde{m}_{jk} J_k \qquad (7)$$

This strategy ensures that, unlike the interfacial concentrations, the linking operators display two types of invariance: firstly, with respect to the choice of E or i_f, and secondly, with respect to the fate of the mass-transported species at the electrode/electrolyte boundary. To incorporate the latter information, a description of what happens to the flux at the boundary is given in the form:

$$J_j = \epsilon_j i_f + \sum \left(\frac{\partial J_j}{\partial \Gamma_k} \right) \frac{d\Gamma_k}{dt} \qquad (8)$$

or

$$J_j = \epsilon_j i_f + \sum \tilde{\delta}_{jk} \Delta \Gamma_k \qquad (9)$$

To reckon with coupling between various species, equations (4)-(9) are written in the matrix form:

$$i_{nf} = \tilde{C}_E \Delta E + \tilde{C} \Delta \Gamma \qquad (10)$$

$$\Delta \Gamma = \tilde{\gamma} \Delta C + \gamma_E \Delta E \qquad (11)$$

$$\Delta E = r_t i_f + r \Delta C \qquad (12)$$

$$\Delta C = \tilde{m} J \qquad (13)$$

$$J = \epsilon i_f + \tilde{\tilde{\epsilon}} \Delta \Gamma \qquad (14)$$

Rangarajan has solved these equations to get

$$E = Z_f i_f = [I - r\tilde{m}\tilde{\delta}(I - \tilde{\gamma}\tilde{m}\tilde{\delta})^{-1}\gamma_E]^{-1}[rt + r(I - \tilde{m}\tilde{\delta}\tilde{\gamma})^{-1}mE]i_f \qquad (15)$$

$$i_{nf} = Y_{nf}(\Delta E) = [\tilde{C}_E + \tilde{C}(I - \tilde{\gamma}\tilde{m}\tilde{\delta})^{-1}\gamma_E]\Delta E + \tilde{C}\tilde{\gamma}(I - \tilde{m}\tilde{\delta}\tilde{\gamma})\tilde{m}\epsilon i_f \qquad (16)$$

where Z_f and Y_{nf} are the impedance and admittance, respectively, of the interface.

These general equations cover *any* specific mode of transport and any geometry of electrode; further, they do not include any restriction at all on the entities undergoing mass transfer or adsorption. Rangarajan has also shown that the general solutions (15) and (16) embrace a hierarchy of special cases, of which two examples are cited below.

2.1. Randles-Ershler Expression (CD Case)

For this special case in which the double layer and adsorption–desorption are considered to play no role in determining the system behavior, $\tilde{\gamma} = 0 = \tilde{\gamma}_E$, and therefore equation (15) reduces to a simple form for the impedance:

$$Z_{CD} = Z_F = r_t + r\tilde{m}\epsilon = r_t + Z_W \qquad (17)$$

where Z_W is the Warburg impedance.

Under linear conditions, the charge transfer resistance, r_t, is given by

$$r_t = \frac{RT}{nFi_0} \qquad (18)$$

Further, the matrices **r**, **m**, and ϵ have the following representations:

$$\mathbf{r} = \left[\frac{RT}{nFC_0^\circ} \quad \frac{RT}{nFC_R^\circ}\right], \tag{19}$$

$$\tilde{\mathbf{m}} = \begin{bmatrix} \dfrac{1}{\sqrt{pD_0}} & 0 \\ 0 & \dfrac{1}{\sqrt{pD_R}} \end{bmatrix} \quad \text{semi-infinite linear diffusion} \tag{20}$$

and

$$\epsilon = -\begin{bmatrix} 1 \\ 1 \end{bmatrix} \left(\frac{n_1}{nF}\right) \tag{21}$$

Hence, by simple matrix multiplication of (19), (20), and (21), the Warburg impedance becomes

$$Z_W = \mathbf{r}\tilde{\mathbf{m}}\epsilon = \frac{RT}{n^2 F^2 C_0^\circ \sqrt{D_0 p}}\left[1 + \frac{C_0^\circ \sqrt{D_0}}{C_R \sqrt{D_R}}\right] \tag{22}$$

These results, it will be noted, have been arrived at by straightforward operator and matrix algebra. As far as the representation of $\tilde{\mathbf{m}}$ is concerned, it is taken from a tabulation (furnished by Rangarajan) covering a host of mass-transport models for various initial conditions and various boundary conditions away from the electrode. It is invariably advantageous to use Laplace transformation methods and to leave the representation of $\tilde{\mathbf{m}}$ in the p domain.

2.2. Frumkin–Melik-Gaikazyan Expression (ABD Case)

For this special case in which the faradaic current is absent, the sought-after admittance, Y_{ABD}, becomes [cf. equation (16)]

$$Y_{ABD} = \tilde{C}_E + \tilde{C}(I - \tilde{\gamma}\tilde{\mathbf{m}}\tilde{\delta})^{-1}\gamma_E$$

Further processing by writing the matrices $\tilde{C}$, $\tilde{\gamma}$, $\tilde{\delta}$, $\tilde{\gamma}_E$, and $\tilde{\mathbf{m}}$, and using matrix algebra yields the following expression for semi-infinite linear diffusion and one species

$$Y_{ABD} \longrightarrow \left(\frac{\partial q_m}{\partial E}\right)_\Gamma p + \left(\frac{\partial q_m}{\partial \Gamma}\right)_E \left(\frac{\partial \Gamma}{\partial E}\right)_C \frac{p}{\Delta_0}$$

which is the Frumkin–Melik-Gaikazyan result for charge separation/equilibrium adsorption–desorption/diffusion.

The purpose of highlighting this contribution of Rangarajan was not to describe past work, but to suggest that the extension, elaboration, and refinement of this systems approach is likely to be a major preoccupation in the electrode process research of the future.

In guessing at possible developments, one must realize that the Rangarajan treatment was necessarily restricted in several ways. Firstly, it was confined to four phenomenological components of the interfacial system, viz., charge separation, adsorption-desorption, charge transfer at electrode, and mass transfer in the electrolyte. Secondly, its most detailed exposition has been in the special case of linear electrochemical systems.

Thirdly, only certain aspects of the general nonlinear case have been handled.[12,13] For instance, it has been assumed that the process of double-layer charging does not interact with the faradaic process and that the faradaic current alone is of interest. Even this faradaic current has been considered to depend linearly on the interfacial concentrations, i.e., the nonlinear case has been restricted to a reaction scheme which will underwrite this linear dependence. A further assumption is that adsorption–desorption is of no consequence and that an indifferent (supporting) electrolyte is present. In short, what has been treated is the nonlinear regime of a generalized Randles–Ershler scheme.

The restrictions described above have not been motivated by a desire "to make life easy," but by the compulsion "to make an impossible life just bearable." These very limitations, however, define the tasks for the coming years and therefore hint at the directions along which the breakthroughs will be achieved. One may well hope for closed analytical expressions, but the chances are that answers will only come through numerical computation.

3. The Phenomenological Components and Their Interactions

A systems approach can only be developed on the basis of an *a priori* knowledge of the phenomenological components and their interactions. In that sense, it is necessary to be equipped with a "model" of the interface.

How does one arrive at such a model? The classical approach is to carry out an experimental study in which systematic variations are made of various interfacial and/or cell parameters (electrical variables, bulk concentrations, etc.) and the emerging data are then processed to suggest a plausible model. The whole procedure is elaborate, tedious, and almost bureaucratic. As a result, the understanding of interfaces becomes so time-consuming that the objective of controlling them often gets postponed. A quickening of the ritual is imperative. Fortunately, this impatience is compatible with the growing power of computers. It seems, therefore, reasonable to predict an increasing tendency to link online computers with electrochemical cells for the discovery of true models of interfaces.

For instance, one can imagine a computer stimulating an interface with a programmed electrical signal and also triggering off the output generated by a test model in response to that same input signal. The comparison of the two outputs—one from the interface (which is best stored for repeated use) and the other from the test model—will reveal the degree of truth of that model. If the model has sufficient credibility, the computer will then move on to other stimuli specially suited to *establish* that model. If, however the test model is patently false, then the computer will scan through its library of memorized models until it finds one with a response identical to the stored response of the interface under study.

One can foresee several versions of such computer-controlled electrochemistry. For example, one can consider a set of electrochemical cells with varying concentrations of some species, or with electrodes for which some characteristic parameter differs in some systematic way. It should cause no surprise if such advances of the future will reduce the determination of models to a spare afternoon's effort. But, these advances will have to be preceded by considerable work in the systems analysis of interfaces and in the development of software and hardware for computers specially designed for coupling to electrochemical cells.

Notwithstanding such advances in cell–computer interfacing, one can see that the analysis of electrical responses alone cannot provide all the answers and that there is a desperate need for developments in experimental techniques, particularly for looking at solid electrodes. In the absence of such developments, electrochemistry will continue to have its present lopsided character wherein the exactness of experiment and theory in the case of mercury is at least an order of magnitude better than that in the case of solid electrodes which form the basis of industrial electrochemistry.

It is not easy to visualize the direction which will lead to a breakthrough with solid electrodes. However, one can predict with fair certainty that it will not be achieved with any one technique. For experience shows that every new technique of studying solid electrodes emerged on the electrochemical scene with great fanfare and promise, but quite soon, hopes were betrayed, and the stage was set for the next "magic" technique. The reasons are simple. The electrode/electrolyte interface is a multifaceted reality, and each technique reveals at best a few of the many facets. Thus, the revelation is never complete, and the knowledge thus derived always inadequate.

The future invites an "assembly-line" approach[14] in which the activity of an electrode/electrolyte interface is "frozen" suddenly—perhaps by freeze drying—and the electrode along with the relevant layer of electrolyte is passed through an assembly line of surface study techniques including perhaps LEED, ellipsometry, microscopy, ATR, ESCA. This procedure must be rigged in such a way that at no time during this passage through the assembly line does the electrode encounter destructive environments. Of course, such an assembly line may well become redundant if the trick of preparing reproducible surfaces is mastered.

4. The Microscopic Components

Thus far, the picture of the future has been associated with the phenomenological components of the interfacial system. To probe deeper, one must involve the particles constituting the interface and their interactions. It is at this particulate level of analysis that the past has faced its most traumatic problems and the future will witness its most spectacular successes. Two particular issues are likely to attract the greatest attention.

The first issue concerns a microscopic treatment of charge transfer. Clearly, if electrochemistry is to achieve a quantum jump beyond the pioneering analyses of Tafel, Butler, Volmer, Gurney, and Frumkin, it must get behind the veil of exchange-current densities, rate constants, transfer coefficients, and symmetry factors. This success at the more fundamental level will be heralded by *a priori* calculations of these electrode kinetic parameters. It is these calculations which one hopes will emerge from a quantum mechanical treatment of the processes.

Here too one can forecast with some assurance an increasing role for the computer. The approach of two simple atoms has already been computer analyzed to reveal the changes of energy underlying their reaction,[15] and one can foresee an increasing capability for this technique. If the not-too-distant future witnesses breakthroughs in approximation methods, then one can imagine mammoth computers handling a larger and larger number of reactant particles. It may then turn out that one can computer analyze the changes in energy when a suitably approximated electrode is approached by an ion associated with a relevant portion of the solvent.

What number of solvent molecules are to be included in the relevant portion of the solvent—only those in the primary solvation sheath as the Gurney–Butler viewpoint would urge, or the whole solvent engaged in lattice-like vibrations? The first view can only be a naive simplification justified for the origins of a subject, and the second view in all its grandiose generality may prove intractable for a microscopic treatment.

The sorting out of such problems will lead to a detailed picture of the course of a charge-transfer reaction—the mechanism by which a reacting particle is energized sufficiently to permit electron tunneling, and the factors on the electrode side and the electrolyte side which influence this mechanism.

An analysis of the "electrode factors" will lead on to an understanding of the catalytic role of an electrode. It is clear that this analysis will lean heavily on solid-state science and the light it throws on the behavior of electrons in metal oxide lattices. This suggestion derives from the fact that even the *noble* metals face an electrolyte with a thin coat of "oxide." With increasing knowledge of the electronic and chemical properties of these oxides and of the methods of preparing them, a stage may come when a base-metal electrode will be covered with an "oxide film" tailor-made to have the right catalytic properties for a particular reaction while possessing the appropriate electronic conduction properties and adequate chemical stability with respect to the electrolyte. Only then will the search for ideal, but inexpensive, electrocatalysts culminate in a triumph which will have far-reaching implications on electrochemical energy conversion and on electrolytic production techniques.

The analysis of the "electrolyte factors" which affect the mechanism of charge transfer leads to the second major issue, viz., a microscopic picture of how adsorbed electroinactive species (ions or molecules) alter the energetics of charge transfer. This picture of the *modus operandi* of surface-active ions and molecules is just emerging.[16–18]

Three types of adsorption-induced effects on electrode kinetics can be distinguished: (1) blocking effects, (2) electrostatic effects, and (3) noncoulombic interaction effects. These effects have been formally separated by Sathyanarayana[18] so that the apparent rate constant, $k_{s,a}$, is related to the true rate constant, $k_{s,t}$, the electrostatic potential, Φ_x, at the reaction site, and a noncoulombic interaction work, $W(\Theta)$:

$$k_{s,a} = k_{s,t}(1 - \Theta) + k'_{s,t}\Theta \exp\left[(\alpha n - z_0)(\Phi_{x(q)} + \Phi_{x(\Theta)})F/RT\right] \exp\left[W(\Theta)/RT\right]$$

where Φ_x has been resolved into a charge-dependent term, $\Phi_{x(q)}$, due to ions in the diffuse layer, and a coverage-dependent term, $\Phi_{x(\Theta)}$, due to particles in the adsorbed layer. This separation leads to several advantages—firstly, the magnitude of each effect may be determined from experiments on $k_{s,a}$ and, in future, from exact theory; secondly, the relative magnitudes of the three effects are bound to be correlated with the structure of the surfactant particle; and thirdly, the conditions required for enhancing (or diminishing) a particular effect of the adsorbate can be specified.

The obvious objective is to achieve a sensitive adsorption-induced control over reaction rates, i.e., large values of $\pm[dk_{s,a}/d\Theta]$, where the positive sign represents an acceleration of reaction rates (electrocatalysis) and the negative sign, a retardation (inhibition). An ideal state of affairs is attained when the sensitivity coefficient, $\pm[dk_{s,a}/d\Theta]$ is virtually zero for all other electrode reactions except the desired one. For a surfactant to achieve the sensitivity and specificity of control just described, it must work in the *electrostatic* mode, because the blocking mode usually results in inhibition and a reduction in the utilization of electrode surface area, and the noncoulombic mode is not as significant at low coverages.

Thus the problem reduces to one of understanding the electrostatic effect of surfactants on reaction rates and correlating these effects with the structure of the surfactants. This understanding involves detailed calculations of $\Phi_{x(\Theta)}$, the micropotential at the reaction site due to adsorbed particles at a coverage of Θ, and it is reasonably sure that the limitations in current theory, for example, restriction to either ions or molecules, to limitingly low coverages, etc., will soon be overcome. Then, one will have detailed, theoretically based selection rules for the choice of surfactants.

Acknowledgments

The author wishes to acknowledge his gratitude to his colleagues in the Electrochemistry Group, Professors S. Sathyanarayana and S. K. Rangarajan, for exposing him to their ideas.

References

1. G. Wills with R. Wilson, N. Manning, and R. Hildebrandt, *Technological Forecasting*, Penguin Books, England (1972).
2. O. Helmer, Analysis of the future: The Delphi method, in: *Technological Forecasting for Industry and Government* (J. Bright, ed.), pp. 116–143, Prentice Hall, Englewood Cliffs, New Jersey (1968).
3. A. V. Bridgewater, Morphological methods: Principles and practice, in: *Technological Forecasting* (R. Arnfield, ed.), pp. 211–250, Edinburgh University Press (1969).
4. J. E. B. Randles, *Disc. Faraday Soc. 1*, 11 (1947).
5. B. V. Ershler, *Disc. Faraday Soc. 1*, 269 (1947).
6. A. N. Frumkin and V. I. Melik-Gaikazyan, *Dokl. Akad. Nauk SSSR 78*, 855 (1951).
7. W. Lorenz and F. Mockel, *Z. Electrochem. 60*, 507 (1956); W. Lorenz, *Z. Electrochem. 62*, 192 (1958).
8. S. K. Rangarajan, *J. Electroanal. Chem. 55*, 297–327 (1974); also in: *Topics in Pure and Applied Electrochemistry* (S. K. Rangarajan, ed.), pp. 7–14, SAEST, Karaikudi, India (1975).
9. S. K. Rangarajan, *J. Electroanal. Chem. 55*, 329–335 (1974).
10. S. K. Rangarajan, *J. Electroanal. Chem. 55*, 337–361 (1974).
11. S. K. Rangarajan, *J. Electroanal. Chem. 55*, 363–374 (1974).
12. S. K. Rangarajan, *J. Electroanal. Chem. 56*, 1–25 (1974).
13. S. K. Rangarajan, *J. Electroanal. Chem. 56*, 27–53 (1974).
14. E. Yeager, personal communication.
15. R. F. Bader and A. K. Chandra, *Can. J. Chem. 46*, 953 (1968).
16. R. Parsons, *J. Electroanal. Chem. 21*, 35 (1969).
17. W. R. Fawcett and S. Levine, *J. Electroanal. Chem. 43*, 175 (1973).
18. S. Sathyanarayana, *J. Electroanal. Chem. 50*, 195–209 (1974).

11

Summary: Electrode Processes

Eric Sheldon

From any association with John Bockris one learns always to expect the unexpected. So perhaps it is out of place for me to feel surprised or disconcerted at being called upon to deal with a subject in which I was last actively involved two decades ago! But the masterly reviews by Conway and Reddy that form the basis of my summary can make two decades seem to vanish as though they had never been, re-creating the sense of immediacy and urgency and significance that underlay all the progress in one's appreciation of electrode processes as the very life-pulse of electrochemistry. Let me, then, try to distil the essence of their remarks, but through the thought process of a practicing physicist rather than along the more established lines of physical chemistry. And I will try to exercise brevity— though perhaps not in as drastic a form as that employed in a still-treasured letter from Bockris urging me to hasten the completion of a manuscript:

> Dear Sheldon,
>> Well?
>>> Sincerely,
>>> J. O'M. B.

to which only one reply was possible:

> Dear Bockris,
>> Yes, thank you.
>>> Sincerely,
>>> E.S.

That was at the time, in the late 1940s and early 1950s, when the detailed understanding of electrode kinetics, and electrode processes in general, came to fruition. The seeds planted by Faraday, Arrhenius, Debye and Hückel, Grahame, and Frumkin, and nurtured by Erdey-Gruz and Volmer, Gurney, Horiuti, and Polanyi, were tended by Bockris and his school into full bloom. The mechanisms of ionic solvation, cathodic hydrogen evolution, and anodic oxygen evolution were elucidated; the general treatment of consecutive reactions was advanced; the tangled mysteries of the double layer— promptly generalized to a multilayer manifold—were unraveled; and the way was paved for ongoing kinetic studies of pure surface processes. In particular, it has meanwhile been possible to trace out the kinetic behavior of chemisorbed intermediates, soluble intermediates in different solutions, and organic electrochemical reactions. Especially over the past decade, good use has been made of *in situ* optical methods such as normal and electron microscopy, spectrophotometry, reflectance techniques, ellipsometry, etc., offering a

---

Eric Sheldon • Department of Physics, University of Lowell, Lowell, Massachusetts.

microscopic insight into macroscopic phenomena. However much is already known, we have still barely begun to understand the intricacies of electrochemical phase growth processes, as in the deposition of metals and the formation of oxides.

Nor do I feel that even now the powerful methods of quantum chemistry and statistical mechanics have been brought into full play. Electron transfer and proton transfer have still not yet been subjected to as detailed a scrutiny as modern quantum techniques would invite, and I submit that this constitutes a challenge at the very forefront of electrode research. Thermodynamics still remains to be exploited further, and, for example, in the domain of diffusion studies the statistical potentialities of a Markov-process theoretical approach have yet to be realized. Considerations of stochastic processes, particularly those pertaining to pseudorandom noise, may well have much to offer toward arriving at new parametric descriptions of electrochemical phenomena. We are beginning to understand the detailed characteristics of simple quantal systems and states—now the latest methods beckon toward the scrutiny of substates.

These thoughts tie in perhaps with those put forward so cogently by Reddy in his survey of the possible future trends in this field. In the place of the time-honored parameters that Conway alluded to, as depicted for instance in potential energy diagrams (e.g., variation of potential, absorption energy or solvation energy, isotherms for reactants and for intermediates), one may adopt the more sweeping morphological approach inherent in systems analysis. As the components in an overall ensemble are subdivided, the subsystems interact and the subprocesses interlock; it is then feasible to study the individual simple classes of coupling in quasi-isolation, as it were, and build up an overview of the interaction scheme from a detailed visualization of its subconstituents. Reddy has deservedly drawn attention to a major preliminary step in this direction by singling out for especial mention the treatment that Nagarajan has latterly put forward toward an all-encompassing grand synthesis of the interplay between different interacting partners at an electrode/electrolyte interface. Nagarajan's scheme is still a restricted one in the sense that it refers to linear electrochemical systems and is confined to a consideration of but four phenomenological components, namely

1. Charge separation
2. Adsorption–desorption considerations
3. Charge transfer at the electode
4. Mass transfer

In a matrix formalism evolved from sets of linear differential equations, the treatment offers every inducement to generalizations that embrace nonlinear systems having more affinity with those actually encountered in practice. Closely akin to this path of development is the expansion of detailed models of the interface under a variety of conditions: the vital role of quantum-mechanical calculations cannot be underestimated here, particularly when abetted by variational, approximational, and simulational techniques that the advent of off-line and on-line computer analysis has rendered possible. A word of caution, though: Computers must not be regarded as the be-all and end-all arbiters in this line of research. I think we are all well aware of the risk that their indiscriminate employment may lead to the submergence of—dare I say it?—the underlying physics, or to go back further, the underlying natural philosophy. How often now do we find our students devoting their uncritical adherence to the ever-spinning magnetic tapes and disks, when we would so much rather have them believe in our hard-won, cherished vision of our science!

Yet I am sure that as we increasingly come to terms with computers, so will we also increasingly make use of offerings from the other scientific disciplines. A couple of examples will suffice: The transfer of charged particles at an interface is a quantum-mechanical tunneling problem at potential barriers not unlike the analogous situation in nuclear physics, and I would remind you that much interest has recently been generated by the realization that fission can involve penetration through a *doubly* humped barrier. Perhaps the methods developed for the quantitative evaluation of this situation may with felicity be applied to the commensurate problem in the electrochemical domain. The other example is an obvious one: The fact that our understanding of processes at solid electrodes still lags significantly behind that for liquid electrodes suggests not only that we could advantageously invoke new aspects of solid-state physics but even borrow from engineering technology the virtues of an assembly-line approach. (Reddy's talk included the suggestion, as you will recall, of freeze-drying the reactants at an appropriate stage of their interaction and passing the electrode through a sequence of diagnostic studies involving nondestructive testing.) We would also need to pursue the recognition of the role played by inactive adsorbed electrospecies, since these can exert blocking effects, electrostatic influence, and noncoulombic perturbations upon the course, rate, and manner of an electrochemical interaction.

With an improved understanding of all these wondrously variegated aspects of this generously multifaceted subject, we may ultimately bring about a new era in electro-chemistry, an era in which the comprehension of electrode processes may enable one to tailor-make more effective electrocatalysts, to devise more efficient storage batteries and fuel cells, to inhibit and virtually to eliminate electrocorrosion, to initiate more productive methods of energy conversion and utilization, and so in myriad ways place more advantageously than heretofore the intrinsic wealth of electrochemistry toward the enhancement of the quality of life. Then, and only then, will electrode processes have taken their rightful place among the order of things, as the nucleus of electrochemistry, as the jewel in the lotus.

12

Some Current Problems on Semiconductor/Electrolyte Interfaces

Karl H. Hauffe

1. Introduction

A large number of catalysts used in industry for the acceleration and the selection of chemical reactions are compound semiconductors, such as ZnO, TiO_2, CoO, CuO, NiO, Cr_2O_3, etc. The attempts at elucidation of the single steps decisive for the mechanism of the reaction occurring often have not been successful. However, today we know that electrons and/or holes become involved in such catalytic reactions. The study of electronic steps becomes complicated due to the absence of an easily accessible reference system, particularly if a gas reaction is catalyzed at a semiconductor surface. In order to get information on the decisive steps during a catalysis, we are interested in the determination of the exchange of electrons and holes between the energy bands and the surface states arising from the adsorption of the reacting molecules on the semiconductor surface.

From the experimental point of view, some problems can be solved if the catalyst is employed as anode or cathode in a suitable electrochemical cell and the reacting species are dissolved in the electrolyte because now the exchange of electrons and holes with the reacting species can be controlled by a reference electrode. By this electrochemical cell technique we can learn a little about the decisive interaction of the electronic species with the reacting molecules and ions adsorbed on the surface of the semiconductor electrode. By means of this technique we can examine the speed and energy of electron capture by oxidizing agents, as for instance oxygen, which does occur with electrons from the conduction band. More powerful oxidizing agents, however, are able to strip electrons from the valence band, a process which is equivalent to hole injection. In a similar way, we can study electron injection from reducing agents, such as hydrogen, carbon monoxide, and many organic compounds. Furthermore, we can investigate the mechanism of reaction of electron–hole pairs, generated by light of sufficient energy, with the reacting species adsorbed on the surface. Particularly, from these reactions new information on the mechanism has been obtained. Stimulated by electrophotographic studies, the sensitization of the electron transfer through the interface of the photoconductor/chemisorbed species by organic dyes, as for instance rhodamine B, bromphenol blue, and azo dyes, has been elucidated. Finally, it remains to be tested to what extent the results obtained from electrochemical measurements can be transferred for the evaluation of the semiconductor/ gas interface. Furthermore, the electrochemical cell technique is a useful tool for devising means to prevent a photocatalytic effect. Such a problem is of outstanding technical importance for the chalking effect of TiO_2-resin coatings.

The electrochemical cell technique with semiconductor electrodes was first introduced for germanium by Brattain and Garrett[1] and for zinc oxide by Dewald.[2] It was

Karl H. Hauffe • Institute of Physical Chemistry, University of Göttingen, West Germany.

employed by Gerischer and Tributsch[3] and Hauffe and his co-workers[4] for the investigation of the sensitization of the charge transfer through photoconductor/electrolyte interfaces, particularly the ZnO/electrolyte interface, by selected organic dyes. Worth mentioning are the stimulating papers by Morrison[5, 7] and Morrison and Freund[6] on surface phenomena associated with the semiconductor, particularly zinc oxide/electrolyte interface.[8, 9] In order to describe some current problems on these electronic reactions, we shall restrict the discussion exclusively to work carried out on zinc oxide. But the same technique and theoretical considerations can be employed on CdS, TiO_2, and organic semiconductor/electrolyte interfaces, as has been reported in the literature.[10]

Apart from the experimental work, theoretical investigations have been carried out on the electron-exchange reactions between energy levels associated with ions both in adsorbed states and in the electrolyte and energy levels in the semiconductor. Besides Gerischer's treatment,[11] which is based on Gurney's model[12] and which is similar to Dewald's theoretical considerations, Marcus[13] and Levich and Dogonadze[14] based their analysis on a continuum theory but with the restriction of very weak interactions between the reacting species. Therefore, they could employ the ordinary perturbation theory, with a limited application. Myamlin and Pleskov[15] gave an excellent survey of the field of semiconductor electrochemistry, both with regard to experiment and theory.

2. Semiconductor/Electrolyte Interface

By introducing a semiconductor as electrode in an electrochemical cell, two main problems have to be considered for the understanding of the electrode processes in the dark and under illumination. The first problem is the structure of the space-charge layer which determines the direction of the charge transfer. The second problem concerns the role of the charge carriers, free electrons or holes, in electrode reactions. Here the situation is still rather complex. In order to more easily understand the results of current voltage and capacity measurements discussed below, it is helpful to introduce a simplified model, represented in Figure 1. As already mentioned, we will consider, in the main, the interface ZnO/electrolyte. As is well known, zinc oxide is an n-type semiconductor with a rather significant exhaustion space-charge layer. This is characterized by a strong bend-

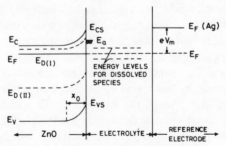

Fig. 1. Schematic representation of the energy levels and the Fermi potentials including energy levels for dissolved species in the system: ZnO|electrolyte|AgCl|Ag. The symbols are as follows: E_F = Fermi potential of zinc oxide, $E_F(Ag)$ = Fermi potential of silver, E_C = energy of the conduction band edge in the bulk, E_{CS} = energy of the conduction band edge at the surface, E_V = energy of the valence band edge in the bulk, E_{VS} = energy of the valence band edge at the surface, V_S = diffusion potential between the surface and the bulk, eV_m = energy difference between the Fermi potentials, E_a = energy of an adsorbed species, $E_{D(I)}$ = dissociation energy of $Zn^x = Zn^· + e'$ (the first donor level), $E_{D(II)}$ = dissociation energy of $Zn^· = Zn^{··} + e'$ (the second donor level), x_O = thickness of the space-charge layer.

ing up of the energy-band edges according to eV_S, where e is the electronic charge and V_S the surface-barrier potential difference. The Fermi potential for ZnO is E_F and for the Ag/AgCl reference electrode it is $E_F(\text{Ag})$. The energy of the conduction band edge E_C is at the surface $E_C + eV_S \equiv E_{CS}$ and of the valence band edge E_V at the surface $E_V + eV_S \equiv E_{VS}$. In contrast to the metal/electrolyte interface, here the Helmholtz double layer shall be omitted since the potential difference is located in the space-charge layer. E_a denotes the energy of an adsorbed species which can be located above or below the Fermi potential. The difference of the Fermi potentials $E_F(\text{Ag}) - E_F = eV_m$ can be measured by a voltmeter.

Reacting species dissolved in the electrolyte can also be assigned an energy level. In particular, redox electrolytes participate in the electrode reactions. In order to decrease the resistance of the electrolyte a high concentration of an inert salt, e.g., KCl, is added. Since these ions do not exchange electrons with the bands of the semiconductor, the notation of their energy levels is not needed. In the following representation we employ Morrison's description,[7] avoiding the complexity of the hybrid model.

The assumption, reasonable at first glance, of describing the system in terms of equilibrium electron transfer near thermodynamic equilibrium for the competing oxidizing and reducing agents in the presence of a redox electrolyte, is often not fulfilled as Morrison pointed out. Another reasonable assumption is the explicit separation of the overall electron exchange into electronic steps and chemical reactions which cannot be represented on the band model. A further assumption introduced by Morrison is that each form of an ion in the solution can be characterized by one energy only.

As can be seen from Figure 1, the surface-barrier potential V_S and also the thickness x_0 of the space-charge layer become smaller when the charge at the ZnO electrode becomes less positive. If an external field with the negative pole on ZnO is employed, the bands become flat with increasing voltage and finally bend downwards (enrichment of electrons). By this technique, one can determine the flat-band potential. In this case the Helmholtz double layer and V_H can no longer be neglected. Generally, a change in applied voltage V_m appears entirely as a change in the surface-barrier potential V_S. When no other reaction occurs on the ZnO surface, the potential V_S is caused by electron capture during chemisorption of oxygen according to

$$O_2(\text{aq}) + e' = O_2^-(\text{ads}) \tag{1}$$

where e' denotes a free electron and (aq) and (ads) the molecule in the aqueous and adsorbed states, respectively.

3. Experimental

For the current voltage and capacity measurements, a typical electrochemical cell is represented schematically in Figure 2. The semiconductor electrode is a ZnO single crystal cut perpendicular to the c axis with the (0001) plane (= Zn^{2+} plane) in contact with the electrolyte. The (000$\bar{1}$) plane (= O^{2-} plane) exhibits an equivalent behavior but the dissolution of the crystal from this plane is faster. As a working electrode a platinum net was installed with a circular opening for the light beam. This passed through the optical window and was obtained from a high-pressure xenon lamp via an interference filter. Sometimes the electrolyte at the anode is separated by a salt bridge from the electrolyte at the cathode. The tube voltmeter TV (input resistance of about 10^{12} Ω) measures the potential difference between the semiconductor electrode and the Ag/AgCl reference electrode. The picoammeter A measures the current passing through the elec-

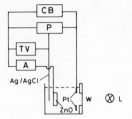

Fig. 2. Schematic representation of the electrochemical cell technique for voltage current and capacity measurements. TV = tube voltmeter, A = picoammeter, P = potentiostat, CB = capacity bridge, W = window for illumination, L = high-pressure xenon lamp.

trolyte and the semiconductor electrode, i.e., across the ZnO/electrolyte interface including the space-charge layer which acts as a barrier to the electron transfer when the ZnO electrode is the anode. Finally, with the capacity bridge we measure the capacity C of the space-charge layer vs. the applied voltage V_S and plot $1/C^2$ vs. V_S according to the Mott-Schottky relation

$$V_S \approx eN_D A^2 \epsilon\epsilon_0/2C^2 \tag{2}$$

obtained from the parallel plate capacity formula

$$C = A\epsilon\epsilon_0/x_0 \tag{3}$$

and the Schottky relation

$$V_S \approx eN_D x_0^2/2\epsilon\epsilon_0 \tag{4}$$

Here N_D denotes the density of charge in the space-charge layer and ϵ the dielectric constant. Equation (2) is fulfilled with ZnO because minority carrier effects do not exist, this is in contrast to a germanium electrode where equation (2) is valid only for a limited range.

The transistor method described in the literature[7] can be employed only if germanium, silicon, and 3/5 compound semiconductors are used as electrodes since these materials can be changed into an n-type or p-type region near the surface by doping. If minority carriers are desired in the n-type zinc oxide, then holes can be produced by illumination with light $\leqslant 380$ nm. Therefore, the photoproduced holes have been used to study hole reactions.

4. Experimental Results

In order to demonstrate the advantage of application of a semiconductor as an electrode in a suitable electrochemical cell, we shall discuss some experimental results with a zinc oxide single crystal operating as anode in the following electrochemical cell

$$+ \text{In}|\text{ZnO single crystal}|\text{electrolyte}|\text{platinum} - \tag{5}$$

where the composition of the electrolyte can be changed in regard to the solvent and the salt additions. As can be seen from Figure 3, the current through the cell with an aqueous electrolyte with KCl as supporting electrolyte is very low in the dark ($\sim 10^{-10}$ A/cm^2), when the zinc oxide electrode is anodically polarized. At cathodic polarization, however, a rather large current of $\geqslant 10^{-4}$ A/cm^2 appears. The constant current which appears both in the dark and under illumination is noticeable when the applied voltage has attained 0.5 V. For these measurements, the electrolyte was composed of 0.5 M KCl and 0.1 M

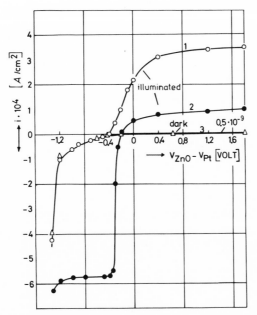

Fig. 3. Voltage–current curves obtained with the electrochemical cell (1) in the dark and under illumination at various electrolytes and with the same light intensity. Curve 2 was obtained with an addition of 10^{-4} mole quinhydrone in an acetate buffered 0.5 M KCl electrolyte. [From Hauffe and Range.[9]]

acetate buffer with a pH = 4.7. This result indicates that the transfer of electrons is energetically much more favorable from ZnO to the chemisorbed species than in the reverse direction. During illumination with 372-nm light, however, the anodic current (i.e., with an anodic polarization of the ZnO electrode) is significantly increased, up to 10^{-4} A/cm² depending on the light intensity, within a small voltage range between 0.5 and 2 V. On the basis of these experiments, it may be concluded that during illumination electron injection occurs from the adsorbed negatively charged species into the ZnO crystal by a recombination process, while the holes in the valence band are generated by the illumination. Because of the potential gradient in the space-charge layer, the generated electron–hole pairs $e' \sim |e|^{\cdot}$ are mainly dissociated: the holes $|e|^{\cdot}$ are moving toward the surface and the electrons e' repelled into the bulk. However, this occurs to a noticeable extent only if the holes have been consumed at the surface by recombination according to:

$$O_2^-(ads) + |e|^{\cdot} \longrightarrow O_2(aq) \qquad (6)$$

or

$$H_2(ads) + |e|^{\cdot} \longrightarrow H_2^+(ads) \qquad (7)$$

The charge transfer under emission and recombination of charge carriers is schematically represented in Figure 4.

According to the current voltage curves in Figure 3, the chemisorption of oxygen should be highly favored in the dark, since the transfer of electrons occurs from the semiconductor to the adsorbing species. A noticeable desorption of oxygen with a simultaneous injection of electrons into the valence band should occur only during illumination when holes are produced. Besides this mechanism, an injection of electrons into the conduction band is also possible when an exothermic reaction occurs.

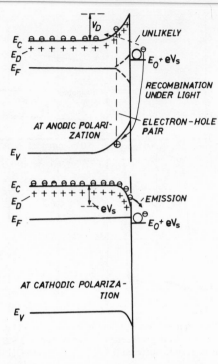

Fig. 4. Schematic representation of electron emission and recombination in the dark and under illumination between the adsorbed species and the electrons and/or holes. E_O represents the energy level of an adsorbed oxygen.

Because of the barrier layer, the current in the dark should be much smaller. Furthermore, at first glance, the change in the dark current under the same experimental conditions for various ZnO crystals is astonishing. This behavior becomes understandable if the dissolution of the ZnO crystal in the electrolyte is considered. Therefore, quantitative dissolution experiments have been carried out.[16] The steady-state dissolution rate of a ZnO single crystal with the (0001) plane in contact with the electrolyte consisting of 0.1 M KCl and 0.1 M acetate buffer (pH = 4.6), which was attained after 20 hr,* amounted to 5×10^{-5} g/cm^2 · hr at 25°C. From this value and knowing the current during the dissolution, the density of donors can be calculated as about 2.5×10^{18} cm^{-3}, in agreement with values calculated by Kiess with another method.[17] From these experiments we may conclude that the dark current depends on both the speed of dissolution of the ZnO crystal and the density of donors in the crystal. The dissolution was increased by illumination under formation of hexagonal pits where the side of the hexagonal pit is parallel to the side of the hexagonal plane of the crystal (Figure 5).

4.1. Some Electronic Reactions at the ZnO Electrode

As an example we choose the decomposition of formic acid on a positively biased ZnO electrode under UV illumination investigated by Morrison and Freund.[18] The overall reaction

$$HCOO^-(ads) \longrightarrow CO_2(aq) + H^+(aq) + 2e'(ZnO) \tag{8}$$

---

*At shorter times the speed of dissolution is higher because of the initial roughness of the crystal surface.

Fig. 5. Photographs of hexagonal pits generated in the (0001) plane of a ZnO crystal in an acetate buffered (pH = 4.6) KCl electrolyte during illumination with UV light: (a) hexagonal pits at higher magnification; (b) the same hexagonal pits, the sides of which are parallel to those of the hexagonal plane.

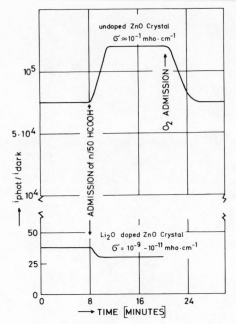

Fig. 6. The increase of the ratio of the photo- and dark current, i_{phot}/i_{dark}, of ZnO in an initially nitrogen flushed electrolyte (composition as in Figure 3) after addition of HCOOH, which is destroyed by oxygen. A Li_2O-doped ZnO crystal exhibits a small decrease of i_{phot}/i_{dark}.

results in a near doubling of the photocurrent when oxygen is absent. As can be seen from Figure 6, the ratio i_{phot}/i_{dark} is doubled when formic acid is added to a nitrogen-flushed KCl electrolyte. In the presence of oxygen, however, the doubling of the current disappears. We propose tentatively that the following mechanism

$$HCOO^-(ads) + e' \sim |e|^{\cdot} \longrightarrow HCOO^*(ads) + e' \qquad (9)$$

can be expected as starting step. This is followed by decomposition of the adsorbed radical $HCOO^*(ads)$ with simultaneous formation of protons and CO_2; an equivalent amount of free electrons is injected into the ZnO crystal (Figure 7):

$$HCOO^*(ads) \longrightarrow CO_2(aq) + H^+(aq) + e' \qquad (10)$$

This step represents an exothermic process supplying the energy necessary for the electron injection into the crystal which is responsible for the above-mentioned doubling of the anodic current. A schematic representation of the main steps during the decomposition of formic acid on an illuminated zinc oxide surface is shown in Figure 7. In the presence of oxygen, step (10) is substituted by the following one:

$$HCOO^*(ads) + O_2(aq) \longrightarrow CO_2(aq) + H^+(aq) + O_2^-(ads) \qquad (11)$$

and

$$HCOO^*(ads) + O_2(aq) + H_2O \longrightarrow CO_2(aq) + 3/2H_2O_2 \qquad (12)$$

As can be seen from the lower part of Figure 6, the anodic photocurrent of a Li_2O-doped zinc oxide crystal contacted with the same electrolyte and flushed with nitrogen was decreased by admission of formic acid. If we may tentatively assume that the Li_2O

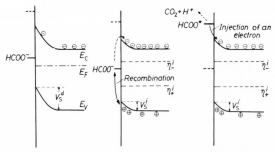

Fig. 7. Schematic representation of the main steps during the decomposition of formic acid on an illuminated zinc oxide surface at 25°C, according to Hauffe.[26] η_+^i and η_-^i denote the quasi-Fermi potentials ($\equiv$ imrefs) for holes and electrons, respectively, during illumination. V_S^d and V_S^i are the surface potentials in the dark and under illumination.

doping causes a p-type disorder according to

$$1/2O_2(\text{gas}) + Li_2O = 2Li|Zn|' + 2|e|^{\cdot} + 2ZnO \tag{13}$$

then the decrease of the current becomes understandable because the following steps are analogous to steps (9) and (10):

$$HCOO^-(\text{ads}) + |e|^{\cdot} + e' \sim |e|^{\cdot} \longrightarrow HCOO^*(\text{ads}) \tag{14}$$

and

$$HCOO^*(\text{ads}) + |e|^{\cdot} \longrightarrow CO_2(\text{aq}) + H^+(\text{aq}) \tag{15}$$

However, further experiments are necessary for the elucidation of this mechanism.

Typical compounds for current doubling are CH_3OH, C_2H_5OH, and As^{3+}. These compounds are two-equivalent chemical species in which the intermediate valence state is unstable so that they change valence by two electrons. In contrast one-equivalent chemical substances, as for instance $[Fe(CN)_6]^{4-}$ and halogen ions, cause no current doubling.[6,19]

We now consider a solvated one-equivalent oxidizing species X, able to capture an electron at the ZnO surface, using the relations deduced by Morrison.[5] One of the problems is the determination of the speed of the electron transfer for the reaction

$$X + e' \xrightarrow{k} X^- \tag{16}$$

and whether the rate or the current density J obeys a first-order law according to

$$J = ekn_S[X] = e\bar{c}\rho n_S[X] \tag{17}$$

i.e., first order in [X] and in the concentration of electrons at the surface, n_S. The rate is formulated in the first term of equation (17) in chemical kinetics where k is the rate constant and in the second term in solid-state parameters where ρ is the capture cross section for electrons and $\bar{c}$ the mean velocity of electrons in the conduction band. In both terms e denotes the elementary charge. The quantity of n_S can be determined by capacity measurement via V_S according to equation (2) and via the expression:

$$n_S = N_D \exp\left(\frac{-eV_S}{kT}\right) \tag{18}$$

Finally, the parameter $\rho[X]$, which may be called the electron reactivity of the oxidizing species, can be evaluated from equation (17).

In Figure 8 the electron reactivity is plotted vs. the redox potential for various one-equivalent oxidizing agents, as for instance Cr^{3+}, Cu^{2+}, Ce^{4+}, $[Fe(CN)_6]^{3-}$, $[Ag(NH_3)_2]^+$, MnO_4^- and $IrCl_6^{2-}$. It was shown that the vanadous and chromous ions which have a positive redox potential inject electrons into the conduction band of ZnO. The same was observed with $[Fe(Cn)_6]^{4-}$ which has an energy level about 0.15 eV below the conduction-band edge.[20]

While Gerischer and Beck[21] introduced a qualitative relationship between the redox potential $E°$ employed in Figure 8 and the energy level of a species, Morrison[5] derived a quantitative relationship for one-equivalent redox couples which yields in first approximation:

$$E_X = E_X° + \text{const.} \tag{19}$$

where E_X is the energy level of a solvated species and $E_X°$ the corresponding redox potential of the couple X/X^-. It is obvious that relation (19) can be improved by introducing the energy of hydration, hydrolysis, and other chemical transformations. The ZnO conduction-band edge was estimated to be at -0.2 ± 0.1 eV on the $E°$ scale of Figure 8. One may expect that electron capture involving energy levels substantially above the conduction-band edge must be ineffective and unfavorable. However, one cannot predict whether the capture cross section for stronger oxidizing species should pass through a peak and decrease. As can be seen from Figure 8, the value of $\rho[X]$ decreases rapidly for species with a redox potential more positive than $[Fe(CN)_6]^{3-}$. This fact can be considered as evidence that the $[Fe(CN)_6]^{3-}$ level is located near the conduction-band edge where a fast electron exchange occurs.

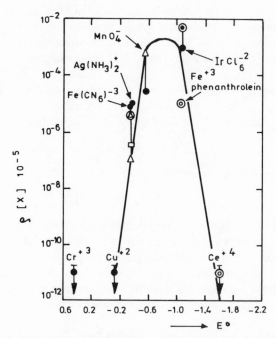

Fig. 8. The variation in electron reactivity (electron-capture cross section) as a function of the redox potential for various one-equivalent oxidizing agents according to Morrison.[5] The value of pH is indicated as follows. $\circ = 1.5$, $\bullet = 4$, $\triangle = 9$, and $\square = 12$.

4.2. Electrophoretic Phenomena

Further information on the mechanism of the electron exchange between a ZnO surface and an adsorbed species has been obtained by electrophoretic measurements of finely dispersed ZnO powder in an insulating solvent, for instance toluene or hydrocarbons, in the dark and under illumination.[22] The dispersion is stabilized by calcium diisopropylsalicylate or by calcium decanate. The calcium ions will be adsorbed on the surface of the ZnO particles while the anions remaining in the solvent encircle the positively charged ZnO grains in the form of a space-charge layer as is schematically represented in Figure 9. If an external voltage is applied to the electrophoretic cell consisting

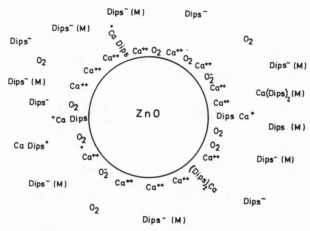

Fig. 9. Model of one Ca-diisopropylsalicylate-stabilized ZnO grain dispersed in toluene, according to Hauffe and Garcia.[22] Dips$^-$(M) = diisopropylsalicylate anion in form of a micelle, Ca(Dips)$_2$ = the undissociated molecule.

of a quartz chamber with two parallel plates, as represented in Figure 10, then an electrophoretic migration is observed in that the positively charged ZnO particles migrate toward the cathode. The velocity of the migrating species can be measured by a microscope under red light. Subsequently, we superpose a pulse of UV light from a xenon high-pressure lamp. The ZnO particle then stops and migrates in the opposite direction. Figure 11 represents such a measurement. The electrophoretic mobility of the particles has been calculated from experimental results as 1.2×10^{-5} in the dark and as 0.5×10^{-5} cm^2/V $\cdot$ sec under illumination.

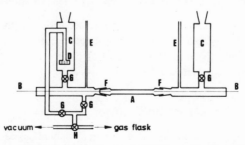

Fig. 10. Schematic representation of the measuring unit [From Hauffe and Volz.[22]] A = quartz measuring chamber, B = platinum electrodes, C = stock bin (glass), D = glass frit, E = capillaries (0.3 mm diameter), F = normal ground junctions, G = Rotaflon stopcocks, H = three-way stopcock.

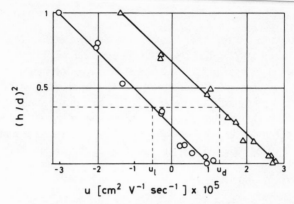

Fig. 11. Linearized form of the parabolic mobility distribution of ZnO particles (0.0005 M Ca-dips) through the depth of the measuring chamber. Dashed lines represent the depth at the zero value of the osmotic flux on the ordinate and the mobility u_d and u_1 in the dark and under illumination, respectively, on the abscissa. [From Hauffe and Volz.[22]]

From these experimental results we may tentatively deduce the following mechanism: Under the assumption that the stabilizer dissolved in toluene is dissociated

$$Ca^-(dips)_2 \rightleftharpoons Ca^-dips^+ (toluene) + dips^- (toluene) \qquad (20)$$

we may assume that the positively charged species are adsorbed on ZnO

$$Ca^-dips^+ (toluene) \rightleftharpoons Ca^-dips^+ (ads) \qquad (21)$$

and

$$Ca^-dips^+ (toluene) \rightleftharpoons Ca^{2+} (ads) + dips^- (toluene) \qquad (22)$$

while the negatively charged anions (dips = diisopropyl salicylate) are surrounding the ZnO particles (Figure 9). Since the reversal in migration under illumination appears only if oxygen is present (before the experiment the dispersion was flushed with oxygen), we suppose the following mechanism:*

$$O_2 (toluene) + e' \sim |e|^\bullet = O_2^- (ads) + |e|^\bullet \qquad (23)$$

$$|e|^\bullet + dips^- (toluene) = dips^x (ads) \qquad (24)$$

and/or

$$|e|^\bullet + toluene = toluene^+ (a \text{ ``solvated'' hole}) \qquad (25)$$

$$dips^x (toluene) + toluene = dips^- (toluene) + toluene^+ \qquad (26)$$

Depending on the extent of this change in charge according to the equations (23)–(26), the surface of the ZnO crystallites has changed its charge from positive to negative. These experiments also indicate the important role of electron exchange with the surroundings when electron-hole pairs are generated.

4.3. Sensitization of the Electron Transfer

The justification for the application of the electrochemical cell technique as a useful tool for the investigation of the electron transfer sensitized by light-excited dyes, as for

*In the absence of oxygen no change in the direction of migration could be detected.

instance rhodamine, orange, and other azo dyes, was confirmed by experiments with zinc oxide resin layers doped with the same dyes or dye mixtures employed in the electrochemical cell.[23] There resulted roughly the same wavelength dependence of the photo-current through the electrochemical cell and the speed of photo-discharge of a charged ZnO-resin layer (electrofax process). In these measurements, the ZnO single crystal operates as an anode in a suitable electrochemical cell. The corresponding dye or dye mixture was dissolved in the electrolyte according to the scheme

$$\text{light} \longrightarrow \text{thin In layer} | \text{ZnO crystal} | \text{electrolyte} + \text{dye} | \text{platinum} \qquad (27)$$

The light from a xenon high-pressure lamp was therefore passed through the rear of the crystal via a monochromator.

As can be seen from Figure 12, morin is a significant sensitizer, particularly in neutral and weakly basic electrolytes. The sensitizing ability of morin for the electron transfer from the electrolyte into zinc oxide was almost destroyed in an electrolyte (90 vol % CH_3OH + 10 vol % H_2O + 10^{-2} mole KCl/liter) with a pH between 3 and 4 established by an acetate buffer. Under optimum experimental conditions, the sensitizing ability of rhodamine B and the azo dye [1 (2-hydroxyphenylazo)-2-naphthol] is only about 30% of that of morin. The strong pH dependence of their sensitizing ability is worth noting. For instance, rhodamine B exhibits an excellent sensitization in acidic electrolytes, while in the basic region the sensitization is poor. In contrast to that, the azo dye exhibits its best sensitizing ability in a neutral or weakly basic electrolyte. The first requirement for sensitization by a dye is its capability for adsorption and the position of the excited energy level (singlet or triplet) relative to the conduction-band edge. As was demonstrated in a more extensive study,[23] the azo dyes with two OH groups in the 0-0' position exhibit a complex formation called chelate complexes with zinc ions, both in a neutral and in a basic electrolyte. Chelate complex formation in an acidic environment is not possible.

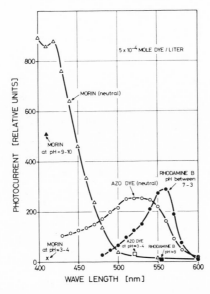

Fig. 12. Wavelength dependence of the photocurrent sensitized by morin, rhodamine B, and an azo dye in 10^{-2} M KCl solution, consisting of 90 vol % methanol and 10 vol % water, at various pH's. Concentration of the single dyes = 5 × 10^{-4} mole/liter each. [From Hauffe et al.[27]]

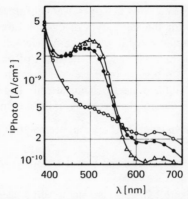

Fig. 13. Spectral sensitization of the anodic photocurrent in the presence of 2×10^{-4} mole/liter 2,2′-azodiphenol in a 90 vol % methanol–10 vol % water solution with 10^{-2} mole/liter $[(C_2H_5)_4]$ Cl (○ pH = 3.0, ● pH = 7.1, and △ pH = 11.0). [From Danzmann and Hauffe.[23]]

To demonstrate the correlation between the optical absorption of a chelate complex in the electrolyte and the remission spectrum of the dyed zinc oxide and the photocurrent sensitized by the dye, we use as an example 2,2′-azodiphenol. As can be seen from Figure 13, the photocurrent is sensitized by 2,2′-azodiphenol only in a neutral or basic electrolyte. Furthermore, the optical maximum of the sensitized anodic photocurrent at 500 nm agrees with the maximum of the remission spectrum (Figure 14) and of the absorption spectrum of the dye chelate, i.e., when zinc salt is added to the electrolyte (Figure 15). In the absence of zinc ions in the dye solution, the maximum appears at 400 nm.

As represented in Figure 16a, the sensitization of the photocurrent by orange II, which exhibits only a weak sensitization (5×10^{-4} mole/dm^3 in an aqueous electrolyte, buffered with sodium acetate + acetic acid to a pH of 4.5), was significantly increased when 0.9×10^{-4} mole/dm^3 rhodamine B was added.[24] Figure 16b shows the rise in photoconductivity through zinc oxide under illumination with 460-nm light (absorption maximum of orange II) when rhodamine B is introduced into the orange II-containing electrolyte. In correlation with this effect, a decrease in the yield of fluorescence of a 0.6×10^{-4} mole/dm^3 rhodamine B solution was observed when orange II was added (Figure 17).[24]

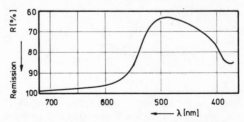

Fig. 14. Remission spectrum of 2,2′-azodiphenol on ZnO, adsorbed from a 10^{-3} M methanol solution, measured on undyed ZnO. [From Danzmann and Hauffe.[23]]

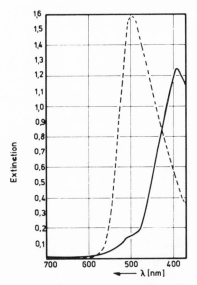

Fig. 15. Wavelength dependence of the absorption spectra of 10^{-4} M 90 vol % methanol–10 vol % water solution of 2,2′-azodiphenol without (—) and with (----) ZnSO$_4$. [From Danzmann and Hauffe.[23]]

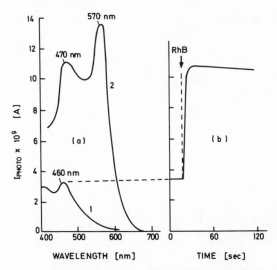

Fig. 16. The increase of the sensitization of orange II by addition of rhodamine B. (a) curve 1, orange II; curve 2, orange II + rhodamine B, 5×10^{-4} mole/dm^3 each. (b) The time dependence of the increase of the photocurrent after addition of rhodamine B. [From Hauffe and Bode.[24]]

These results can be discussed under the assumption that the azo dye orange II and rhodamine B form an associate adsorbed on zinc oxide and that during the illumination with 460-nm light, the azo dye will be excited and transfer its energy to rhodamine B (RhB), which injects an electron into ZnO:

$$\text{orange(ads)} + h\nu \longrightarrow \text{orange*(ads)} \tag{28}$$

$$\text{orange*(ads)} + \text{RhB(ads)} \longrightarrow \text{orange(ads)} + {}^+\text{Rhb}^-\text{(ads)} \tag{29}$$

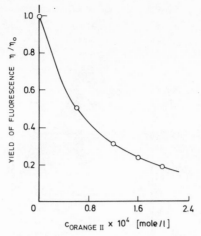

Fig. 17. The decrease of the yield of fluorescence of orange II with increasing amounts of rhodamine B. [From Hauffe and Bode.[24]]

$$^{+}RhB^{-}(ads) + A^{-}(aq) \longrightarrow RhB^{-}(ads) + A(aq) \qquad (30)$$

$$RhB^{-}(ads) \longrightarrow RhB(ads) + e' \ (ZnO) \qquad (31)$$

where A^{-} represents an anion, e.g., Cl^{-}, OH^{-}, Ac^{-}. This mechanism of a combined energy and electron transfer sensitization in the sequence (28)–(31) seems to have general validity. The existence of the assumed associated species is supported by the enhancement of the adsorption of rhodamine B in the presence of orange and by the significant decrease in the yield of fluorescence of rhodamine B due to increasing concentrations of orange II. This decrease in fluorescence follows a mechanism which is quite different from that caused by addition of halide ions.

Another example of the influence of a binary dye system on its sensitizing ability is represented in Figure 18 where the photocurrent in an equimolar mixture (5×10^{-4}

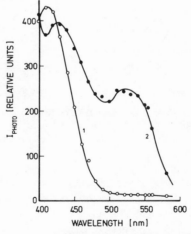

Fig. 18. Wavelength dependence of the photocurrent in an equimolar mixture (5×10^{-4} mole dye/liter each) of morin and 1 (2-hydroxyphenylazo)-2-naphthol in a basic (curve 2) and in a neutral (curve 1) electrolyte (10^{-2} M KCl in 90 vol % methanol–10 vol % water).[27]

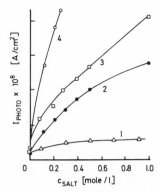

Fig. 19. Sensitization of the anodic photocurrent by a 10^{-4} mole/liter aqueous rhodamine B solution (pH = 4.6) as a function of the concentration of various salts:[25] 1, KNO_3; 2, KCl; 3, KBr; and 4, KI.

mole/dm^3 of each dye) of morin and 1(2-hydroxyphenylazo)-2-naphthol in a basic (curve 2) and in a slightly acidic electrolyte (90 vol % methanol + 10 vol % H_2O + 10^{-2} mole/dm^3 KCl) is plotted vs. the wavelength of absorbed light. While the sensitizing ability of morin in the acidic electrolyte is still significant, that of the azo dye is completely destroyed.

It has been found that halogen ions are able to increase the sensitizing ability of rhodamine B.[25] The increase in the sensitized photocurrent becomes larger in the sequence Cl$^-$, Br$^-$, and I$^-$ (Figure 19). KNO_3 has no influence on the sensitization. These results are correlated with the decrease in the yield of fluorescence of rhodamine B in the presence of halogen ions (Figure 20). The promoting influence of the halide ions, which we call cosensitizers, on the sensitization of the anodic photocurrent by rhodamine B and on the quenching of the fluorescence have prompted us to propose a mechanism

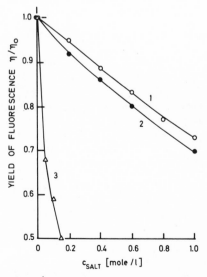

Fig. 20. Fluorescence yield of a 10^{-4} mole/liter aqueous rhodamine B solution as function of halide salt concentration (pH = 4.6): 1, KCl; 2, KBr; and 3, KI.

based on an electron transfer from the quenching A^- ion (Cl^-, Br^-, and I^-) to the excited sensitizer molecule $^+RhB^-$ according to

$$A^-(aq) + {}^+RhB^-(aq) \longrightarrow RhB^-(aq) + A(aq) \tag{32}$$

with the subsequent step

$$A(aq) + RhB^-(aq) \longrightarrow A^-(aq) + RhB(aq) \tag{33}$$

for quenching of the fluorescing dye in a homogeneous solution, or

$$A^-(aq) + {}^+RhB^-(ads) \longrightarrow A(aq) + RhB^-(ads) \tag{34}$$

and

$$RhB^-(ads) \longrightarrow RhB(ads) + e' \; (ZnO) \tag{35}$$

for sensitization of the photocurrent. On the basis of this mechanism, the correlation between fluorescence and sensitization is obviously in contrast to the mechanism of quenching the fluorescence of rhodamine B by orange II.

Without doubt, research on semiconductor/electrolyte interfaces will supply further stimulating results which will help to elucidate rather complex surface phenomena, for instance catalytic reactions, sensitization of electron transfer and semiconductor electrochemistry. It is obvious that these relations demonstrated with zinc oxide may be employed on cadmium sulfide, titanium dioxide, and other broad-band semiconductors.

Acknowledgment

Some of these results have been obtained by investigations sponsored by the Deutsche Forschungsgemeinschaft and the Fonds der Lehemischen Industrie.

References

1. W. H. Brattain and C. G. B. Garrett, *Bell Syst. Techn. J. 34*, 129 (1955).
2. J. F. Dewald, *Bell. Syst. Techn. J. 39*, 615 (1960).
3. H. Gerischer and H. Tributsch, *Ber. Bunsenges. Phys. Chem. 72*, 251 (1968); H. Tributsch and H. Gerischer, *Ber. Bunsenges. Phys. Chem. 73*, 251 (1969).
4. K. Hauffe and J. Range, *Z. Naturforsch. 23b*, 736 (1968); K. Hauffe, H. J. Danzmann, H. Pusch, J. Range, and H. Volz, *J. Electrochem. Soc. 117*, 993 (1970).
5. S. R. Morrison, *Surface Sci. 15*, 363 (1969).
6. S. R. Morrison and T. Freund, *J. Chem. Phys. 47*, 1543 (1967); *Electrochim. Acta 13*, 1343 (1968).
7. S. R. Morrison, Surface phenomena associated with the semiconductor/electrolyte interface, *Prog. Surface Sci. 1*, 105 (1972).
8. H. Gerischer, *J. Electrochem. Soc. 113*, 1174 (1966).
9. K. Hauffe and J. Range, *Ber. Bunsenges. Phys. Chem. 71*, 690 (1967).
10. See for instance W. Mehl and J. M. Hale, Insulator electrode reactions, *Adv. Electrochem. Electrochemical Eng. 6*, 399 (1967).
11. H. Gerischer, *Z. Phys. Chem, N.F. 26*, 223, 325 (1960); *27* 48 (1961); see also *Adv. Electrochem. Electrochemical Eng. 1*, 139 (1961).
12. R. W. Gurney, *Proc. R. Soc. London 134A*, 137 (1931).
13. R. A. Marcus, *Annu. Rev. Phys. Chem. 15*, 155 (1964).
14. V. G. Levich and R. R. Dogonadze, *Dokl. Acad. Nauk SSSR 133*, 158 (1960).
15. V. A. Myamlin and Y. V. Pleskov, *Electrochemistry of Semiconductors*, Plenum Press, New York (1967).

16. K. Hauffe and O. Haeggqwist, *Z. Phys. Chem.* N.F. *85*, 191 (1973); H. Erbse, K. Hauffe, and J. Range, *Z. Phys. Chem.* N.F. *74*, 248 (1971).
17. H. Kiess, *J. Phys. Chem. Solids 31*, 2379, 2391 (1970).
18. S. R. Morrison and T. Freund, *J. Chem. Phys. 47*, 1543 (1967).
19. W. Gomes, T. Freund, and S. R. Morrison, *J. Electrochem. Soc. 115*, 818 (1968); *Surface Sci. 13*, 201 (1968).
20. S. R. Morrison, *Surface Sci. 13*, 85 (1969).
21. H. Gerischer and F. Beck, *Z. Phys. Chem. N.F. 13*, 389 (1957).
22. K. Hauffe and H. Volz, *Ber. Bunsenges. Phys. Chem. 77*, 967 (1973).
23. H. J. Danzmann and K. Hauffe, *Ber. Bunsenges, Phys. Chem. 79*, 438 (1975).
24. K. Hauffe and U. Bode, *Disc. Faraday Soc. 58*, 281 (1974).
25. U. Bode, K. Hauffe, Y. Ischikawa, and H. Pusch, *Z. Phys. Chem. N.F. 85*, 144 (1973).
26. K. Hauffe, *Rev. Pure Appl. Chem. 18*, 79 (1968).
27. K. Hauffe, H. Pusch, J. Range, and D. Rein, Dye sensitization of zinc oxide by means of the electrochemical cell technique, in: *Current Problems in Electrophotography* (W. F. Berg and K. Hauffe, eds.), p. 178 ff Walter de Gruyter, Berlin (1972).

13

Corrosion—The Past

E. C. Potter

1. Electrochemical Basis of Corrosion

One has only to contemplate that the pocket flashlight and the portable transistor radio are commonly driven by the Leclanché dry cell to realize that metallic corrosion is electrochemical in origin. Roughly 10 mg of zinc are electrochemically oxidized every second a small flashlight is operated. If we assume that one quarter of the world's population owns a "transistor" and has it switched on for an average of one hour a day, then some 30 metric tons of zinc are dissolved daily in this way. Clearly, metallic corrosion can amount to electrochemistry on a grand scale.

Notwithstanding the indubitably electrolytic course of corrosion in various primary batteries, it was long questioned whether the most familiar manifestations of immersed metallic corrosion are electrochemical. The doubts were profound enough for researchers to prove specifically that the immersed rusting of iron is quantitatively faradaic. This was accomplished by Hoar with Evans[1] in 1932, and Agar's more elaborate experiments[2] with zinc in 1939 finally secured the electrochemical basis for immersed corrosion. The proof that atmospheric corrosion (i.e., in moist air) can also be electrochemical by virtue of electrolysis within aqueous surface films only about 30 μm in thickness was made by Rosenfeld.[3] In apparent contrast, the dry high-temperature oxidation of a metal is not electrochemical in the usual sense, but an electric current must negotiate an ionic oxide growing on its parent metal in order to sustain the formation of fresh oxide ions from the surrounding oxygen molecules. The field-assisted enchancement of this process, akin to an electrolysis, was described by Lacombe and his associates.[4] Thus, the three principal categories of metallic corrosion have all been shown to have an electrochemical basis, although, for the purposes of this paper, custom will be observed in according only immersed corrosion (including underground corrosion) a full connection with the science of electrochemistry.

2. Potential/pH Diagram

Only the precious metals, with silver and copper, occur native. While this indicates that the remaining metals have long since reached terrestrial equilibrium in the combined state, nevertheless the accumulated bulk of certain metals that man has liberated or blended is now destined to remain as metal virtually forever. For example, the metals aluminum, titanium, chromium, tantalum, and their separate alloys (including austenitic stainless steel) are permanent not only in air but often also underground and in fresh and saline waters. These facts are inconsistent with the electromotive series, which has been and sometimes still is invoked to explain how metals at one end of the series are base and

E. C. Potter • CSIRO, Division of Process Technology, North Ryde, Australia, 2113.

corrodible while others at the opposite end are noble and corrosion-resistant. The series fails because aluminum is base yet is not attacked by fresh water, while gold is noble but is dissolved by certain cyanide solutions. In addition, the fixed intermediate position of hydrogen in the series is inappropriate and only a proportion of metal displacement reactions happen in the way predicted by the series.

These anomalies and difficulties found some resolution through Pourbaix[5] who, in 1938, recalled that the equilibrium potential of a metal in an aqueous solution depends not only on the activities (or concentrations) of the relevant metal ions in solution but often also on the hydrogen ion activity (as conveniently expressed by pH). Out of this essentially thermodynamic approach grew the now familiar potential/pH diagram. The Second World War prevented these diagrams from becoming widely known, and it was not until April 1946 that Pourbaix lectured on them at Cambridge University. Subsequently Agar translated Pourbaix's book, which was published in English[6] in 1949. The diagrams then caught on, so much so that Pourbaix (aided by an international group of collaborators and under the auspices of CEBELCOR*) spent a large part of the 1950s assembling potential/pH diagrams for all the chemical elements. A glimpse of the metals side of this task was presented to the 1st International Congress on Metallic Corrosion[7] in 1961, and the full *Atlas of Electrochemical Equilibria in Aqueous Solutions*[8] appeared soon afterward.

Although the potential/pH diagram has been generally available for almost 30 years, it has been developed chiefly by a select group of devotees, and the time has probably not yet arrived for a balanced assessment of its impact on corrosion science. Certainly, however, the potential/pH diagram is attractive educationally for summarizing in easy visual style a diversity of metals performance in aqueous conditions. For each metal a domain of absolute stability (immunity, as conferred by cathodic protection) is revealed, but the other two types of domain, corresponding to corrosion and passivity, cannot necessarily be interpreted realistically. For example, if a metal finds itself in a corrosion domain of its potential/pH diagram, there is no indication of the rate of the corrosion or of the form it will take; furthermore there is no guarantee that a passivity domain delineates conditions where the oxide film has the proper adhesion and physical structure to be protective. Attempts to incorporate such information in the diagrams begin to mar their initial simplicity and attraction.

Of the many examples that illustrate the value or limitations of potential/pH diagrams, two will be included here. The first shows how convenient the *Atlas* is to survey and select those metals that are resistant to a certain aqueous environment, namely pure water at 25°C. In Figure 1, by Pourbaix and his associates,[9] the unhatched areas are the theoretical domains of immunity and passivation, and the hatched areas similarly indicate corrosion. The central band (shaped like a parallelogram) common to all the diagrams and sloping from upper left to lower right defines the region of stability of water. If, therefore, the rest potential of the metal on the vertical line at pH 7.0 lies inside the stability band for water at an unhatched spot, then that metal is suitable at 25°C for the storage of pure water without significant contamination. Figure 1 comprises the diagrams for the 13 metals (excepting 5 platinum-group metals) that meet the conditions set. The practical man will particularly note the confirmation of the favorable position of aluminum, and the laboratory technician will be convinced of the wisdom of the block tin condenser even though the diagram applies strictly only for 25°C. The metallurgist will notice the absence of suitable alloys from the display.

*Centre Belge d'Étude de la Corrosion.

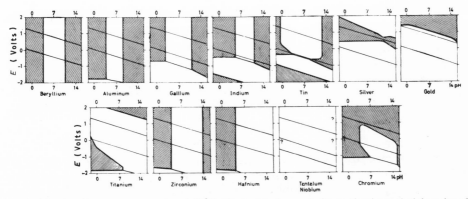

Fig. 1. Metals resistant to pure water at 25°C. The hatched areas indicate the theoretical domains of corrosion. The unhatched areas indicate the theoretical domains of immunity and passivation. [From M. Pourbaix and G. Govaerts.[9]]

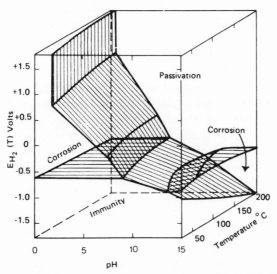

Fig. 2. Potential–pH–temperature diagram assuming passivation by films of Fe_3O_4 and Fe_2O_3, taking $\bar{S}^{\circ}_{HFeO_2}$ for 25°C = 10 e.u. [From Townsend, Figure 6.[10]]

The other example demonstrates that potential/pH diagrams may be constructed for temperatures far from 25°C. Figure 2 reproduces* Townsend's calculated diagrams for iron[10] as a three-dimensional display covering temperatures up to 200°C. Here the usefulness of the display is more illusory since [as has been pointed out by Ashworth[11]] the pH scale is itself temperature dependent in that the neutral point falls from 7.0 to below 6.0 in the temperature range considered and the composition of the solution is an important factor through the dissociation constant of the solute. For example, assuming ionic activity coefficients are unity, the pH of a decinormal, fully dissociated acid remains 1.0 at all temperatures. However, for the same description of alkali, it is pOH that remains 1.0 at all temperatures, causing the corresponding pH to fall from 13.0 at 25°C to about 10.5 at 200°C. Indeed, in appreciating the pH scale we should view pH 13.0 at

*Some obvious errors in the original diagram have been corrected.

200°C with the same curiosity as pH 15.5 at 25°C. In consequence not too much significance need be attached to the enlargement with rising temperature of the small wedge at
the right of the front face in Figure 2. This is the domain of stability of the soluble
species $HFeO_2^-$ formed when iron corrodes in strongly alkaline solution, a situation with
some practical bearing on the safety of high-pressure steam generators.

3. Potential/Current Diagram

It is, of course, extremely useful to know if a metal is immune to corrosion in given
circumstances using the potential/pH diagram, but practical situations most often subject
metals to corrosive risks and the *rate* of corrosion then becomes of great importance.
From about the turn of the century the pioneer corrosion workers gave considerable
attention to corrosion-rate measurements, but the advent of polarization curves in corrosion science waited until Evans and Hoar[1] had demonstrated that the current in an
ordinary corrosion cell was worth equating to the metal loss. The vogue then started for
diagrams with two polarization curves, separate for anodes and cathodes on the corroding
metal, each being plotted as a straight line against linear scales of potential and current.
The anodic and cathodic lines were seen to approach each other with increasing current,
and at their intersection a common potential and a maximum corrosion current were
reached, a situation freely identified with the practical corrosion cell where the cathodes
and anodes would be virtually short-circuited. Depending on the influence of the exposure conditions on the steepness and positions of the individual polarization lines, there
arose a classification of corrosion processes depending on whether they were controlled
by the cathodic, the anodic, or by both reactions.[12,13] Likewise a separation of inhibitors into cathodic or anodic types emerged. Since, however, the separate polarization
lines were near to inaccessible to observation in many practical cases, the so-called Evans
diagrams had to be predominantly schematic and have become more an influence in corrosion education than in research.

This situation changed radically with the acceptance of the potentiostat in the corrosion laboratory about 20 years ago. The deliberate, controlled anodization of the whole
of a corrosion specimen then began to have additional meaning, particularly for the study
of passivity. One of the pioneers was Edeleanu,[14] who helped to make the semilogarith-
mic anodic polarization curve of a passive metal so well known that in a decade it was
used in the emblem of the 3rd International Congress on Metallic Corrosion. Edeleanu
showed how to combine this curve with the corresponding cathodic curve so that the
intersections explained how passivity would sometimes be unreliable. In doing this he was
tacitly using the concept of mixed potential introduced by Wagner and Traud[15] in 1938
and since invoked at some time by almost every electrochemist working in corrosion
science. It is difficult to combine justice and brevity when reporting the importance of
the anodic polarization curve to the advance of corrosion science in recent years, but two
examples show something of this influence.

A characteristic of the anodic passivation curve is that the dissolution rate of an
already active metal rises sharply to a maximum as the metal potential is made more
positive. Further anodic movement of the potential causes an abrupt decrease of the
dissolution rate to the low plateau associated with the passive state. The peak dissolution
rate corresponds to the critical current density for passivation, which depends chiefly on
the metal and the composition and concentration of the solution. In some cases the
passive state may persist over one volt or more if anodization is extended, but at sufficiently positive potential the transpassivation region is entered and the passivity succumbs

to the highly oxidizing environment. As an example, Kolotyrkin[16] found that a specimen of chromium immersed in normal sulfuric acid at room temperature was dissolving at a rate close to 10 A/m². Upon shifting the rest potential of the metal by at least 0.5 V in the anodic direction, a peak dissolution rate of about 200 A/m² (the critical current density for passivation) was passed through before the dissolution rate dropped to about 5×10^{-4} A/m² as the metal became passivated. This diminution of the ordinary corrosion rate of chromium by a factor of 20,000 upon passivation is obviously all the difference between failure and success with this metal in the acid solution, and the opportunity therefore arises to procure protection by deliberately putting the potential of the chromium at a value (such as +0.1 V on the hydrogen scale) where passivity is assured. To do this the chromium is combined externally with a suitable cathode in the same acid solution, and a battery in a control circuit keeps the potential of the chromium anodic in the passive range. This method, which bears some procedural resemblance to *cathodic* protection, has been called *anodic* protection[17] and has been successfully applied industrially in a number of instances where the metal is capable of passivation.[18] Arguably, anodic protection is the only completely new anticorrosive principle that has emerged in the period under review, and notably it has been devised as a direct result of electrochemical kinetics research.

A second example of the influence of the anodic polarization curve in corrosion science arises from the work of Kolotyrkin and his associates on the effect of solution composition. Nickel immersed in normal sulfuric acid at ordinary temperature dissolves slowly at a rest potential not far from zero on the hydrogen scale. The addition to the acid of different types and concentrations of oxidizing agents shifts the rest potential to more positive values, and the corresponding effect on the dissolution rate can be directly observed and converted into the equivalent current density. In Kolotyrkin's work with Bune,[19] the rest potential was shifted over a range up to +1.4 V on the hydrogen scale by selection of the oxidizing agent from among ferric sulfate, ceric sulfate, potassium

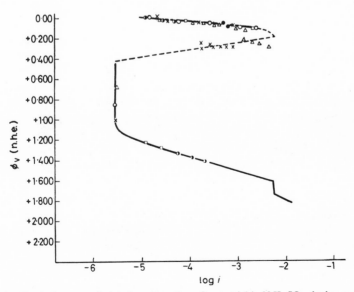

Fig. 3. Steady dissolution rate of nickel as a function of potential in N H_2SO_4 during anodic polarization (the curve), and without polarization in presence (at different concentrations) of: o, $Fe_2(SO_4)_3$; ×, $K_2Cr_2O_7$; Δ, H_2O_2; □, $KMnO_4$; ◑, $Ce(SO_4)_2$; •, O_2. [From Ya. M. Kolotyrkin.[19]]

dichromate, potassium permanganate, dissolved oxygen, and hydrogen peroxide. Afterward, the derived potential/current trace, arising solely from chemical adjustment of the system (no polarization), was seen to be the same as that produced by polarization alone in the pure sulfuric acid (Figure 3). To use Kolotyrkin's elegant words:

> These results testify first of all to the absence of any difference in principle between anodic and chemical passivation, and consequently to the fact that the ions and molecules of the oxidizing agent do not participate directly in the formation of the passive film, since if they did, it would be hard to explain the absence of a specific influence of the type of the oxidizing agent on the kinetics of the anodic dissolution of the metal.

4. The Tafel Line in Corrosion Science

The modern story of electrochemical kinetics pivots around the Tafel line in one way or another. The electrochemist entering corrosion science sooner or later commits himself to probing the meaning of Tafel lines for corroding metals. In this connection, the great Russian tradition in electrochemical kinetics established by Frumkin and a succession of collaborators was the genesis of present-day thinking. Around 1940 (but the work did not become generally known until after 1945) Kolotyrkin,[20] of Frumkin's school, investigated why the Tafel line observed for nickel in dilute mineral acid, instead of proceeding asymptotically to the reversible (hydrogen) potential as the meter current approached zero, levelled off to a (rest) potential some tens of millivolts cathodic to the expected reversible potential. It was known, of course, that this distortion of the Tafel line was connected with the dissolution of the nickel in the electrolyte, but Kolotyrkin made some painstaking experiments to disentangle the electrolytic events quantitatively.

He set the nickel potential at a number of values on either side of the rest potential (reversing the meter current as required), and at each potential he measured volumetrically the rate of hydrogen evolution. These volumetric rates gave him the electric currents for hydrogen evolution and, on plotting these with the corresponding potentials, he traced the Tafel line that would have been observed but for the interfering effect of the nickel dissolution. On subtracting this hydrogen-evolution Tafel line from the "distorted" Tafel line given by the ordinary fully electrical procedure, Kolotyrkin obtained the Tafel line for the dissolution of nickel and only needed to read off the current on this line at the observed rest potential to estimate the corrosion rate of nickel immersed in the dilute mineral acid he had selected. Nowadays he would have saved himself much work by extrapolating the straight section of the ordinary cathodic Tafel line back to the rest potential and identifying the current at this conjunction with the corrosion rate of the nickel.

Although the newcomer to corrosion-science literature may be excused for missing the connection, almost 20 years later Stern and Geary[21] were able to elaborate on Kolotyrkin's approach, because advances in electrochemical kinetics had by then produced a detailed two-term expression for the whole Tafel line starting from the reversible potential.[22] By applying the full Tafel expression to the case where the $M \rightleftharpoons M^{z+}$ reaction was balanced by (say) the $2H^+ \rightleftharpoons H_2$ reaction, and then making some simplifying assumptions, they obtained a convenient and convincing expression for the metallic corrosion current. The expression is,

$$i_{corr} = (1/2.303r) \left[b_a b_c / (b_a + b_c) \right], \tag{1}$$

where i_{corr} is the corrosion current, r is the polarization resistance at the corrosion (rest) potential, and b_a and b_c are the numerical Tafel slopes for anodic and cathodic reactions, respectively. In many cases i_{corr} is reasonably insensitive to the values of b_a and b_c, so that the essence of the method is the measurement of r from the linear polarization region close to the rest potential. Actually, the empirical use of polarization–resistance measurements to estimate corrosion rates of metals in aqueous solutions had been foreshadowed by Bonhoeffer and Jena[23] and had already been practiced by Skold and Larson.[24] In a sense what Stern and Geary accomplished was the quantitative expression of the constant of proportionality between polarization resistance and corrosion rate.

Response to the Stern–Geary equation has been mixed since its introduction in 1957. On one hand more rigorous treatments of its theoretical basis have been published, leading to unwelcome complexity; on the other hand many practical corrosion workers trust the original equation (1) and the corrosion rates calculated using it. The position is much in need of assessment, but the following develops a brief but broad opinion.

Oldham and Mansfeld[25] have examined theoretically the effect on the reliability of the Stern–Geary equation of making the original assumptions. This examination, which amounted to balancing four partial electrode reactions instead of two as Stern and Geary did, shows that this latter scheme suffices provided the two exchange-current densities are not too close. Otherwise the corrosion rates calculated from equation (1) may be many times in error, for which circumstance Mansfeld and Oldham[26] suggest an extra term for equation (1) to avoid most of the discrepancy. Other authors, notably Barnartt[27] and also Matthews,[28] have discussed the conditions controlling the length of the linear departure of potential from the rest value as meter current is raised, and there is agreement that the slope (polarization resistance) at the rest potential is the proper value to take. Thus, from the point of view of principle and derivation, the Stern–Geary equation appears a sound simplification of a more complex general equation. Admittedly the simplified equation can yield wrong results under certain coincidental conditions, but there seems no evidence that such conditions have intervened vitally in a practical case.

The other consideration affecting the utility of the Stern–Geary equation is the severely practical one of checking its results against those of a direct method such as weight loss. A recent demonstration of such scrutiny is due to Driver and Meakins,[29] who have estimated the corrosion rates of steel in dilute mineral acid with and without the addition of selected organic corrosion inhibitors. Table 1 includes a representative selection of the results, showing good concordance among corrosion rates estimated from

Table 1. Comparison of Stern–Geary Method with Other Methods for Estimating the Corrosion Rate of Steel in Mineral Acid at 20°C [Extracted from Driver and Meakins[29]]

Acid	Inhibitor	Inhibitor concentration	Tafel slopes (mV)		Corrosion rate (A/m^2)		Corrosion inhibition (%)	
			b_a	b_c	Tafel intersection	Stern–Geary equation	Stern–Geary	Weight loss
1 M HCl	A^a	nil	74	82	1.38	1.37	–	–
		10^{-5} M	64	75	0.24	0.25	82	83
		10^{-3} M	71	72	0.027	0.028	98	96
0.5 M H$_2$SO$_4$	B^b	nil	81	97	1.65	1.57	–	–
		10^{-5} M	110	90	0.51	0.54	66	75
		10^{-3} M	235	81	0.038	0.037	98	96

^{a}A = n-dodecylpyridinium bromide, $C_{12}H_{25}NC_5H_5Br$.
^{b}B = N, N'-diethylthiourea, $C_2H_5NH \cdot CS \cdot NHC_2H_5$.

Table 2. Estimated Corrosion Rate of Copper in Aerated 0.1 M NaCl at 20°C

Tafel slopes (mV)		Polarization resistance (Ω)		Corrosion rate (A/m^2)		
b_a	b_c	r_a	r_c	Tafel intersection	Stern–Geary (using r_a)	Stern–Geary (using r_c)
0.063	0.24	6.5×10^4	3.7×10^4	2.02×10^{-2}	3.04×10^{-2}	5.34×10^{-2}

(1) weight loss, (2) the Stern–Geary equation, and (3) extrapolation of the Tafel line back to the rest potential.

In Driver and Meakin's work the cathodic reaction was the evolution of hydrogen and the corrosion product was soluble, so that there was a good opportunity to observe cathodic and anodic Tafel lines without profound alteration to the metal surface. In many other practical cases, however, the cathodic reaction is the electroreduction of dissolved oxygen, and the final corrosion product progressively alters or obscures the corroding metal surface by deposition, depending on the number of coulombs passed during observation of Tafel lines and the order in which the various anodic and cathodic excursions are carried out. An example, given by Nancarrow (personal communication), is the corrosion of copper in aerated 0.1 M sodium chloride at 20°C, where the observed polarization resistances, r_a and r_c, for respective anodic and cathodic departures from the rest potential, differ substantially, indicating that the practical procedure itself influences the course and rate of corrosion. The uncertainties of the Stern–Geary method in this case are appreciated from Table 2.

It appears, therefore, that the Stern–Geary equation should be used with much circumspection, particularly in the common practical case of solid deposition on the corroding surface. However, the situation is not so self-defeating that the observer should always prove first by independent measurement that the Stern–Geary equation is suitable to use with the system of interest. Much the same opinion has been expressed by Stern and Weisert.[30]

5. Concluding Remarks

At the start of the period under review the study of corrosion science was still the province of a relatively small number of teams, chiefly in universities and official laboratories. Today, however, corrosion science is organized and its study widespread; many nations have at least one society or organization devoted to corrosion science and protection, and about 10 international journals specialize in corrosion-science publication. In North America the National Society of Corrosion Engineers and in Europe the European Federation of Corrosion are large and active media for the promotion of every aspect of corrosion science, and the International Congress on Metallic Corrosion has convened six times in the past 15 years. Corrosion science has become a huge subject, now covered by selective rather than comprehensive literature.

Indeed, this summary of corrosion-science development during the past 30 years has been selective, even allowing for a curtailment to the electrochemical sector of the subject. However, rather than compile a detailed catalog of often disjointed progress, the author has chosen to present milestones on the trail as he himself perceives it.

References

1. U. R. Evans and T. P. Hoar, *Proc. R. Soc. 137*, 343 (1932).
2. J. N. Agar; see U. R. Evans, *J. Iron Steel Inst. 141*, 221 P (1940).
3. I. L. Rosenfeld, *Atmospheric Corrosion of Metals*, 238 pp. Engl. trans. publ. by National Assoc. of Corrosion Engrs., Houston, Texas (1972, original Russian edition, 1960).
4. F. Schein, B. le Boucher, and P. Lacombe, *Compt. Rend. 252*, 4157 (1961); P. Desmarescaux and P. Lacombe, *Compt. Rend. 256*, 5133 (1963).
5. M. Pourbaix, *Métaux Corros. 14*, 189 (1938).
6. M. Pourbaix, *Thermodynamics of Dilute Aqueous Solutions, with Applications to Electrochemistry and Corrosion* (J. N. Agar, trans.) 136 pp., Edward Arnold, London (1949).
7. M. Pourbaix and G. Govaerts, *Proc. 1st International Congress on Metallic Corrosion*, p. 96 Butterworth, London. Editorial note signed by L. Kenworthy (1962).
8. M. Pourbaix, *Atlas of Electrochemical Equilibria in Aqueous Solutions*, Pergamon Press, New York (1966).
9. M. Pourbaix and G. Govaerts, *Proc. 1st International Congress on Metallic Corrosion*, p. 98 Butterworth, London. Editorial note signed by L. Kenworthy (1962); M. Pourbaix, *Atlas of Electrochemical Equilibria in Aqueous Solutions*, p. 72, Pergamon Press, New York (1966).
10. H. E. Townsend, *Proc. 4th International Congress on Metallic Corrosion*, p. 477, National Association of Corrosion Engineers, Houston, Texas. Ed. N. E. Hamner (1969), *Corros. Sci. 10*, 343 (1970).
11. V. Ashworth, *Proc. 4th International Congress on Metallic Corrosion*, p. 486, National Association of Corrosion Engineers Houston, Texas. Ed. N. E. Hamner (1969).
12. U. R. Evans, *An Introduction to Metallic Corrosion*, 2nd ed., p. 69 et seq, Edward Arnold, London (1963).
13. H. H. Uhlig, *Corrosion and Corrosion Control*, 2nd ed., p. 50 et seq, Wiley, New York (1971).
14. C. Edeleanu, *J. Iron Steel Inst. 188*, 122 (1958).
15. C. Wagner and W. Traud, *Z. Elektrochem. 44*, 391 (1938).
16. Ya. M. Kolotyrkin, *Proc. 1st International Congress on Metallic Corrosion*, p. 11 Butterworth, London. Editorial note signed by L. Kenworthy (1962).
17. C. Edeleanu, *Nature 173*, 739 (1954); *Metallurgia 50*, 113 (1954).
18. H. H. Uhlig, *Corrosion and Corrosion Control*, 2nd ed., p. 227, Wiley, New York (1971).
19. Ya. M. Kolotyrkin and N. J. Bune, *Z. Phys. Chem. 214*, 264 (1960); also Ya. M. Kolotyrkin, *Proc. 1st International Congress on Metallic Corrosion*, p. 13 Butterworth, London. Editorial note signed by L. Kenworthy (1962).
20. Ya. M. Kolotyrkin and A. N. Frumkin, *Dokl. Akad. Nauk SSSR 33*, 445 (1941).
21. M. Stern and A. Geary, *J. Electrochem. Soc. 104*, 56 (1957).
22. J. O'M. Bockris, *Modern Aspects of Electrochemistry*, Chapter IV, p. 180 et seq, Butterworth, London (1954).
23. K. F. Bonhoeffer and W. Jena, *Z. Elektrochem. 59*, 151 (1951).
24. R. V. Skold and T. E. Larson, *Corrosion 13*, 139 (1957); T. E. Larson, R. V. Skold, and E. Savinelli, *J. Am. Water Works Assoc. 48*, 1274 (1956).
25. K. B. Oldham and F. Mansfeld, *Corrosion 27*, 434 (1971).
26. F. Mansfeld and K. B. Oldham, *Corros. Sci. 11*, 787 (1971).
27. S. Barnartt, *Corros. Sci. 9*, 145 (1969).
28. D. B. Matthews, *Aust. J. Chem. 28*, 243 (1975).
29. R. Driver and R. J. Meakins, *Br. Corros. J. 9*, 227 (1974).
30. M. Stern and E. D. Weisert, *Proc. Am. Soc. Test. Mater. 59*, 1280 (1959).

14

Corrosion—The Future

Einar Mattsson

Abstract

The continuous growth in world population together with the expected decline in high-grade ores and energy resources will necessitate a more efficient utilization of metal-based construction materials. In addition, in the more advanced technology developed to meet the future needs of the community, equipment will be developed where failures are highly disruptive or even unacceptable, either because of health hazards, e.g., in nuclear reactors, or because shutdowns would be prohibitively expensive. These advances must lead to an augmented demand for materials of increased corrosion resistance and for better methods of corrosion control, protection, and prevention.

Better corrosion resistance can be achieved via improved materials selection and better design, both of which can be effected on the basis of current knowledge. Thus, engineers will need to be able to recall relevant information from corrosion literature in a more convenient way. Computer-based retrieval of data will hence be commonplace, and the stored information must be more complete and better classified than today; international cooperation seems appropriate to achieve this end.

Improved corrosion resistance will in many cases require research on the mechanism of corrosion processes. In this work, electrochemical methods including the recent electrode impedance technique will certainly play an important role. Special attention will have to be paid to the following phenomena: stress corrosion cracking, passivity, atmospheric corrosion, and wet corrosion at high temperatures. Of great interest among metallic construction materials are high-strength–low-alloy steels, aluminum, titanium, magnesium, manganese, and metal matrix composites.

In advanced equipment where corrosion failures are unacceptable, continuous corrosion control may be essential so that protective measures can be effected in good time. These control methods must be nondestructive and often must involve remote sensing of the corrosion rate, e.g., in closed systems or at great depths in marine environments or in the ground. In this context the polarization-resistance method might be useful. An electrochemical technique for the determination of atmospheric corrosion rates should also be available. Other possibilities are acoustic methods which are capable of detecting stress corrosion cracking at an early stage or light emission techniques which can reveal localized corrosion before any real damage has accrued.

When control methods have indicated that corrosion has begun, it is clearly vital to arrest the attack as quickly as possible before failure can ensue; electrochemical protection, whether anodic or cathodic, is likely to be important in this respect because this technique is often easy to apply and control. The alternative method of corrosion pre-

Einar Mattsson • Director of Research, Swedish Corrosion Institute, Drottning Kristinas väg 48, S-114 28 Stockholm, Sweden.

vention by inhibitors often seems risky today. It is sometimes even uncertain whether a commercial inhibitor will prevent or actually accelerate corrosion! Considerable efforts are needed to bridge the great gap existing between scientific work conducted in the laboratory and actual service applications.

At present, the most common method of corrosion prevention is by application of organic or inorganic coatings, antirust painting being preponderant. To achieve more rational development work in this area, better fundamental knowledge of the corrosion of painted metal and a quick, reliable test method for evaluation of anticorrosion paints are required. A serious drawback of antirust painting is that a good result necessitates a costly pretreatment of the surface and dry, well-controlled conditions during application. New techniques which would permit painting of steel without extensive prior cleaning and under moist conditions (preferably under water), would be a considerable step forward. Among inorganic coatings aluminum is mentioned as an interesting alternative to zinc for protection of steel.

1. Introduction

Metallic corrosion under wet conditions is generally electrochemical, occurring in corrosion cells at the metal surface. The corrosion field includes a vast number of phenomena and problems, and it would be quite impractical in a survey like this to aim at a complete coverage of the whole field. The author has chosen to touch on some aspects of probable future developments which have appeared during the course of his work. As this has for nearly two decades been concerned mainly with applications in the field of corrosion, the survey will deal with corrosion from a practical point of view.

2. Corrosion and Its Importance to the Community

The deterioration of metallic engineering materials in service is reckoned these days to cost about £25 ($50) per capita per year in industrialized countries,[1-3] i.e., about $15 billion in the U.S.A. alone. Even then, important consequences of corrosive attack such as factory shutdown, decreased efficiency of equipment, and accidents, which are all difficult to cost-assess, are not included. The future importance of corrosion is probably

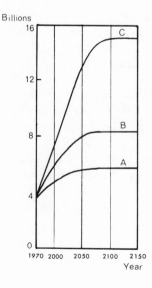

Fig. 1. The growth in world population according to three alternative projections.[4]

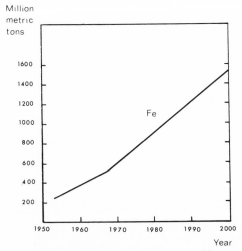

Fig. 2. World requirements for steel.[1]

best considered within the broader perspective of how man and his requirements will develop.

The world population, today about 4 billion, is likely to increase within the next few decades before stabilizing at a level which will be determined by equilibrium with the available resources of the earth.[4] Different opinions exist as to the rate of growth and the level of stabilization (Figure 1), but the latter should lie in the range of 6–15 billion. The increases in population and in the per capita consumption of materials must necessarily imply a steadily increasing output from the metal-producing industries. Estimated increases in the consumption of steel, aluminum, copper, and zinc are presented in Figures 2 and 3.

Actually, the earth's metallic resources seem sufficient to satisfy the demands of mankind in the foreseeable future since exploitation has so far been confined merely to the surface of certain areas of our planet and vast areas of the outer shell and the immense volumes deeper within remain practically untouched. Also, the water of the seas and the enormous tracts of ocean bed have the potential to supply very great quantities of different metals (Figure 4).

In spite of this, the growing consumption of metals will lead to increased prices. There are several reasons for this, one being the increased production cost of metals from low-grade or less-accessible ore bodies and another the anticipated rise in energy prices.

This rise in metal prices will have great consequences for corrosion metallurgists since it will necessitate a greater economy of utilization of metals together with an extension of the service life of metallic constructions. For example, a residential house in Sweden, which may today have a service life of 60 years, is likely to become more expensive to erect in the future because of improvements in standards with much better heat insulation; for such costly dwellings a much longer service life is essential—300 years has been mentioned. Similarly, the service life of motor cars, which is now on average 14 years in Sweden, will tend to rise, probably because of increases in material prices and improved standards (greater cost),[6] so that cars might in future have to last for a generation or longer. All in all, the throwaway lifestyle so prevalent today must be replaced by a preserve and reuse policy. In this context, the role of corrosion protection and prevention is obvious.

Superimposed on the long-term development, fluctuations in the availability of

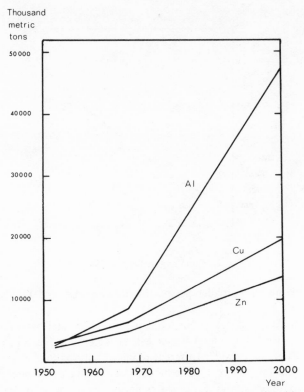

Fig. 3. World requirements for aluminum, copper, and zinc.[1]

metals can be envisaged which are connected with the uneven distribution of resources over the earth's crust and may have political undertones. For example, the price of aluminum could rise precipitously in certain situations since the useful ore bauxite occurs in only a few restricted areas, e.g., Jamaica and Australia.[7] In these circumstances, consumers could have problems until methods for the production of aluminum from more common minerals, e.g., clay, have been developed and set in industrial operation; then, the corrosion metallurgist might have to help in developing alternative techniques with unconventional materials replacing those which have become unobtainable. A similar situation could arise for chromium. The chromium ores occur mainly within the Eastern bloc, Rhodesia, and South Africa.[7]

Another important task for the corrosion metallurgist will be to improve the service performance of metallic components. The consumer will consider to an increasing extent not only the initial cost outlay but also maintenance charges and even the expense involved in recycling material after service. For example, one organization buying a great many motor cars has insisted on having written into the purchasing contract that the absolute price must be related to the results of corrosion checks on delivery. Advanced equipment, where corrosion failures will be either very disrupting or entirely unacceptable, is likely to become commonplace. For example, corrosion failure of a crucial component in a nuclear reactor could be disastrous for the local environment. In such cases, corrosion control and prevention is of the utmost importance.

In addition, man's concern about the environment will have implications for the corrosion field. One aim will be to prevent contamination of the environment with harmful corrosion products or corrosion protection agents. To further avoid the pollution of lakes

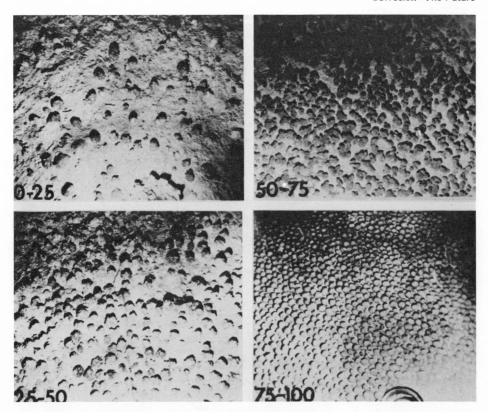

Fig. 4. Mineral nodules on the bottom of the Pacific; composition: 25–30% Mn, 1.0–1.5% Ni, 0.5–1.0% Cu, and 0.25% Co.[5]

and rivers, industries will have to recirculate processing water of different types in closed systems, and as a consequence corrosive conditions can be expected due to the accumulation of dissolved matter.

3. The Application of Already Developed Anticorrosion Techniques

It has been estimated that about one fourth to one third of the cost arising from materials deterioration through corrosion can be saved by applying techniques which are already developed.[1,2] The authorities of several countries are today considering how this substantial saving can be effected. In the U.K. for instance, centers are now being established with the sole aim of advising about corrosion prevention on the basis of existing knowledge. Such establishments already exist in some places, e.g., in Sweden, and an increase in their number and an improvement in their efficiency can be anticipated in the future. Of course, corrosion education also has an important role in this context, and great attention is being paid to this within the European Federation of Corrosion.[8]

A necessary prerequisite for the efficient application of established anticorrosion measures is a convenient access to the relevant information already published on these techniques. Nowadays, the corrosion metallurgist has at his disposal a variety of handbooks and several abstracting journals which cover the corrosion literature more or less completely. However, a literature survey of any special corrosion problem can, even with these sources, become very expensive if carried out manually. Considerable improvements

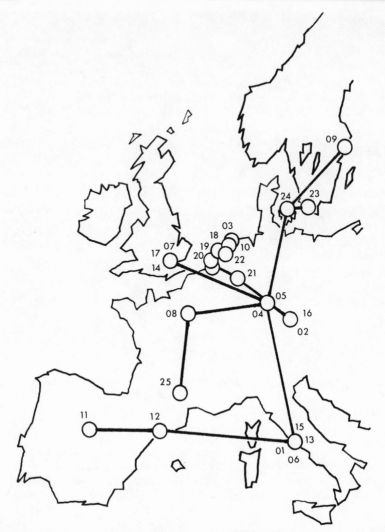

Fig. 5. European network of literature banks as of November 15, 1974; 1, 6, 13, 15, ESRO, Frascati-Rome; 2, 16, ZLDI, Munich; 3, 10, ESRO, Nordwijk; 4, 5, ESRO, Darmstadt; 7, 14, 17, TRC, DTI, DRIC, London; 8, ESRO, Paris, 9, KTH, Stockholm; 11, INTA, Madrid; 12, CIDC, Barcelona; 18, 19, NOCI, SHELL, den Haag; 20, UNILEVER, Vlaardingen; 21, PHILIPS, Eindhoven; 22, TH, Delft; 23, UB, Lund; 24, DTB, Copenhagen; 25, CNES, Toulouse.[10]

in efficiency can be obtained with computer-based retrieval of information from litera-ture banks based on properly selected search terms.[9] In Western Europe, such a lit-erature search can be made at terminals which have access to a network of interconnected literature banks, in turn coordinated from a center at ESRO, Frascati in Rome (Figure 5). These literature banks give adequate coverage of the recent corrosion literature, but older material is less well represented.

A considerable development of computer-based literature retrieval, however, is re-quired and can be anticipated. One of the improvements which can be expected is that the banks will cover at least the modern corrosion literature more or less completely after a few years, since virtually all current contributions are being fed in. However, a reappraisal of the classification techniques employed is needed, since today a considerable number of different classification systems are used.

Concurrent with the development of literature banks, an emergence of materials data banks, including corrosion data, can be envisaged. This possibility is being considered in several countries. The data banks might be operated in a similar way to the literature banks, but it seems difficult to find a classification system which is so simple that computerized retrieval can be carried out without extensive training. International cooperation seems required to achieve and operate both the literature and the data banks.

4. Corrosion Properties of Metallic Materials

Electrochemical techniques of various types employed in Prof. Bockris's group over the years will also be used in the future to investigate the corrosion properties of metallic materials and the mechanisms of corrosion processes. Prof. Conway has given a survey of these techniques in the session on electrode processes. In addition more recent methods, like *electrode impedance measurements*, might be applied to study the complicated multistep electrochemical reactions often involved in the corrosion processes.[11, 12] Further, purely physical methods like LEED (low-energy electron diffraction), Augér spectroscopy, and ESCA (electron spectroscopy for chemical analysis) will be used for the study of oxide coatings, etc.[13–15]

Among corrosion processes of interest in the past particular mention should be made of stress corrosion cracking, which has intrigued corrosion scientists since the beginning of this century (Figure 6). Due to the drastic and destructive effects of this type of corrosion, it is highly desirable that this phenomenon be clarified in order that rational counter measures can be taken.

Another phenomenon which has received much attention in the past is the passivity of stainless steels and its breakdown on localized attack such as pitting and crevice corrosion in the presence of chloride ions. Now that physical methods (LEED, ESCA, etc.) have become available for the study of very thin surface coatings as a complement to the electrochemical techniques, a breakthrough can be expected in research on these phenomena which are of great importance for the application of stainless steels in sea water.

The atmospheric corrosion of metals should also be mentioned as an old subject on which substantial progress is desirable. During the next decade, hopefully, the relation between the corrosion rate and the climatic parameters will be established, so that the atmospheric corrosion of metallic components on a building can be predicted from meteorological data and chemical analysis of the air and deposition.

An area which has more recently come into focus is corrosion and crud formation in nuclear power plants and other steam plants. Here wet conditions exist over a range approaching the critical temperature ($374°C$), while most work on the electrochemical properties of metals is valid for temperatures of around $25°C$ and is not applicable to the high-temperature conditions found in these plants. In recent years survey reports have been published on the thermodynamics of certain systems occurring in power stations[16, 17] (Figure 7). Work on electrode kinetics at these temperatures, however, still remains to be done. The electrochemist will have to overcome great experimental difficulties in carrying out electrode kinetics work in autoclaves in the high-temperature region mentioned, a first problem being to find a good reference electrode.

One recent technical development in this area is operation under so called neutral drift conditions (neutrale Fahrweise). By controlling water conditions (pH 7–8, con-

Fig. 6. Stress corrosion cracking of brass, investigated since the beginning of the century.

ductivity 0.10–0.15 μS/cm) and maintaining a controlled content of dissolved oxygen in the water (0.1–0.2 mg/kg), it has proved possible to passivate carbon steel so that it can be used to replace stainless steel in certain vessels, heat exchangers, and condensate lines.[19,20] Such modification has been estimated to cut the capital investment required by about £1,000,000 ($2,000,000) for a nuclear power plant of moderate size. The fact that this development was possible without detailed knowledge of the high temperature electrochemical properties of the materials indicates that great improvements can be anticipated from the systematic study of these properties.

During the next few decades many new construction materials will be developed either to save material or to replace conventional materials which have become scarce. Corrosion metallurgists will of course have to contribute to this development work by investigating the corrosion properties of the new materials.

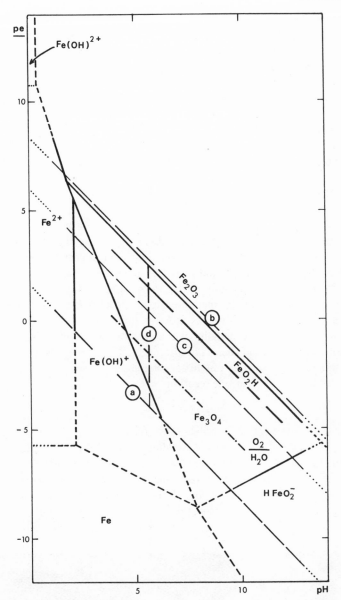

Fig. 7. pe-pH$_T$ diagram for the system Fe–H$_2$O at 350°C, 170 atm.[18]

The efforts to save on material will result in a trend toward higher strength alloys, as these will allow thinner sections to be used in structural components. For example, the carbon steel with a yield strength of 25 kp/mm^2 used for bulk material could partly be replaced with high-strength–low-alloy steel (HSLA).[7] Already today such steels can be produced with a yield strength up to 70 kp/mm^2, and in construction with thinner sections the corrosion resistance may become the limiting factor rather than the strength, which generally determines the dimensions in carbon steel structures. The corrosion metallurgist will have the task of providing these HSLA steels with satisfactory corrosion resistance, if possible by alloying additions (aluminum could be one possibility), but otherwise through protective coatings.

Metals which, due to their uneven distribution, may become scarce for periods in some parts of the world and require replacement are chromium and aluminum. Chromium is a very important alloying constituent in stainless steels, and it would be very difficult to find a substitute for it in these alloys, although aluminum might offer certain possibilities. If aluminum becomes scarce for some period, titanium would be an alternative as it has interesting and very favorable corrosion properties. The fact that titanium is too expensive today for more general application is not likely to prevent its use in the future. Another possible substitute for aluminum is magnesium as this is available in sea water in vast quantities. A more general use of magnesium alloys, however, will require substantial achievements from the corrosion metallurgist in improving their corrosion resistance (Figure 8).

In the future metallurgists may have to use one metal which is not much in evidence today: manganese. One reason for this possibility is that mineral nodules which occur on the sea bottom, and which in the future may become an important metal resource, contain in addition to copper, nickel, and cobalt as much as 25% manganese (Figure 4). A situation could be expected analogous to the one existing for the case of nickel at the beginning of this century. Then the metallurgists at INCO had to develop a technology based on the good corrosion properties of nickel in order to create a market for the nickel available. In a similar way the metallurgists of tomorrow may have to develop a manganese technology. One should then take into account that manganese is an electronegative metal, the standard electrode potential being about -1.1 Volt (against the standard hydrogen electrode) for the electrode reaction

$$Mn^{2+} + 2e^- \rightleftharpoons Mn$$

Thus manganese can offer cathodic protection to steel.

It can also be foreseen that corrosion metallurgists will have to investigate and employ new types of metallic construction materials, i.e., the metal matrix composites.[21] Within reasonable limits these may be tailor-made in accordance with the desires of the designer. Good corrosion resistance, possibly restricted to a surface zone, may be combined with favorable qualities with respect to such properties as strength, stiffness, fatigue life, electrical and thermal conductivity, or density. Materials of this type are aluminum

Fig. 8. Fragments from aircraft component of magnesium alloy, completely mineralized through corrosion during 28 years exposure to sea water in the Baltic.[3]

reinforced with boron or carbon and titanium reinforced with silicon carbide. Metal matrix composites may, for example, come into use in connection with undersea exploration for deep-diving submarines and undersea structures, where a high compressive strength, coupled with good resistance to corrosion by sea water, is needed. Other applications can be expected in the chemical industry for piping, storage tanks, etc. Here high strength coupled with good resistance to corrosion in a variety of aggressive agents will be required. Further, metal matrix composites with good high-temperature properties are being investigated for use in jet engines and turbines. The cost of composites is rather high today, but it is expected to decrease with the degree of automation that can be introduced into the manufacturing process.

5. Corrosion Control

The purpose of corrosion control is to establish and to limit the corrosion taking place on a construction in practical use. The importance of corrosion control is likely to increase considerably, as its use to monitor the corrosion situation in vital parts of advanced constructions becomes more widespread. In corrosion control nondestructive methods should be used, and the corrosion rate should be measured without significant disturbance to the conditions at the exposed metal surface. Further, the technique of measurement should be capable of giving an instantaneous value of the corrosion rate.

Corrosion processes proceed with the production of various types of "emission" which can be used for determination of the corrosion rate. Such emissions include electric current, light, and in certain cases acoustic waves. Today methods which depend on the electric current, i.e., electrochemical techniques, predominate, and they are likely to continue to do so.

Among electrochemical techniques, the polarization-resistance method is currently used for corrosion control and its importance will probably grow in the future. The principle is well known[22] and based on the fact that in the vicinity of the free corrosion potential the relation between current density and activation polarization is linear. The slope R_p of the linear polarization curve is called the polarization resistance. It is related to the corrosion current density i_{corr} by the following equation:

$$R_p = \frac{\Delta U}{i} = \frac{\beta_a \beta_c}{i_{corr} 2.3(\beta_a + \beta_c)}$$

Where ΔU is the activation polarization caused by the current density i, and β_a and β_c are the Tafel constants for the anodic and cathodic electrode reactions. Thus, the polarization resistance is an inverse measure of the corrosion rate. The three electrodes of the measuring cell, i.e., the measurement, counter, and reference electrodes, are generally made from the same metal as the construction to be controlled. The electrodes may be combined in a probe which is placed in the system in question. The polarization-resistance technique can be used to follow the corrosion rate or the corrosivity in a closed system such as a reactor vessel or a pipeline. Corrosion control can be carried out without any disturbing opening up of the system, and the corrosivity can be recorded in the control room of the plant in the same way as temperature, pressure, etc., are recorded now. This technique can also be used in cases where it is difficult to gain access to the corroding surface, e.g., on offshore equipment deep in the sea. Another possible application might be to control the corrosivity of sea water around ships, so that the length of time spent in polluted and corrosive sea water is minimized, and the use of such water for filling tanks, etc., is avoided.

Fig. 9. Atmospheric corrosion testing at Bohus Malmön, Sweden.

Even for atmospheric corrosion testing an electrochemical technique is being developed as a complement to the traditional long-term field tests of panels on racks under atmospheric conditions (Figure 9). It is often difficult to draw conclusions from the conventional field tests because the many parameters of the climate vary in such a complicated manner. Further, these tests require such long exposure times that interest in the results has usually declined when they are finally obtained. The electrochemical technique mentioned originates from the Soviet scientist Tomashov and his colleagues[23] and has been further developed by Kucera and the author at the Swedish Corrosion Institute.[24] Measurements are carried out with an electrochemical cell, one variant of which is schematically shown in Figures 10 and 11. The cell may be considered as a model of the corrosion cells acting at a metal surface, but so designed that the corrosion current

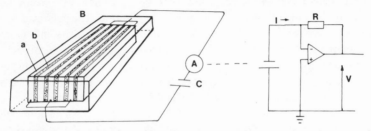

Fig. 10. Electrochemical device for measurement of atmospheric corrosion: A, zero resistance ammeter, circuit shown to the right; B, electrochemical cell (a = electrodes, b = insulators); C, external emf.

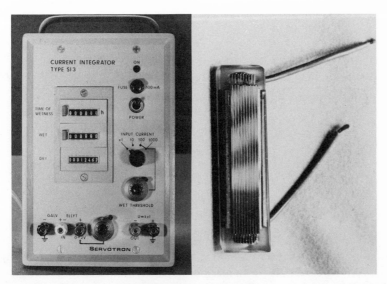

Fig. 11. Electrochemical cell for atmospheric corrosion testing (right); and integrator for accumulation of cell current (left).

can be measured. The cell has two electrode systems, each containing the metal to be studied. One electrode system acts as the anode, the other as the cathode. The electrolyte consists of a film of moisture formed on the cell surface during exposure. An external emf is applied as the driving force, and the current may then be continuously recorded or accumulated using an integrator. When the cell is exposed out-of-doors or in a climate chamber, the cell current varies with the exposure conditions in the same way as the atmospheric corrosion rate and can be used as a measure of this rate (Figure 12). There are many possible applications for this research tool. It may be used, for instance, to

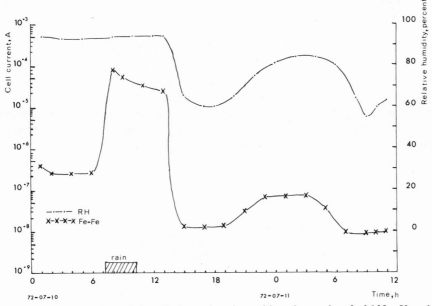

Fig. 12. Cell current in electrolytic cell of type iron–iron with an imposed emf of 100 mV, and the relative humidity in Stockholm over the period July 10–11, 1972.

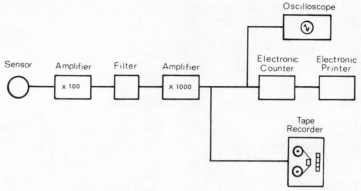

Fig. 13. Stress-wave-analysis technique for detection of stress corrosion cracking.[25]

study atmospheric corrosion in the hollow spaces in the body of a car under different garaging and driving conditions, or to measure variations in the corrosion rate on different parts of a building. Of course, this measurement technique is not restricted to atmospheric corrosion, but can also be applied to other cases of electrochemical corrosion.

There is also an urgent need for a nondestructive method by which the occurrence and propagation of stress corrosion cracking in a construction can be detected. One interesting possibility is to use acoustic emission, since stress corrosion cracking proceeds with the emission of stress waves. Such a stress-wave-analysis technique (SWAT) is already being employed, and the equipment used is shown schematically in Figure 13.[25] The stress waves are picked up by a sensor and amplified, while noise is filtered away. The signals are then collected in an electronic counter and recorded on a digital printer or alternatively stored on tape for later display and analysis. The acoustic emission accompanying stress corrosion cracking has been shown to give more detailed information on the crack-growth process than is otherwise available, and this technique might in the future prove applicable to the nondestructive detection of stress corrosion cracking in structures such as nuclear reactors.

As many corrosion processes are also accompanied by the emission of light, i.e., photons, another nondestructive way of determining the corrosion rate is to measure the light emission from the corroding surface. A team of scientists at the Chalmers University of Technology in Gothenburg is investigating the possibility of using this technique both for dry corrosion, such as oxidation at high temperatures, and for localized attack under wet conditions.[26] The method gives the instantaneous corrosion rate and appears to be very sensitive, so further interesting developments along these lines may be expected.

6. Corrosion Protection

Corrosion can be prevented by a number of means, of which the following will be commented upon here: (1) electrochemical corrosion protection, (2) use of corrosion inhibitors, and (3) use of protective coatings of an organic or inorganic nature.

6.1. Electrochemical Protection

Electrochemical corrosion protection operates by the maintenance of the electrode potential of the object at a level where the corrosion rate is low. Cathodic protection, i.e., controlling the potential in the cathodic region, is widely used today, e.g., for pipelines and ships' hulls. Anodic protection, which functions by the control of the potential in the anodic region at a level where passivation is obtained, is used mainly for titanium and for stainless steel in contact with H_2SO_4.

No fundamental changes in cathodic protection are foreseen, but a considerable increase is likely in the number of applications and practical refinements will continue to be made. Thus, the control of electrochemical protection is likely to be improved. In chemical plants, for instance, with many units such as heat exchangers and reaction vessels, where corrosion conditions are changing continuously, the protective current may be controlled potentiostatically or through polarization-resistance measurements with a computer guiding each specific unit in accordance with its corrosion conditions. Even if such an improvement is not possible for economic reasons today, it might be realistic in the future.

6.2. Corrosion Inhibitors

A corrosion inhibitor is a chemical substance which, when added in small concentrations to an environment, effectively checks, decreases, or prevents the reaction of the metal with the environment. Most inhibitors are used in liquid systems, but vapor corrosion inhibitors which prevent corrosion in the atmosphere also exist.

Today corrosion inhibitors are of great practical importance, being extensively employed, for instance, in the petroleum industry. However, their use seems to be confined mainly to industries which have their own inhibitor expert who is able to select a suitable product among the legions of inhibitor chemicals that have been proposed and who can find out how to use it. A typical technically well-educated engineer who is looking for a corrosion inhibitor may easily get lost in the jungle of scientific literature on the subject. In many cases the information he can get on commercial inhibitor products is poor. Often he cannot even be certain whether they will arrest or accelerate corrosion in his case. Thus, there seems to be a lack of communication between the scientists and the users. In the future the users should be provided with easily understandable, well-organized information on corrosion inhibitors, and they should also be given access to simple (electrochemical?) methods for evaluation. Then corrosion inhibitors might become an effective tool more generally employed in the fight against corrosion.

When using corrosion inhibitors, health and pollution aspects will also have to be taken into account. Inhibitors containing chromate and/or phosphate have been widely used up to now. In the future, however, the use of these inhibitors will probably be forbidden in cases where their toxicity and unfavorable influence on the environment are significant.

6.3. Organic Coatings

Antirust painting is particularly important. It accounts for a major proportion of the total corrosion budget. In the same way as for an unprotected surface, the corrosion of a painted metal surface seems to proceed by the action of an electrochemical corrosion cell. Here the cathodic reaction is reduction of the oxygen permeating through the paint film to the metal surface. The paint film which includes absorbed water, constitutes an essential part of the electrolyte. Thus the electrode kinetics on a painted metal surface would seem to be much more complicated than in plain metal–solution systems. It is now time that the mechanism of corrosion on painted metal be clarified, and it would be of great value if electrochemists were to devote more attention to this difficult subject.

Knowledge of the mechanism of corrosion on painted metal would provide a good basis for the development of a rapid test method for anticorrosion paints. Such a method is urgently needed, as even in this case the conventional field tests require too long exposure times and the accelerated test methods now being employed are often far from representative of the exposure in practice. In fact a quick and reliable method for the evaluation of anticorrosion paints will be almost essential for rational development work in this field.

Today a very well cleaned and dry metal surface is required before painting, and dry conditions are necessary during the paint application to obtain long-lasting protection (Figure 14). These requirements lead to great costs and inconveniences. For the maintenance painting of hulls, which is carried out regularly at 1–2-yr intervals, ships have to be dry-docked. The total cost of such a treatment for a 200,000-ton vessel is of the order £100,000 ($200,000) including dry-docking, paint, labor, and loss of earnings. There are indications that such anticorrosion painting may in the future be carried out in a more rational and less costly way, without dry-docking, through underwater painting.[27] This treatment could be carried out in special stations where the ship is driven past stationary units in which fouling is removed, the surface prepared for painting and the paint applied, all under water. One possibility is to apply the paint by spraying under an air shield. The paint used should be water repellent and should skin-harden very quickly. Techniques and systems for the anticorrosion painting of moist and even rusty surfaces may also be hoped for during the next decades. Then, a considerable part of the expense and inconvenience which today arises from anticorrosion painting would be eliminated.

In anticorrosion painting one important development should be mentioned which will radically change the types of paints that may be used in the future. This development is dictated by the authorities who now require a drastic decrease in health hazards in this area. Paints with unhealthy solvents, like aromatic or chlorinated hydrocarbons, will probably have to be replaced with solventless or water-based paints. Further, substitutes

Fig. 14. Anticorrosion painting of hull under shelter with controlled temperature and humidity conditions at Kockums Ship Yard in Malmö.

might be required for certain pigments, like chromates and red lead, now being used extensively as corrosion inhibitors in antirust paints. The reason is that these pigments may have a toxic effect. Even zinc powder, used in antirust paints to give galvanic cathodic protection to steel, may cause health problems on subsequent welding. In the search for a replacement, manganese might be considered, as this metal is also electronegative with respect to iron. Also certain types of binders in anticorrosion paints, such as epoxy and polyurethane resins, are seriously criticized because of a tendency to cause skin allergy.[28]

6.4. Inorganic Coatings

Among the inorganic materials used as coatings for corrosion protection, zinc is predominant and extensively used on galvanized steel. With the increase in air pollution by sulfur compounds during recent decades, zinc coatings have lost much of their protective quality. To illustrate the strong influence of pollution, it may be mentioned that the corrosion rate of zinc in the polluted urban atmosphere of Stockholm is about 5 μm/yr as compared to only 0.8 μm/yr in the rural atmosphere 90 km outside Stockholm.[29] It is clear that this will significantly affect the life of the 30-μm-thick zinc coating existing on standard quality galvanized steel sheet. In the search for an alternative coating material, aluminum comes into consideration as it is well-known for its good corrosion resistance in air polluted with sulfur compounds.[30] Another advantage of aluminum in this context is its low density (2.7 g/cm^3 as compared to 7.1 g/cm^3 for zinc), which means that the quantity (weight) of aluminum in the coating is small. In light of these considerations it is remarkable that aluminized steel has not hitherto been widely used, although its possibilities have been known for several decades. Lately, however, interest in this material seems to have been awakened and in the future one may expect a shift from galvanized to aluminized steel.

7. Conclusion

In conclusion, after all this gazing into the crystal ball, it seems appropriate to quote a wise man—a politician—who said: "It is difficult to prophesy—especially about the future." The present author is inclined to agree.

References

1. Material Needs and the Environment Today and Tomorrow, final report of the National Commission on Materials Policy, June 1973, Library of Congress Card Number 73-600202, Washington, D.C. (1973).
2. Survey of Corrosion and Protection in the United Kingdom, report of the Committee on Corrosion and Protection (T. P. Hoar, chairman), 129 pp, Her Majesty's Stationery Office, London (1971).
3. E. Mattsson, Elektrokemi och korrosionslära, Bulletin No 56, 2nd ed., Swedish Corrosion Institute, Stockholm (1974).
4. Norges resurssituasjon i global sammenheng. Norges Offentlige Utredninger 1974: 55, Universitetsforlaget, Oslo (1974).
5. G. Hambraeus, Svensk mineralförsörjning i globalt perspektiv. Paper presented at "Hindersmässan", Örebro 1975-02-01.
6. Weak Points of Cars, AB Svensk Bilprovning, Stockholm (1974).
7. From Abundance to Scarcity. Is that the Materials' Story? Product Engineering (N.Y.) 45, 28 (1974).

8. Corrosion Education Manual, Second International Edition. European Federation of Corrosion, Working Party on Corrosion Education (E. B. Bergsman, chairman). Swedish Corrosion Institute, Stockholm (1974).

9. Z. Gluchowicz, Selective dissemination of information and retrospective searches. Computer-based documentation services from RIT Royal Institute of Technology Library, Stockholm (1973).

10. T. Hedman, Royal Institute of Technology Library, Stockholm, personal communication.

11. R. Knoedler and K. E. Heusler, *Electrochim. Acta 17*, 197 (1972).

12. C. Gabrielli and M. Keddam, *Electrochim. Acta 19*, 355 (1974).

13. C. Leygraf, A Technique for Surface Characterization Based on Low Energy Electron Scattering and its Application to the Primary Oxidation of Iron, dissertation at the Royal Institute of Technology, Department of Physical Chemistry, Stockholm (1973).

14. C. Leygraf, G. Hultquist, S. Ekelund, and J. C. Eriksson, *Surface Sci. 46*, 157 (1974).

15. I. Olefjord, Applications of ESCA in Materials Technology, dissertation at Chalmers University of Technology, Department of Engineering Metals, Gothenburg (1974).

16. D. Lewis, *J. Inorg. Nucl. Chem. 33*, 2121 (1971).

17. D. Lewis, *Ark. Kemi 32*, 385 (1971).

18. D. Lewis, AB Atomenergi, Studsvik, Sweden, personal communication.

19. R. K. Freier, VGB-Speisewassertagung p. 11 (1969).

20. R. K. Freier, VGB-Speisewassertagung p. 8 (1970).

21. L. W. Davis and S. W. Bradstreet, *Metal and Ceramic Matrix Composites*. Cahners Publishing, Boston (1970).

22. M. Stern and A. L. Geary, *J. Electrochem. Soc. 104*, 56 (1957).

23. N. D. Tomashov, G. K. Berukshtis, and A. A. Lokotilov, *Zavod. Lab. 22*, 345 (1956).

24. V. Kucera and E. Mattsson, ASTM Special Technical Publ. 558, p. 239 (1974).

25. C. E. Hartbower, W. G. Reuter, C. F. Morais, and P. P. Crimmins, ASTM Special Technical Publ. 505, p. 187 (1972).

26. S. Andersson, B. Kasemo, and I. Marklund, Chalmers University of Technology, Gothenburg, Sweden, personal communication.

27. Symposium on Protective Coatings for Underwater Application, City of London Polytechnic, London (1972).

28. U. Ulfvarson, Synpunkter på kemiska risker i färgindustrin och industrier som förbrukar färg, I Bindemedel, II Pigment och lösningsmedel, III Industrier med användning av färg- och lack-produkter, *Faergoch Lack 20*, 3, 6–8 (1974).

29. E. Mattsson and V. Kucera, 2nd International Symposium on Modelling the Effect of Climate on Electrical and Mechanical Engineering Equipment, p. 90, Liblice, CSSR, (1971).

30. H. P. Godard, W. B. Jepson, M. R. Bothwell, and R. L. Kane, *The Corrosion of Light Metals*, Wiley, New York (1967).

15

The Future of Electro-Organic Chemistry

James W. Johnson

It seems that much of the growth that can be expected in the future for electrochemistry in general will be associated with the field of electro-organic chemistry. As organic technology makes ever-increasing contributions to societal needs, corresponding pressures on electro-organic technology will develop. These will arise for various reasons, among which are: (1) In the future, increasing proportions of energy available for chemical processing will probably be in the electrical form. Efforts to use this energy in the most economical way will give electrochemical methods obvious advantages if they can be adapted. (2) The costs of energy will cause existing chemical processes to be examined in terms of the production of electricity as a by-product if the processes were carried out electrochemically. This could be viewed as the operation of a fuel cell, primarily for the reaction products rather than the electricity. (3) Environmental aspects will be of prime consideration. Electrochemical processes are frequently alternatives that inherently contribute less to noise, thermal, and chemical pollution than the corresponding nonelectrochemical processes. (4) The versatility of electrochemical processing will also give added weight for its consideration. Frequently the changing of process variables, such as potential, temperature, concentration, etc., will give a different product. Such a change by nonelectrochemical means would most often be accomplished only with changes in the process. The conservation of capital and efficiency of use of established operations would also be very attractive.

Since organic chemistry can affect almost all aspects of electrochemical processing, the various components basic to these processes will be looked at separately.

1. Electrodes

The use of organic materials, most commonly graphite, as electrodes has long been a practice. In recent years, new forms, such as "vitreous" carbon,[1] have been developed that have special properties, e.g., biologically implantable, low porosity, high erosion resistance, high mechanical strength, etc. The work of Panzer and Elving[2] compares the behavior of pyrolytic graphite and glossy (or vitreous) carbon electrodes in various systems and illustrates variations that can be attained by processing. Hand *et al.*[3] report on the satisfactory performance and convenience of carbon-cloth electrodes for preparative electrolyses. Undoubtedly future developments will also involve deposition by various techniques of other materials on carbon substrates that will impart special properties. An example of this is the reduction of oxygen on a semiconductor surface as reported by Jahnke and Schonborn.[4] Polymeric iron phthalocyanine deposited on carbon was, under certain conditions, a better electrocatalyst than Pt.

In general semiconductors as electrocatalysts appears to be a fruitful area for further

James W. Johnson • Chemical Engineering Department, University of Missouri, Rolla, Missouri.

investigations. The type of charge carriers (holes or electrons) should exert an important influence on the type of charge-transfer reaction that occurs at an electrode.

Composites are also a group of materials that should receive consideration beyond strictly mechanical considerations. Surface variations, even on a macroscale, can often show unexpected results.

The use of organic redox reference electrodes (e.g., chloranil)[5] may also see some revived interest in the future due to their solvent-independent potentials. This would be of special concern to researchers doing mechanistic studies who are interested in comparing current–potential relationships in different media.

2. Solvents

The influence of the solvent used during an electrolysis can be varied and of great importance. Among items to consider are: (1) It may be a source of reactants (e.g., H or O from aqueous solutions). (2) It must either dissociate or dissolve a solute that will dissociate to form a significant number of charge-carrying species in order to conduct current. (3) It must be relatively stable over the potential range to be used. (4) It may be used to alter the structure of the electrical double layer. (5) It may be used to alter the nature of reactants or products. (6) It must be separable from the product. Coupled with these are also economic and safety factors.

More and more solvents other than H_2O are being found to meet special requirements; the majority are organic compounds. Examples of these are acetic acid for acetoxylations; methanol for metoxylations, reductions, and Kolbe-type reactions; and acetone, pyridine, alcohols, dimethyl sulfoxide, propylene carbonate, as well as numerous others, for the electrowinning and electropurification of metals. As a better understanding of the electrochemical nature of these solvents evolves, their use in commercial processes will also increase.

3. Electrolytes

In general, the solubility of inorganic salts in organic solvents is considerably smaller than in water, thus making charge transport through the solution more difficult. This situation can often be improved by using organic compounds as the supporting electrolyte. Examples are tetraethyl- and tetrabutylammonium salts, acetates, carbonates, potassium methoxide, and tetraphenylborate. These compounds can also be used to change the structure of the electrical double layer which may influence the course of a reaction. For example, ethyl bromide in propylene carbonate is reduced to ethane at a Pb cathode when the supporting electrolyte contains Li^+, K^+, NH_4^+, or Ca^{2+}. If the supporting electrolyte contains R_4N^+ or R_3S^+, tetraethyl lead is produced.[6]

4. Membranes

Membranes or diaphragms used in electrochemical cells are often classified as permeable or ion exchange. Many organic materials are included in each classification. Membranes become of prime importance when anolyte and catholyte mixing must be prevented to conserve reactants or products. This is also true for processes using large currents and/or which have low conductivity electrolytes where power consumption can be decreased by decreasing the electrode spacing. Such savings are often possible only if suitable membranes are available. This area seems to get very little primary attention, as pointed out by MacMullin and Fioshin[7] and emphasized by the small number of articles appearing in the literature. Techniques, such as reported by Weininger and Holub,[8] of preparing membranes of extremely fine pore size could have a profound effect on electro-chemical processing and should be encouraged.

5. Processing

Electrochemical processing will ultimately benefit from advances and new developments in the preceding areas. The benefits will not be incurred automatically, however, as initially they will be associated with a single process and then applied or incorporated individually into other processes through improvisation or of necessity.

The electrochemical processing of organic materials is an area of industrial chemistry where advancement will be assisted by the versatility of many items of processing equipment. Many cells can possibly be adapted for different products by changing electrodes, solvent, diaphragms, potential, temperature, etc., as well as reactants in a given unit. (This would be a logical point to initiate standardization practices with equipment and components.) Flow-through-type cells would seem to be best suited for this. The electrolyte outside the cell could readily have adjustments made in reactant concentration, pH, temperature, etc., as well as have products removed. Eventually this should lead to standardized processing procedures that would be applicable for a large number of products.

Some areas that appear particularly fruitful for study that would involve organic materials in electrochemical processing are discussed below. They were compiled by a brief perusal of the literature and are intended to neither be complete nor to have ascertained the most significant areas.

5.1. Metal Processing

Numerous processes have been described by which metals are deposited cathodically from electrolytes containing the organometallic compounds. These are usually associated with metal-refining processes[9] of active metals such as Be, Zn, Cd, and Al, as well as Na, K, and Mg. The similarity of the processes for the different metals suggests that one facility and several electrolyte components might serve for refining several different metals.

A desirable aspect of these processes is the removal in the anode sludge of impurities in the metal being refined that do not form stable alkyl derivatives. This suggests a possible application for electrowinning metals directly from their ores if they possess sufficient conductivity. The production of lead from galena would be a candidate for such use. The primary incentive in this case would be to eliminate the SO_2 produced by conventional smelting processes that contributes to air pollution.

5.2. Carbon Dioxide Reduction

The reduction of carbon dioxide (or carbonate ions) is an area that deserves more study in conjunction with nonaqueous media. The production of carbon monoxide, formic acid, formate, formaldehyde, etc., with reasonable efficiencies and low costs could have a profound influence on syntheses in the organic chemical industry.

5.3. Anodic Processes

Most organic compounds can be anodically oxidized to some extent. One problem that generally plagues these processes, especially when beginning with common hydrocarbons, is the reaction selectivity. Often a desired product has a greater reactivity than the initial reactants, thus leading to a complex mixture of reaction products. With this qualification in mind, several types of anodic processes can be described.

One type is substitution reactions involving C–H bonds.[10] Examples of substituents are halides, CN, OH, OCH_3, OAc, NO_3, etc. While few of these reactions have been studied extensively, most have been found to be dependent on the substrate, electrode, electrolyte, and solvent.

Another type is the complete oxidation to CO_2. Compounds to which this applies are mainly the more elementary hydrocarbons (alkanes, alkenes, etc.) and their partially oxygenated products (alcohols, aldehydes, etc.). Current thinking is that unique properties of a very limited number of electrodes (e.g., Pt) are essential for these reactions. This has been an area under intense investigation in recent years in conjunction with fuel-cell research.[11] Satisfactory long-term economic performance has yet to be demonstrated, but the desire for more efficient use of natural resources will keep this an active area of interest. The more promising aspects of fuel cells now seem to be for on-site generation devices (as opposed to small portable ones) where conventional electrical transmission systems are not practical or where electrical power can be generated as a by-product of electrochemical processing. The work done in this area has contributed significantly to the application of new instrumental techniques and electrode kinetic theory to organic processes. Hopefully this will continue.

Partial oxidation of aromatic compounds and couplings (apart from polymerization) are also types of anodic reactions that can be used in special circumstances. Although some have limited usefulness, they do lend themselves to theoretical studies in that the coulombic efficiency for a single product is often quite high and the number of steps in the reaction sequence is relatively small.

The entire area of anodic processes is one that should greatly benefit from systematic mechanistic studies.

5.4. Cathodic Processes

The major effort in electro-organic chemistry has probably been expended in this area, although much of it is oriented toward syntheses. Reaction intermediate and product identification, as opposed to electrode kinetic parameters, have been the basis of most mechanism postulations. Types of reactions which can be mentioned here are hydrogenation of multiple C–C bonds, cleavage, coupling, dehalogenation, etc.[12-14] Coupling reactions appear to be the most likely candidates for further commercial applications. Some success has already been realized in this area by Baizer in which adiponitrile, a monomer used in nylon production, is produced from acrylonitrile.[15]

This is also an area made to order for persons with academic interests in electrode kinetics and where meaningful contributions can be made. Much preliminary screening and results of tedious, time-consuming analytical work have been reported which allow one to do a study that might be precluded by either time requirements or availability of special equipment.

5.5. Electrogenerated Reagents

The use of electrogenerated reagents (egr) would appear to meet several specialized needs in electro-organic chemistry. Types of reactions that can be effected in this manner are oxidations, reductions, additions, substitutions, couplings, cleavages, etc. The reaction of the organic species with the egr may be either heterogeneous or homogeneous. Examples of electrogenerated reagents are amalgams, halogen atoms (and molecules), hydrogen atoms (and molecules), and redox systems (both inorganic and organic). Some advantages of egr are: (1) allows taking the reagent to the substrate in lieu of bringing the substrate to the electrode, (2) may allow regeneration of the reagent where it is expensive or disposal is difficult, (3) may allow generation of the reagent during times of low-load electrical use and storage for later use, and (4) may allow a selectivity that cannot be attained through reaction at an electrode.

Solvated electrons are a special case of electrogenerated reagents.[16] Their use is a relatively new method for carrying out reduction reactions. The existence of these species for periods of time sufficiently long for them to be considered reactants is primarily associated with nonaqueous polar solvents which also have low proton activity. Solvents reported to solvate electrons include ammonia, diglyme, amines, and amides. LiCl is frequently the electrolyte, although tetrabutylammonium salts have been used. Reductions of aromatic compounds, alkenes, acetylenes, and aliphatic amides have been reported. Differences in the stereography of the products have been noted compared to those from electrocatalytic hydrogenations. The solvent also appears to have other important influences. Further studies in this area are certainly warranted and could result in preparative methods for special organic chemicals.

5.6. Electropolymerization

The polymerization of numerous monomers (methyl methacrylate, acrylonitrile, styrene, fluoroolefins, etc.) by electrochemical means can be readily demonstrated. The results of many studies have been reported.[17] Although the major role of the electrode appears to be that of an initiator, some distinctive characteristics in structure and composition of polymers produced in this manner have been noted. The initiation step can be accomplished by activating either the monomer or some other species. Careful consideration must be given to both the solvent and electrolyte if the reaction is to be sustained. Once the polymerization reaction has been initiated, much of the subsequent reaction can be described in terms of traditional polymer chemistry.

Controlling the production rate of the activating species (by control of current) gives some control of the size and size distribution of the polymer molecules. The variation in the types of activating species and the orienting influence of the electric field around the electrode are thought to influence the structure. The possibility of producing new types of polymers by electrochemical means has also been suggested.

Due to the large quantities of polymers produced and their ever expanding use, any efficiencies effected by electrochemical processing could be translated into equally large savings. This appears to be one of the industrial frontiers for electrochemistry.

5.7. Pollution Control

The increasing attention that must be given to pollution control by all industries is favorable to electrochemistry. In general, electrochemical alternatives to traditional processes are less polluting or produce pollutants that are easier to contain. For example, Bockris and Reddy have pointed out that if electricity were generated from hydrocarbons by direct energy conversion rather than going through a thermal cycle, not only would an increased efficiency in the use of the fuel be realized, but numerous atmospheric pollutants would not be released.[18] Also, the use of electrochemical methods to clean up pollution should not be overlooked. The ability of electrode reactions to efficiently remove appreciable amounts of dissolved materials present at low but objectionable levels is paralleled by very few other methods. Electrochemical processes can also often operate continuously with very little interference to established routines.

The ability to electrochemically sterilize water contaminated with living organisms, as developed by Stoner, is another interesting recent development.[18] The many possibilities presented here will undoubtedly stimulate a great deal of activity in this area.

5.8. Corrosion

Although this is the last topic presented here, it is by no means the last nor the least associated with electro-organic chemistry. The annual losses due to corrosion are tremendous and generally irrecoverable when measured on scales of either money, energy, or manpower. Advances in this area could have vast implications. Electro-organic chemistry has a place here mainly from the widespread use of organic inhibitors, of which there is certainly much remaining to be learned. The subject is further complicated as synergistic effects prevent components of inhibitors from being studied separately to assess their effects. This is also an area that will continue to receive attention.

In summary, with appropriate efforts by the electro-organic chemical community, future opportunities appear unlimited. As with most opportunities, though, they must be won by dedicated labor in both the academic and applied sectors.

References

1. F. C. Cowlard and J. C. Lewis, *J. Mater. Sci. 2*, 507 (1967).
2. R. E. Panzer and P. J. Elving, *J. Electrochem. Soc. 119*, 864 (1972).
3. R. Hand, A. K. Carpenter, J. C. O'Brien, and R. F. Nelson, *J. Electrochem. Soc. 119*, 74 (1972).
4. H. Jahnke and M. Schonborn, paper delivered at the 19th CITCE Meeting, Detroit, Michigan (September 26, 1968).
5. D. J. G. Ives and G. J. Janz, *Reference Electrodes*, p. 314, Academic Press, New York (1961).
6. R. Galli and F. Olivani, *J. Electroanal. Chem. 25*, 331 (1970).
7. R. B. MacMullen, *J. Electrochem. Soc. 120*, 135C (1973); M. Y. Fioshin, in: *Progress in Electrochemistry of Organic Compounds* (A. N. Frumkin and A. B. Ershler, eds.), pp. 287–318, Plenum Press, New York (1971).
8. J. L. Weininger and F. F. Holub, *J. Electrochem. Soc. 117*, 340 (1970).
9. A. Brenner, in: *Advances in Electrochemistry and Electrochemical Engineering* (C. W. Tobias, ed.), Vol. 5, pp. 205–248, Interscience, New York, (1967).
10. L. Eberson, in: *Organic Electrochemistry* (M. Baizer, ed.), pp. 447–468, 781–804, Marcel Dekker, New York (1973).
11. B. J. Piersma and E. Gileadi, in: *Modern Aspects of Electrochemistry* (J. O'M. Bockris, ed.), Vol. 5, pp. 47–175, Plenum Press, New York (1966).
12. R. Dietz, in: *Organic Electrochemistry* (M. Baizer, ed.), pp. 253–277; Marcel Dekker, New York (1973); M. R. Rifi, pp. 279–314; H. Lund, pp. 315–345; L. G. Feoktistov and H. Lund, pp. 347–397; M. M. Baizer, pp. 399–411, 679–704; L. Eberson, pp. 413–428; L. Horner, pp. 429–444; L. Horner and H. Lund, pp. 729–779.
13. A. P. Tomilov *et al.*, *Electrochemistry of Organic Compounds*, pp. 137–304, Halsted Press, New York (1972).
14. A. B. Ershler, in: *Progress in Electrochemistry of Organic Compounds* (A. N. Frumkin and A. B. Ershler, eds.), pp. 203–239, Plenum Press, New York (1971).
15. M. M. Baizer, *J. Electrochem. Soc. 111*, 215 (1964).
16. H. Lund, in: *Organic Electrochemistry* (M. Baizer, ed.), pp. 839–851, Marcel Dekker, New York (1973).

17. G. Parravano, in: *Organic Electrochemistry* (M. Baizer, ed.), pp. 947–974, Marcel Dekker, New York (1973); A. P. Tomilov *et al.*, *Electrochemistry and Organic Compounds*, pp. 498–509 and references cited therein, Halsted Press, New York, (1972).
18. J. O'M. Bockris and A. K. N. Reddy, *Modern Electrochemistry*, Vol. 2, pp. 1350–1400, Plenum Press, New York (1970).
19. Work of Dr. Glenn Stoner, Materials Science Department, University of Virginia, Blacksburg, Virginia.

16

Summary: Electro-Organic Chemistry

Hendrik Keyzer

Drs. Menzies and Johnson have managed to condense the vast reaches of past and future of electro-organic chemistry admirably into the short time available. This summary is necessarily a paraphrase of their statements. However, I wish to add several comments, for some of which I am deeply indebted to Dr. Manuel M. Baizer.

Over a number of decades, areas of chemistry have tended to specialize to such a degree that each exhibits a vocabulary intelligible only to acolytes. This is obvious at organic seminars, where physical chemists are lulled to sleep by the flash of curly arrows and the jingles of SN_1, SN_2 incantations, whereas the organic chemist at physical chemical seminars watches with glazed expression as the blackboard is whitewashed with differential equations and quantized hallucinations. The disparity has been a matter of concern to the creative scientist who has always been compelled to reach out across a multitude of disciplines to weld the fragments he finds into a useful whole. The modern practitioner of electro-organic chemistry shows a tendency to be an *uomo universale* because his field involves synthesis and analysis of compounds, determination of structures, catalysis, electron-transfer processes, membrane phenomena, polymer chemistry, biological oxidation–reduction events, and engineering features. Organic electrochemistry is apposite to kinetics, reaction mechanisms, solvation, solid-state effects, and a host of other chemical aspects.

To many classical organic chemists a vast spectrum of challenging problems has opened up which can be solved with relatively simple techniques and comparatively inexpensive apparatus. Dr. Menzies has mentioned electropolishing, plating, and film formation, to which may be added radical scavenging, decarboxylative and reductive coupling, allylic substitution, reductive cyclization, and carbonyl preparation.

Ancient chemistry began to take form out of an assembly of apparently unrelated facts largely mineralogical in nature and based on technological need. In the middle of the 19th century the discovery of the periodic table allowed incisive predictions to be made, giving chemistry an impetus of grandiose proportion. With the discovery of the nature of carbon and its bonds and the function of carbon in living processes, branches of science sprang up which appealed to the chemist because this element yielded compounds capable of systematic classification. To an extent this caused the remaining 90% of the periodic table to suffer some neglect until technological pressure again encouraged research (basically electrical in nature) on elements such as germanium, silicon, and a variety of other metals which, amongst other things, led to the invention of remarkable transducers and the moronic but invasively powerful analog of the human brain, the computer.

Technological pressure has also introduced the synthesis of polymers which replace

---

Hendrik Keyzer • Chemistry Department, California State University, Los Angeles, California.

and complement natural products. Since these compounds can be tailored, it is no surprise that effort has been expended not only to replace the inorganic materials of our technology but to improve upon them, e.g., the room-temperature organic superconductor dream. On a more general level, polymers promise to diminish man's dependence on resources, limited or inaccessible for whatever reason. Electroinitiated polymerization is receiving a great deal of attention now, and although this branch of science is still embryonic, there is mounting evidence that the initiation and termination steps can be finely controlled and that unusual polymers can be synthesized more economically electrochemically than by any other method.

Dr. Baizer has pointed out that recurrent patterns of publications describing a complete research project in electro-organic chemistry involves the polarographic study of compounds in anhydrous media in the presence of proton donors. Presumed intermediates receive attention. Cyclic voltammetry is applied to gauge the reversibility of electrode reaction. Electron-spin resonance of possible radical intermediates is studied. Electron-density maps are determined so that further reactions may be designed. The compounds are then subjected to large-scale electrolyses, the products isolated and characterized, and their structure and genesis rationalized on the analytical electrochemical data obtained.

The scope of today's organic electrochemistry includes: extension of the range of useful solvents and electrolytes so that previously intractable organic compounds can engage in electron transfer at an electrode, generation of solvated electrons, elucidation of electrode processes, electrode matrices incorporating catalysts, electro-oxidative flow systems, electrolysis of substrates in thin-layer cells, simultaneously coupled organic syntheses at both electrodes, fluidized electrode systems, organic complex characterization, and so on.

In the future Dr. Johnson envisages the electrowinning of metals, new membrane technology, vitreous carbon electrodes, and the electrogeneration of reagents. The possibilities are extensive. One application of electro-organic chemistry pointed out by Bockris excites major interest, i.e., the conversion of organic sewage with high efficiency to the starting materials required by bacteria to produce protein, a process which not only contributes to pollution abatement but may ameliorate pressing food problems.

The extension of the philosophy of organic chemistry with electrochemistry is one of the most incisive ways to face the materials and energy conservation demands of the generation which follows ours and to which we have the profoundest of moral obligations in terms of giving life and the maintenance of acceptable qualities of life.

17

Aluminum Extractive Metallurgy—Trends

Nolan E. Richards

All primary aluminum is still made by the Hall–Heroult process of 1886. The alumina is extracted from bauxite through the Bayer process of 1888. This should not be construed as a complete lack of creativity on behalf of the technologists supporting the extraction of aluminum—there have in fact been great strides made in automation, control, efficiency, energy and materials conservation, and evaluation of alternative processes. In the last decade, as distinct from the preceding period, there has been a swift application of evolving research and development to actual operations. This has been an important factor in restraining production costs of aluminum.

Operation-oriented research focuses on the factors of:

1. Current efficiency (reducing or suppressing the extent of the back-reaction $CO_2 + Al \rightarrow CO + Al_2O_3$)

2. Reducing any element of the gross cell potential; potential drop in anodes, cathode cavities, electrolyte and anodic overpotential; through improved understanding of the electrochemistry and electrode reactions

3. Stabilizing and optimizing the conditions for the delicately balanced interface; molten electrolyte (density 2.15), supernatant and in chemical equilibrium with molten aluminum (density 2.30)

4. Improving the dissolution and uniformity of concentrations of aluminas, in electrolytes that can be selected as optimum for a particular operation

5. Prolonging the life of the cells and all mechanical equipment ancillary to the cell in the destructive environment of phase changes at 700–950°C, and stress effects of a finite sodium activity on the graphitic domain in carbonaceous materials

6. Collecting and recycling all volatilized or entrained material from systems enclosing the cells

7. Accommodation to frequent changes in the properties of the industry's raw materials

8. Development of basic and subsequent practical technology that makes existing processes and facilities obsolete.

Electrowinning aluminum is not a space-intensive process. However, within the liquid aluminum cathode, the interaction of magnetic fields (intensity approximately 100 G with orthogonal current vectors on the order of 0.5 A/cm²) can produce metal motion deleterious to the productivity of each cell. Stability of liquid aluminum in most modern cells, and hence current efficiency, have improved because of better, if imperfect, understanding of magnetohydrodynamics. This has culminated in more advantageous placement and configuration of the electrical bus bar systems around the series of cells and understanding of how to regulate the depth of the cathodic aluminum.

Nolan E. Richards • Reduction Research Division, Reynolds Metals Company, Sheffield, Alabama 35660.

Even before the public acknowledgment of an "energy crisis," both the alumina refining and the alumina reduction plants were making concerted efforts in reducing unit power consumption—their economic survival dictated this! For example, in the 1950s, it took approximately 18.7 kWh to produce 1 kg of aluminum. Today, in the same cell, carrying 55% more current (actually setting up increased ohmic potentials and electrode overpotentials), the typical unit power consumption is 17.6 d.c. kWh/kg aluminum. In modern United States plants (which tend to operate at higher anodic current densities than similar plants in Europe), this figure lies in the range of 14.7 to 16.5 dc kWh/kg.

Cell design and improved carbonaceous materials are only partially responsible for this. Through improved knowledge about electrolysis—in particular, control of alumina concentration in the cryolite-based electrolyte—we have increased the proportion of time that cells are operating at near optimum conditions: viz., at as low a voltage as practical, at a temperature 10–15 degrees above the liquidus, and with uniform faradaic yield. Not so long ago it was common practice to withhold alumina from the cell so that "anode effects," indicative of moving up the polarographic wave from oxygen ion to the discharge of fluoride ions, would occur. This "purge," or really fix on a minimum alumina concentration was honestly accepted as essential for continued trouble-free operation. And so it was then. But now, control of the rate at which alumina is added to the cell through the scheduled periodic additions of several hundred kilograms or, at shorter intervals, several kilograms of alumina results in more uniform alumina concentration. Achieved by mechanical means, this control of alumina not only gives lower cell voltages, it also permits lower operating temperatures because the composition is kept closer to that of the eutectic. Concomitantly, the vapor pressure of fluoride salts and cumulative material being recycled through fume-collection systems are reduced.

Even further advantage has been taken of these colligative properties of electrolytes recently. Hall actually taught the practice to which I allude in his second patent. A combination of experimentation, patient evaluation, and operation around process-model contours has culminated in a serious industry-wide evaluation of proportions of lithium fluoride in the electrolyte for aluminum.

Classically, calcium fluoride was used as a freezing-point depressant. Examination of the properties conferred on cryolite systems, so well characterized for the industry by research teams led by Grjotheim, Matiasovsky, Baimakov, and others, shows that of the few fluorides besides CaF_2 that could be used, LiF is claimed to be superior, closely followed by MgF_2 then CaF_2.

An example of how an aluminum producer has used basic electrochemistry to improve the sensitivity of control of alumina electrolysis, and hence production, is through development and use of an alumina concentration meter. This meter comprises a probe immersed in a cryolitic electrolyte and a current programmer and readout system. The principle is one of limiting current density for the discharge of oxygen-containing fluoride ions on a carbonaceous anode—together with the sharp increase in voltage—to a small, but finite amount of fluoride ions discharge when O anions are exhausted from the double layer. Solid-state logic programs any number, typically 20, of levels of current density from 2 to 25 A/cm^2 to the boron nitride-insulated anode (which can be easily resurfaced). As soon as the anode effect is initiated, the current supply is automatically stopped, and the reading of the step retained on a light-emitting diode system. This corresponds to a specific concentration of *dissolved* Al_2O_3. The calibration can be expanded or compressed, but the ultimate, practical capability of useful resolution of the system is about ±0.2 wt % Al_2O_3.

This electroanalytical tool has been used to characterize various alumina addition practices to industrial cells, to determine current efficiency as a function of Al_2O_3 concentration, and to educate operators in the area of improving their means and judgment in maintaining more uniform Al_2O_3 concentration in the electrolyte.

Through using costly salts in some of its total capacity (an annual ton of aluminum requires an inventory of approximately 0.07 tons of electrolyte), the industry will have to develop methods for recovering lithium values from that electrolyte absorbed into the carbonaceous lining of the cathodes. Presently, all of the sodium cryolite values are recovered. The existing chemical process will have to be modified, perhaps as has been suggested by King, to reclaim the lithium values economically.

"Black and dirty," carbon is the heart of our present extraction process, the anode. The prebaked and self-baking anodes now being fabricated have improved or optimized properties of electrical conductivity, strength, density, and compromised chemical and electrochemical reactivity to the oxygen in air and electrolyte, respectively. Quality of the ingredients, pitch and coke, plus the finished carbon electrodes, is monitored by aluminum producers carefully and constantly.

The economic battle might not be won if it were not for the extent of monitoring operating parameters that is actually practiced in reduction plants today. For instance, it is not unusual to find the electrolyte of each cell analyzed weekly for molar ratio, NaF/AlF_3, and all individual cationic components except Al and Na, and the temperature of the electrolyte determined at a similar rate. In special studies, the anode overpotential can be measured on the operating anode against a reference electrode, the current efficiency inferred from continuous infrared, or discrete chromatographic analyses of samples of the gases evolving from the anode. Continuous monitoring of an effective "cell resistance," R_c, derived from the total cell potential, V:

$$R_c = (V - E_k)/I$$

where E_k is a rest potential and I, the current, is being used in various ramifications as a means of anticipating anode effects before the actual limiting anodic current density is reached.

Finer and more consistent control of the multitude of cells in an alumina reduction plant has been attained through the routine manual or automated acquisition and analysis of such data on the parameters.

Implementing this practice a step further, plants are using different degrees of closed-loop control to regulate interelectrode distance of each cell. More sophisticated control systems, coupled with an electronic data processor such as that at National Southwire (described by Murphy), can execute many process operations—breaking the crust, adding the alumina, regulating the anode position, and anode effect suppression. Process and research people are progressing slowly along a learning curve of evolutionary control. In the not-too-distant future it will be technically feasible to use closed-loop control for the optimization and operation of individual cells consistent with maintenance of a given maximum or fixed power demand.

Realizing that the published price for fluoride (as HF) is about $0.32 (U.S.) per pound, it can be appreciated that *without* the pressures from the Environmental Protection Agency, aluminum extractors have been diligent in conserving and recovering all fluoride compounds. In the past, nearly all the fluoride and some hydrocarbon effluent was recovered in four main areas: the stacks of the carbon-baking furnaces, from air drawn through hoods over the cells, from air exhausting from potrooms through roof vents, and from the linings of spent cells. Many aluminum plants recovered cryolitic values from the cell gases where, after dilution by ambient air, the components are CO_2, CO, $NaAlF_4$, and HF. This is achieved through wet scrubbing with alkaline solutions or dry scrubbing by sorption on alumina particles. Other aluminum plants had what we call integrated cryolite recovery plants in which spent cell linings were processed for recovery and formation of essentially stoichiometric cryolite. There was and still is a rejected product called "black mud," comprising about equal proportions of alumina and carbon and minor amounts of calcium, silicon, and iron. Retention of this material is a solid-waste problem. The industry is anxious to eliminate this, for both aesthetic and economic reasons. Practical alternate routes to recovery of the increasingly valuable high-purity carbon and separation of the alumina for reuse in the process are needed.

For several years, aluminum producers have been striving in the assembly of technology, capital, *and* engineering equipment (and raw materials!) to meet the standards for emissions being imposed. Several viable alternatives have evolved after critical evaluation. Common to all, you will find a primary hooding duct and fan system. Then there are two classes of systems to remove the gaseous and particulate fluorides: (1) Dry systems, which by differing mechanical means, depend upon absorption of the fluorides on metallurgical grade alumina. This material is recycled to the cells. Final removal of dust may involve baghouses or electrostatic precipitators. (2) Slightly alkaline solutions are used in wet scrubbing systems which coordinate nicely with wet electrostatic precipitators. Both the flow and underflow from the wet systems can be melded into conventional cryolite recovery systems. We may expect practical problems as we progress towards "zero emission." For instance, there has been inadequate time to establish optimum materials of construction and corrosion rates for critical components in wet electrostatic precipitators. Catalytic burners or converters may be a preferred method to eliminate residual hydrocarbons from the effluent of carbon-baking furnaces or the ventilation withdrawn from cells with self-baking Soderberg electrodes. There are opportunities for electrochemistry in control of aluminum-plant effluents, e.g., electrochemical deposition of insoluble fluoride complexes at parasitic electrodes.

Meanwhile, aluminum plants are very active with their internal programs for *in situ* improvement of existing and evolving systems expected to meet technologically feasible standards for emissions. This includes developing people and technical expertise for monitoring and analyzing environmental problems.

Since the evaluation of the subhalide process, new processes for the recovery of aluminum have included a chloride electrolysis and recycle system (called the Alcoa process), the reduction of aluminum chloride with manganese (called the Toth process), and newer aspects of the carbothermic reduction of alumina (Reynolds and Showa Denko).

Details of the Alcoa process are, of course, not known. The principles, chlorination of alumina and electrolysis of $AlCl_3$-based electrolytes, have been studied sporadically since the work of Bunsen in 1852. Alcoa is to be recognized for its apparent satisfactory resolution of the engineering and materials of construction in the production and transport of aluminum chloride and design of cells with the option of bipolar electrodes which can be arranged vertically. This configuration is a space-conserving factor that the aluminum industry would welcome. As I understand it, the basic flow diagram of the Alcoa process is:

$$3CO\uparrow$$

$$+$$

$$Al_2O_3 + 3Cl_2 + 3C \longrightarrow 2AlCl_3 \xrightarrow{680-720°C} 2Al + 3Cl_2$$

$$Al \begin{vmatrix} M_ICl \\ M_{II}Cl \end{vmatrix} Cl$$

There have been prior disclosures in German patents showing that the temperature for the Oerstad process could be lowered by traces of $NaAlCl_4$. Implementation of this or similar beneficial chemistry in the chlorination of alumina, the use of a vertical bipolar cell, and a permanent anode could conceivably reduce the unit energy requirements for aluminum by 20–30%.

Not knowing the state of Alcoa's progress, it would be presumptuous to anticipate what technology may be lacking. From long association with molten salt systems, however, one can anticipate that considerable research will be required to resolve the role of oxychlorides in the chloride systems, the details of dependence of partial pressures on composition and temperature, and the long-term effects of chlorides permeating the primary lining materials and other materials of construction.

Dr. Toth's process involves the interaction of Mn, produced in a modified blast furnace,

$$MnO_2 + C \longrightarrow Mn + CO_2$$

with gaseous aluminum chloride,

$$Al_2O_3 + 3C + 3Cl_2 \longrightarrow 3CO + 2AlCl_3$$

Viz.,

$$3Mn + 2AlCl_3 \longrightarrow 2Al + 3MnCl_2 (1)$$

after which the $MnCl_2$ is recycled to another furnace to recover the MnO_2 and chlorine

$$MnCl_2 + O_2 \longrightarrow MnO_2 + Cl_2.$$

This approach, suitable for containment, is still under development by Applied Aluminum Research Corporation. The industry is awaiting their further disclosures, particularly in connection with a convincing energy and cost appraisal, the chemistry and physical conditions for the displacement reaction, and assurance of the purity and form of the aluminum extracted. It appears that additional study on the thermodynamics of the $Mn/AlCl_3$ system may be needed to better judge the levels of Mn left in the aluminum product. In more recent disclosures, the temperature of the final step to aluminum has been described as about 300°C. Particulate aluminum is difficult to coalesce into metallurgically useful products. Thus, if this is to be an element of the final process, demonstration of the means to form billets, etc., would be a desirable objective.

Showa Denko, proposing that aluminum of negligible carbide content can be recovered by sequential extraction through temperature fluctuations in a mass of $Al-Al_4C_3$, has revived the old dream of an "aluminum blast furnace." The arc-furnace reduction of an alumina–carbon charge is directed towards production of aluminum only,

$$Al_2O_3 + 3C \longrightarrow 2Al + 3CO,$$

but the formation of aluminum carbide has generally proved unavoidable. Their contribution has been the claim that the temperature dependence of the equilibria between Al and Al_4C_3 can be exploited to recover a product containing significantly less Al_4C_3 than one would expect.

The same materials, C and Al_2O_3, in different stoichiometric ratios, are used by Kibby (Reynolds Metals) in first preparing an intermediate, aluminum monoxycarbide (Al_2OC), in a prereduction furnace. This Al_2OC is vaporized in a heating zone such as a controlled arc at a controlled rate so as to produce aluminum vapor and CO. The aluminum is recovered by condensation at a sufficiently low temperature that the vapor pressure of the liquid aluminum is less than the partial pressure of aluminum. In this system for the intensive production of aluminum, energy is recovered from both the CO generated in pre-reduction and distillation stages and the hydrogen used to depress the partial pressure of aluminum in the condenser. Reynolds does not disclose full details on materials of construction. A greatly improved data base for the thermodynamics of the Al–O–C system and the physical properties of the aluminum oxycarbides is needed in support of processes like these two.

The areas of research really needed to support alternative processes to the Hall–Heroult process include:

1. Impact of $AlCl_3$ on the physical properties and strength of a range of structural materials such as steels, Inconels, composite metals

2. Equilibria of $AlCl_3$ in reaction with selected metals as a function of temperature, pressure, and chemical means of modifying activities

3. Kinetics and equilibria of solid–solid, solid–gas reaction in the Al–C–O system

4. Free-radical reactions and frozen equilibria resulting from plasma reactions as suggested by R. K. Rains in 1970

5. Means of improving the design and materials of construction of the cathodes of existing electrolytic cells.

In addition to the implications of these longer-range programs, problems of another nature have descended upon the aluminum industry—raw materials supply. Wide publicity has been given to the unilateral action desired by Jamaica in increasing threefold the revenue to the government from bauxite mines. This has focused attention on an already well-known fact, that 80% of the United States' alumina requirements are imported. Methods of recovering alumina from domestic clays, laterites, and anorthosites have been practiced on a pilot scale by the industry and the U.S. Bureau of Mines alike. These were more costly than the Bayer process, but with the rapid trends to increasing costs of overseas bauxites, they will be competitive. Consequently, the industry, already involved in demonstration plants capable of utilizing large holdings of aluminous ores, will be alert to the alternatives needed to avert economic impasses on raw materials.

Other raw materials becoming more difficult to secure with the preferred specifications or purity are metallurgical cokes and pitches for anode fabrication and anthracite coal for cathode construction. The industry has provided for such contingency by developing technology for solution-extraction-refining processes for coal they own. These can yield high-purity cokes and pitches suitable for replacing those derived from petroleum and coal tar. And again, as prices increase and the supply and cost of long-term contracts for energy become less certain, processing that which was once uneconomical becomes increasingly competitive.

18

Future Prospects for Electrodepositing Metals from High-Temperature Inorganic Melts

D. Inman and D. E. Williams

Within the broad brief of "electrowinning and depositing nonaluminum metals," this contribution concentrates on future prospects for electrodepositing metals, particularly refractory metals, from high-temperature inorganic melts. Some justification for this approach would seem to be necessary.

Thus, with regard to the past, a search of the literature convinces one that it is completely impossible to do justice to earlier work on a broad front, without the review becoming a mere bibliography. Several specialist reviews[1-3] are, however, available, dealing with the very large number of processes which have been researched, developed, and sometimes (but rarely) put into commercial practice in the long history of electrolytic metallurgy.

We have decided to concentrate on electrodeposition from melts for several reasons. First, high-temperature inorganic melts are good solvents for refractory source materials which may be difficult to treat by hydrometallurgical processes. Many refractory waste materials are attractive secondary sources of the common metals. Second, it may be possible to reduce the number of stages in an overall extraction process by dissolving the source material directly in a high-temperature inorganic melt prior to electrodeposition. The resulting simplification of the extraction process could be particularly significant now that energy costs are often the dominant factor in process economics.

Of course, many elements, in particular the alkali, alkaline earth, lanthanide, actinide, and refractory (transition) metals, can only be electrodeposited in pure form from melts (or, in some cases, other nonaqueous solvents). Also, many of the apparently well-established aqueous electrowinning processes are easily vitiated by the codeposition of hydrogen. However, the advantages of extraction processes based on electrodeposition from melts, over the alternative pyrometallurgical processes, are frequently theoretical or speculative only, so that the problems posed by these systems provide particularly exciting challenges for the future.

In our modern situation, technological progress involves ecological and sociological wisdom and political and economic reality as well as the initial scientific advance. We refer to fundamental research undertaken within this total context as "mission-oriented."

In the present article, we have considered future prospects under the following main headings:

1. What are the advantages of carrying out electrodeposition operations from high-temperature inorganic melts?

D. Inman and D. E. Williams • Department of Metallurgy and Materials Science, Imperial College of Science and Technology, London SW7 2BP, England.

2. What are the main problems preventing the more general application of electro-deposition processes based on inorganic melts?

3. What are the generally desirable objectives of research work in this area?

4. What means of process control do we have at our disposal?

5. Some examples of the mission orientation of fundamental principles are given.

6. Some technological areas where further development seems particularly promising are considered.

1. Advantages of High-Temperature Inorganic Melts

High-temperature inorganic melts offer the following general advantages as solvents for electrochemical operations:

1. The solvent melts themselves exhibit wide ranges of electroinactivity—that is their decomposition voltages, which as a first approximation can be related to the Gibbs free energies of formation of the melts from their constituent elements, are large. In practice, however, these "theoretical" decomposition voltages may be reduced in two ways. On the one hand, because of the high solubilities of metals in their own halide salts (e.g. Ca in CaF_2), the cations in alkaline and alkaline earth metal halide melts often deposit at less cathodic potentials than those expected for the deposition of the appropriate metal in its standard state. (This can be explained quite easily on the basis of the Nernst equation.) On the other hand, the oxy anions in, e.g., molten alkali and alkaline earth metal nitrates, sulfates, borates, and carbonates, are reduced at less cathodic potentials than are the metal ions of these solvents. Nevertheless, the experimental decomposition voltages are still relatively large, compared with those of other solvents.

2. They are good (ionic) conductors of electricity. Again, it is necessary to modify this statement in some cases. For example, in the case of Ca dissolution in molten CaF_2, the resulting metal–molten salt solution exhibits electronic conductivity and this has important consequences for the electroslag remelting process used in steel refining.[4] Again semiconductivity can arise when a metal exhibits more than one valence state (e.g. Fe^{2+} and Fe^{3+} in silicate slags), or when sulfide ions are present in solution.

3. They are powerful solvents for inorganic materials. This, of course, can be a severe disadvantage when one is trying to obtain a suitable container.

4. Voltage losses due to polarization are often small. This is not as generally true as might have been expected *a priori* for high temperatures. Although desirable from the point of view of energy conservation, there are certain circumstances, e.g., in the electro-plating of refractory metals (see below), where the presence of a non-diffusion-controlled stage in the charge-transfer process proper *or* the coupling of a slow reaction to the charge-transfer process probably prevents the formation of dendrites.

5. Chemical reactions between metal deposits and solvents and between metal ions and solvents are generally absent. For example, refractory metals cannot in general be deposited from aqueous solutions for one or more of the following reasons. (a) Their low oxidation states reduce water and their high oxidation states oxidize water. (b) Oxy cations, which can give rise to oxide deposits, are often formed. (c) The metals often become coated with oxide films. (d) The metals often electrodeposit at potentials negative to those for hydrogen evolution.

These sorts of reactions are generally absent in halide melts (cf. the stability of, for example, Ti^{2+} and Ta^{5+}). On the other hand, problems such as the loss of volatile solutes and the formation of insoluble clusters can arise unless careful attention is paid to the composition of the solvent melt.

2. The Main Problems

Why have not more of the processes which have been conceived in laboratories been developed commercially? Economic factors (the high capital cost of an electrowinning plant) are obviously playing a part. At the scientific level, the main problems that one can see at the moment may be considered under four general headings: separation of the metal deposits from the melt after electrolysis, keeping the metal solute stable in the solution, special problems arising from oxide solutes, and materials problems.

We will briefly describe these problems and subsequently, in Sections 3 and 4, we will discuss the present understanding and the future prospects for their solution.

2.1. Separation of Metal Deposits from Melts

Many laboratory methods for the electrodeposition of high-melting-point metals give a dendritic or powdery product.[3] If the metal is formed as dendrites or powders at the cathode, then its subsequent separation from the solidified solvent is a process analogous to the beneficiation of a crude ore. Expensive and wasteful leaching, grinding, and flotation operations are required. A successful electrowinning process requires the metal to be formed as a liquid or as a massive, coherent, and continuous deposit on the cathode.

2.2. Stability of Solutes

This type of complication is particularly marked with transition-metal compounds. The disproportionation or polymerization of lower valency states and the volatilization of higher valency states are the chief worries. Disproportionation reactions lead to powdery deposits, and polymerization reactions can give insoluble materials. Normally, halide melts are used, and rigorously pure melts and a strictly controlled atmosphere must be maintained to avoid the formation of insoluble oxides or oxohalides, either as precipitates or as films covering the electrodes.

Problems can arise through the oxidation of the solute by the anode product (e.g., chlorine). This in turn makes necessary cells employing ceramic diaphragms which, if they are too porous, allow too much mixing of anolyte and catholyte, and if they are not porous enough, introduce high electrical resistance. In addition, the porous ceramics are more easily corroded by the melts than are the bulk materials.

2.3. Special Problems Arising from Oxide Solutes

In many cases, the most generally available solutes are oxides (e.g., alumina in aluminum winning), and these pose special problems, even assuming that molten solvents in which they are sufficiently soluble are available. These special problems are of two types. First, the formation of oxy ions can lead to oxidic rather than metallic cathodic deposits. Second, at carbon or graphite anodes the formation and consequent dissolution of CO and CO_2 in the melts can give rise to carbides following the electroreduction of the gases at the cathodes.[5]

2.4. Materials Problems

Problems associated with the choice of materials of construction, their corrosion, and the hydrolysis of melts by moisture are all particularly severe at high temperatures. This of course is one reason why high-temperature processes in general are avoided if possible.

Much research in this area has been conducted in recent years on the materials science of the bulk materials,[6] the improvement of melt purification and handling in inert atmospheres, and the science behind the interactions of melts and metals.[7]

3. Generally Desirable Objectives

In view of the problems described above, it is possible to formulate some of the generally desirable objectives of any mission-oriented fundamental research work in this field, and these are given below. Each of course involves a very wide area of research. We will also give here a brief comment on experimental methods.

3.1. Form of Deposits

Barton and Bockris[8] have shown that dendritic deposits are formed when the rate of the electrode reaction is controlled by the rate of mass transfer of the reactant to the electrode surface. Thus, in order to avoid the formation of dendrites, melts in which the rate of electrodeposition is *not* diffusion controlled should be devised. This is a difficult task because in general, at the high temperature of the melts, chemical and electrochemical reactions are reversible (see Section 2). At present, it seems possible to introduce irreversible steps into the electroreduction reaction sequence by a suitable choice of the anionic composition of the solvent (see Section 4). Improvements in cell design, to increase the rate of mass transfer, are also needed.

3.2. Stability of Baths

One must devise melts to minimize the disproportionation of the lower valence states, the formation of insoluble lower valence states, and the volatilization of higher valence states. The cationic composition of the solvent appears to be the most important parameter here (see Section 4). There is still much to be learned about the chemistry and thermodynamics of transition-metal solutes in inorganic melts.

3.3. Oxide Solutes

As previously stated, one problem is to obtain a melt in which oxides are sufficiently soluble, without the consequent formation of stable oxy cations. Thus, when UO_2Cl_2 is dissolved in a halide melt, the stable cation UO_2^{2+} is formed and cathodic reduction gives UO_2.[9] On the other hand, in the aluminum electrowinning bath, evidence exists for the formation of stable oxy anions (e.g., $AlO_2F_2^{3-}$), which decompose and yield aluminum metal by the cathodic reduction of the positive component.

A particularly clearcut example is that of vanadium. Although lithium–vanadium bronzes appear to result from the electrolysis of V_2O_5 in LiCl–KCl,[10] vanadium metal can apparently be produced by the electrolysis of the same solute in $CaCl_2$.[11] Again, the cationic composition of the solvent appears to be the most important parameter. Qualitatively, it appears that the interaction between the cationic component of the solvent and the oxide ions competes with the interaction between the vanadium ion and the oxide ions.

A present trend[12] seems to be to chlorinate the metal oxide source material before electrolysis. However, there can be severe materials problems associated with handling chlorine and metal chloride vapors at high temperatures.

Furthermore, concerning the problem of carbide formation, it should be possible to devise electrodes which will effect the separation of oxygen from melts, without reaction to produce CO and CO_2. Much research in recent years has concerned the solid, oxide-ion-conducting electrolytes based on zirconia, which could perhaps be used as ion-selective anodes for this purpose.[13]

The investigation of new electrode materials could be a fruitful area for future research. Perhaps it would be possible to devise an ion-selective anode for chlorine? Such anodes would provide an attractive alternative to the use of porous diaphragms, but obvious problems would arise from the heavy currents generally employed in electrolytic operations and the corrosion of materials by the melts.

3.4. Materials Problems

In order to minimize corrosion and hydrolysis and, incidentally, to conserve energy, it is desirable to employ melts which remain molten to as low temperatures as possible. Ideally, of course, one needs a solvent which is molten at room temperature, possesses all the desirable properties of the melts mentioned above, and none of the deleterious properties!

3.5. Experimental Methods

Rigorous purification to ensure the quality of the system (melts and gases) is needed. For example, the presence of traces of moisture or oxygen will give rise to oxide films on electrodes of metals such as Ti or Ta. The problem is particularly marked for melts containing LiCl. Procedures involving bubbling HCl through the melt, evacuation, bubbling of argon, and pre-electrolysis are used. Even so it is still not certain that impurity effects are absent.

Many studies in high-temperature electrochemistry have used the chronopotentiometric method[2] because of its simplicity. However, the usefulness of this technique for studying complex reactions is rather restricted. When disproportionation and polymerization reactions occur, the results become exceedingly confusing. Also, the technique relies on the establishment of well-defined diffusion conditions, and this is difficult to attain with high-temperature systems. Further development in experimental techniques is needed. Indeed, one must be careful to establish the appropriate independent variables for the system. These are frequently potential and time, not current.

Most electrochemical methods are ambiguous when it comes to establishing the exact nature of the intermediates and products in an electrode process. Indeed, the chemistry of these molten systems—the nature of the compounds and oxidation states involved—is frequently not well known. The extension of the spectroscopic methods, which have proved so powerful in the study of electrode reactions at room temperature, to high-temperature systems would be a major development.

4. Control of Processes

The following modes of control over molten salt processes for electrodepositing metals are available: solvent anion composition (first-order effect), solvent cation composition (second-order effect), concentration of the solute, temperature, applied potential, applied current density, and acid–base, redox, and complexation reactions.

Most of these modes of control are well known to aqueous solution electrochemists, but because so many high-temperature extractive processes are apparently dirty, empirical, and relatively unsophisticated, the high degree of design sophistication and control that can be attained is often not appreciated.

In the following, we will illustrate, in turn and by examples, the effect of each of these control parameters. We will try to show the utility of simple qualitative theories, for example the concept of soft and hard acids and bases (SHAB), in systematizing a confused mass of data. We will also illustrate the importance of analogies to aqueous solution chemistry.

4.1. Solvent Anionic Composition

This is a first-order effect, as is clearly exemplified by the marked differences between chlorides and fluorides as electroplating baths for refractory metals.

The all-fluoride baths, developed by Senderoff and Mellors at Union Carbide,[3,14] were remarkably successful in producing coherent, structurally stable deposits, but in most cases plate quality deteriorated in the presence of chloride and other foreign anions, and powders or dendrites were produced. The cationic composition of the melt was not critical, but control of the solute valence in an intermediate state was critical.* The process was general for the refractory metals, (Ti, Zr, Hf, V, Nb, Ta, Cr, Mo, W) except for Ti.

On the other hand, only molybdenum could be plated in a coherent form from an all-chloride bath.[15,25] The overall reaction was irreversible. Addition of fluoride in small amounts made the reaction reversible[16] and would thus presumably result in a reduction of the quality of the plate, although this point has not been tested experimentally. The quality of the deposit was critically dependent on the cationic composition of the solvent in this case.[3]

In the fluoride baths the formation of coherent deposits was attributed to an irreversible electrochemical step, and in the all-chloride bath to a slow chemical step—dissociation of a dinuclear cluster—preceding the electrochemical reduction.

In terms of the SHAB concept, the "hard" fluoride ion forms strong complexes with the higher oxidation states of the solute. The lower oxidation states are unstable and tend to disproportionate, producing powdery electrode deposits. The highest oxidation states form neutral stable compounds which volatilize from the melt. In the intermediate oxidation states, stable and highly charged "hard" anionic species are formed, which do not volatilize from the melt and are electroreduced irreversibly. The SHAB concept leads one to speculate that the "softer" chloride ion can displace fluoride from these intermediate valence complexes, giving rise to "softer" complexes which can interact more strongly with the electrode and hence be electroreduced more easily. The study of electrode–solute interactions and the electrical double layer in molten salt solvents is important in this respect. Adsorption effects have been tentatively identified.[16]

In the molybdenum case, the "softer" chloride ion favors the formation of lower oxidation states and polymeric clusters. The "hard" fluoride ion breaks up these clusters. In an extreme case, tantalum, insoluble polymers seem to be produced during electroreduction in chloride melts.[17] These effects are also dependent on the cationic composition of the solvent.

---

*For example, in the case of niobium a preliminary electrolysis is required to reduce the niobium [initially in the (V) state] to a mean valence not exceeding 4.2.

4.2. Solvent Cationic Composition

Although the solvent cations exert only a second-order effect within any particular anion system, their effect can be quite important, and even critical, especially in respect to the thermal stability of melt solutions. Thus, the partial pressures of Ti(IV), Zr(IV), and Hf(IV) above lithium chloride solution approach 10^3 torr, but above CsCl solution, the vapor pressures are only of the order of 1 torr, at $1000°K$.[18] Tantalum(V) shows similar behavior. Good enhancement of thermal stability can be obtained by using as solvent melts those containing cesium chloride.[17]

The SHAB concept is useful here, too. It interprets the effects in terms of the modification of the "hardness" or "softness" of the solvent anions by interaction with the solvent cations. Thus, chloride ion is "harder" in cesium chloride solvent than in lithium chloride. A nice analogy can be drawn with the relative reactivities of the halide ions in water and in dipolar aprotic solvents. The halides behave as much "harder" reagents in the nonaqueous solvents.[19]

The harder the anion, the stronger is its interaction with solutes in high oxidation states. In other words, the interaction of the solvent anions with the solute is affected by the polarization of the solvent anions by the solvent cations. This effect has indeed been identified as a major factor affecting the activity coefficients of the components of binary alkali halide melts.[20]

The effects of solvent cation composition on the polymerization and disproportionation of lower oxidation states can be similarly understood. Thus, in the case of tantalum, disproportionation and polymerization of lower valence states are important phenomena in CsCl solvent, but in KCl solvent or in LiCl–KCl eutectic solvent, polymerization, at least, appears to be a much less important phenomenon.[17] It is noteworthy that the most stable compounds of tantalum are with Rb or Cs as the cation and that solid polymeric compounds of the closely related element niobium have so far been prepared only with Rb or Cs cations.[21]

The binary aluminum chloride–alkali chloride melts offer interesting possibilities because of the range of cation–anion interactions that can be obtained in them. In terms of the SHAB idea, the anion "hardness" can be varied over a wide range by varying the cation composition. The variation of coordination geometry of transition-metal solutes with composition that is observed[22] can be related to this. Further study of electrode reactions in these solvents should prove fruitful. However, although these melts would seem to offer certain advantages as solvents for metal deposition, for example very low melting temperatures, they possess the major disadvantage of a relatively low decomposition voltage. The same objection applies to melts containing zinc chloride.

4.3. Solute Concentration

In most cases, concentrated solutions are advantageous for the enhancement of mass-transfer rates whereas dilute solutions are advantageous from the point of view of thermal stability. However, if polymerization occurs, but without the formation of an insoluble species, more concentrated solutions can be advantageous from this point of view as well.

Thus, in the case of Mo, the vapor pressure of $MoCl_3$ in alkali chloride solution decreases with increasing concentration.[15] This can be related to the formation of dinuclear complexes at higher concentration, and as we have seen, has the added advantage of introducing a slow chemical step, dissociation of the cluster, into the overall electrode process.

4.4. Temperature

The temperature limits are set at the lower end by the liquidus temperature of the solvent and the solubility of the solute. The upper limit is set by volatilization of the solute, degradation of the deposit, and corrosion of the container. From virtually every point of view, the lower the temperature of operation is, the better.

4.5. Electrode Potential

For most electrodeposition purposes, it is better to control selectivity rather than rate (i.e., to control the thermodynamics rather than the kinetics), and thus potentiostatic rather than galvanostatic control is desirable. The application of potentiostats to electrochemical studies in molten media is facilitated by the low ohmic resistance of these media.

From the industrial point of view, the development of higher power potentiostats is needed. An awareness of the meaning and importance of electrode potential, as distinct from cell voltage, and of the principles of reference electrode design is also needed.

4.6. Applied Current Density

Control of the current density rather than the electrode potential has usually been applied, both in industrial and laboratory high-temperature electrochemical operations, because the necessary electrical equipment is more straightforward.

Control of the rate of deposition is important in determining the balance between dendrite and powder formation in a mass-transfer-limited deposition process. Control of the rate is also critical in the formation of surface (diffusion) coatings, in which one must balance the rate of deposition with the rate of diffusion of the metal deposited into the substrate.

4.7. Acid–Base, Redox, and Complexation Reactions

As high-temperature melts and their mixtures have been increasingly investigated in recent years, many phenomena identical *or* similar to those well known for lower-temperature solvents have been repeatedly observed. In particular, complex ion formation and acid–base reactions have been identified and characterized. For the bulk reactions, the Lux–Flood acid–base definition (base = acid + O^{2-}) has been known for many years, but it is only relatively recently that its consequence for electrode reactions involving oxy anions have been looked into.[7] One noteworthy feature has been the increasing awareness of the part played by superoxide and peroxide ions when the simple oxide ion is exposed to oxidizing environments.[23] Redox phenomena are also well known and, in particular, the role played by certain melts in stabilizing unusual valence states has attracted a lot of interest.[24]

Examples of the use of these phenomena in dissolving, stabilizing, and separating melt solutes will be given below.

5. Some Illustrations of the Mission Orientation of Fundamental Principles

5.1. Electrodeposition of Molybdenum

In 1954 Senderoff and Brenner[25] showed that it was possible to obtain electrodeposited molybdenum plates from chloride melts. Further noteworthy work was that of Senderoff and Mellors,[15] who first identified the role played by dinuclear complexes. The work of Inman and Spencer[26] has also been published. Salient features which are related to the control of the processes have been given earlier and are summarized below.

1. The thermal stabilities of the baths were shown to increase as the sizes of the alkali metal cations of the solvent melts were increased. The melt NaCl–KCl (20 mole%:80 mole%) employed at approximately 750°C was found to be particularly advantageous in this respect.

2. The removal of the solute ($MoCl_3$) by volatilization was found to decrease as its concentration increased. This was related to the formation of multinuclear complexes.

3. The overall rate of electrodeposition in chloride melts was shown to be controlled by the rate of a coupled chemical process, viz.

$$Mo_2^{6+} \longrightarrow 2Mo^{3+} \quad \text{(slow)}$$

$$Mo^{3+} + 3e^- \longrightarrow Mo \quad \text{(fast)}$$

The dinuclear species is probably wholly *or* partly chloride bridged.

4. The rate of electrodeposition apparently increased as fluoride ions were added to the chloride melts.

How is it possible to apply these results? First, we can see how to produce the optimum bath, from the thermal stability point of view, on the basis of scientific principles. Second, we can see that it is desirable that this bath be operated in the regime where the rate of metal deposition is controlled by the rate of the coupled chemical reaction and that (in this case at least) fluoride ions be excluded.

5.2. Chromium Extraction from Chromite Ores Using Chloride Melts

Processes, of course, are already available for this purpose, but over the years much attention has been paid to the research of a more economical method for the production of chromium metal. Although most authors concerned themselves with the production of chromium itself from a pure feedstock, Fleck and Wong[27] investigated the possibilities of extracting chromium directly from its chromite ore (main components: ferrous oxide and chromic oxide). Two different melts were used to promote dissolution: The first was a mixture of $Na_4P_2O_7$, NaCl, and $NaPO_3$, and the second a mixture of $Na_2B_4O_7$ and Na_2CO_3. In both cases the chromium content of the product was low (maximum 50.7%). The major impurity was, of course, iron, which was not separated from the chromium prior to electrolysis.

In devising a method for the extraction of chromium from its chromite ores, redox, acid-base, complexation, and electrode reactions in LiCl–KCl at 500°C, were combined to achieve the dissolution and separation of the metallic components of the mineral. The details of this work have already been published,[28] so only a summary will be given here.

Cr_2O_3 as well as FeO can be dissolved in the chloride melt by bubbling SO_3. In the latter case, the SO_3 acted as an oxidizing agent as well as an acid. The Fe(III) compounds which were formed exhibited some instability. A synthetic chromite (Cr_2O_3 + FeO) was then suspended in the melt. In this case, SO_3 bubbling led to the preferential dissolution of the FeO. Only around 3% of the Cr_2O_3 dissolved in the presence of FeO. It could be dissolved subsequently by a similar SO_3 treatment after the separation of the iron.

The observation that Cr_2O_3 remained practically unattacked, whereas FeO was completely soluble when the mixed oxides were treated with SO_3, could be explained on the basis of the following facts: (1) FeO is a more basic oxide than Cr_2O_3; (2) Cr_2O_3 is less reactive than FeO; and (3) SO_3 is preferentially used as an oxidizing agent for Fe(II) ions.

Al_2O_3 is the main impurity in natural chromite ores, but this was not included in the synthetic chromite experiments in order not to obscure the reactions associated with the main components. Aluminum ions can, however, be preferentially complexed by fluoride ions, and in this way their cathodic electrodeposition potential is "well clear" of that for chromium. The other important component of the chromite ore is MgO. Mg(II) deposits at much more negative potentials than chromium.

The main negative feature of this investigation was the difficulty encountered in trying to obtain good electroplates of chromium in the presence of sulfate ions in the chloride melts. (It had earlier been shown that good electroplates were obtainable from pure chloride melts.) This mainly arose because SO_3 produced anodically from sulfate ions is reduced at less cathodic potentials than Cr(II) ions, the intermediate in the reduction of Cr(III) to metal. Some improvement was effected by using a separate anode compartment. However, chromium metal can react with sulfate anions if they are present at high enough concentrations. Some methods for eliminating the sulfate anions formed by the SO_3 treatment were suggested in the original paper.

The main reason for discussing the work above has been to illustrate the way in which fundamental principles can be directly mission oriented. Although these principles are well known in relation to aqueous electrochemistry, this is not so for molten salts where, almost of necessity, many technological approaches have been *ad hoc* in conception, only subsequently being rationalized and explained in scientific terms.

6. Some Promising Areas for Technological Development

One's view of the areas awaiting development tends to be rather subjective. Our main criterion in selecting the areas below is that they are all ripe for the application of fundamental principles in the ways illustrated above for molybdenum and chromium.

6.1. Recovery of Metals from Waste Materials and Unusual Ores

In order to conserve materials and energy resources, there is a general trend towards reclaiming metals from waste materials. In addition, ore sources are becoming leaner, and hitherto unused types of ore may have to be employed. It is difficult to extract metal values from materials such as pyrometallurgical slags, flue dusts, clays, and refractory rubble using conventional room-temperature solvents. On the other hand, processes utilizing molten salts and slags can be conceived to deal with this problem.

For example, in a process which we are currently researching, one would dissolve refractory material in a molten oxide solvent, extract metals with an immiscible halide melt, and recover the metals by electrodeposition from the halide. We have outlined above (Section 4) some of the means that are available for obtaining selectivity in such a process. One disadvantage, from the energy-conservation point of view, is the necessity of operating at high temperature, but there are many situations where waste heat from other operations could be utilized. The processing of slags and dusts being produced by an existing extraction plant is one such situation.

6.2. Electrochemical Process Engineering at High Temperatures

Very little systematic attention has been paid to the systematic aspects of cell design, such as scale-up, heat transfer, and mass transfer. These are properly the concern of electrochemical engineering.

Richardson[29] has recently drawn attention to the wide disparity in the rates of a range of metallurgical processes. Among these, the electrolytic winning of aluminum is noteworthy for its slow throughput per unit of reactor volume (0.00005 ton $m^{-3} \cdot min^{-1}$). This may be compared with the processing rates in the iron blast furnace, (0.004 ton $m^{-3} \cdot min^{-1}$). The important factors are the rate of mass transfer to the reacting interface and the area of this interface.

In recent years there has been a great deal of attention to these factors in aqueous electrochemistry, and a number of cell designs have been evolved.[30] Extension of these designs into high-temperature electrochemistry is needed. The development of bipolar electrode arrays for aluminum electrowinning[12] is encouraging in this respect. It is important to make metallurgical processes go as quickly as possible in order to obtain high outputs from small reactors, which tend to waste less heat and have lower capital costs.

6.3. Direct Reduction of Sulfide Ores to Metals

The electron is a powerful and controllable reducing agent. It is a much more powerful reducing agent than hydrogen, and therefore, in principle, direct electrolytic reduction could be used to advantage in place of existing processes which employ roasting to oxidize sulfides to oxides prior to hydrogen reduction, e.g., Mo production from molybdenite (MoS_2).

The anodic dissolution of a metal sulfide in an aqueous medium is gradually gaining importance as a method for the simultaneous recovery of the metal and elementary sulfur.[31] Sulfur is produced from the sulfide anode, releasing metal ions into the solution. Metal is electrodeposited at the cathode. The alternative process, in which the sulfide is made the cathode, forming the metal and releasing sulfide ions to the solution, is very slow in aqueous media.

Both processes seem feasible using a molten electrolyte with a molten sulfide matte as one electrode, and the problems associated with the fabrication of sulfide electrodes are avoided. With molten electrolytes, difficulties arise in the anodic process because of the production of polysulfides, and in the cathodic process because the presence of sulfide ions in the melt gives rise to semiconduction and hence to a reduction in the current efficiency. However, promising developments can be expected in the future.

7. Conclusion

Clean electrolytic processes have much to offer in the preservation of the quality of the environment, whether or not they are acceptable in conventional economic terms. We have discussed some of the possibilities, advantages, and problems associated with the electrodeposition of metals from high-temperature inorganic melts, and we look forward with interest to the future of this exciting field.

Acknowledgments

We should like to thank Mr. G. F. Warren for making available his results on the electrochemistry of tantalum in alkali halide melts, in advance of publication.

This work was supported by the S.R.C. and the Wolfson Foundation.

References

1. C. A. Hampel, *Encyclopaedia of Electrochemistry*, Reinhold, New York (1964).
2. D. Inman *et al.*, Electrochemistry, *Chem. Soc. Spec. Per. Rep. 1*, 166 (1970); *2*, 61 (1972); *4*, 78 (1974).
3. S. Senderoff, *Metall. Rev. 11*, 97 (1966).
4. M. E. Peover, *J. Inst. Metals 100*, 97 (1972).
5. P. J. Bowles, *Advances in Extractive Metallurgy*, p. 600, The Institution of Mining and Metallurgy, London (1968).
6. F. S. Martin *et al.*, *High Temperature Chemical Reaction Engineering: Solids Conversion Processes* (F. Roberts, R. F. Taylor, and T. R. Jenkins, eds.), p. 215, Institution of Chemical Engineers, London (1971).
7. D. Inman and N. S. Wrench, *Br. Corros. J. 1*, 246 (1966).
8. J. L. Barton and J. O'M. Bockris, *Proc. R. Soc. London A268*, 485 (1962).
9. R. G. Robins, *J. Nucl. Mater. 3*, 294 (1961).
10. H. A. Laitenen and D. R. Rhodes, *J. Electrochem. Soc. 109*, 413 (1962).
11. O. Watanabe, *J. Electrochem. Soc. Jpn. Overseas Ed. 34*, 91 (1966); *Chem. Abstr. 66*, 25421p (1967).
12. J. G. Peacey and W. G. Davenport, *J. Met. 26*(7), 25 (1974).
13. B. Marincek, *Schweiz. Arch. Angew. Wiss. Tech. 33*, 395 (1967), quoted by T. H. Etsell and S. N. Flengas; *Chem. Rev. 70*, 339 (1970).
14. S. Senderoff and G. W. Mellors, *J. Electrochem. Soc. 114*, 586 (1967).
15. S. Senderoff and G. W. Mellors, *J. Electrochem. Soc. 114*, 556 (1967).
16. D. Inman, R. S. Sethi, and R. Spencer, *J. Electroanal. Chem. 29*, 137 (1971).
17. G. F. Warren, personal communication.
18. S. N. Flengas and P. Pint, Can. Metall. Q. 8, 151 (1969).
19. A. J. Parker, *Chem. Rev. 69*, 1 (1969).
20. J. Lumsden, *Thermodynamics of Molten Salt Mixtures*, p. 70, Academic Press, London (1966).
21. A. Broll, A. Simon, H. G. von Schnering, and H. Schafer, *Z. Anorg. Allgem. Chem. 367*, 1 (1969).
22. H. A. Oye and D. M. Gruen, *Inorg. Chem. 4*, 1173 (1965); C. A. Angell and D. M. Gruen, *J. Inorg. Nucl. Chem. 29*, 2243 (1967).
23. P. G. Zambonin, *Anal. Chem. 43*, 1571 (1971).
24. C. R. Boston, in: *Advances in Molten Salt Chemistry* (J. Braunstein, G. Mamantov, and G. P. Smith, eds.), Vol. 1, p. 129, Plenum Press, New York (1971).
25. S. Senderoff and A. Brenner, *J. Electrochem. Soc. 101*, 16 (1954).
26. D. Inman and R. Spencer, *Advances in Extractive Metallurgy and Refining*, p. 413, The Institute of Mining and Metallurgy, London (1974).
27. D. C. Fleck and M. M. Wong, U.S. Pat., 3,126,327 (1964).
28. D. Inman and J. C. L. Legey, *Physical Chemistry of Process Metallurgy* (J. H. E. Jeffes and R. J. Tait, eds.), p. 69, Institute of Mining and Metallurgy, London (1974).
29. F. D. Richardson, *Trans. Inst. Min. Metall. 84A*, p. 19, Bulletin No. 824 (1975).
30. R. W. Houghton and A. T. Kuhn, *J. Appl. Electrochem. 4*, 173 (1974).
31. F. Habashi, *Minerals Sci. Eng. 3*(3), 3 (1971).

19

Mineral Processing—A Review of the Current State of Fundamental Knowledge

Thomas W. Healy

1. Introduction

Mineral processing, as practiced to upgrade ore from a mineral deposit into a salable concentrate from which the metal can be extracted, is now a sequence of complex operations which are controlled and optimized by the cooperative efforts of engineers and chemists. Whenever those operations involve water as the dispersary phase, the chemistry of the unit operation is necessarily and inextricably electrochemical.

Mineral processing ceased to be an art when it was first understood that the fundamental control of most processes is based on understanding the electrochemistry of the mineral–water system. It is important to stress, however, that with a few notable exceptions, the art of mineral processing was practiced for many centuries before man began to grasp the science! However, our ability to treat lower and lower grade ores and to recycle materials now depends most definitely on our ability to understand the fundamental electrochemistry of mineral processing operations.

The aim of this chapter is to present a summary of the principles of mineral processing as we now understand them and to show how intelligent application of such principles now and in the past has enabled mineral processing to change from an art to a science.

Since the aim is to demonstrate how electrochemical principles have been applied to mineral processing, it is necessary to be selective and to single out a few operations for consideration. The unit operation of froth flotation, more than any other mineral processing operation, made possible the present multimillion tons/day operations all over the world, and it is therefore appropriate that it form the basis of the chapter. Other operations such as pelletizing, thickening, and comminution could also be considered. Froth flotation is that unit operation which is used to separate one kind of particulate solid from another through their selective attachment to air bubbles in an aqueous suspension. The volume[1] to commemorate the 50th anniversary of froth flotation shows very clearly how the vast mineral development of the United States, Canada, Australia, Africa, and the U.S.S.R. began with the introduction of froth flotation. Each day in the United States, for example, the unit operation of froth flotation is responsible for the treatment of more than one million tons of materials. The source of most of the common base metals is the flotation concentrates of metal sulfides, PbS, $CuFeS_2$, ZnS, etc., and continuous flotation produces large tonnages of these materials economically by fairly standard techniques. In recent years, however, there has been an increasing need by modern technology for the so-called nonmetallic minerals, including the ceramic oxides, silicates,

Thomas W. Healy • Department of Physical Chemistry, University of Melbourne, Parkville 3052 Victoria, Australia.

and clays; phosphate fertilizers; iron ores; high-temperature refractories; nuclear minerals; and others. Consequently, there is an increasing need to provide these important raw materials by such methods as selective flotation separation. In the United States in 1960, for example, phosphate materials made up one third of the total 21 million tons of all flotation concentrates produced.[1] This was almost twice the tonnage of any other single material produced by flotation treatment.

In all that is to be considered, we shall continually stress the nature and properties of the mineral–aqueous solution interface. Herein is the basis of the assertion that mineral processing science *is* electrochemical science, since the interface is composed of charged sites, adsorbed ions and dipoles, and an electrical double layer with its attendant electrical potential profile. It is usual to consider sulfide and nonsulfide mineral processing as two separate but related topics. This makes engineering sense in that usually sulfide mineral processing involves recovery of the sulfide from a low-grade ore (i.e., high throughput, low retention, high rejection) while nonsulfide processing usually involves a high-grade ore (i.e., somewhat higher throughput, low rejection, high retention). It also, as we shall see, makes electrochemical sense in that sulfide mineral–water systems are E_h-pH (redox–hydrolysis) controlled systems, while nonsulfide systems are more simply pH (hydrolysis only) controlled systems.

Thus we shall first examine the "state of the science" of nonsulfides, i.e., their fundamental interfacial electrochemistry, adsorption of ions and molecules at the interface, and the properties that are changed by adsorption. Then we shall look at sulfide surfaces and stress the ways in which they differ from the simpler nonsulfide minerals.

2. Nonsulfide Mineral Chemistry

Flotation science and technology has two main areas of concern: (1) finding proper physicochemical conditions for achieving appropriate selectivity between mineral species, and (2) determining conditions for optimum process kinetics. In this paper the principles upon which the selective separation of particulate solids can be achieved by froth flotation will be developed primarily in terms of the surface chemistry of nonmetallic mineral systems. Particular emphasis will be placed on the principles involved in making solid particles selectively hydrophobic. Factors that control the kinetics of bubble-particle attachment will not be discussed, since they involve less electrochemical factors than the act of alteration of the surface to a hydrophobic (i.e., air avid) condition.

It is appropriate at this stage to summarize some of the basic definitions of flotation science that may not be immediately known to electro- and surface chemists. The common flotation technology terms[2] are as follows.

A *collector* is the surface-active agent (e.g., fatty acid, alkyl xanthate or thiophosphate, long-chain sulfate, sulfonate, amine) that is added to the flotation pulp where it is selectively adsorbed on the surface of the desired mineral to render it hydrophobic. An air bubble preferentially adheres to such a surface and concentrates the mineral into the froth layer at the top of the flotation cell.

A *frother* is again a surface-active agent (e.g., short-chain alcohol, aromatic alcohol, ethylene oxide surfactant), added to the flotation pulp primarily to stabilize the froth in which the hydrophobic minerals are trapped. When long-chain soaps or amines are used as collectors, a frother may not be necessary.

An *activator* is a material added to promote selective adsorption of the collector on a particular mineral. Metal ions, for example, can activate the anionic surface of the mineral quartz so that anionic collectors (e.g., alkyl sulfonates) can adsorb and induce flotation;

again, copper ions can promote collector adsorption on the usually unreactive ZnS surface.

A *depressant* is a material added to prevent the collector from adsorbing on a particular mineral in the mixture in the flotation pulp. Metal ions, for example, may act as depressants by preventing adsorption of a cationic surfactant on an otherwise anionic surface of a particular mineral. Sodium silicate can depress minerals that tend to respond to anionic collectors.

Laboratory flotation studies can, at the simplest level, involve measurement of the contact angle at the air–solution–mineral interface[3] or the bubble "pick-up" where a bubble is pressed onto a bed of particles.[2] These static tests of hydrophobicity can be supplemented by flotation in a glass model of a flotation cell[4] or in the so-called Hallimond tube apparatus developed by Hallimond[5,6] and perfected by Ewers[3] and Fuerstenau.[7] The value of the Hallimond tube is that the experiments can be done in the presence of collector only, whereas frother must be added in float-cell replicas. Vacuum flotation studies in which air or gas saturated mineral pulp is subjected to partial vacuum to form bubbles appears to measure "incipient flotation"[8] and is useful to delineate the limits of flotation–nonflotation in a given system. The reader is referred elsewhere[2,9] for a comprehensive analysis of experimental techniques including the thermodynamics of contact angle measurements in flotation systems.

2.1. The Electrical Double Layer

In the case of ionic solids such as AgI, Ag_2S, $BaSO_4$, CaF_2, $CaCO_3$, and others, a surface charge can arise when there is an excess of one of the lattice ions at the solid surface. Equilibrium is attained when the electrochemical potential of these ions is constant throughout the system. Those particular ions which are free to pass between both phases and therefore are able to establish the electrical double layer are called *potential-determining ions* (p.d.i.). In the case of AgI, the potential-determining ions are Ag^+ and I^-. For a solid such as calcite, $CaCO_3$, the potential-determining ions are Ca^{2+} and CO_3^{2-}, but also H^+, OH^-, and HCO_3^- because of the equilibria between these latter ions and CO_3^{2-}.

For oxides, hydrogen and hydroxyl ions have long been considered to be potential determining,[10,11] although the mechanism as to how pH controls the surface charge on oxides is still speculative. It is well known that oxides possess a hydroxylated surface in the presence of water, and we can generate the surface charge by sets of thermodynamically equivalent equilibria such as

$$AH_2^+ \rightleftharpoons AH + H^+ \qquad (K_+) \qquad\qquad (1)$$

$$AH \rightleftharpoons A^- + H^+ \qquad (K_-) \qquad\qquad (2)$$

where K_+ and K_- are surface dissociation constants of the proton acid groups of the metal oxide surface. Thus in the case of the oxide, and also the many aluminosilicate minerals, changing the pH of the solution will markedly affect the magnitude and sign of the surface charge. Similarly, change in the activity of say Ag^+ (or I^-) for AgI will alter the charge of the AgI–water interface. There is a solution condition—equivalent to the electrocapillarity maximum—where the surface has equal numbers of positive and negative groups, and this unique point of zero charge (p.z.c.) is a most valuable interfacial parameter in mineral systems. Table 1 summarizes a selection of p.z.c. values of salt-type and oxide-type minerals; the listing follows earlier reviews[2,12,13] to which the reader is referred to locate exact pretreatment and other experimental conditions.

Table 1. Point-of-Zero-Charge Values of Various Nonsulfide Minerals

Material	p.z.c.	Material	p.z.c.
Barite ($BaSO_4$)	pBa 6.7	Cassiterite (SnO_2)	pH 4.5
Calcite ($CaCO_3$)	pH 9.5	Zirconia (ZrO_2)	pH 4.0
Fluoroapatite ($Ca_5(PO_4)_3(F, OH)$)	pH 6	Rutile (TiO_2)	pH 5.8–6.7
Fluorite (CaF_2)	pCa 3	Hematite (natural Fe_2O_3)	pH 4.8–6.7
Hydroxyapatite ($Ca_5(PO_4)_3OH$)	pH 7	Hematite (synthetic Fe_2O_3)	pH 7.5–8.8
Scheelite ($CaWO_4$)	pCa 4.8	Goethite ($FeOOH$)	pH 6.8–7.2
Silver iodide (AgI)	pAg 5.6	Corundum (α-Al_2O_3)	pH 9.1
Silica gel (SiO_2)	pH 1–2	Boehmite ($AlOOH$)	pH 7.8–8.8
α-Quartz (SiO_2)	pH 2–3	Magnesia (MgO)	pH 12

The p.z.c. refers, quite explicitly, to a measurement wherein the charge of the surface is being determined. For example, the surface charge ($\sigma_0 \mu C/cm^2$) is conveniently described in terms of adsorption densities (Γ mol/cm^2) of p.d.i., thus for AgI,

$$\sigma_0 = eF(\Gamma_{Ag^+} - \Gamma_{I^-}) \qquad (3)$$

or for oxides,

$$\sigma_0 = eF(\Gamma_{H^+} - \Gamma_{OH^-}) \qquad (4)$$

Now this charge, balanced by the charge of the rest of the electrical double layer, gives rise to a potential–distance profile out from the surface. For AgI, fashioned in the form of an electrode (e.g., Ag/AgI), we can measure a potential difference between solid and solution, and the solution condition corresponding to zero potential is referred to as the isoelectric point (i.e.p.). In the absence of specific adsorption, i.e., indifferent electrolyte only being present, then the p.z.c. and i.e.p. are identical.

For most oxides, silicates, etc., we are unable to fashion an electrode, and we must rely on the classical electrokinetic potential or zeta potential to give us a measure of potentials between solid and solution. Again, but this time for slightly different reasons, if supporting electrolyte only is present, the point of zero zeta potential with respect to p.d.i. and the i.e.p. and the p.z.c. are identical.

If we return to the reversible electrode system of Ag/AgI, we are well aware that the potential difference measured (E) is given by

$$dE = \frac{-2.303RT}{F} d \log a_{Ag^+}$$

$$= -59.2 \, d(pAg) \text{ at } 25°C \text{ for } E \text{ in mV} \qquad (5)$$

This classical Nernstian response is real, but a problem arises immediately as to how to relate dE to surface charge given by equation (3) above. By convention, but in reality in terms of an ion–dipole model of Lange,[14, 15] we split E into an outer potential (ψ_0) and a chi potential difference and write

$$d\psi_0 = -59.2 \, dpAg$$

where ψ_0 is the potential difference between the plane of surface charges and bulk solution (considered as zero potential). The quantity ψ_0 so defined does not require connection between the surface and the bulk crystal and the attached Ag metal. It exists for an isolated AgI particle dispersed in water, a condition where we cannot conveniently measure ψ_0. In no way is it being suggested that the Nernst equation as we know it is incorrect, but rather it is suggested that uncritical use of the Nernst equation to describe the potential difference, and its change with pAg, at an isolated AgI-solution interface, is an assumption and one worthy of investigation.

Turning to oxides—where as noted we cannot usually fashion an electrode—the use of equation (5) in the form

$$d\psi_0 = -59.2 \, (d\text{pH}) \tag{6}$$

is an assumption for which *no* evidence is available to support it. Levine and Smith[16] and later Yates *et al.*[17] have shown that for an oxide for which the charge is given by equation (4), the potential of the surface wherein resides the fundamental charge is given by a more complex expression where the variation of ψ_0 with pH is dependent on ionic strength and surface-group dissociation constant.

The simplest way of illustrating the magnitude and significance of the ψ_0–pH departure from Nernstian behavior is to consider the variation of ψ_0 with pH or ionic strength at various values of the parameter ΔpK where

$$\Delta pK = pK_- - pK_+ \tag{7}$$

i.e., the relative acid strength of the second to the first dissociation constants of the diprotic surface acid. This is shown in Figure 1 for an amphoteric surface at a pH of 3 units above the pH of the p.z.c. and as a function of ionic strength. The deviation at $\Delta\text{pH} = 3$ (i.e., 59.2 × 3 mV) from the Nernstian behavior is both ΔpK and ionic-strength dependent. As ΔpK increases, the deviation from Nernst behavior increases so that by ΔpK of about 6, instead of 59 mV/pH unit, we observe 30–35 mV/pH unit at 10^{-3} M. Silica appears to have a pK of 6–8; FeOOH, 4–5; TiO$_2$, 3–4; Al$_2$O$_3$, 2–3; etc.[18–20]

Continuation of the quantitative analysis of the structure of oxide and other non-sulfide mineral–solution interfaces will need to consider what now appears to be a general non-Nernstian potential–pH relationship. Fortunately, the use of electrical double-layer theory in the past has usually been restricted to the significance of the p.z.c. or i.e.p. There is good experimental evidence, for example, that the p.z.c. itself is related in a general way to the average electrostatic field of an oxide lattice.[20,21] The field, which is the vector sum of simple coulombic forces acting on an electron (as a probe) located at fixed distance above the lattice, increases as the valence of the cation of the lattice in-

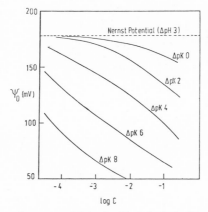

Fig. 1. The variation of the surface potential (ψ_0) with 1 : 1 electrolyte concentration (C) for oxide surfaces of various ΔpK values for 3 pH units above the p.z.c. The so-called Nernstian response (independent of C) is shown for comparison.

creases and as the density (ions/unit volume) increases. Empirically it is found that the pH of the p.z.c. increases with increasing field. The basis of this experimental correlation clearly bears rethinking. Curiously those oxides with low p.z.c. show an adsorption sequence of $Cs^+ > Li$, whereas oxides with a p.z.c. greater than pH 4 show $Li^+ > Cs^+$.[22]

2.2. Surfactant Adsorption of Oxide–Water Interfaces

The total free energy of adsorption of species ranging from simple small organic or inorganic ions and molecules, through surfactants, all the way to complex polyelectrolytes, must of necessity be broken down into component parts, viz.,

$$\Delta G^\circ_{ads} = \sum_j \Delta G^\circ_j \tag{8}$$

where ΔG°_j may be contributions to adsorption identifiable as separate independent processes. Following an earlier treatment,[23] equation (8) may be given by

$$\Delta G^\circ_{ads} = \Delta G^\circ_{elec} + \Delta G^\circ_{solv} + \Delta G^\circ_{hb} + \Delta G^\circ_{vdw} + \Delta G^\circ_{chem} \tag{9}$$

where ΔG°_{elec} is the $+/-$, or coulombic, free energy change and is given by $z_\pm e\psi_\delta$. ΔG°_{elec} is negative when the adsorbing ion is opposite in sign to the sign of the net surface charge. Thus cation adsorption is, in coulombic terms, spontaneous above the pH of the p.z.c., while anion adsorption is spontaneous below the pH of the p.z.c. The potential ψ_δ is the electrostatic potential at a plane distance δ out from the surface, and it is at this plane where the particular ion adsorbs. If the adsorbing species were a dipole, then ΔG°_{elec} in such a case is always negative and is given by the product of the electric field at the interface and the dipole moment.

The second term in equation (9) is the change in free energy of the hydration sphere of an ion as it adsorbs and becomes partly desolvated in the process. For adsorption on minerals that are insulators this ΔG°_{solv} term is always positive but can be near zero or slightly negative for adsorption of ions on a semiconducting mineral, such as rutile, which has a dielectric constant of the same order of magnitude as that of water. ΔG°_{solv} is negative for ion adsorption at metal–electrolyte interfaces.

The third and fourth terms will be considered in more detail in subsequent sections. It is sufficient at this stage to identify them as contributions to adsorption due to hydrophobic interactions (ΔG°_{hb})[24] and London dispersion of Van der Waals interactions (ΔG°_{vdw}). The final term ΔG°_{chem} includes specific "chemical" interactions ranging from hydrogen bonding to covalent bonding.

This method of analysis of data on the adsorption of organics at a mineral–water interface is illustrated by reference to the data of Figure 2. The system is sodium dodecylsulfonate adsorbing at the alumina–aqueous electrolyte interface.[2] Shown as a function of the equilibrium concentration of the surfactant is the adsorption density, the zeta potential of the alumina colloid, and the change in wettability of the alumina. The wettability is directly related to the work of adhesion (W_A) and the contact angle (θ) at the alumina–water–air interfaces as

$$W_A = (1 + \cos\theta)\gamma_{AW}$$

where γ_{AW} is the air–water interfacial tension and θ is measured through the water phase. The initial rise in adsorption density is paralleled by no change in zeta potential, indicating that surfactant ions are exchanging in the diffuse layer and are not entering the inner part of the double layer. Then at a particular concentration termed the "hemimicelle

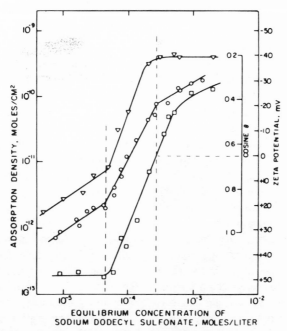

Fig. 2. The adsorption density of sodium dodecylsulfonate on alumina, the zeta potential of alumina, and the contact angle on alumina as a function of the equilibrium concentration of sodium dodecyl-sulfonate at pH 7.2 and ionic strength of 2×10^{-3} N at $24 \pm 1°C$ controlled with NaCl: ($\triangledown$) contact angle, ($\circ$) adsorption density, ($\square$) electrophoretic mobility. [Reprinted with permission from Fuerstenau and Healy.[2]]

concentration" (HMC), the adsorption and the magnitude of the zeta potential both increase sharply; there is also a sharp increase in the contact angle. Adsorption eventually reduces the zeta potential to zero, and the adsorption density continues at a less marked rate of increase. If the marked increase in adsorption at the HMC is identified as the onset of hydrophobic interactions (ΔG°_{hb}) then at zero zeta potential

$$\Delta G^\circ_{ads} = \Delta G^\circ_{hb}$$

If hydrophobic effects are considered as the product of the number of —CH_2— groups in the chain (n) times a hydrophobic energy per CH_2 group (ϕ_{hm}), then at zero zeta potential

$$\Delta G^\circ_{ads} = n\phi_{hm}$$

since

$$\Delta G^\circ_{ads} = -RT \ln K_{ads}$$

where K_{ads} is the adsorption equilibrium constant. Then from the Stern model of the electrical double layer, in terms of the Grahame approximation valid at low coverage,

$$RT \ln [\Gamma_0/(2rC_0)] \simeq n\phi_{hm} \qquad (10)$$

where Γ_0 and C_0 are the adsorption densities and equilibrium concentration at zero zeta potential, respectively, and r is a constant; the adsorption density at zero zeta potential is constant at constant pH and supporting electrolyte concentration for an homologous series of surfactants. Thus if C_0 is plotted as a function of n, a straight line results, the slope of which yields ϕ_{hm}, the hydrophobic contribution per CH_2 group to the total free energy of adsorption.

In seeking a general understanding of the relation between the structure of an organic collector and its adsorption properties, it is found that one can separate out hydrophilic functions from hydrophobic functions. Examples of hydrophilic centers are polar head groups, double bonds, and ether linkages, while examples of hydrophobic functions include methyl and methylene groups and aromatic or heterocyclic centers.

For the process of transferring a polar group out of water to an interface, the free-energy change is positive and must be recovered by, say, a negative electrostatic free energy (ΔG°_{elec}) of interaction with the charged surface. In contrast, *the system* will lower its free energy when hydrophobic centers are removed from water into a hydrophobic environment. Such a hydrophobic environment is provided by a micelle interior, a hemimicelle of adsorbed organics at a solid–liquid interface, by hydrocarbon–water solutions, or by hydrocarbon liquid.

Considerable advance has been made recently by Lin and Somasundaran[25] who summarized many of these hydrophobic-group free-energy changes. They considered the measured changes in solubility, micelle formation, and adsorption of many organics as a function of the alkyl chain length of the organic. The known linearity (Traube's rule), of log (property) vs. chain length yields a hydrophobic energy per —CH_2— residue. A selection of hydrophobic energy values for various processes is given in Table 2.

The system gains increasing amounts of energy/CH_2 as a CH_2 residue is transferred from solution to micelle, to hemimicelle, to hydrocarbon solution to pure hydrocarbon liquid.

As first pointed out by Fuerstenau and Modi,[26] measurement of the sign of the charge on oxide mineral surfaces is necessary in order to ascertain the fundamental response of minerals to flotation with anionic or cationic collectors. An example of the dependence of flotation on the sign and magnitude of the surface charge is presented in Figure 3, which is based on the experiments of Iwasaki *et al.*[27] on the electrokinetic and flotation behavior of goethite ($FeOOH$).

The behavior of numerous mineral–collector systems similar to those shown in Figure 3 have been published, including the flotation of alumina,[26, 28] hematite and magne-

Table 2. Free Energy Changes per CH_2 Group for Various Transfer Processes Involving Organics[25]

Transfer of alkyl chains		Organic reagents	Total free-energy change per CH_2 group (ϕ_{CH_2}) (kT)
From	To		
Aqueous solution	Ionic micelles	Fatty acid soaps, alkyl sulfates, alkylammonium, and trimethyl ammonium halides	−0.61 to −0.69
Aqueous solution	Hemimicelles at S–L interface	Alkylammonium acetates, alkylammonium sulfonates	−0.71 to −0.98
Aqueous solution	Nonionic micelles	Oxyethylene hydrocarbons	−1.10 to 1.22
Aqueous solution	Hydrocarbon solution	Long-chain ($n = 8$–16) fatty acids	−1.39
Aqueous solution	Hydrocarbon liquid	Alcohols ($n = 4$–10)	−1.34 to −1.39
		Octane	−1.17
		Alkyl amines ($n = 10$–14)	−1.53

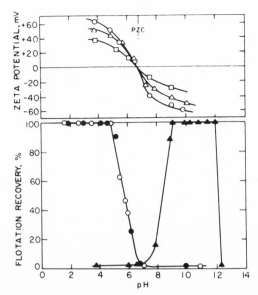

Fig. 3. The dependence of the flotation of goethite (FeOOH) on surface charge. The upper curves show the zeta potential as a function of pH at different concentrations of NaCl, indicating the p.z.c. to be at pH 6.7: (○) 10^{-4} M, (△) 10^{-3} M, (□) 10^{-2} M. The lower curves are the flotation recoveries in 10^{-3} M solutions of dodecylammonium chloride (○), sodium dodecylsulfate (●), or sodium dodecylsulfonate (▲). [After Iwasaki *et al.* (1960b); reprinted with permission from Fuerstenau and Healy.[2]]

tite,[27] ilmenite,[29] aluminosilicates,[30] and rutile.[29] All of the above studies were conducted with collectors with 12 carbon atoms. From a technical point of view, increasing the length of the hydrocarbon chain decreases the selectivity of the collector in making flotation separations.

As for other types of minerals, an early study of collection of minerals with amine salts by Taggart and Arbiter[31] showed that the contact angle on barite with dodecylammonium chloride increased with increasing SO_4^{2-} concentration and decreased with increasing Ba^{2+} concentration in solution. These results can be interpreted in terms of the double-layer model since Ba^{2+} and SO_4^{2-} are potential determining in this system.

In summary, separation of nonsulfide minerals can be achieved by finding conditions where minerals may be oppositely charged so that a cationic or anionic collector adsorbs only on the desired mineral. For example, in considering the flotation separation of a mixture of quartz and rutile, the two minerals are oppositely charged between approximately pH 3 and 6. Within this pH range quartz can be floated from rutile with an alkylammonium collector or, on the other hand, rutile can be floated from quartz with a sulfonate collector. Above pH 6, however, both minerals are negatively charged, and a separation cannot be achieved with these collectors.

In nonsulfide flotation the process of activation, i.e., preconditioning of a surface to allow adsorption of a particular surfactant, is most frequently achieved by control of the adsorption of metal ions and in particular, the hydrolyzable metal ions. The example shown in Figure 4 is taken from the comprehensive work by M. C. Fuerstenau *et al.*[32] We note that the quartz, with its increasingly negative charge above pH 2–3, would not adsorb the anionic alkylsulfonate collector.

The dramatic changes in surface properties induced by metal ions in these metal-activation systems can best be appreciated by considering the accompanying zeta-potential

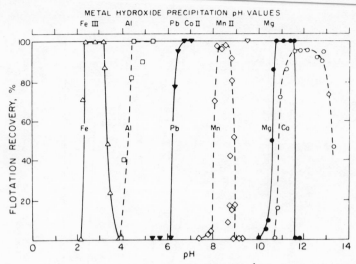

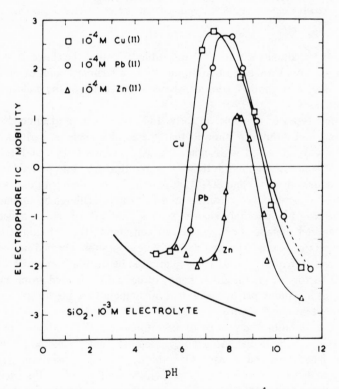

Fig. 4. The effect of cation hydrolysis in the activation (2×10^{-4} M) of quartz for flotation with a sulfonate (6×10^{-5} M) as collector. [After Fuerstenau *et al.* (1963), reprinted with permission from Fuerstenau and Healy.[2]]

Fig. 5. Electrophoretic mobility behavior of SiO_2 in solutions of 10^{-4} M copper(II) or zinc(II). The solid line at the bottom of the figure is for SiO_2 in the absence of added metal ions. [Reprinted with permission from Healy.[33]]

changes. An example is shown in Figure 5 where the zeta-potential–pH behavior of quartz in the presence of Cu(II), Pb(II), or Zn(II) is shown.[33] The unique ability of such metal ions, and those responsible for the activation effect shown in Figure 4, to induce (– to +) followed by (+ to –) changes in the sign of the zeta potential of a negative substrate is closely allied to their hydrolysis behavior. Some have suggested that one particular mononuclear hydrolysis product such as $PbOH^+$ is specifically (i.e., in an electrochemical sense) adsorbed, while others have invoked polynuclear complexes such as $Al_8(OH)_{20}^{4+}$; although it is difficult to understand the massive (– to +) zeta reversal with a species like $PbOH^+$ and impossible to generalize polynuclears to those systems which do not form then neither adsorbate allows the subsequent higher pH (+ to –) reversal without desorption, and desorption is not observed experimentally. These limitations led James and Healy[34] to suggest that the low pH (– to +) reversal was due to surface nucleation-precipitation of the hydroxide of the metal; once so coated, the substrate behaves as the metal hydroxide, and the system switches abruptly and accordingly to give the normal i.e.p. of the hydroxide. Experimentally the pH of zero zeta potential as part of the high pH (+ to –) reversal does indeed correspond to the pH of the i.e.p. of the metal hydroxide. Metal activation in flotation is seen in this model of James and Healy as a surface-coating rather than an ion-adsorption process.

3. Sulfide Flotation

The realization that electrochemical processes were inextricably involved in sulfide mineral flotation came first with the realization[35] that oxygen was required to promote "adsorption" of a collector such as potassium ethyl xanthate at the PbS (galena)–water interface. Prior to that time, uptake of the organic collector $C_2H_5CSSO^-$ to produce a hydrophobic (air-avid) surface was thought to occur by a double decomposition reaction with xanthate (X^-) as, e.g.,

$$PbS + 2X^- \rightleftharpoons PbX_2 + S^=$$

If surface lead xanthate were more insoluble than surface lead sulfide, abstraction of xanthate would occur and the surface would therefore be in a flotable condition. Similarly the inability of ethyl xanthate to float ZnS (sphalerite) was thought to be due to the fact that zinc xanthate (surface) was more soluble than surface zinc sulfide.

With the wisdom of hindsight it is easy to challenge this model; it is suspicious, to say the least, to find that an *ethyl*, two-carbon surfactant is able to induce sulfides to float in contrast to the nonsulfides where collectors are C_{12}–C_{16} in chain length. This suggests, for example, that either sulfides are inherently close to an air-avid condition before xanthate adsorption and/or that the mineral-water interfacial energy is lowered by a dramatic chemisorption process involving electron sharing or transfer at the interface.

3.1. Electrochemical Models of Sulfide Surface Reactions

There is now a wealth of evidence that various surface-controlled processes applied to sulfide minerals are electrochemical in nature. Amongst the most spectacular evidence is, for example,

1. The direct relationship between the E_h-pH states and conditions of the model sulfur–oxygen–water system and the acid dissolution or leaching of sulphides;[37, 38]
2. The changes in contact angle of an air bubble at the PbS–water interface in the

presence of ethyl xanthate when the PbS sample was fashioned in the form of an electrode of an electrochemical cell;[37]

3. The finding that the normally unreactive ZnS will abstract xanthate if irradiated with UV light of a wavelength corresponding to the band gap of ZnS;[39]

4. The fact that diethyldithiophosphate inhibits oxidation reactions at the copper sulfide (Cu_2S) water interface *and* promotes air avidity by formation of Cu(I) and Cu(II) thiophosphate surface complexes.[40]

We can summarize the essential electrochemical nature of sulfide mineral–organic collector interaction as

$$X^- \longrightarrow X_{ads} + e \quad \text{and/or} \quad 2X^- \longrightarrow X_2 + 2e$$
$$\text{oxidation}$$

where X^- and X_2 are xanthate and dixanthogen, respectively, and X_{ads} represents chemisorbed xanthate.

$$O_2 + e \longrightarrow O_2^-, \text{etc.} \quad \text{and/or} \quad H_2O + e \longrightarrow OH^- + \tfrac{1}{2}H_2$$
$$\text{or} \quad O_2 + 4H^+ + 4e \longrightarrow 2H_2O$$
$$\text{reduction}$$

The solid, in acting as an electrode, provides the pathway for the redox electron transfer. We must still account for the hydrophobic condition which results from such a short two-carbon (or four-carbon, if dixanthogen is the active species) chain. It is fair to say that there is evidence for chemisorbed xanthate and dixanthogen and even for elemental sulfur, an inherently hydrophobic material; the sulfur can be thought to be generated at pH values below about pH 8 by reactions[41,42] such as

$$S^{2-} \longrightarrow S^\circ + 2e$$

Future research must attempt to reconcile the now well-established electrochemical reactions at sulfide–water interfaces and the manifest ease with which many sulfide minerals can be made hydrophobic.

Other reactions at sulfide–water interfaces must also contain an electrochemical component. Among intentional or inherent activators we could list

$$Cu^{2+} \longrightarrow Cu^+ + e \quad Ag^+ \longrightarrow Ag + e \quad Fe^{3+} \longrightarrow Fe^{2+} + e$$

When we consider leaching reactions, again they are controlled by electrochemical processes. A particularly interesting example[39] is the dissolution of ZnS where the dissolution is found to be catalyzed by UV light corresponding to the absorption edge of ZnS. It is thought to be due to an expression of the reaction of electron–hole pairs generated by the light with interstitial zincs.

$$Zn_i^+ + e/p \longrightarrow Zn^{2+} + e$$

Again we need reactions such as

$$O_2 + e \longrightarrow O_2^-, \text{etc.}$$

to provide the reduction half of the redox couple.

4. Conclusions

Wet processing of minerals has now attained a state where future advances will depend on careful control of electrochemical processes at mineral–water interfaces. The demands to treat ultralow-grade ores and increasing amounts of recycled material and to limit pollution levels in reject effluent can no longer be regarded as an extension of the present art; application of the general principles of electrochemistry to these problems presents many new challenges.

Acknowledgment

The author acknowledges the generous support of the Australian Mineral Industries Research Association (AMIRA) which made this work possible.

References

1. *Froth Flotation—50th Anniversary Volume*, AIME, D. W. Fuerstenau, ed., New York (1962).
2. D. W. Fuerstenau and T. W. Healy, in: *Adsorptive Bubble Separation Processes* (R. Lemlich, ed.), Chapter 6, Academic Press, New York (1972).
3. K. L. Sutherland and I. W. Wark, *Principles of Flotation*, Aus.I.M.M. Melbourne, 1955.
4. M. C. Fuerstenau, *Eng. Min. J. 165*, 108 (1964).
5. A. F. Hallimond, *Min. Mag. 70*, 87 (1944).
6. A. F. Hallimond, *Min. Mag. 72*, 201 (1945).
7. D. W. Fuerstenau *et al.*, *Eng. Min. J. 158*, 93 (1957).
8. P. Somasundaran and D. W. Fuerstenau, *Trans. AIME 241*, 102 (1968).
9. A. M. Gaudin, Flotation, McGraw-Hill, New York (1957).
10. F. F. Aplan and D. W. Fuerstenau, "Principles of Non-metallic Flotation," in Ref 1, p. 170.
11. H. R. Kruyt, *Colloid Science*, Vol. 1, Elsevier, Amsterdam (1952).
12. G. A. Parks, *Chem. Rev. 65*, 177 (1965).
13. T. W. Healy and D. W. Fuerstenau, *J. Colloid Sci. 20*, 376 (1965).
14. H. Lange, *Z. Elektrochem. 55*, 76 (1951).
15. R. Parsons, in: *Modern Aspects of Electrochem.* (B. E. Conway and J. O'M. Bockris, eds.), Vol. 1, p. 103, Butterworth, London (1954).
16. S. Levine and A. L. Smith, *Disc. Faraday Soc. 52*, 290 (1971).
17. D. E. Yates, S. Levine, and T. W. Healy, *J. Chem. Soc. Faraday Trans. 70*, 1807 (1974).
18. A. L. Smith, in: *Dispersion of Powders in Liquids* (G. D. Parfitt, ed.), 2nd ed., pp. 86–131, Elsevier, London (1973).
19. G. R. Wiese *et al.*, *Electrochemistry*, M.T.P. Publications, London 1975.
20. R. J. L. Wright and R. J. Hunter, *Austr. J. Chem. 26*, 1183 (1973).
21. T. W. Healy, A. P. Herring, and D. W. Fuerstenau, *J. Colloid Interface Sci. 21*, 435 (1966).
22. F. Dumont and A. Watillon, *Disc. Faraday Soc. 52*, 352 (1971).
23. T. W. Healy in: *Organic Compounds in Aquatic Environments* (S. J. Faust and J. V. Hunter, eds.), Chapter 9, Marcel Dekker, New York (1971).
24. G. Nemethy and H. A. Scheraga, *J. Chem. Phys. 36*, 3382 (1962).
25. I. Lin and P. Somasundaran, *J. Colloid Interface Sci. 37*, 731 (1971).
26. D. W. Fuerstenau and H. J. Modi, *Adaptation of New Research Techniques to Mineral Engineering Problems*, MIT Press, Cambridge (1956).
27. I. Iwasaki, *Trans. AIME 232*, 383 (1965).
28. H. J. Modi and D. W. Fuerstenau, *Trans. AIME 217*, 381 (1960).
29. H. S. Choi *et al.*, *Can. Min. Met. Bull. 60*, 217 (1967).
30. J. S. Smolik *et al.*, *Trans. AIME 235*, 367 (1966).
31. A. F. Taggart and N. Arbiter, *Trans. AIME 169*, 266 (1946).
32. M. C. Fürstenau *et al.*, *Trans. AIME 226*, 449 (1963).
33. T. W. Healy, *Proc. Aust. I.M.M. Conference W.A.*, 1973, p. 477, Aus. I.M.M. Melbourne (1972).
34. R. O. James and T. W. Healy, *J. Colloid Interface Sci. 40*, 53 (1972).
35. N. I. Plaksin and S. V. Bessonov, *Proc. II Int. Min. Proc. Congress, London 3*, 361 (1957).
36. E. Peters and H. Majima, *Can. Met. Q. 7*, 111 (1968).
37. R. Tolun and J. A. Kitchener, *Trans. I.M.M. 73*, 313 (1964).
38. R. Woods, *J. Phys. Chem. 75*, 354 (1971).
39. D. R. Dixon, R. O. James, and T. W. Healy, *Trans. AIME 258*, 81 (1975).

40. S. Chander and D. W. Fuerstenau, *Electroanal. Chem. Interfacial Electrochem.* *56*, 217 (1974).
41. R. K. Clifford and J. D. Miller, paper presented at AIME Annual Meeting, Dallas, Texas (1974).
42. T. W. Healy, M. S. Moignard, and D. R. Dixon, *A. M. Gaudin Memorial Symposium*, AIME (1976, in press).

20

Electrochemistry and Mineral Processing—
The Future

D. F. A. Koch and B. J. Welch

1. Introduction

The processing of minerals from ore to the finished product requires many stages. Initially, exploration is required to find an ore body, it has then to be mined, and finally processed. Between the mining and processing stages there is often a beneficiation step to increase the quality of the ore, followed by transportation to the processing site. Electrochemistry has played a role in all these areas, although in some cases this role has only been identified quite recently.

The future development of the mineral industry will require an increasing attention to technology which is compatible with society's demands for environmental protection. Thus, the concentrations and total quantities of both gaseous and liquid effluents which result from processing will need to be monitored and minimized. A more efficient utilization of our resources will be required which would entail the development of the most economic techniques for treating lower grades of ore and more efficient use of energy in its treatment. More sophisticated exploration techniques would also be required to enable deeper-seated ore bodies to be discovered. The attainment of these future needs could be aided by the application of electrochemical principles and techniques, and in this paper we would like to discuss these under the headings relevant to the mining industry, viz., exploration, mining, beneficiation, and processing.

2. Exploration

Most ore bodies which are currently being mined were discovered by prospectors who observed indications on the surface which led to a more detailed study of the area. More recently, physical techniques (e.g., electromagnetics, induced polarization, and spectral reflectance) have assisted in the process of discovery. However, the chemical transformations of an ore body which occur after its formation are little understood, and because increased knowledge could assist in our interpretation of surface and subsurface mineralization, some effort has recently been directed in this area. In the case of a sulfide ore body we have in most cases an electrochemical cell with an electronic conductor (sulfides), electrolyte (ground water), and reactants and products. Although this thought was proposed in 1830,[1] an elegant application has only recently been made.[2,3]

Application of Pourbaix-type diagrams to mineral systems have been developed by Garrels and Christ[4] to establish the thermodynamic conditions under which mineral deposits may be formed. Experimental measurements of Eh and pH have recently related

D. F. A. Koch • Division of Mineral Chemistry, CSIRO, Melbourne, Australia. *B. J. Welch* • School of Chemical Technology, University of New South Wales, Australia.

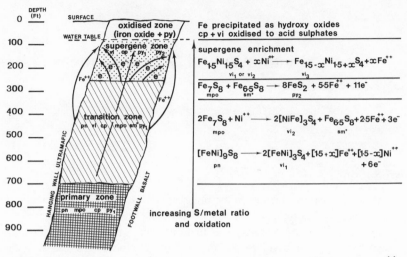

Fig. 1. Electrochemical model for the *in situ* oxidation of a nickel sulfide ore body.[2]

laboratory findings to the composition of an ore body which is being currently mined. A study of the system Ni, S, H_2O, O_2 has led to a representation of the compositional variation of an ore body by an electrochemical model which is shown diagrammatically in Figure 1.[2]

From this it can be seen that the oxidation of the ore occurs below the water table with the transfer of electrons to the air–water interface by the electronically conducting ore body. The cathodic reaction of this interface is the reduction of oxygen. The oxidation leads to a series of oxidation products which depend on the depth and consequently Eh. This is an interesting application of electrochemistry which could in the future predict the type of mineralization to be expected in depth from a knowledge of the structure of the ore body.

A knowledge of the form in which minerals can be transported could be useful in predicting the occurrence of an ore body, and the concentrations of chlorine and sulfur complex ions of the heavy metals in aqueous solution can explain the formation of some ore bodies.[5,6]

3. Mining

Problems associated with the oxidation of sulfides have been encountered in mining. The mining operation allows access of air to a previously enclosed ore body, and heating may result in an ultimate temperature in excess of 600°C and copious evolution of sulfur dioxide. It has been shown that the processes which cause the initial rise in temperature require water and air and produce sulfur. They may be expressed as two electrochemical reactions

$$MS \longrightarrow M^{+n} + ne + S \tag{1}$$

$$O_2 + 4H^+ + 4e \longrightarrow 2H_2O \tag{2}$$

Sulfur has been identified as an oxidation product[7] and pyrrhotite (FeS) as the mineral sulfide with sufficient reactivity to initiate the temperature rise. This greater

Fig. 2. Coarse granular electrodeposit of ZrB_2.

reactivity of pyrrhotite over other commonly occurring sulfides is borne out by electrochemical measurements.[8]

The reactions shown above may be used to advantage in another aspect of a mining operation, viz., the use of a backfill to consolidate stopes once an area has been mined out. The oxidation of pyrrhotite can result in the formation of dissolved Fe^{3+}, and the reprecipitation of the iron can cause accretion of a blend of waste rock (silicate) and flotation tailings (sulfide) to produce a cement of sufficient strength to enable mining operations to continue around the filled area.[9] The overheating phenomenon may be avoided by the correct choice of fill composition so that voids are not present and access of air is inhibited.

These are current applications, and the reactions are only now being understood. Greater application of our increasing knowledge of sulfide electrochemistry in this area may be expected to utilize a material which would otherwise be an embarrassing waste. Underground mining is currently the most costly factor in a mineral-treatment operation, and a decrease in costs is quite possible.

The mining operation can also be improved with better equipment that enables safer and speedier procedures to be used. One area where electrochemistry can play an important role is in the preparation of drill tips that are spark-free, thus minimizing the chance of igniting the ore. Research by Schlain et al.[29] has made considerable progress in this area. This approach may in the future be extended even further to the direct deposition of hard abrasive materials by application of fundamental aspects of electrocrystallization.[30] For example, an extreme range of deposits of ZrB_2 on nickel can be made from fused-salt electrodeposition.[31] The types include an extremely well-bonded and hard, rough-edged deposit as would be desired (Figure 2).

4. Beneficiation

Once the ore is mined it is crushed and ground and then usually concentrated by gravity, electromagnetic or electrostatic separation, froth flotation, or a combination of several of these techniques. Flotation results from the attachment of air bubbles to a mined surface which is made hydrophobic. The process by which a hydrophobic surface is produced in sulfides has recently been shown to be electrochemical in nature,[10,11] consequently it can be studied both experimentally and theoretically by utilizing the advances in electrode kinetics made by Bockris and his co-workers.[32] The overall process involves two electrochemical reactions, one involving xanthate ion which is oxidized and chemisorbed and the other the cathodic reduction of oxygen. Linear-sweep voltammetry has shown several oxidation states of xanthate on sulfide and metal surfaces[11] (Figure 3), and these oxidations have been associated with the occurrence of flotation[12] (Figure 4).

The half-reactions may be written

$$R-O-C\overset{S^-}{\underset{S}{\lessgtr}} \longrightarrow \left[R-O-C\overset{S}{\underset{S}{\lessgtr}}\right] + e \qquad (3)$$

$$2\left[R-O-C\overset{S}{\underset{S}{\lessgtr}}\right] \longrightarrow R-O-C\overset{S}{\underset{S-S}{\lessgtr}}-R-O-C\overset{S}{\underset{S}{\lessgtr}} \qquad (4)$$

Actual measurements of Eh in operating plants[13,14] and laboratory experiments show that the mixed potential mechanism, where reactions of the type in equations (3) and (4) are associated with the cathodic half-reaction for oxygen reduction, can be applied to flotation plant behavior. Eh measurements have also been used to control the oxidizing leaching of uranium ores.[16]

To date the empirical development of the flotation process has achieved remarkable success in concentrating sulfides. Typical results for lead, zinc, and copper sulfides in some Australian plants are shown in Table 1. Very high efficiencies are obviously being obtained.

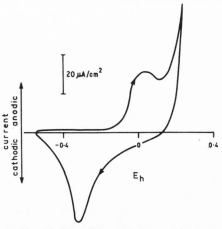

Fig. 3. Voltammogram for the reactions of ethyl xanthate on a galena electrode. Sweep rate 10 mV/sec; 0.1 M borate at 25°; 10^{-2} M ethyl xanthate.[11]

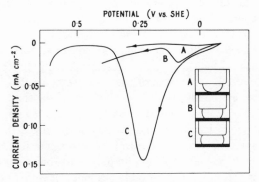

Fig. 4. Voltammograms showing the reduction of dixanthogen in 0.05 M borate (pH 9.2).[12] A: 0.07 monolayers, contact angle 10°; B: 1.0 monolayer, contact angle 60°; C: 7.0 monolayers, contact angle 75°.

In the future, flotation will be applied, if possible, to more "difficult" ores which currently cannot produce the results shown in Table 1. The problems are twofold. First, the physical problem of mineral liberation from the gangue material has to be achieved, and second there will be a need for new and specially selective reagents for some of the more complex ores. Flotation reagents will continue to be empirically tested but our increasing knowledge of the electrochemical reactions are already leading to a rationality in

Table 1. Concentration of Metal Values by Flotation

	Concentration of metal		
Ore type	In ore	In concentrate	Recovery
PbS with Zn and Fe sulfides and silicates	13% Pb	73% Pb	97% Pb
ZnS with Fe sulfides and silicates	10% Zn	52% Zn	87% Zn
$CuFeS_2$ with Fe sulfides and silicates	1% Cu	26% Cu	90% Cu

our examination of new reagents, as for example the correlation between the reversible potential for collector oxidation and flotation performance.[15]

5. Processing

Once beneficiated, the ore has to be processed to produce, in most cases, a metal. Changes in processing techniques are occurring partly as a result of the concern about pollution by effluents and also because of the need to minimize energy consumption. Most sulfides are smelted, and objections have been raised to the associated evolution of sulfur dioxide. Quite stringent requirements have been applied particularly in the U.S.A. (the U.S. Environmental Protection Administration Primary Standard is 0.13 ppm SO_2 daily). Also associated with many sulfides are toxic minor impurities such as arsenic, and their distribution during processing will be of increasing importance.

The problem of gaseous sulfur effluents may be solved by two approaches: first, by using pyrometallurgical processes which produce sulfur dioxide of a sufficiently high concentration so that it can be converted to sulfuric acid or elemental sulfur at an acceptable cost, and second by the use of a hydrometallurgical route which may oxidize the bulk of the sulfur only to elemental sulfur or another solid form (e.g., an inert iron sulfide).

On the pyrometallurgical side, flash smelting is receiving increasing application for the treatment of copper and nickel sulfides,[17] in which a gas containing approximately 10% sulfur dioxide is produced which can readily be converted to sulfuric acid. Both the sulfide and a phase such as lime, alumina, or silica are added as fine particles at the top of a shaft furnace. The reaction occurs during the passage down the shaft and then as the slag and matte phases are equilibrated at the bottom of the shaft. The equilibrium between matte and slag is electrochemical in nature because the slags are primarily ionic and the mattes are good electronic conductors. Consequently, the transfer of species across the matte–slag interface could well involve electron transfer. Although some of the charge-transfer reactions at this type of interface have been considered,[18] the fate of minor impurities is only now being seriously examined because of their possible discard in a polluting manner. Arsenic, for example, could be tied up in an inert form in a slag, but its distribution between the matte and slag phases is not favorable. Suitable complexing agents and careful adjustment of the oxidation potential may lead to an increased transfer to the slag phase.

Emanating from various pioneering studies to elucidate the structure of liquid silicates,[33] we have now developed a good understanding of structural aspects of ionic melts generally. This understanding has included the tendency of species to coordinate and form complexes, and therefore it is somewhat surprising that there has been no systematic attempt to utilize this knowledge to influence trace metal–melt equilibrium in extractive metallurgy. Not only pollution, but also quality control, considerations will stimulate future application of our present knowledge of complexing in ionic liquids during mineral processing. We have already established the feasibility of utilizing the coordinating characteristics during a pilot plant study of de-bismuthizing lead using a molten caustic electrolyte.[34] In this modification of the Dittmer process the bismuth ions are coordinated in an ionic form within the electrolyte.

Hydrometallurgical processes do not produce gaseous sulfur dioxide, and the air oxidation of a metal sulfide in an aqueous suspension could, in the simplest case produce metal ions in solution and elemental sulfur by the half-reactions (1) and (2). The treatment of copper sulfides has received the greatest attention, and some of this work has been recently reviewed.[19] The direct electrolysis of sulfides to form a metal at the cathode and sulfur at the anode has been used commercially for nickel[20] and copper,[21] and other direct electrochemical processes are at the pilot-plant stage.[22,23] There are two basic processing techniques used. One makes use of a cast massive anode made from a matte (molten sulfide) which is then cooled and electrolyzed in an aqueous electrolyte at temperatures between 20 and 90°C,[20,21] where the anode reaction in the case of nickel is

$$NiS \longrightarrow Ni^{2+} + S + 2e$$

and the cathode reaction is

$$Ni^{2+} + 2e \longrightarrow Ni$$

The other type of process treats the sulfide in a powdered form, and the most probable route of reaction is via the electrochemical regeneration of a dissolved oxidant.[22,23]

Molten salt systems can be electrolyzed in a similar manner, for example the Halkyn process[39] for electrolysis of PbS dissolved in molten $PbCl_2$ can be represented by the anode reaction

$$PbS \longrightarrow Pb^{2+} + S + 2e$$

and at the cathode

$$Pb^{2+} + 2e \longrightarrow Pb$$

This process has been operated in a pilot plant, but encountered problems due to build-up of impurities.

Impurities are always present in the feed for metallurgical processing, and in pyro-metallurgical operation the major impurities are discarded either with the slag or as a gaseous effluent. As we have already mentioned, the role of minor impurities in these operations will receive increasing attention. In a hydrometallurgical operation impurities have to be removed, and they can be discarded in the solid residue after leaching or as a precipitate from the solution. The high current efficiencies for zinc electrolysis, for example, are only obtained after careful solution purification, and in this case minor impurities are largely removed in a ferric hydroxide precipitate. A recent paper shows that a readily filterable goethite (FeOOH) can be obtained which carries down with it cations and anions.[24]

A process recently developed at the kinetic scale in our laboratories for the treatment of chalcopyrite concentrates[25] shows how impurities can be distributed in a hydro-metallurgical–electrochemical process. The most commonly occurring copper sulfide is chalcopyrite ($CuFeS_2$), and this is more difficult to oxidize than those with less iron.[8] Consequently a preliminary heat treatment at 300–400°C was used to produce idaite (Cu_5FeS_6) and pyrite (FeS_2):

$$5CuFeS_2 + 4S \longrightarrow Cu_5FeS_6 + 4FeS_2$$

The idaite is readily oxidized, while pyrite is the least reactive of the common sulfides,[8] and consequently a controlled oxidation using cupric chloride solution selectively dissolves the copper-containing mineral, leaving the major portion of the iron as an inert material which could be an acceptable waste. The leaching half-reactions may be written as

$$Cu_5FeS_6 \longrightarrow 5Cu^+ + Fe^{2+} + 6S + 7e$$

Table 2. Distribution of Impurities During CuCl$_2$ Leach at 90°C[26]

Element	% Dissolved
Cu	98
Fe	12
Zn	100
Mo	nil
As	21
Au	nil
Ag	84
Bi	91
Te	80

and

$$Cu^{2+} + e \longrightarrow Cu^+$$

The metal impurities are distributed in the leaching step as shown in Table 2.

The low solubility of Fe and Mo is probably due to the high oxidation potential of these sulfides, while the sulfides of Bi, Cu, Zn, and Ag are readily oxidized. Au will not be oxidized under these conditions, and the As results vary considerably with the ore type. A compound of the type Cu$_3$AsS$_4$ may be readily oxidized, while FeAsS could be relatively inert.

As in most hydrometallurgical processes, complete selectivity in leaching cannot be obtained, and purification steps will be needed to remove Fe, Zn, etc.; this would be a combination of an iron precipitation as referred to above and crystallization from a bleed-off.

The example was given in some detail to illustrate future problems which will have to be solved if hydrometallurgy is to supplant pyrometallurgy. There is abundant scope for the electrochemist and colloid chemist to contribute to improved processes which may meet our environmental needs.

There will also be the need to treat more dilute solutions as an effluent or a main stream to recover some of the valuable materials. The application of fluidized-bed electrolysis, where the larger surface area enables high diffusion-controlled currents for metal deposition,[27] may well be a future development. Cementation techniques, where a more noble metal is deposited on a less noble metal (e.g., Cu on Fe) by galvanic coupling, are also likely to receive wider attention from the electrochemists[28] with a view to extending their applications.

The increasing requirement for optimum energy usage and optimum recovery of the desired product is highlighting the need for more sophisticated methods of process control. In all cases some form of sensor is required to provide an input into a computer-control system. Recent advances in the potentiometric use of ion-selective electrodes have enabled us to envision its application in aqueous systems. However, further advances are needed if a high degree of selectivity is to be obtained, and an increasing understanding of the reactions may help in this direction. Recent work has associated a series of potentials on copper sulfides with discrete two-solid-phase systems[35] (Figure 5). Furthermore there is no reason from a theoretical standpoint why high-temperature, ion-selective electrodes cannot be developed to aid mineral processing.

Voltammetric methods have now advanced to a point where they can be considered for on-stream analysis, and they have the special advantage that preconcentrations (e.g., in

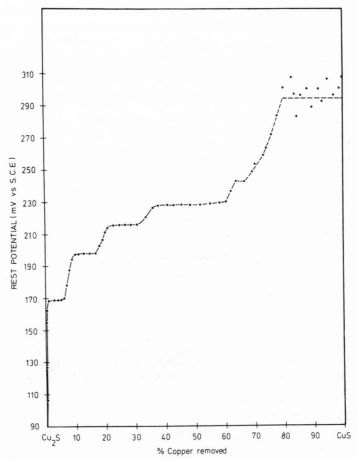

Fig. 5. Reversible potential of a copper sulfide electrode in 10^{-2} M $CuSO_4$ at 25°C as a function of electrode composition.[35]

a hanging drop) can be achieved during the analysis thus enabling us to monitor trace impurities.

At present the future role of molten salts in electrochemical processing of minerals is somewhat clouded. The numerous advantages of molten salts that have continuously been proposed—such as high conductivity of electrolyte and faster electrode kinetics because of higher temperature involved—are in practice countered by less attractive features. Probably the worst aspect has been the low current efficiencies when compared to aqueous electrolysis. From an overview of published work, one can generally blame the low efficiencies on the strong solvation properties of molten salts. Interaction between metal and molten electrolyte often results in "dissolution of the metal" to form a readily oxidized species, the oxidation proceeding by both a de-electronation at the anode as well as a chemical reaction with the anode product.[40] Another important cause of lower current efficiency which is particularly pertinent to mineral processing arises from water being introduced to the molten salt system.

While the solubility of gases is low[36] (typically 10^{-6} mole/cm^3), their dissolution is often exothermic and therefore the solubility increases with temperature. The problem of water dissolving is compounded by its hydrolysis reactions. For example in chloride melts this gives rise to dissolved HCl. We have found[37] that cathodic reduction of the proton proceeds via the following mechanism:

$$H^+ + e = H_{ads}$$

$$H_{ads} + H_{ads} = H_2 \text{ (dissolved)}$$

The overall reaction product (H_2) is soluble in the melt at the concentration levels involved. In these fluid, highly agitated electrolytes, transport of the dissolved H_2 to the counter electrode is rapid. Oxidation of the hydrogen completes the cyclic process which will continuously contribute to reduction in efficiency.

This example highlights the future need for stringent quality control in mineral processing if the metal is to be electrowon in fused salt environment at a minimum energy level.

Despite the above disadvantages that arise from the solvent properties of molten salts, their selectivity for dissolution of oxides and sulfides[38] may in the future be turned to advantage in leaching partly oxidized ores. For example, molten lead chloride will preferentially dissolve both lead oxide and lead sulfide and form solutions of high concentration amenable to electrowinning.

References

1. R. W. Fox, *Phil. Trans. 2*, 309 (1830).
2. M. R. Thornber, *Newcastle Conference 1972*, pp. 51–58, Australasian I.M.M., (1972).
3. E. H. Nickel, J. R. Ross, and M. R. Thornber, *Econ. Geol. 69*, 93–107 (1974).
4. R. M. Garrels and C. L. Christ, *Solutions, Minerals and Equilibria*, Harper and Row, New York (1965).
5. H. C. Helgeson, *Complexing and Hydrothermal Ore Deposition*, Pergamon Press, New York (1964).
6. J. O. Nriagu and J. M. Anderson, *Chem. Geol. 7*, 171–183 (1971).
7. G. M. Lukaszewski, *Mining and Petroleum Technology* (M. J. Jones, ed.), pp. 803–819, Institution of Mining & Metallurgy, London (1970).
8. D. F. A. Koch, *Modern Aspects of Electrochemistry*, No. 10 (J. O'M. Bockris and B. E. Conway, eds.), pp. 211–237, Plenum Press, New York (1975).
9. G. M. Lukaszewski, *Symposium on Mine Filling*, pp. 87–96, Australasian I.M.M., Melbourne (1973).
10. R. Tolun and J. A. Kitchener, *Trans. Inst. Min. Met. 73*, 313 (1964).
11. R. Woods, *J. Phys. Chem. 75*, 354 (1971).
12. J. R. Gardner and R. Woods, *Aust. J. Chem. 26*, 1635–1644 (1973).
13. K. A. Natarajan and I. Iwasaki, *Trans. Soc. Min. Eng. AIME 254*, 323 (1973).
14. D. A. J. Rand, unpublished work.
15. G. Winter and R. Woods, *Separ. Sci. 8*, 251–257 (1973).
16. K. A. Natarajan and I. Iwasaki, *Min. Sci. Eng. 6*, 35 (1974).
17. R. Malmström and T. Tuominen, *Advances in Extractive Metallurgy and Refining*, p. 529, Institution of Mining and Metallurgy, London (1972).
18. F. R. Richardson, *Physical Chemistry of Melts in Metallurgy*, Vol. 2, p. 495, Academic Press, New York (1974).
19. K. N. Subramanian and P. H. Jennings, *Can. Met. Q. 11*, 387 (1972).
20. W. W. Spence and W. R. Cook, *Trans. Can. Inst. Min. Met. 67*, 257 (1964).
21. T. Popademitriou and J. R. Grasso, *Engelhard Ind. Tech. Bull. 10*, 121 (1970).
22. P. Kruesi, E. S. Allen, and J. L. Lake, *Can. Min. Metall. Bull. 66*, 81 (June, 1973).
23. B. J. Scheiner, R. E. Lindstrom, D. E. Shanks, and T. A. Henrie, *U.S. Bur. Mines Tech. Progr. Rept. No. 26* (1970).
24. P. T. Davey and T. R. Scott, *Trans. Inst. Min. Metall. 84C*, 83 (1975).
25. K. J. Cathro, *Proc. Aust. Inst. Min. Metall. No. 252*, pp. 1–11 (1974).
26. K. J. Cathro, in *Extractive Metallurgy of Copper*, Vol. II, p. 776 (J. C. Yannopoulos and J. C. Agarwal, eds.), AIME, New York (1976).

27. J. A. E. Wilkinson and K. P. Haines, *Trans. Inst. Min. Metall. 81C*, 157 (1972).
28. G. P. Power and I. M. Ritchie, *Proc. R. Aust. Chem. Inst. 42*, 39 (1975).
29. D. Schlain, F. X. McCawley, and C. Wyche, *J. Electrochem. Soc. 116*, 1227 (1969).
30. J. O'M. Bockris and G. A. Razumney, *Fundamental Aspects of Electrocrystallization*, Plenum Press, New York (1966).
31. E. J. Frazer, K. E. Anthony, and B. J. Welch, *J. Electrodeposition Surface Treatment 3*, 169 (1975).
32. J. O'M. Bockris and A. K. N. Reddy, *Modern Electrochemistry*, Plenum Press, New York (1970).
33. J. O'M. Bockris, J. A. Kitchener, S. Ignatowicz, and J. W. Tomlinson, *Disc. Faraday Soc. 4*, 265 (1948).
34. H. J. Gardner and W. T. Denholm, *South Australia Conference 1975*, Australasian I.M.M., Melbourne (1975), pp. 197–201.
35. D. F. A. Koch and R. J. McIntyre *J. Electroanal. Chem. 71*, 285 (1976).
36. G. J. Janz, *Molten Salts Handbook*, Academic Press, New York (1967).
37. N. Q. Minh and B. J. Welch, *Aust. J. Chem. 28*, 965 (1975).
38. B. J. Welch, P. L. King, and R. A. Jenkins, *Scand. J. Metall. 1*, 49 (1972).
39. A. R. Gibson and S. Robson, *Br. Pat. 448, 328* (1936).
40. P. J. Barat, T. Brault, and J. P. Saget, *Light Metals*, Vol 1, p. 19, Metallurgical Society of AIME, New York (1974).

21

Summing Up: Mineral Processing

A. T. Kuhn

Let us admit that two lectures can only touch on the vast subject that is mineral processing. A week's symposium would hardly do the subject justice. What can one say in so
short a space? First, it is clear that from now on, mineral processing will get harder and
harder. Politics will become increasingly important, and just as we have seen the take-
over of oil extraction from overseas investing companies in the Middle East, it is clear that
local governments will demand greater financial and decision-making participation in the
working of any ore body. Second, environmental considerations have already left their
mark on approaches to mineral extraction. In the electrochemical context, we have only
to think of the infamous "red mud" which is produced during aluminum extraction at
the bauxite to alumina stage, and the dumping of which is said to threaten the Mediter-
ranean. But last of all, we know that the ore bodies we work, wherever they be, will be
poorer, leaner and more contaminated than they were. In spite of the jest about the
"Electrochemical Mafia," let us make no claims in the first of these three problem areas.
But as to the others, electrochemistry is increasingly being seen as a cleaner way of processing. Because high energies can be transmitted from one molecule to another at ambient temperatures, we can avoid the problems of fumes which have proved intractable
in so many cases. And at the same time, if we apply our energy intelligently, perhaps we
can extract substances from their ores more efficiently than by other means. In looking
at the problem, we should think about first chemical, then physical mineral processing
techniques.

1. Chemical Processing

All over the world metals are recovered from their ores by leaching and electrowinning. If we look at the cells in which this metal deposition takes place, we shall see that
they scarcely differ from those of half a century ago. Still, they are tanks with flat electrodes hung in them. Now, one feels, the first fruits of the electrochemical engineer's
thinking are beginning to percolate through to the mining and extraction companies. In
Zambia, Constructors John Brown, Ltd. collaborating with various mining companies, have
erected two-meter-square, fluidized-bed, copper-winning cells. They are considering the
recovery not only of copper but also cobalt, in such cells. I believe the experiment, in the
case of copper, at least, will not succeed. This is not to argue that the status quo should
be preserved. Simply, one feels that (as Pickett has pointed out) the very high current
densities across the diaphragm in such cells result in higher power consumption. In the
case of the copper, the result is approximately a doubling of the power consumed as the
cell voltage goes from about 4 to 9 V. Of course like any cell voltage figure, this can be

---

A. T. Kuhn • Department of Chemistry and Applied Chemistry, University of Salford, Salford M5
4WI, Lancashire, England.

pushed one way or another by changing current density. But the overall point remains true. In the case of copper, this means that the cost element due to power consumption increases from about 12 to 25% of the mine price (once again, we must be careful in our definitions), and this can only be offset by reducing the capital cost of the cell (expressed in terms of cost per annual recovery capacity). At the same time, the whole question of "starter" beads must be considered and additional factors such as the point that particulate metals are much more likely to pick up impurities in handling and remelting than slab cathodes. Of course, as workers in the field have pointed out, the presence of an ion-exchange diaphragm means that all sorts of options are now open to the plant designer. Catholytes can be circulated quite independently, and streams from several recovery loops can be mixed. Even more important is the greater ease with which higher pH catholyte liquors can be obtained. But what has happened is that (in the main) people have leaped straight from the grandfather clock to the quartz-crystal chronometer. Is the latter the correct solution for every problem? I think that the electrowinning cell of the future will be found somewhere between the extremes of simplicity and sophistication offered by the two cells mentioned above. By all means let us incorporate ion-exchange membranes and let us enhance the mass-transfer regime too. Harvey and co-workers at Kennecott have shown one way, using air-sparged mass-transfer enhancement, and I have referred in my own talk to those workers and cited references to them. There, too, we have discussed the ideas in which solvent extraction is combined with electrowinning especially when ores are poor. Whatever the cell, it will need an inert anode, and in our chapter on electrocatalysis we have touched on this vital question.

The classical "Tainton" 1% Ag + Pb anode is now virtually extinct. New alloys of lead, many of which were recently discussed at a symposium on the electrochemistry of lead, offer greatly reduced corrosion, sometimes with improved oxygen overvoltage. Looking even further forward, we must ask whether one day—as several patents already suggest—ores may be ground and metals extracted in the cell without a separate and external leaching process. Lastly, we must remember that when fossil fuels do become really scarce (either because of exhaustion of fuels or because of pricing themselves out of the market), nuclear or solar or wind power will hopefully provide us with electricity which can be directly hooked into our existing electrochemical technology.

Electrorefining does not appear to be an area in which dramatic developments can be expected soon, although the worth of molten-salt refining processes, for example for lead and alloys, is not to be overlooked. We must watch for developments (research is now proceeding) to fuse the two discrete processes of winning and refining to produce an acceptably pure metal in a single electrochemical operation. This, one supposes, is an idea symmetrical with that in which electrorefining is made one with the manufacturing process, using high-rate electroforming.

2. Physical Processing

Here I would like to refer to my own discussion on electroflotation where it can be seen that the Russians (if no others) are extremely serious about this process as a means for separating mixed ores prior to leaching. All the remarks made there in a primarily environmental context apply in the field of processing. At the moment, emphasis is still on a definite "unit process" electroflotation. But can we envision, either separately or as part of another process, the idea of "electroconditioning" of an ore? I think that many experts in the field of electroflotation would argue that we already have this, even if it has been arrived at by somewhat empirical routes. And of course we have earlier heard about the electroconditioning, if one wishes to call it so, of the xanthate flotation agents.

In summary, I feel that a great future awaits electrochemical processing of minerals. It has been suggested that electrocatalysis is an exciting field for the scientist. Where cell design is concerned, there is a temptation to feel that this phase is already past and that the next phase rests with the industrial chemical engineer to pick up these ideas that have relevance to a large-scale set-up and engineer them into a cell which must, perforce, be a compromise between cost, performance, and simplicity.

22

High-Temperature Electrolytes—The Past Thirty Years

John D. Mackenzie

1. Introduction

However one attempts to be restrictive, the term "high-temperature electrolytes" must still cover an extremely large number of solid and liquid systems. It is obviously impossible to present a comprehensive state-of-the-art review corresponding to only 30 minutes of lecture. In preparing this review and the lecture, as one who has had over 20 years of active involvement in high-temperature electrolytes, I have asked myself a simple but difficult question: "What really have been the most exciting and important developments in these past 30 years?" In any such attempts to be selective, it is also difficult to be totally unbiased when there has been personal involvement. I have tried to be fair. However, to those colleagues whose important contributions to the field of high-temperature electrolytes have not been identified in this review I can only say that it has not been intentional.

I have arbitrarily considered "high temperatures" to be any temperature above 100°C. Any solid or liquid which conducts an electric current primarily through its ions is an electrolyte. Only those which are relatively conductive will be considered here. The three groups of materials discussed below are: molten salts of high fluidity, glass-forming systems of low fluidity, and highly conductive solid electrolytes.

2. Molten Salts

Thirty years ago, molten salts were already well recognized by some as highly conductive ionic liquids. NaCl, for instance, was considered to be ionized on fusion to yield mobile Na^+ and Cl^- ions. Indeed Faraday had studied the electrical properties of some 50 molten salts. However, they were still not relatively widely known, experimental results were sparse, and theoretical treatments lacking. During the last 30 years, an extremely large amount of research has been carried out. J. O'M. Bockris and his co-workers have made many important contributions in this period. Practically every available experimental tool has been utilized and adapted for high temperature use. A colossal quantity of physicochemical data now exist. "Molten salts" have become an established and important branch of chemistry. Although industrial exploitation of this vast body of knowledge is still limited, practical applications are increasing. Most of the published research is summarized in a number of books.[1-10] The large amount of experimental data generated mainly in the U.S.A. and in the U.S.S.R. has led to the establishment of a Molten Salts Data Center at the Rensselaer Polytechnic Institute, Troy, New York, in 1965 under the direction of G. J. Janz. Almost all available data can be obtained from reports and the *Molten Salts Awareness Bulletin* published by this Center.

John D. Mackenzie • School of Engineering and Applied Science, University of California, Los Angeles, California 90024.

2.1. Direct Structural Studies

X-ray diffraction, neutron diffraction, Raman spectra, and infrared spectra measurements have been made on a number of fused salts, mainly halides and nitrates. Despite the experimental uncertainty, such results have furnished some important information on the structure of molten salts. X-ray and neutron diffraction data on halides are summarized in Table 1. It is apparent that on melting, the average separation distances between cations and anions are decreased. Secondly, the coordination numbers for nearest neighbors in the melt are appreciably less than those in the crystal. The second-nearest-neighbor distances on the other hand appear to be slightly larger than the corresponding crystal values. There thus appears to be a high degree of short-range order in the melt. The molar volumes of the melt are some 15–20% greater than those of the corresponding crystals at their melting temperatures. Some kind of a "quasi-lattice" model with "holes" in the melt are therefore likely, although a more detailed picture is lacking. That such results are "average" values, that detailed structures must vary from salt to salt, and that the details are uncertain obviously contribute to the difficulties in theoretical calculation of properties of fused salts.

Distinct bands have been observed in the vibrational spectra of halides, nitrates, and chlorates.[12] Assignments of such vibrational frequencies have been made and the symmetry of the species determined. A quasi-lattice structure is again indicated. In addition, complex anions and molecular species have been identified. These include $ZnCl_3^-$, $CdCl_3^-$, $GaBr_4^-$, $ZnCl_2$, $CdCl_2$, and $HgBr_2$. Indeed, one of the most significant achievements in

Table 1. Nearest-Neighbor and Next-Nearest-Neighbor Distances and Coordination Numbers in Molten and Solid Alkali Halides from X-Ray and Neutron Diffraction[11]

| | | Nearest neighbors | | | | Next nearest neighbors | | | |
| | | Liquid | | Solid | | Liquid | | Solid | |
	Method	r (Å)	C.N.	r (Å)	C.N.	r (Å)	C.N.	r (Å)	C.N.
LiF	X	1.95	3.7	2.10	6	3.0	8	2.97	12
LiCl	X	2.47	4.0	2.66	6	3.85	12	3.76	12
		2.45	3.5			3.80	8.3		
LiBr	X	2.68	5.2	2.85	6	4.12	12.8	4.03	12
LiI	X	2.85	5.6	3.12	6	4.45	11.3	4.41	12
NaF	X	2.30	4.1	2.40	6	3.44	9	3.39	12
NaCl	X	2.80	4.7	2.95	6	4.2	9	4.17	12
NaI	X	3.15	4.0	3.35	6	4.8	8.9	4.74	12
KF	X	2.7	4.9	2.80	6	3.86	9	3.96	12
KCl	X	3.10	3.7	3.26	6	–	12	4.61	12
	N	3.10	3.5	3.26	6				
RbCl	X	3.30	4.2	3.41	6	–	12	4.82	12
CsCl	X	3.53	4.6	3.57	6	4.87	7.1	5.05	12
CsBr	X	3.55	4.6	3.86	6	5.4	8.3	4.46	6
	N	3.55	4.7		8	5.2	7.9		
CsI	X	3.85	4.7	4.08	8	5.5	7.2	4.72	6

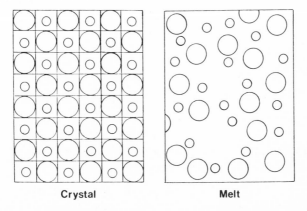

Crystal   Melt

Fig. 1. Two-dimensional simplified representations of crystalline and molten NaCl.

the past 30 years has been the identification of a large number of complex ion molecular groups in molten salts. Recently, Monte Carlo calculations have been made to obtain radial distribution functions.[13] The structural features obtained are similar to those furnished by X-ray diffraction. The type of simplified two-dimensional representation of the structure of molten NaCl first presented by Bockris and Bloom[14] is shown in Figure 1. This is considered to be satisfactory at present.

2.2. Transport

Molten salts are mostly fluid and very conductive liquids. The electrical conductivity of fused halides, for instance, is about 1 ohm^{-1} cm^{-1} at the melting temperature. The viscosity is in the order of centipoises. Exceptions fall into two ends of a spectrum. Beryllium and zinc halides are highly viscous and fairly insulating. Mercury halides, on the other hand, are fluid but insulating. The viscous melts are considered to be "polymeric" whereas the mercury salts are more "molecular" and "covalent." A great deal of conductivity measurements have been made. Over fairly wide ranges of temperatures, the equivalent conductivity apparently obeys the Arrhenius equation

$$\Lambda = A_\Lambda \exp\left(-E_\Lambda/RT\right) \tag{1}$$

where A_Λ is an empirical constant and E_Λ the activation energy. E_Λ for a number of salts in Table 2 is seen to vary from 2 to 6 kcal/mole. E_κ is approximately 1 kcal/mole less than E_Λ.

The viscosity η generally also obeys an Arrhenius relation,

$$\eta = \eta_0 \exp\left(E_\eta/RT\right) \tag{2}$$

where η_0 is an empirical constant and E_η the activation energy for viscous flow. As expected, E_η is seen to be larger than E_Λ in Table 2.

The absolute-reaction-rate theory has been widely applied to fused salts. The entropy of activation for conductance, ΔS^*, evaluated from

$$\Lambda = 5.18 \times 10^{18}\, Z_i(D+2)l_i^2 \exp\left(\Delta S^*/R\right) \exp\left(-\Delta H^*/RT\right) \tag{3}$$

furnishes some interesting information. Here, Z_i is the charge on the ion conducting, l_i is half the migration distance, and D is the dielectric constant of the melt. For all the alkali

Table 2. Comparison of Activation Energies
for Equivalent Conductance (E_Λ), Ionic
Diffusion (E_D), and Viscous Flow (E_η)
for Some Molten Salts[a]

	E_Λ	E_{D^+}	E_{D^-}	E_η
NaCl	2.99	7.14	7.43	9.31
TlCl	3.61	4.54	4.56	
NaNO$_3$	3.22	4.97	5.08	4.04
AgNO$_3$	2.90	3.73	3.84	3.62
KNO$_3$	3.61	5.53	5.76	
CaCl$_2$	5.29	9.9	8.86	12.0
NaCl$_2$	4.14	12.17	10.62	26.3
PbCl$_2$	4.09	7.76	7.74	6.76

[a] Values in kcal/mole.

halides, and some alkaline earth halides, ΔS^* is approximately -6 e.u./mole. These are considered to be totally ionized. For the highly viscous polymeric melts of $ZnCl_2$ and $BeCl_2$, ΔS^* is $+70$ e.u./mole and $+300$ e.u./mole, respectively at the melting temperature and decreases when the temperature is raised. Measurements of transport numbers have also furnished interesting results although the soundness of the experimental techniques has frequently been questioned. The observed transport numbers for the cations in halides and nitrates vary from 0.6 and 0.9 at temperatures near the melting point.

Measurements of diffusion coefficients have shown that the Nernst–Einstein equation

$$D_i = RT \, \Lambda_i / |Z_i| F^2 \tag{4}$$

is not obeyed.[15] Here Λ_i is the ionic equivalent conductance, and F is the faraday. The ratio of $\Lambda_{calc}/\Lambda_{obs}$ varies from 1.1 to 1.9 for the nitrates and the halides. More so, the activation energy for ionic diffusion E_D, be it for cation or anion, is appreciably larger than E_Λ, as shown in Table 2. Bockris and Hooper[16] have attributed this difference to the presence of large excess volume in the melt which may result in the formation of paired cation–anion vacancies. This will contribute more significantly to mass transport but less to electrical transport. Hence Λ_{calc} is larger than Λ_{obs}. Quantitative explanation of the results in Table 2 must await further theoretical developments. Contrary to the relation between diffusion and conductivity, the Stokes–Einstein relation which relates diffusion to shear viscosity is apparently applicable to salt melts.[15]

A number of theories have been developed to explain transport properties and applied to molten salts. The so-called significant-structure theory which postulates that vacancies or holes confer gaslike properties on a certain fraction of ions was applied to alkali halides.[17] Calculated values of conductivity and viscosity apparently agree well with observed values. Thus for KCl at 1100°C, the observed viscosity and conductivity are 1.00 cP and 2.30 ohm^{-1} cm^{-1}, respectively. The calculated values are 1.03 cP and 1.67 ohm^{-1} cm^{-1}. The hole theory of Frenkel and Furth has been applied by Bockris and co-workers[18,19] to calculate mean hole radius which is of the same magnitude as ionic radii. Rice has developed a statistical mechanical theory from which ionic diffusion coefficients and viscosity have been calculated.[20] Calculated values of D_i and η at 1100°C for alkali halides are in good agreement with observed values. Angell and Moynihan have applied the Gibbs and Adam theory to many low-melting fused salts.[21] At present, it is difficult to conclude which of these many theories is most appropriate for molten salts.

Certainly, further significant progress in this area is difficult without additional direct structural information.

Measurements of ionic conductivity of molten-salt mixtures have revealed some interesting results. For an ideal mixture, the equivalent conductance should be given by

$$\Lambda = X_1 \Lambda_1 + X_2 \Lambda_2 \qquad (6)$$

It appears that few mixtures are ideal.[1] The majority show negative deviation from additivity, for example, the mixtures of $PbCl_2$ with alkali halides.[1] Conductivity results together with molar volume studies suggest that nonideal behavior is the result of complex-ion formation. Thus, large ions such as $PbCl_3^-$, $PbCl_4^{2-}$, and $PbCl_6^{4-}$ are indicated, and these have low mobilities. A few systems show positive deviation from ideality, for example, $CaCl_2 + MgCl_2$.

2.3. Thermodynamics and Other Studies

Much of the existing experimental data are to be found in the book by Lumsden.[3] In addition, the National Standard Reference Data System Series (NSRDS), published by the Molten Salts Data Center under G. J. Janz are another valuable source of data. The Temkin model[22] has been used widely over the years to calculate thermodynamic functions. Recent theoretical work has been reviewed by Braunstein.[23]

Activity coefficients calculated for mixtures have shown both ideal ($CdCl_2 + AgCl$ and $PbCl_2 + LiCl$, for instance) and nonideal ($CdCl_2 + KCl$, for instance) behavior.[1] Nonideal behavior is attributed to the presence of complex ions. Thus

$$CdCl_2 + KCl = KCdCl_3 \longrightarrow CdCl_3^-$$
$$CdCl_2 + 2KCl = K_2CdCl_4 \longrightarrow CdCl_4^{2-}$$
$$CdCl_2 + 4KCl = K_4CdCl_6 \longrightarrow CdCl_6^{4-}$$

The formation of such complex ions is supported by results of many independent studies. These include phase-diagram considerations, emf of cells, vapor pressure, calorimetry, cryoscopy, compressibility, density, molar refractivity, and surface tension.[1] Molar volume studies, for instance, showed large deviation from ideality for the system $CaCl_2 + MgCl_2$ as found from activity coefficient measurements. Cryoscopy studies also indicate

Table 3. Some Systems with Complex Ions[1]

System	Experimental method	Complex ions postulated
$KCl + CdCl_2$	Conductance, molar volume, Raman spectra, emf, vapor pressure, cryoscopy, heat of mixing	$CdCl_3^-$ $CdCl_4$ $CdCl_6^{4-}$
$KCl + ZnCl_2$	Conductance, Raman, emf	$ZnCl_3^-$ $ZnCl_4^{2-}$
$KCl + PbCl_2$	Conductance, molar volume, transport no., surface tension, emf, vapor pressure	$PbCl_3^-$ $PbCl_4^{2-}$ $PbCl_6^{4-}$
$KF + NaF + ZrF_4$	Vapor pressure, IR	ZrF_5^-

the presence of $CdCl_4^{2-}$ ions. Thermochemical measurements indicate the presence of $MgCl_4^{2-}$ in $CaCl_2 + MgCl_2$ mixtures. Some of the complex ions revealed by various methods are shown in Table 3.

An outstanding study was the determination of compressibility by Bockris and Richards.[19] They measured the velocity of ultrasound through molten halides and calculated the isothermal and adiabatic compressibility. The results were used to obtain important structural information such as free volume.

2.4. Molten Salts as Solvents

Up to some 30 years ago, most studies on electrolytes were based on aqueous systems. The liquid range of water is only 100°C. Most molten salts, on the other hand, have very large liquid ranges. The liquid ranges (boiling temperature–freezing temperature) for NaF and NaCl, for instance are 710°C and 650°C, respectively. It is thus natural that a large number of investigations have been reported where molten salts have been used as solvents. Perhaps the most interesting group of solutes is metals. Other nonmetallic elements, gases, oxides, water, and organic substances are also soluble in fused salts. An excellent summary of the behavior and application of molten salts as solvents is to be found in a recent book by Charlot and Tremillon.[4]

Transition metal halides form many interesting complex ions when they are dissolved in fused salts.[24] A summary of such complex ions, the structures of which have been revealed by ultraviolet and visible absorption spectra, is shown in Table 4.

Numerous metals are soluble in their halide melts to form *electronically* conducting solutions.[25] At high enough temperatures, alkali metals are soluble in all proportions. Solubility generally decreases with the addition of another salt. Some of these solutions are intensely colored. Bredig[25] has shown that the metals are not in the form of colloidal suspensions. For many systems, electrons introduced by the metal are in shallow traps and these yield highly conductive solutions. Other mixtures, on the other hand, contain deeper traps. Such traps are believed to be diatomic molecules such as Na_2, diatomic molecule ions such as Hg_2^{2+}, Cd_2^{2+}, Ca_2^{2+}, and monomeric ions of metal in lower than normal valence state such as B_1^+ and Nd^{2+}. Attempts have been made to treat such solutions in terms of color centers as for the F centers of solid salts. Detailed understanding of the structures of such melts and the mechanisms of electronic conduction are lacking at present. Some solubility information on a number of metals is given in Table 5.

In so-called displacement solubility, for instance,

$$M_1 + M_2 X \rightleftharpoons M_1 X + M_2$$

metal of the original halide may be precipitated and a new halide formed. An electrochemical series of metals in molten salts has been established by Markov and Delimarskii.[26]

Gases dissolved in molten salts may or may not lead to reaction. For instance, CO_2, He, Xe, Ne, and Ar in halides do not react. However, halogens in halides may lead to reactions.[27] Thus, Cl_2 is soluble in $LiCl + KCl$, $PbCl_2$, and $AgCl$. The diffusion coefficient of chlorine in these melts is in the order of $1-4 \times 10^{-4}$ cm^2/s over the temperature range of 400–600°C. Such high diffusion coefficients are not understood and have been attributed to the probable formation of Cl_3^- ions which can produce a chain-reaction-type diffusive process.

Many oxides are soluble in alkali halides,[4] e.g., CaO, BaO, CdO, Hg_2O, Cu_2O, and PbO. These dissociate into O^{2-} ions and cations which are stabilized by the Cl^- ions. ZnO,

Table 4. Some Complex Ions in Fused Salts from Absorption Spectra[24]

Ion	Medium	Temperature (°C)	Geometry
$(TiCl_6)^{3-}$	LiCl–KCl	400	Distorted octahedral
$(VO)^{2+}$	LiCl–KCl	400	
Ti^{2+}	$AlCl_3$	227	Distorted octahedral
$(VCl_4)^-$	$CsAlCl_4$	800	Distorted tetrahedral
$(VCl_6)^{3-}$	LiCl–KCl	400	Octahedral
$(VCl_6)^{4-}$	LiCl–KCl	400	Octahedral
$(VCl_4)^{2-}$	LiCl–KCl	1000	Distorted tetrahedral
$(CrCl_6)^{3-}$	LiCl–KCl	400	Octahedral
$(CrF_6)^{3-}$	LiF–KF–NaF	640	Octrahedral
Cr^{3+}	Li_2SO_4–Na_2SO_4–K_2SO_4	600	Distorted octahedral
$(ReCl_6)^{2-}$	LiCl–KCl	4450	Octahedral
$(CrCl_4)^{2-}$	LiCl–KCl	400	Distorted tetrahedral
Cr^{2+}	$AlCl_3$	227	
$(MnCl_4)^{2-}$	LiCl–KCl	400	Tetrahedral
Mn^{2+}	$AlCl_3$	227	Octahedral
$(FeCl_4)^{2-}$	LiCl–KCl	400	Tetrahedral
$(CoCl_4)^{2-}$	LiCl–KCl	400	
$(CoF_6)^{4-}$	LiF–NaF–KF	300	Octahedral
$(NiF_6)^{6-}$	LiF–NaF–KF	500	Octahedral
$(PdCl_4)$	LiCl–KCl	400	Tetragonal
$(PtCl_4)$	LiCl–KCl	400	Tetragonal
$(CuCl_4)^{2-}$	LiCl–KCl	400	Distorted tetrahedral

Table 5. Solubility of Some Metals in Molten Salts[4]

Metal	Molten salt	Solubility (mole %)	T°C
Alkali	Alkali halides	total	1000
Al	$AlCl_3$	10^{-7}	230
Al	AlI_3	0.02	230
Sb	$SbCl_3$	0.02	270
Sb	SbI_3	3.1	270
Bi	$BiCl_3$	23	270
Bi	$BiBr_3$	40	270
Bi	BiI_3	65	450
Sn	$SnCl_2$	0.003	500
Sn	$SnBr_2$	0.009	500
Pb	$PbCl_2$	0.006	500
Pb	PbI_2	0.05	500
Ga	$GaCl_2$	2	230
Ga	$GaBr_2$	9	230
Ga	GaI_2	40	230

MgO, NiO, etc., are only sparingly soluble. V_2O_5, P_2O_5, and B_2O_3 dissolve to form complex oxide anions such as VO_3^- and BO_2^-.

2.5. Some Applications of Molten Salts

A detailed discussion of the various applications of molten salts is beyond the scope of this review. Secondly, the past 30 years is considered to be only the beginning of the industrial exploitation of molten salts and thus this section must of necessity be short. Perhaps the most important "new" development (i.e., other than the manufacture of metals through complex melts) has been the molten-salt breeder reactor at Oak Ridge. This was started in 1960 and made critical in 1965. The fuel was uranium and thorium fluoride in a molten mixture of BeF_2, LiF, and ZrF_4. Technical feasibility was proven by 1969. Although the future of such reactors is still questionable, it has been and still is an important and active application of molten salts. A pilot plant was successfully operated using molten mixtures of $NaNO_2$, $NaNO_3$, and KNO_3 as a heat-transfer system by the American Hydrotherm Corp., New York, in 1971. Other applications include fuel cells based on molten carbonates, energy-storage systems, high-energy batteries, gasification of coal, and electrolysis of $AlCl_3$ to give Al.

Perhaps an area which shows the greatest promise is the use of molten salts as a reaction medium. For instance, halogenation reactions can be carried out for the recovery of uranium from fuel elements as gaseous UF_6. The mixture comprising KF, ZrF_4, and AlF_3 is a good solvent for UF_4. When F_2 is passed through the melt, UF_4 is converted to UF_6 and then recovered as a pure gas. The displacement solubility reaction of metal in fused salts can result in the formation of pure metals or anhydrous salts. For example,

$$2U + 3PbCl_2 = 2UCl_3 + 3Pb$$

Examples of other possible applications of this type are to be found in the book by Charlot and Tremillon.[4]

2.6. Summary of Status of Molten Salts

1. In the past 30 years, molten salts have grown from items of curiosity to be an established field of chemistry.

2. The use of all available experimental tools has been brought to bear to study molten salts.

3. As a result, a large body of accurate and useful experimental data exists. These data have been critically evaluated and assembled for easy utilization.

4. Significant structural information has been obtained and theoretical models generated. The confirmation of the existence of large numbers of complex ions in the melt is a most important advance.

5. Molten salts are excellent solvents and hold promise for chemical reaction studies and applications.

6. The large body of knowledge obtained to date has set the stage for future applications of molten salts, although practical uses are still limited at the moment.

3. Glass-Forming Systems

Prior to the 1950s, glass formation was generally considered to be limited to molten oxide systems containing a fairly high concentration of so-called glass-formers. These are, for instance, SiO_2, B_2O_3, and P_2O_5. Their derivatives are the silicate, borate, and phosphate glasses, respectively. In the past 20 years, significant progress has been made in the understanding of why and how liquids can be cooled to give solid glasses.[28-31] In fact, all liquids can be cooled to give glasses providing nucleation and/or crystal growth can be suppressed. Even metallic glasses have since been prepared. Since we are concerned with high-temperature electrolytes I shall confine my discussion primarily to molten oxides, their mixtures, and the behavior of their glasses at elevated temperatures. There is little doubt that scientifically this group of high-temperature electrolytes has been extremely interesting and challenging. Technologically, because most of the glasses in use today derive from silicates, borates, and phosphates, this group is of obvious importance.

Glasses have been known and exploited for many hundreds of years. However, these past 30 years have been a remarkable period for scientific progress as well as new industrial developments. Perhaps before we commence to review some of the significant developments, we should first answer the question: "What is a glass?" The relationships between crystal, liquid, supercooled liquid, and glass are best illustrated in an idealized volume–temperature diagram as shown in Figure 2. If crystallization is somehow suppressed when a liquid is cooled to temperatures below the melting point T_m, a supercooled liquid is formed. Further cooling results in increasing viscosity. When the viscosity reaches approximately 10^{14} P the relaxation times for molecular transport become longer than typical experimental times. The attainment of equilibrium volume through contraction is no longer possible. Thus, a metastable solid with liquid-like disordered structure results. This is termed a glass.[32] The temperature range corresponding to viscosities of about 10^{13}–10^{15} P is the so-called glass-transition temperature, T_g. For common oxide glasses, this is approximately 500°C. At temperatures even 100–200°C below this, the solid glass is fairly conductive (resistivity about 10^5 Ωcm for instance). Electrolysis readily occurs and the transport number of the cations, principally Na^+, is unity. Ionic conductivity increases with increasing temperature. At temperatures above 1000°C, the resistivity can be less than 1 Ωcm.[33] These oxide glass-forming systems are thus in every sense high-temperature electrolytes.

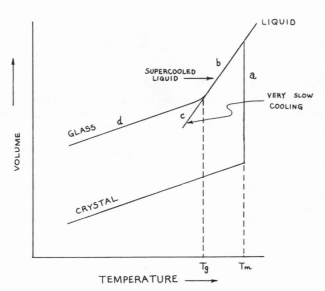

Fig. 2. Volume–temperature relationship of an idealized glass-forming system.

3.1. Transport in Molten Oxides

When sodium carbonate is mixed with silica and heated, the carbonate first decomposes. A chemical reaction then takes place which may be formally represented by

$$-\overset{|}{\underset{|}{Si}}-O-\overset{|}{\underset{|}{Si}}- + Na_2O \longrightarrow -\overset{|}{\underset{|}{Si}}-O^- \underset{Na^+}{\overset{Na^+}{}} {}^-O-\overset{|}{\underset{|}{Si}}-$$

It is seen that a "bridging" oxygen is destroyed to yield two "nonbridging oxygens." The sodium ions are electrostatically attracted to these nonbridging oxygens. The polymeric structure of SiO_2 is progressively destroyed by the alkali oxides. Viscosity of the melt decreases and ionic conductivity increases with increasing alkali content. The measurements of ionic conductivity, transport number, and viscosity of silicate melts by Bockris and co-workers were started in the late 1940s and have contributed greatly to the structural understanding of such electrolytes and their resultant glasses.

Some typical values of conductivity of mixed oxides are shown in Table 6. Conductivity–temperature behavior is generally described satisfactorily by the usual Arrhenius equation. For both silicates and borates, the activation energy for conductance of alkali systems, E_K is about 10 to 20 kcal/mole and apparently does not vary much with composition. For the alkaline earth systems, E_K decreases from about 50 kcal/mole to less than 20 kcal/mole when M_2O is decreased from 20 to 60 mole %. The structures of such melts are thus dependent on the nature of the cations. By the application of transition-state theory, Bockris and co-workers were able to evaluate the free energy of activation for ionic conduction, ΔG^*. When this is plotted against the so-called "ion-oxygen attraction," defined by

$$I = 2ze^2/r^2 \tag{6}$$

where z is the valence of the cation, e is the electronic charge, and r is the cation-oxygen separation distance, an interesting relationship results. ΔG^* for Al^{3+} and Ti^{4+} is much higher than for the alkali and alkaline earth ions. This gave rise to the concept that such ions can compete with Si^{4+} ions in the melt to give complex anions. Such ideas have become important in subsequent understanding of phase separation in glasses.

Viscosity measurements in molten silicates confirm the break-up of the silicate network.[43,44] When alkali and alkaline earth oxides are added to molten silica, the viscosity and activation energy for viscous flow both drop drastically. The thermal expansion of binary silicate melts also shows very large increase when M_2O exceeds about 10–20 mole

Table 6. Electrical Conduction in Binary Liquid Oxides[34]

System and composition	T (°C)	Specific conductivity (ohm^{-1} cm^{-1})	Conduction mechanism	Reference
$Li_2O \cdot SiO_2$	1300	0.1	Ionic	35, 36
$CaO \cdot 2SiO_2$	1650	0.2	Ionic	36, 37
$PbO \cdot SiO_2$	1000	0.1	Ionic	38
$1.2MnO \cdot SiO_2$	1350	1.0	Ionic	36, 37
$19FeO \cdot SiO_2$	1400	125	10% Ionic	39
$2.75CoO \cdot SiO_2$	1450	–	72% Ionic	40
$2.5CoO \cdot SiO_2$	1440	–	95% Ionic	40
$Na_2O \cdot 2B_2O_3$	1000	1.3	Ionic	41
$CaO \cdot 2B_2O_3$	1200	0.1	Ionic	42
$PbO \cdot B_2O_3$	1000	0.5	Ionic	38

%.[45,46] The conductivity, viscous flow, and molar volume studies led Bockris and co-workers to postulate "discrete anion" structures in molten silicates. Such discrete anions are the result of the break-up of the polymeric network of silica. Since the melts form glasses on cooling, it is likely that such large anions are present in the solidified glass structures as well. This concept is fundamentally different from the so-called "random network theory" of the 1930s. Although the detailed structures of the large anionic groups are uncertain, the Bockris school has demonstrated that discrete anions must exist. Such theories have contributed significantly to the understanding of structures and behavior of solid glasses.

Viscosity and density studies of molten borates revealed their structural differences from the silicates. This has been attributed to the ability of the boron ion to attain both three- and fourfold coordination whereas the silica ion invariably has fourfold coordination.[33]

Much of the theoretical progress made was only possible because of the successful development of high-temperature physicochemical techniques.[47] Experimental studies of reaction of refractories with molten oxides, for instance have been important in the electric melting of glass in which the molten silicate becomes a resistive element between molybdenum electrodes.

Although it appears that much progress has been made in the past 30 years concerning transport properties of molten silicates, much remains to be done. One of the most interesting properties of binary silicates is the so-called "mixed alkali effect." In a series of binary silicate melts of the composition $30M_2O \cdot 70SiO_2$, for instance, the equivalent conductance of mixed alkali oxide systems (e.g., $15Li_2O \cdot 15K_2O \cdot 70SiO_2$) exhibits marked negative departure from ideal behavior.[48] Viscosity isotherms also show such large negative departure from linearity.[49] Not only is the individual behavior of conductivity and viscosity difficult to explain, the apparent *opposite* behavior of conductance and viscosity is certainly an interesting mystery at present.

3.2. Transport in Solid Glasses at Elevated Temperatures

The two most interesting topics are certainly the mixed alkali effect and the interaction of solid glasses with fused salts.

The mixed alkali effect mentioned above for melts actually becomes much more pronounced when the temperature is decreased.[50] For alkali silicate glasses of the composition $M_2O \cdot 2SiO_2$, for instance, the mixed alkali glass has resistivities some three orders of magnitude greater than those of the binary systems (Figure 3).[51] The activation energy for ionic conduction also increases sharply when glasses contain two different alkali ions. An excellent review of this behavior has been published by Isard.[52] This anomalous property is apparently found also in alkali germanates, borates, and borosilicates. Alkaline earth ions apparently do not show such behavior. Although many theories have been proposed to explain the mixed alkali effect, no single one appears to be entirely satisfactory. The complexity of oxide glass structure is certainly manifested by this most interesting anomaly.

The electrolytic nature of solid glasses at elevated temperatures and its interaction with fused salts has been widely studied in the past 20 years. When a silicate glass containing alkali ions (for instance, Li^+ ions) is placed in a molten salt such as $NaNO_3$ at 350°C, an ion-exchange process commences.[53] The Li^+ ions on the surface of the glass migrate into the fused salt. Na^+ ions diffuse into the glass to replace the Li^+. Since the volume occupied by the Li^+ ion is much less than that required by an Na^+ ion and since

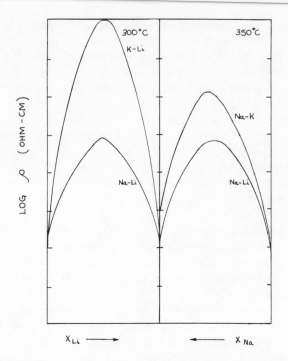

Fig. 3. The mixed-alkali effect in glass. [From Lengyel and Boksay.[51]]

the glass is held at temperatures below T_g (i.e., rigid), a compressive stress is created on the surface. In time a thick compressive layer is developed. Glasses are well-known to break fairly easily under tension. The compressive layer thus results in strengthening of the glass. Modulus of rupture values, normally some 10,000–20,000 psi, has been increased to over 100,000 psi by this method. Scientific understanding in this area lags behind industrial exploitation. How the structure of the glass affects ion exchange and the simultaneous stress relaxation due to volume viscosity during ion exchange is not clearly understood at present.

Oxide glasses are not the only ionically conducting glasses. Sulfide glasses are normally thought to be electronic conductors. However, Plumat has shown that glasses of the composition GeS_2–Na_2S are actually ionic conductors with Na^+ as the principal current carrier.[54] More recently, GeS_2-based glasses containing Ag and As_2S_2-based glasses containing Ag have also been shown to be purely ionic conductors.[55]

In general, cations are the only current carriers in oxide glasses. At very high temperatures and for oxide glasses of extremely high resistivity, oxygen ion motion has been suggested. However, the only proven anionic conductivity in glasses concerns halide ions. Recently, PbO–SiO_2 glasses containing some 30 mole % of F, Cl, Br, and I were studied by Schultz and Mizzoni.[56] At temperatures between 250 and 350°C, electrolysis measurements showed unambiguously that the halide ions are the principal current carriers.

Oxide glasses are not all ionically conducting, of course. When some 10–15% of a transition metal oxide such as Fe_2O_3 is present, a mixed ionic–electronic conductor results.[57] Higher concentrations of transition metal oxides in glasses usually lead to purely electronic conductors.

3.3. Structure, Kinetics, and Phase Separation

Direct structural studies of molten oxides and solid glasses have not resulted in much understanding of these complex materials.[58] X-ray diffraction, in particular, has been disappointing. For silicates, the only certainty is that the SiO_4 tetrahedron is the basic building unit in practically all systems studied. How such SiO_4 tetrahedra are linked together is not clear. This was the status in 1945. It is still the status today. With simple B_2O_3 glass, for instance, the situation was even worse. The early work of Warren suggested that BO_3 triangles were the basic building units. The addition of alkali oxides to B_2O_3 results in the formation of BO_4 tetrahedra. Apparently so much uncertainty exists that subsequent workers who repeated the X-ray diffraction studies suggested that BO_4 tetrahedra actually were the basic units in B_2O_3. Addition of alkali oxides results in the possible formation of BO_3 triangles! This controversy was finally resolved through the use of NMR absorption which unambiguously showed that glassy B_2O_3 is made up of BO_3 triangles. Alkali oxides react with B_2O_3 to yield BO_4 tetrahedra.[59] Thus the early work of Warren was finally confirmed. The ratio of four-coordinated boron to three-coordinated boron for the same molar concentration of metal oxides in B_2O_3 can be different for various oxides. At present it is not certain why this is so and how the BO_3 and BO_4 groups are linked together. Infrared absorption studies have not been fruitful in that unambiguous interpretation of the data is not possible.[60]

In summary then, direct structural studies indicate that short-range order exists in glass. This is exemplified by building units such as the SiO_4 tetrahedron, BO_4 tetrahedron, and BO_3 triangle. No description of longer-range interactions is possible.

Glasses are metastable solids. At low temperatures, the free-energy barrier to crystallization is significant and hence glasses can be considered stable. The relationship between diamond and graphite at room temperature and one atmosphere is similar to the glass–crystal relationship. When a glass is heated up to the supercooled liquid region, however, transformations can easily occur. This is the temperature region between the liquidus temperature or melting point and the glass-transition temperature, in other words, when the viscosity is between 1 and 10^{13} P. At these high temperatures, the glass, or truly, the supercooled liquid, may phase separate into two liquids in two different ways. Alternatively, it may crystallize. Both phase separation and crystallization have yielded important industrial products. The two processes are sometimes related. In the past 30 years, a great deal of scientific effort has been devoted to the understanding of phase separation and the measurement of kinetics of crystal growth.

In one form of phase separation, one phase which is a minor constituent segregates in the form of dispersed droplets. Classical nucleation and growth theories are fairly satisfactory in describing such a process.[61] The nucleation and growth of the droplets are governed by diffusion and viscosity. In the other form, the two phases appear continuous and interpenetrate one another. The volume fractions of the two separating phases are approximately equal. This is generally referred to as "spinodal decomposition." Phase separation originates from small compositional fluctuations throughout the supercooled liquid. No nucleation step is necessary.[62] In an isotropic system, spinodal separation leads to an interconnected structure. The simplest way to picture such a phase separated liquid is to imagine a system containing intimately mixed white and red noodles. The void space is then squeezed out so that the entire volume is occupied by the two interwoven noodles, each representing a liquid phase. The diameter of the noodles would vary from 10 Å to 100 Å. Such structures have been revealed by electron microscopy.

Crystallization of supercooled oxide melts is again depicted to be a two-step process. Nucleation precedes crystal growth. The kinetics have been correlated to diffusion and viscous flow. Good agreement is frequently obtained between observed and calculated crystallization rates. From scientific understanding, controlled crystallization with the aid of nucleation catalysts has resulted in important industrial products.[63,64] This is described further below.

3.4. Metastability of Supercooled Oxide Melts and Important Applications

Molten oxides at high temperatures are good solvents. Silver halides, for instance, can be dissolved in molten borosilicate melts. On cooling to temperatures below the liquidus, molten silver halide droplets first separate out. Their sizes can be controlled to less than 50 Å in diameter. Subsequently small crystallites of silver halides are formed and dispersed in the solid glass which remains transparent. On exposure to light with wavelength from 4000 to 6000 Å, the glass darkens. On removal of the irradiation source, the darkness disappears. These are called photochromic glasses.[65] The darkening and fading mechanism is complex.

Phase separation in oxide glasses containing Ti^{4+}, Zr^{4+}, and P^{5+} ions probably results in large complex oxy anions formed by these cations. Such polyvalent ions are good nucleation catalysts and have been widely utilized in glass compositions to give glass ceramics.[64] Noble metals such as gold can be dissolved in molten glasses. On cooling they form colloidal particles which are also good nucleation catalysts for the preparation of glass ceramics. The uniformity of the distribution of catalysts in the melt and the control of their concentration permits the formation of very fine-grained glass ceramics of high mechanical strength. Some glass ceramics are even transparent to visible light despite their practically complete crystallinity.

Spinodal decomposition in borosilicate glasses results in the formation of a practically pure SiO_2 phase interconnected within an alkali borate phase. The latter is easily leached out by hot water or dilute acids. A porous silica glass with pore diameters between 10 and 20 Å is formed.[66] Many interesting applications have been developed with these porous glasses. On subsequent heating, the porous structure can be collapsed to give a low-expansion silica glass, commonly known as Vycor. The ionic nature of molten glass is an important factor in controlling the morphology of the phase-separated glass and hence its practical applications. The nature of the anionic structure of oxide melts is of particular importance. When separation of the two phases occurs and the glass is cooled, large stresses can be generated within the phase and/or at the interface during leaching. Further fundamental studies of the structure of molten oxides will no doubt be beneficial to such practical applications.

The solubility of gases and water in molten oxides has been the subject of many investigations because of its importance in the production of glass. Detailed considerations are beyond the scope of this review. One interesting recent study, however, is worthy of mention. Many gases apparently are highly soluble in molten glasses at elevated pressures.[67] The gases can be trapped if the melt is allowed to cool under pressure. As much as 5% solution is common. When hydrogen is dissolved (0.02–2%) the resultant glass has a remarkable degree of resistance to radiation damage by X rays, γ rays, and neutrons. Normally the glass would color on irradiation. This has been attributed to the destruction of the color centers generated from irradiation by the hydrogen present in the glass.

3.5. Slags

Slags are closely related to glasses. Much of the knowledge acquired in the past 30 years has originated from the study of molten oxides regarded as slags. Detailed discussions on this topic are not possible in this review. The beginning of this era of understanding can justifiably be identified with a meeting in 1952, held at the Imperial College.[68] Much of the work since then has originated from Imperial College under F. D. Richardson. An excellent review of this field is to be found in two recent volumes authored by Richardson.[69]

3.6. Summary of the Status of Glass-Forming Systems at High Temperatures

1. Molten oxides of the common glass-forming type, i.e., silicates, borates, etc., are ionic liquids which can be highly conductive.
2. The addition of alkali and alkaline earth oxides to oxides with polymeric structures such as SiO_2 and B_2O_3 results in the break-up of the network. Large discrete anionic groups are then formed.
3. Techniques such as X-ray diffraction have been relatively unproductive in the elucidation of the structures of this group of high-temperature electrolytes.
4. The ionic nature of the melt has been exploited in electric melting of glass. The ionic nature of the supercooled liquid has led to the ion-exchange strengthening of glass.
5. The most important scientific advances were made on the supercooled liquid, especially in the fields of phase separation and crystallization. Even more important practical applications have been developed, the most prominent of which is the manufacture of glass ceramics.
6. Scientific understanding is still lagging behind industrial practices.

4. Solid Crystalline Electrolytes of High Conductivity

Although most oxides are ionic solids, the binding energy between cations and anions is usually very large. Normally then, appreciable conductivity is only observed at high temperatures. Conductivity can be ionic and/or electronic. Experimental values of the mobilities of electrons and holes in ionic compounds are some 100–1000 times larger than the mobilities of ions.[70] Thus ionic conductivity should predominate only when the concentrations of holes and electrons are very small. Alternately, the mobilities of ions must be unusually large. In this section, only those electrolytes with ionic transport numbers close to unity will be considered. Secondly, at the temperatures of practical importance for each electrolyte, the resistivity must be less than approximately 10 Ω cm. Thirdly, practically all my remarks will be confined to ionic conductivity and diffusion. These electrolytes can be conveniently divided into three groups. They are: (1) the defect-stabilized oxides having the calcium fluoride structure, e.g., calcia-stabilized zirconia (CSZ), (2) the β-alumina family, and (3) the cation disordered compounds of the silver halides and chalcogenides type.

These highly conducting electrolytes have been called "superionic conductors" by Rice and Roth.[71] Compared to the fused salts and oxide melts discussed in the two previous sections, information on this third class of materials was practically nonexistent 30 years ago. During this period they have been extensively studied because of their promise in practical applications and their interesting physicochemical properties.

4.1. Defect-Stabilized Oxides

The group includes the following:

$$CaO \cdot M'O_2 \ (M' = Zr, Hf, Th, Ce)$$

$$ZrO_2 \cdot M''_2O_3 \ (M'' = La, Sm, Y, Yb, Sc)$$

$$ThO_2 \cdot Y_2O_3$$

These materials possess the cubic fluorite structure. All are oxygen-ion conductors. Their structures are discussed in detail elsewhere.[72] The oxygen ions are able to migrate easily because of the presence of large amounts of oxygen ion vacancies. For instance, for $CaO \cdot ZrO_2$(CSZ), some 15–25 mole % of CaO can be in solid solution. The presence of Ca^{2+} in lattice positions normally occupied by the Zr^{4+} results in the formation of oxygen ion vacancies for electrical neutrality. Charge is now easily transported and so CSZ can be termed an extrinsic anionic conductor. For a solid solution with about 15 mole % CaO, the electrical resistivity at 1000°C is about 10 Ω cm.[73] The electrical resistivity is independent of oxygen partial pressures for the range of $1-10^{-22}$ atm over a wide temperature range.[74] Transport number for electrons is apparently less than 0.002. The diffusion coefficient of O^{2-} ion is about 10^{-7} cm²/s at 1000°C, whereas D for Ca^{2+} is about 10^{-17} cm²/s for $Zr_{0.85}Ca_{0.15}O_{1.85}$.[75] Conductivity calculated from diffusion of O^{2-} ions by the Nernst–Einstein relation agrees very well with observed values over a wide range of temperature. The activation energies for conductivity and O^{2-} diffusion are practically identical.[72] It is evident that such a doped oxide is essentially an oxygen-ion conductor.

One interesting feature for these oxides is the variation of conductivity at any temperature with dopant concentration. For CSZ, $ZrO_2 \cdot Y_2O_3$, $ThO_2 \cdot Y_2O_3$, $ThO_2 \cdot CaO$, and $HfO_2 \cdot CaO$ conductivity first increases with increasing dopant concentration, reaches a maximum value and then decreases.[75] This has been attributed to vacancy ordering, vacancy clustering, and dopant–vacancy interactions.[70] Another interesting observation is the linear increase of the activation energy for conduction with dopant concentration.[73] E^* at 9 mole % CaO is 20 kcal/mole and reaches 40 kcal/mole at 22 mole % CaO, with no inflection at around 15 mole % where the conductivity maxima are located. This has been attributed to the effects of the large size of Ca^{2+} ion which is appreciably larger than the Zr^{4+} ion (0.99 Å vs. 0.79 Å).

CSZ and $Y_2O_3 \cdot ZrO_2$ have been successfully used as the solid electrolytes in fuel cells operating at about 1000°C on hydrogen and oxygen.[76] Such cells have been considered for coal-burning power plants.[77]

4.2. The β-Aluminas

The sodium β-aluminas, with a $Na_2O:Al_2O_3$ ratio of 1:5 to 1:11, have been known for a long time as useful refractory bricks. "Monofrax" refractories are well known to glass-melting tank designers. They are highly inert and can be used to temperatures well in excess of 1000°C. It was not until the 1950s that workers in the Ford Motor Co. Scientific Laboratory announced to bewildered ceramists that the β-aluminas can be highly ionically conducting and that they hold great promise as solid electrolytes in high-energy-density batteries. At room temperature, for instance, the resistivity of single crystals of β-alumina can be as low as 100 Ωcm and at 300°C, 10 Ωcm. Since then many β-aluminas have been reported in which other monovalent cations can replace sodium and Ga^{3+} and Fe^{3+} can replace Al^{3+}. Mixtures of alkali ions (Na^+ and Li^+, for example) and

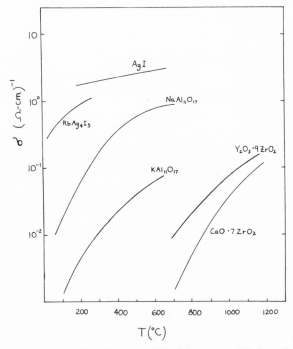

Fig. 4. Ionic conductivity of some high-temperature solid electrolytes.

alkali and alkaline earth ions (Mg^{2+} and Na^+, for example) give a more stable β-alumina than with only Na^+ and can have higher ionic conductivity (Figure 4).[70, 71]

In $NaAl_{11}O_{17}$, which has a hexagonal layer structure, the sodium ions are located in planes perpendicular to the c axis. The planes are separated from one another by blocks consisting of four layers of Al^{3+} and O^{2-} ions. The Na^+ ions have high mobilities in the plane but are immobile along the c axis. The sodium ions can be ion exchanged in appropriate molten salts to give other β-aluminas such as $AgAl_{11}O_{17}$.

Sodium–sulfur batteries have been designed to operate at about 300°C.[78] The biggest problems are the difficulties of preparing dense, conductive, strong polycrystalline β-alumina and the sealing of the electrolyte to other components of the battery. A great deal of engineering development is being conducted at present.

4.3. The Nonoxides

This group is comprised of MAg_4I_5 (M = NH_4, K, or Rb), Ag_6IWO_4, α-AgI, α-Ag_3HgI_4, $Ag_4HgSe_2I_2$, etc.[71] They were discovered independently in 1966 and 1967 by Bradley and Greene[79] and Owens and Argue.[80] These materials are not really high-temperature electrolytes since their ionic conductivities are so high (0.2 Ωcm, for example) that the most likely applications are probably for batteries to be operated at room temperature. The activation energies for ionic conduction are 2–6 kcal/mole. The cationic transport number is unity. α-AgI, however, is only conductive at temperatures above 150°C since it undergoes a transition to a fairly insulating β-modification which is stable at room temperature. The mobile cations, e.g., $16Ag^+$ in $RbAg_4I_5$ are distributed somewhat randomly over 72 available sites. These Ag^+ ions can be conveniently regarded as being in "liquid-like" states,[71] and hence can migrate rapidly. A review of the structures and conduction mechanisms of these solids has been published by Wiedersich and Geller.[81]

4.4. Theories on Conduction

Theoretical progress has been mainly in the form of small improvements of the simple "hopping" model.[82, 83] VanGool suggested a somewhat new mechanism of conduction for these highly conductive ionic solids by postulating a momentary adjacent domain structure.[82] Ionic transport occurs through movement of the walls between domains. The most innovative approach in this period, in this author's opinion, is the model proposed by Rice and Roth.[71] The model assumes that in these compounds there is an energy gap ϵ_0 above which ions of mass M, belonging to the conductive species, can be thermally excited from localized ionic states to free-ion-like states in which an ion can now migrate rapidly with a velocity v_m and energy $\epsilon_m = \frac{1}{2}Mv_m^2$. On account of the interaction with the rest of the solid, such an excited free-ion-like state will have a finite lifetime τ_m. On the basis of a postulated Boltzmann transport equation for the thermal occupations of the various free-ion-like states, equations were derived for ionic conductivity, thermal conductivity, and thermoelectric power.

It is difficult to conclude at present which model is the most appropriate one for the above three groups of highly conductive solids. We do know, however, that one necessary condition for fast ion transport is an excess of available sites in the lattice for the conductive ions and that the activation or excitation energy must be relatively small. It will be of great interest and challenge to prepare new "superionic conductors" through structural understanding.

Acknowledgments

I am grateful to J. O'M. Bockris who not only first introduced me to materials discussed in this review, but guided and stimulated me continuously these past 20 years. I have relied heavily on the excellent book by H. Bloom[1] for the section on molten salts. The advice of G. J. Janz is much appreciated. The support of the Directorate of Chemical Sciences, AFOSR through Grant 75-2764 is gratefully acknowledged.

References

1. H. Bloom, *The Chemistry of Molten Salts*, W. A. Benjamin, New York (1967).
2. G. J. Janz, *Molten Salts Handbook*, Academic Press, New York (1967).
3. J. Lumsden, *Thermodynamics of Molten Salt Mixtures*, Academic Press, New York (1967).
4. G. Charlot and B. Tremillon, *Chemical Reactions in Solvents and Melts*, Pergamon Press, New York (1969).
5. S. Petrucci, ed., *Ionic Interactions*, Vol. 1, Academic Press, New York (1971).
6. S. Petrucci, ed., *Ionic Interactions*, Vol. 2, Academic Press, New York (1971).
7. G. Mamantov, ed., *Molten Salts, Characterization and Analysis*, Marcel Dekker, New York (1969).
8. J. Braunstein, G. Mamantov, and G. P. Smith, eds., *Advances in Molten Salt Chemistry*, Vol. 1, Plenum Press, New York (1971).
9. M. Blander, ed., *Molten Salt Chemistry*, Wiley-Interscience, New York (1964).
10. B. Sundheim, ed., *Fused Salts*, McGraw-Hill, New York (1964).
11. H. A. Levy and M. D. Danford, in: *Molten Salt Chemistry* (M. Blander, ed.), p. 109, Wiley-Interscience, New York (1964).
12. D. W. James, in: *Molten Salt Chemistry* (M. Blander, ed.), p. 507, Wiley-Interscience, New York (1964).
13. J. Krogh-Moe, T. Ostvold, and T. Forland, *Acta Chem. Scand. 23*, 14 (1969).

14. J. O'M. Bockris and H. Bloom, in: *Fused Salts* (B. Sundheim, ed.), p. 35, McGraw-Hill, New York (1964).

15. C. T. Moynihan, in: *Ionic Interactions* (S. Petrucci, ed.), Vol. 2, Chapter 5, Academic Press, New York (1971).

16. J. O'M. Bockris and G. W. Hooper, *Disc. Faraday Soc. 32*, 218 (1962).

17. W. C. Lu *et al.*, *J. Chem. Phys. 49*, 797 (1968).

18. T. Emi and J. O'M. Bockris, *J. Phys. Chem. 74*, 159 (1970).

19. J. O'M. Bockris and N. E. Richards, *Proc. R. Soc. London A241*, 44 (1957).

20. S. A. Rice, *Trans. Faraday Soc. 58*, 499 (1962).

21. C. A. Angell and C. T. Moynihan, in: *Molten Salts, Characterization and Analysis* (G. Mamantov, ed.), p. 315, Marcel Dekker, New York (1969).

22. M. Temkin, *Acta Physicochim.*, URSS *20*, 414 (1945).

23. J. Braunstein, in: *Ionic Interactions* (S. Petrucci, ed.), Vol. 2, Chapter 4, Academic Press, New York (1971).

24. S. L. Holt, in: *Ionic Interactions* (S. Petrucci, ed.), Vol. 1, Chapter 8, Academic Press, New York (1971).

25. M. A. Bredig, in: *Molten Salt Chemistry* (M. Blander, ed.), p. 367, Wiley-Interscience, New York (1964).

26. B. F. Markov and Yu K. Delimarskii, *Electrochemistry of Fused Salts*, Sigma Press, Washington, D.C. (1961).

27. J. D. VanNorman and R. J. Tivers, in: *Molten Salts, Characterization and Analysis* (G. Mamantov, ed.), p. 509, Marcel Dekker, New York (1969).

28. D. Turnbull and M. H. Cohen, *J. Chem. Phys. 29*, 1049 (1958).

29. J. D. Mackenzie, ed., *Modern Aspects of the Vitreous State*, Vol. 1, Butterworths, London (1960).

30. J. D. Mackenzie, ed., *Modern Aspects of the Vitreous State*, Vol. 2, Butterworths, London (1962).

31. J. D. Mackenzie, ed., *Modern Aspects of the Vitreous State*, Vol. 3, Butterworths, London (1964).

32. J. D. Mackenzie, in: *Modern Aspects of the Vitreous State* (J. D. Mackenzie, ed.), Vol. 1, Chapter 1, Butterworths, London (1960).

33. J. D. Mackenzie, in: *Modern Aspects of the Vitreous State* (J. D. Mackenzie, ed.), Vol. 1, Chapter 8, Butterworths, London (1960).

34. J. D. Mackenzie, *Advances in Organic and Radiochemistry*, Vol. 4, p. 293, (1962).

35. J. O'M. Bockris *et al.*, *Trans. Faraday Soc. 48*, 75 (1952).

36. J. O'M. Bockris *et al.*, *Trans. Faraday Soc. 48*, 536 (1952).

37. J. O'M. Bockris *et al.*, *Disc. Faraday Soc. 4* 265 (1948).

38. J. O'M. Bockris and G. W. Mellors, *J. Phys. Chem. 60*, 1321 (1956).

39. J. Inouye, J. W. Tomlinson, and J. Chipman, *Trans. Faraday Soc. 49*, 796 (1953).

40. T. Baak, *Acta Chem. Scand. 8*, 166 (1954).

41. L. Shartsis *et al.*, *J. Am. Ceram. Soc. 36*, 319 (1953).

42. L. Shartsis *et al.*, *J. Am. Ceram. Soc. 36*, 35 (1953).

43. J. O'M. Bockris and D. C. Lowe, *Proc. R. Soc. London A226*, 423 (1954).

44. J. O'M. Bockris, J. D. Mackenzie, and J. A. Kitchener, *Trans. Faraday Soc. 51*, 1734 (1955).

45. J. O'M. Bockris, J. W. Tomlinson, and J. L. White, *Trans. Faraday Soc. 52*, 299 (1956).

46. J. W. Tomlinson, M. S. R. Heynes, and J. O'M. Bockris, *Trans. Faraday Soc. 64*, 1822 (1958).

47. J. O'M. Bockris, J. L. White, and J. D. Mackenzie, eds., *Physicochemical Measurements at High Temperatures*, Butterworths, London (1959).

48. R. E. Tickle, *Phys. Chem. Glasses 8*, 101 (1967).

49. J. P. Poole, *J. Am. Ceram. Soc. 32*, 230 (1949).

50. B. Lengyel, *Glastech. Ber. 18*, 177 (1940).

51. B. Lengyel and Z. Boksay, *Z. Phys. Chem. 204*, 157 (1955).

52. J. O. Isard, *J. Non-Cryst. Solids 1*, 235 (1969).

53. M. E. Nordberg *et al.*, *J. Am. Ceram. Soc. 47*, 215 (1964).

54. E. R. Plumat, *J. Am. Ceram. Soc. 51*, 499 (1968).

55. Y. Kawamoto *et al.*, *J. Am. Ceram. Soc. 57*, 489 (1974).

56. P. C. Schultz and M. S. Mizzoni, *J. Am. Ceram. Soc. 56*, 95 (1973).

57. J. D. Mackenzie, in: *Modern Aspects of the Vitreous State* (J. D. Mackenzie, ed.), Vol. 3, Chapter 5, Butterworths, London (1964).

58. S. Urness, ref. in: *Modern Aspects of the Vitreous State* (J. D. Mackenzie, ed.), Vol. 1, Chapter 29, Butterworths, London (1960).

59. P. J. Bray and A. H. Silver, in: *Modern Aspects of the Vitreous State* (J. D. Mackenzie, ed.), Vol. 1, Chapter 5, Butterworths, London (1960).

60. N. F. Borrelli *et al.*, *Phys. Chem. Glasses 4*, 11 (1963).
61. J. J. Hammel, *J. Chem. Phys. 46*, 2234 (1967).
62. J. W. Cahn, *J. Chem. Phys. 42*, 93 (1965).
63. L. L. Hench and S. W. Freiman, eds., *Advances in Nucleation and Crystallization in Glasses*, Columbus, Ohio, The American Ceramic Society (1971).
64. P. W. McMillan, *Glass Ceramics*, Academic Press, New York (1964).
65. W. H. Armistead and S. D. Stookey, *Science 144*, 150 (1964).
66. M. L. Hair and M. E. Nordberg, *J. Am. Ceram. Soc. 52*, 65 (1969).
67. S. P. Faile and D. M. Roy, *Bull. Am. Ceram. Soc. 47*, 401 (1968).
68. F. D. Richardson, *The Physical Chemistry of Melts*, Institute of Mining and Metallurgy, London (1953).
69. F. D. Richardson, *Physical Chemistry of Melts in Metallurgy*, Vols. 1 and 2, Academic Press, New York (1974).
70. W. L. Worrell, *Bull. Am. Ceram. Soc. 53*, 425 (1974).
71. M. J. Rice and W. L. Roth, *J. Solid State Chem. 4* 294 (1972).
72. P. Kofstad, *Nonstoichiometry, Diffusion, and Electrical Conductivity in Binary Metal Oxides*, Wiley-Interscience, New York (1972).
73. R. E. Carter and W. L. Roth, in: *EMF Measurements in High-Temperature Systems*, (C. B. Alcock, ed.), Elsevier, London (1968), p.125.
74. E. C. Subbarao, in: *Nonstoichiometric Compounds*, (L. Mandelcorn, ed.), Chapter 5, Academic Press, New York (1964).
75. C. B. Alcock, in: *EMF Measurements in High-Temperature Synthesis*, (C. B. Alcock, ed.), p. 109, Elsevier, London (1968).
76. D. H. Archer *et al.*, in: *Fuel Cell Systems*, (G. J. Young and H. R. Linden, eds.), Chapter 24, American Chemical Society, Washington, D.C. (1965).
77. R. L. Zahradnik *et al.*, in: *Fuel Cell Systems*, (G. J. Young and H. R. Linden, eds.), Chapter 25, American Chemical Society, Washington, D.C. (1965).
78. C. C. Liang, in: *Applied Solid State Science*, (R. Wolfe, ed.), Vol. 4, p. 95, Academic Press, N.Y. (1974).
79. J. N. Bradley and P. D. Greene, *Trans. Faraday Soc. 62*, 2069 (1966).
80. B. B. Owens and G. R. Argue, *Science 157*, 308 (1967).
81. H. Wiedersich and S. Geller, in: *Chemistry of Extended Defects in Nonmetallic Solids*, (L. Eyring and M. O'Keefe, eds.), p. 629, North-Holland, Amsterdam (1970).
82. W. VanGool, ed., *Fast Ion Transport in Solids, Solid State Batteries and Devices*, North-Holland, Amsterdam (1973).
83. J. W. Patterson, in: *Electrical Conductivity in Ceramics and Glass* (N. M. Tallan, ed.), Chapter 7, Marcel Dekker, New York (1974).

23

High-Temperature Electrolytes—The Future

J. W. Tomlinson

1. Introduction

At one time restricted to salt melts, glasses, and slags, the field of high-temperature electrolytes now ranges over a far wider variety of fluids from organic anion–cation systems melting below 100°C to highly conducting refractory solids and supercritical electrolyte solutions bordering on dense plasmas. The ability to vary experimental conditions over extended ranges of temperature and pressure has shown that many classes of substances formerly only known as semiconductors or insulators become electrolytes at the right density and temperature.

A crystal ball has not proved useful in compiling this prescient review nor does it seem likely that the seeds of the future necessarily lie in the past. What follows is an idiosyncratic attempt to assess the potential of present events, rather than to present a research program in anticipation of a multimillion dollar benefaction. At the outset, the current political "weltschmertz" suggested a prudent prescription for consolidation, but in the course of writing caution has been replaced by all the old enthusiasm which the author experienced on first entering the field of high-temperature ionics as a colleague of John Bockris in 1947.

2. Retrospect

Prior to the nineteenth century, the history of chemistry is almost synonymous with the history of high-temperature electrolytes because of the great concern of the alchemists and their predecessors with pyrochemical and pyrometallurgical operations involving fluxes and fusions of all kinds. These essential tools in the search for the *lapis philosophorum* acquired great mystical and metaphysical significance which still surrounds them in some measure today.

Their scientific study was begun by Davy at the beginning of the nineteenth century and exploded with Faraday into a century of activity throughout Europe. On February 18, 1833, Faraday, whose laws of electrolysis were based mainly on experiments with molten salts, writes in his diary:

> 300. *Chloride of Silver*—most beautiful—fuzed by lamp on glass—conducts well—intense action—deep descoloration at P. Pole and apparently chloride platina formed. Metallic silver at N. Pole and on lifting up pole drew out a wire of silver, the metal being reduced as it went on and communication being completed through it. Might in this way draw out silver bar at rate of $\frac{1}{2}$ inch per second quite regular.[1]

This epitomizes the research which followed—being concerned with electrical conduction and with the electrowinning and electrorefining of metals always in mind. The

---

J. W. Tomlinson • Professor of Physical Chemistry, Victoria University of Wellington, New Zealand.

impetus of these technological aims continued into the early decades of the twentieth century and the record of this golden age of high-temperature electrochemistry is recorded in now famous books by Drossbach,[2] Lorenz,[3] Allmand,[4] and Mantell,[5] among many others.

The renaissance of high-temperature electrochemistry coincided with the entry of John Bockris into the field 30 years ago, and it is again significant that the motivation for the greater part of the extensive research effort which followed has been technological, deriving from the metallurgical, glass, and atomic-energy industries and the space and energy programs. Richardson's *Physical Chemistry of Melts in Metallurgy*[6] records a great part of the work and reference to much of the rest can be had through the books of Bockris and Reddy,[7] Mamantov,[8] Petrucci,[9] Sundheim,[10] Blander,[11] Bloom,[12] and Lumsden.[13]

The clear implication is that we must look at the technology of the future to determine the course of high-temperature electrolyte science, but technological and scientific advance have not necessarily been coincidental in the past, and we may usefully attempt to supplement Bockris's own appraisal of some newer concepts in electrochemical science[14] with some comments on high-temperature electrolytes. A most perspicacious survey of recent developments, with which this commentary attempts to avoid undue overlap, has been given by Clarke and Hills.[15]

3. Prospect

The unique features on which the particular properties of high-temperature electrolytes depend are their high charged-particle density and high particle mobility. The mobilities of ions in ionic melts at their melting points are comparable to those in aqueous solution, about 10^{-3} cm^2/s · V, corresponding to diffusivities of about 5.10^{-5} cm^2/s. Since the particle densities for ionic melts are about 20 times those of molar aqueous solutions, their conductivities are correspondingly higher. This is reflected in the customary current densities for electrolytic processes which, in the case of aqueous solutions, vary from 50 to 500 A/m^2 compared with 5000–50,000 A/m^2 for a molten-salt process. The newer solid-state fast ion conductors[16] are being designed with conductivities approaching those of ionic fluids (Figure 1), and their use has revolutionized the development of high-energy-density batteries, of which the outstanding example is the sodium–sulfur cell using a β-alumina electrolyte.[17] It is significant that the Arrhenius activation energies of the fast ion conductors[18] are comparable with those of the alkali halides[19] and nitrates[20] at constant volume, 10–15 kJ/mole. The ion mobilities are therefore approaching within a third to a half of the theoretical maximum corresponding to the thermal velocity. In these circumstances, significant improvements in conductivity can only be achieved by increasing the particle density. It is surprising that no liquid or solid-state hydrogen-ion conductors have yet been developed. Using low-melting ionic hydrides, a rechargeable battery incorporating solid-state hydrogen-ion conductors might have power characteristics an order of magnitude better than those currently employing sodium-ion electrolytes.

Many of the engineering difficulties associated with high-temperature electrolytes are materials problems arising from their reactivity. These can to some extent be minimized by eliminating dissolved oxygen and water which are responsible for grain boundary attack of metals and ceramics, but ultimately the objective must be to develop electrolytes with specific structures which retain high selective ion mobility while at the same time remaining inert to suitably chosen container materials.

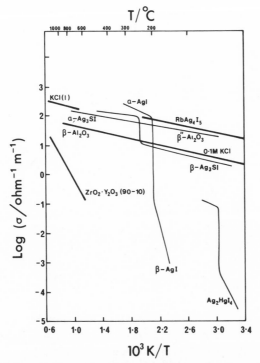

Fig. 1. Conductivity of some highly conducting ionic solids, as a function of temperature (redrawn from reference 81 with additions).

4. The Ionic Liquid State

There are two fundamental problems in describing the liquid state; one is to characterize the molecular interactions and the other is the statistical problem of relating the macroscopic properties of the system to the microscopic or molecular properties. With the exception of the liquid rare gases, the simplest dense fluids are to be found among the high-temperature electrolytes, and these substances, particularly among them the alkali halides, have therefore attracted the attention of theoreticians. The alkali halide ions are isoelectronic with the rare gases, and successful calculations of their thermodynamic properties have been made with a variety of semiempirical theories[21-26] based on simple models, using a pair-wise additive interaction potential which can be simply derived from spectroscopic data and the properties of the corresponding solid. A complication is the effect of the long-range coulombic forces which in principle introduces a requirement for "many-body" formulations in the statistical treatment. In extending the theory beyond ions of the inert gas type, asymmetry of the micropolarizability is a major complication which may be present even though the ions are monatomic.[27]

The approximate kinetic theory of transport in simple fluids developed by Kirkwood and Rice and Allnatt[28] has proved interesting in its application to ionic melts,[29] especially in clarifying the essential role of the coulombic potential which fixes the system volume and local structure but makes little contribution to the dissipation of energy and momentum in transport, which is largely determined by the short-range forces. From a theoretical point of view, the molten organic salts $R_4N^+ \cdot BF_4^-$ are of particular interest because the large cation size increases the steepness of the repulsive potential on an ion size scale, and the melts may therefore be approximated by simple hard-sphere or square-

**Table 1. Activation Volumes for
Self-Diffusion in Liquid Rare
Gases[32] and Conductivity in
Alkali Halides[31]**

ΔV_D (cm³/mole)		ΔV_Λ (cm³/mole)	
Ar	32	6.2	KCl
Kr	41	9.1	RbBr
Xe	66	13.5	CsI

well potential models. In the same context the possibility of varying the size of the quasi-spherical R_4N^+ ions has already proved valuable[37] and some interesting comparisons between the transport and thermodynamic properties of $R_4N^+BR_4^-$ salts with those of their isoelectronic, electrically neutral hydrocarbons have been made, e.g., $Bu_4N^+BR_4^-$ and Bu_4C.[30] The thermodynamic data show that at the same volume and temperature, the pressure of the salt (charged molecules) is 1–2 kbar less than that of the nonelectrolyte (uncharged molecules) and that this is largely due to the increase of $(\partial U/\partial V)_T$ by the coulombic field. The viscosity of the salt is 5–10 times that of the nonelectrolyte at the same temperature and particle number density; this may be ascribed to the increased dissipation of momentum arising from the greater repulsive interactions at the reduced intermolecular separation caused by the coulombic field. For the same reason, the values of ΔV_κ for KCl, RbBr, and CsI[31] are an order of magnitude less than those of ΔV_D for the iso electronic liquids Ar, Kr, and Xe (Table 1).[32]

All of these measurements are inconsistent with transition-state, hole- and free-volume theories of transport, and the current tendency is increasingly to use theories, based on small step diffusion models,[33] deriving from the theory of Kirkwood and Rice and Allnatt.[28] Thermal activation and volume-dependent theories remain, however, valuable in qualitative analyses of the gross changes in behavior associated with changes in density described later. It may be noted in passing that although there are fundamental objections to treating transport processes in liquids in terms of the transition-state theory[34] and there is ample evidence that jumps over barriers requiring energy of activation are not involved,[35] expressions involving terms of the form $e^{-E_a/kT}$ may have entered a new era of respectability as a result of the quantum statistical mechanical treatment of activated rate processes by Wassam and Fong.[36]

Among the more recent theories based on simple physical models which involve a new approach is that of Barton and Speedy.[37] They use the notion of Bernal[38] that the structure of a liquid is best expressed in terms of a distribution of coordination number $g(z)$, and they relate molecular motion to the number of molecules of coordination number zero N_0 arising from local fluctuations in the microstructure. An "L" well potential is used which combines a shallow square well for the long-range part of the intermolecular potential with a deeper square well for the short-range part. The theory accounts qualitatively for the general variation of E_p (see 6.2) with T[39] for ionic migration in liquids referred to later and quantitatively for the temperature and volume dependence of the conductivity of the series of organic salts $R_4N^+BF_4^-$ (R = n-butyl to n-heptyl).

For the solid-state fast-ion conductors Rice and Roth[40] suppose that ions of mass m are thermally excited with an energy ϵ_m from ionic states across an energy gap ϵ_0 into a conduction band where they are propagated freely in the applied field with a velocity $V_m = (2\epsilon_m/m)^{1/2}$ and energy ϵ_m for a lifetime τ_m. Expressions are derived for the electric and thermal conductivity and the thermoelectric power, e.g.,

$$\sigma = \frac{(Ze)^2}{3kT} \ nV_0 l_0 e^{-\epsilon_0/kT}; \ l_0 = V_0 \tau_0; \ V_0 = (2\epsilon_0/m)^{1/2}$$

where n is the density of potentially mobile species. The term ϵ_0 can be obtained from experimental values of $\sigma(T)$ and hence V_0, l_0, and τ_0 can be calculated. For a variety of fast-ion conductors, the calculated parameters are consistent with their lattice structures, and for a series of substituted silver halides values obtained from thermoelectric power data and from conductivity are in close agreement. Developed for solids with a population of liquid-like ions, the theory should be equally applicable to liquids, and the authors have applied it with success to self-diffusions in liquid metals. For KCl at 1200 K we obtain $\tau_0 = 2.10^{-13}$ s and $l_0 = 1.2$ Å. The advantage of the theory over the conventional hopping model is that l_0 and τ_0 are independently accessible whereas α and γ, the jump distance and frequency, are not. It also does not involve the untenable concept of activation over a structural potential-energy barrier. An extension to cover the volume and temperature dependence of conductivity is clearly desirable.

It is significant that in the solid super ion conductors, the Arrhenius activation energy is very sensitive to ion size for a given structure. Flygare and Huggins[41] have calculated minimum energy paths along tunnels in the α–AgI structure using a pair potential including coulomb, exponential repulsion, and polarization terms, and they find a sharp minimum for $0.8 < r_+ < 0.85$ as shown in Figure 2, which is the size expected for Ag^+. Experimental verification of this behavior has been obtained in the case of the alkali metal and silver cations in β-alumina by Pizzini and Bianchi.[42] Their data are shown in Figure 3 and give remarkable confirmation of the highly specific character of ionic size even in these open-lattice structures. Needless to say such a minimum in E_A is not found

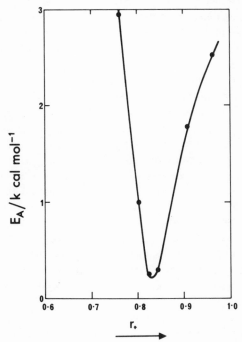

Fig. 2. Calculated activation energy for cation transport in AgI as a function of cation radius. [From Flygare and Huggins.[41]]

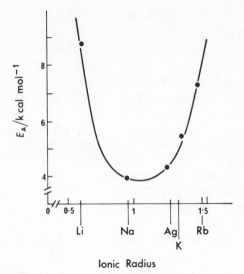

Fig. 3. Arrhenius activation energies for ionic self-diffusion in β-alumina as a function of the ionic radius of the substituent mobile ion. [From Pizzini and Bianchi.[42]]

for a series of molten alkali metal halides with a common anion, indicating the completely different "many-event" character of the migration process in liquids.

5. Computer Simulation

Computer experiments based on both the Monte Carlo and molecular dynamics techniques, have been carried out on ionic liquids.[43,44] Using a straightforward Born-Mayer-Huggins pair potential with an exponential repulsion term and dipole–dipole and dipole–quadrupole dispersion terms, excellent agreement is obtained for the thermodynamic properties of the simpler alkali halides, as well as reasonable values for their diffusion coefficients. Failure in the case of potassium and cesium fluorides and lithium iodide is attributed to the omission from the pair potential of a term for the polarization energy, which would be substantial when the ions differ greatly in size.

Impressive as these results are, the greatest power of the method may well lie in its ability to produce results inaccessible by experiment and to permit an examination of the assumptions of theories of transport. As an example of the former, the mean radial distribution function $g_m(r)$ can easily be resolved into the components $g_{12}(r)$, $g_{11}(r)$ and $g_{22}(r)$ as shown in Figure 4, whereas the extraction of these quantities from diffraction data, although possible in principle, is in practice extremely difficult. In contrast to mixtures of rare-gas molecules, the first peaks of $g_{11}(r)$ and $g_{22}(r)$ coincide even when the ions are as dissimilar in size as Li^+ and I^-, reflecting the importance of the coulomb forces in producing a charge-ordering effect and in determining the local structure. An unexpected feature of the $g(r)$ is the appearance of a small but significant percentage of like ions in the first coordination shell (for KCl, 7% at 1045 K increasing to 15% at 2874 K), this indicates the need for a revision of the basis for evaluating coordination numbers and that the values obtained from the experimental $g_m(r)$ may be erroneously low.[43] It is also striking that whereas $g_m(r)$ shows no order beyond the weak second peak, the resolved components of the third and fourth peaks are quite distinct (see Figure 4). Instantaneous configuration from molecular dynamics simulations show large and persistent

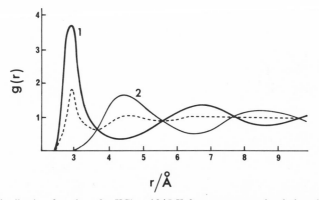

Fig. 4. Radial distribution functions for KCl at 1045 K from computer simulation: $1, g_u(r); 2, g_l(r);$ ---$g_m(r); u$ = unlike ions; l = like ions; m = mean. [From Woodcock and Singer.[43]]

voids of at least ionic size which have been described as geometrical rather than thermodynamic in that they are inaccessible to thermally activated ions.[44] The hole theory[45] may thus be said to have been confirmed insofar as holes are there but not insofar as they play no part in any transport mechanism. Another remarkable result is that the structure of liquid BeF_2, including the characteristic tetrahedral coordination of F^- around Be^{2+} ions, may be simulated with a symmetrical coulomb potential without any need to invoke directional valence forces.[46]

From a molecular dynamics simulation the motion of individual particles may be followed in detail, and in this model diffusion does not take place by means of discrete, thermally activated translations but by a succession of local configurational rearrangements.[44] Smedley and Woodcock[47] have used such a simulation for argon and KCl to examine the basis of the Kirkwood–Rice–Allnatt transport theory. In both cases the velocity autocorrelation function shows a negative region corresponding to frequent short-range reversals of trajectory, and the behavior of the force autocorrelation functions at long times suggests that successive encounters are not statistically independent. In the case of liquid KCl, the slow decay of the coulomb force autocorrelation function, compared with those for the short-range forces, confirms the contention of Rice[48] that the coulomb forces do not contribute significantly to the dissipation of energy and momentum. Other refinements in the assumptions of the theory are indicated, in particular the need to evaluate friction coefficients arising from the significant cross-correlations between hard and soft forces. The many-body character of interactions in liquids implies a many-event mechanism as a model for transport.

Woodcock and Singer[43] conclude that the value of semi-empirical theories based on physical models "lies in their predictive power rather than in the reality of the underlying models." There is, of course, never any "reality" in a physical model, and high-temperature electrolytes have offered and will continue to offer a stimulus to the development of physical models which have the advantage of an analytical rather than a numerical solution.

Computer simulation holds great promise for the advancement of the analysis of the thermodynamic properties of mixtures of high-temperature electrolytes. Coupled with the comprehensive sequence of measurements of heats of mixing by Kleppa[49] and with the aid of the corresponding-states theory of Reiss,[50] the technique seems likely to reduce the description of chemical potentials of the components of even complex mixtures to a matter of routine, if lengthy, computation, limited only by economics and progress in computer technology.

6. Other Aspects of Ionic Migration

The application of thermodynamic thinking to ionic transport in high-temperature electrolytes has had two results of significance for the future.

6.1. Irreversible Thermodynamics

First, the use of irreversible thermodynamics, leading to phenomenological descriptions of transport properties,[51] although it has had little theoretical impact, has cleared the air surrounding several examples of muddled thinking, in particular the transport number muddle[52] and the applicability of the Nernst–Einstein relation. But further, it has led to a very simple and clear analysis of electrorefining in molten-salt systems where the current is limited by mass transport across a hydrodynamic boundary layer.[53] Four years after it was reported that the refining effect could be characterized by a refinability parameter $\beta = \gamma_c(K)^{1/Z_B}/R^*$, where γ_c is the activity coefficient of the impurity metal in the cathode, K is the equilibrium constant for the exchange reaction involving the impurity, Z_B is the charge on the refined metal cation and R^* is a friction coefficient ratio, a pilot-plant process for the electrorefining of lead in fused chloride electrolytes was described.[54] Values of R^* estimated from theory for bismuth silver and copper in lead were 10^{-8}, 0.16, and 2.7, respectively, and these correlated well with the observed retention of these impurities in the anode which can be classed as excellent, good, and poor. The development of refining theory and the acquisition of the basic data, particularly to permit the evaluation of R^*, will clearly be of fundamental importance for a future economy in which recycling of metals plays a major role.

6.2. Volume and Structure

Second, and more far-reaching, is the recognition of the importance of volume in determining the properties of electrolytes or potential electrolytes. Initially the constant-volume principle* focused attention on the desirability of eliminating the effect of volume on transport properties so that the effect of temperature could be isolated. The familiar isobaric temperature coefficient of, for example, conductance,

$$E_p = -R[\partial \ln \Lambda/\partial(1/T)]_p$$

is formally separable into two terms thus:

$$E_p = E_v + (\pi + p)\Delta V$$

where $E_v = -R[\partial \ln \Lambda/\partial(1/T)]_v$; π is the internal pressure, $(\partial U/\partial V)_T$; and ΔV is the so-called activation volume, $-RT(\partial \ln \Lambda/\partial p)_T$. This analysis is quite independent of any transport theory, and E_v and ΔV may be functions of V and T, respectively. However, the transition-state theory and most theories based on a simple physical model allow an identification of E_p, E_v, and ΔV in terms of parameters of the model and predict properties for these parameters which are not observed experimentally. For example, for the alkali nitrates[20] and the salts $R_4N^+BF_4^-$ ($R = C_4$–C_7),[37] E_v and ΔV are strongly de-

---

*For a comprehensive review and an extensive bibliography of earlier work see Barton.[55]

pendent on V and T so that Λ cannot be expressed as a separable function of these variables of state. The Λ, V, and T data for the alkali halides[19] are similarly inconsistent with, for example, the hole and free-volume theories. The greater diagnostic power of pressure–temperature data is of immense value, and with superior experimental techniques which permit the simultaneous measurement of the conductivity and volume as a function of temperature and pressure,[37] the direct evaluation of Λ as a function of the preferred variables of state V and T is possible with greatly increased precision. If the intention is the advancement of theory rather than the mere acquisition of data, it is doubtful whether it will be worthwhile making measurements of the transport properties of ionic fluids, especially conductance, in any other way.

Beyond the implication of the constant volume principle lies an even more fundamentally significant effect of volume. The suggestion[39] that the temperature variation of liquid transport properties could be predominantly determined by restricted free volume or reduced configurational entropy at lower temperatures, by the temperature variation of mobility at intermediate temperatures and by structural factors at higher temperatures (Figure 5), has been amplified[20] and widely applied to molten-salt systems.[55] The significant change as temperature increases is, of course, one of increasing volume or decreasing density, and the effect of density on structure is particularly interesting.

Many instances have been reported of ionic melts whose temperature coefficients of conductance decrease at one atmosphere pressure as temperature increases and in some cases finally become negative.[56] The mercury, cadmium, bismuth, and indium halides are examples, and isobaric conductance maxima were reported for a number of salts which display Arrhenius behavior at lower temperatures. In the case of mercuric iodide, the conductivity decreases with rising temperature over the whole liquid range. Grantham and Yosim[56] suggested that these results could be accounted for by assuming that, with increasing temperature as the liquid approached the critical point, the normal increase in ionic mobility is offset by the formation of molecules, i.e., by increased ion association at higher temperatures. Cleaver and Smedley[57] showed that the behavior of the mercuric halides could be entirely accounted for by the displacement of ionization equilibrium of the type

$$2HgX_2 \rightleftharpoons HgX_3^- + HgX^+$$

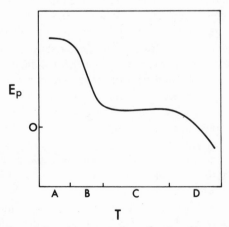

Fig. 5. Hypothetical variation of E_p with temperature showing the complete range of behavior: A, solid-like region near the glass transition temperature; B, volume-dependent range; C, Arrhenius region; D, structure-dependent region.

From measurements over the range 260-370°C and 1-1000 bar they obtained negative volumes of activation for conductivity which corresponded closely to the volumes of ionization calculated with the aid of the Born equation. Their conclusions were confirmed by Darnell and McCallum[58] by measurements at 5.4 and 20.5 kbar. They showed that at these elevated pressures the salts behaved like normal ionic fluids, their conductivity decreasing with increasing temperature over the whole range. Bannard and Treiber,[59] in a more detailed investigation of mercuric iodide up to 3.75 kbar and 850°C, showed that the negative temperature coefficient became more positive with increasing pressure, passing through zero at 2.5 kbar; the volume of activation appeared to approach zero at 5-6 kbar.

The conductivity σ may, of course, be written

$$\sigma = F \sum_i N_i |Z_i| U_i = \frac{\Lambda}{V}$$

where Λ is the molar conductivity, V is the molar volume, and N_i the number of moles of ions of type i, charge Z_i, and mobility U_i per mole of salt. This old expression emphasizes the composite effect of temperature on conductance, first to alter the mobility of the charge carriers and second to change the number density of the charge carriers by shifting the position of ionization equilibria. Thus we may expect that molecular liquids may well become ionic at sufficiently high densities and vice versa. It is of no consequence that temperatures may be required which take the system well into the critical region. For bismuth chloride where T_c = 905°C, Treiber and Tödheide[60] have shown that the degree of ionization varies from 10^{-5} to 1 from 1 bar to 4 kbar. In the supercritical region, where the density may be varied continuously from gaslike to fluid-like values, from 0.15 to 3.0 g/cm^3, the molar conductance changes by almost five orders of magnitude from 5×10^{-3}-2×10^2 Ω^{-1} cm^2/mole. This remarkable transition from an insulator to an ionic conductor is only one of a series of such examples. In the case of water the classic shockwave experiments of Hamann and Linton,[61] confirmed by the static experiments of Holtzapfel and Franck,[62] show that at 1000°C and about 100 kbar (where the density reaches 1.5-1.7 g/cm^3) the conductivity approaches unity due to the increase in the ionization product by more than 12 orders of magnitude over the value under standard conditions (Figure 6). Such values of 10^{-2}-10^{-1} moles^2/dm^6 are consistent with the enthalpy and volume change for the ionization process and indicate that water should be a completely ionized fluid if compressed to a density of about 2.0 g/cm^3 at high supercritical temperatures.[63] Its molar volume and behavior would then be comparable to that of fused sodium hydroxide.[64] Similar static conductance measurements on fluid ammonia up to 600°C and 40 kbar, in which conductivities of 10^{-2} Ω^{-1} cm^{-1} were reached,[65] lead to an estimated value of the ionization product of 4×10^{-4} moles^2/dm^6, about 10^{18} times the normal value. Supercritical ammonia should be fully ionized at a density of about 1.7 g/cm^3. In contrast to these polar fluids which at their critical points are still substantially un-ionized, ammonium chloride at its critical point of 882°C and 1635 bar appears likely to be fully ionized. The conductance of the liquid has been measured[66] along the vapor-pressure curve from the triple point at 520°C where Λ = 85 Ω^{-1} cm^2/mole to about 30°C below the critical temperature where Λ = 250 Ω^{-1} cm^2/mole.

Treiber and Tödheide[60] have pointed out that these substances may be arranged in order according to the reduced density ρ/ρ_c at which they become fully ionized as follows: NH$_4$Cl $\simeq$ 1; BiCl$_3$ $\simeq$ 3.2; HgX$_2$ $\simeq$ 4; H$_2$O $\simeq$ 6.3; NH$_3$ $\simeq$ 7.0. They suggest that the distinction made between ionic melts and weakly conducting polar fluids such as ammonia and water under normal conditions is not fundamental, but only one of degree.

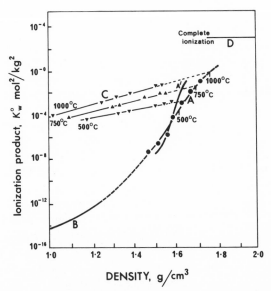

Fig. 6. Ionization product of water as a function of density: A, A', values calculated from shock conductivities; B, curve calculated from low density data; C, static isotherms from Holzapfel and Franck.[62] [From Hamann and Linton.[63]]

The dielectric properties of dense water[64] at high temperatures suggest that it is a good ionizing solvent. Furthermore, since its viscosity in the supercritical region is 10–20 times less than under standard conditions, ionic mobilities may be expected to be high. There is an extensive literature[67] on the behavior of dilute electrolytes up to 800°C and 4 kbar, and some investigations have been made as high as 1000°C and 12 kbar.[64] Equivalent conductances up to an order of magnitude greater than those at room temperatures have been observed at water densities between 0.4 and 0.8 g/cm^3, and at constant density the conductance passes through a maximum value at temperatures close to 300°C. Concentrated solutions also show conductance maxima at temperatures close to this value.[68]

The interpretation of all these observations in terms of the effects of the independent variables, volume and temperature, on the concentration of charge carriers and their mobility has extended our understanding of the nature and range of high-temperature electrolytes in a quite remarkable way. Whether a substance is classed as an electrolyte or an insulator depends entirely on the temperature and density considered, and the possibility of investigating the properties of dense fluids over a wide range of intermolecular separations offers new prospects for clarifying the respective roles of coulombic and short-range interactions in determining particle mobilities. No current quantitative theories of ionic migration have been extended over the range of conditions which are presently being covered, and we are still essentially at the stage of offering essentially qualitative, descriptive explanations of the phenomena.

It is difficult to assess the technological implications of these new results since extreme conditions of temperature and pressure seem impractical, at least under static conditions. Perhaps explosive processes may prove to be less ridiculous than at first sight, and systems involving controlled standing shockwaves are almost acceptable. At this point speculation leads the field of high-temperature electrolytes towards plasma physics.

7. New Fluids

Mackenzie[69] has commented on the solvent properties of molten salts for metals, gases, oxides, and organic substances. The potential reactivity and catalytic properties of, for example, solutions of metals in molten salts, where at high temperatures the components are frequently completely miscible, is largely undeveloped. A full evaluation of the effects of this solubility on the mechanism of electrodeposition in such systems has also not yet been made. Charge transfer may take place in the electrolyte itself, and this provides an interesting theoretical alternative to the more familiar passage of an electron from a surface state in a metal to a level associated with an ion in solution.

Applying a polarization technique (previously used with solids) to the LiCl–KCl eutectic, Heus and Egan[70] were able to compute the components of the conductivity due to electrons, electron holes, and ions as a function of the lithium metal activity. At 700°C, σ_e and σ_h can have values approaching unity, and the authors show the need to assess with care the suitability of this familiar eutectic for use in potential and polarographic measurements. They also indicate the source of problems in measurements of the double-layer capacity and the implications of their findings for electrowinning and electrorefining techniques where electronic conduction may limit the degree of purification possible. The possibility of producing n- and p-type liquid-state semiconductors by simply varying the composition is also attractive.

However, by extending conditions into the high-pressure region many more interesting possibilities emerge. The limited range of salts which are completely miscible with water at temperatures below about 200°C has been summarized by Braunstein,[71] who distinguished five composition regions:

1. Debye–Hückel limiting law range

$$x_{salt} < 10^{-3}$$

2. Extended Debye–Hückel theory range

$$10^{-3} < x_{salt} < 5 \times 10^{-2}$$

3. Hydrate melts

$$5 \times 10^{-2} < x_{salt} < 3 \times 10^{-1}$$

4. Incomplete hydration sheaths

$$2 \times 10^{-3} < x_{H_2O} < 7 \times 10^{-1}$$

5. Gas solutions in molten salts

$$x_{H_2O} < 2 \times 10^{-3}$$

Braunstein emphasized the advantages to be gained from broadly based studies over the whole composition range especially for our understanding of the theory of concentrated solutions.

At somewhat higher temperatures, salts such as the alkali and alkaline earth halides become completely miscible with water. Sourirajan and Kennedy,[72] for example, have reported T, p composition equilibria in the system NaCl–H$_2$O up to 1240 bar, from 250 to 700°C, and over the complete composition range. Such data are of great value in developing the theory of the origin of geothermal and volcanic structures and would be essential in evaluating the potentialities of tapping deep-seated terrestrial thermal power sources. The use of highly reactive, highly fluid, highly conductive liquids of this type for industrial processes has not yet been evaluated, although the small solubility of minerals in superheated steam is a problem well known to boiler-corrosion engineers.

At still higher pressures, substantial regions of miscibility exist in such systems as carbon dioxide–water, benzene–water, ethane–water, and argon–water. Franck[73] has discussed these interesting solutions and points out that it is possible to have mixtures of liquid-like density over the complete composition range at pressures of about 2 kbar and above the corresponding critical temperature. These systems are of great theoretical interest in that they may exhibit a lower critical endpoint and, in some cases, gas–gas immiscibility. The necessary conditions of temperature and pressure are realizable on the industrial scale, but as far as I am aware so far there has been little, if any, attempt to explore the potentialities of these and similar fluids as reaction media. It seems likely that the range of new high-temperature organic, inorganic, and mixed electrolytes will be greatly extended in the near future; this, coupled with the effect of increased density in ionizing polar molecules, can hardly fail to yield some interesting technological possibilities.

8. Reaction Media

The use of high-temperature electrolytes as inorganic reaction media is very well known. Slags, mattes, glasses, fluxes, electrolytes, heat-transfer media, nuclear fuels, and fuel-element processing systems have been highly developed technologically and the subject of scientific investigation for many decades. Some unfamiliar conditions of temperature and pressure in which new reactions may be discovered have already been mentioned. The use of fused salts as organic reaction media has, on the other hand, become the subject of systematic investigation only comparatively recently.

The field has recently been extensively reviewed by Sundermeyer[74] and by Gordon,[75] who distinguish between homogeneous reactions and high-temperature processes in which a reactant gas phase is in contact with an inorganic melt in which it is relatively insoluble. The first category of reactions has been limited by the lack of miscible systems, but the extension of the available ranges of temperature and pressure mentioned in the previous section may introduce new variety into this area. Research in the second category has been more fruitful, and Gordon lists modification of the reactivity of nucleophilic reagents, acids, and bases and electrophilic catalysts; the appearance of new redox systems; and electrochemical and other modes of generation of reactive intermediates among foreseeable developments. Gordon[75] draws particular attention to the advantages of fused organic salts which are low melting, good solvents for organic compounds, susceptible to continuous structural variation, and in fact are reactants in many of the known homogeneous organic–fused salt reactions. Among the more useful criteria for assessing the "polarity" of the solvent, is the cohesive energy density or internal pressure, which has the added advantage of being well correlated with mutual solubilities and other liquid properties.[76]

A wide variety of reactions is discussed by Gordon[75] and Sundermeyer,[74] but those in which the melt acts as a catalyst seem to offer the greatest scope for development. As alternatives to packed or fluidized beds of solid material, fused salts have the advantages of high thermal conductivity which prevents local overheating and a continuously renewable surface much less susceptible to poisoning and blocking. The production of vinyl chloride by the oxidative chlorination of ethane in a $ZnCl_2/KCl$ or $CuCl_2/KCl$ melt is a good example of a reaction where better yields are obtained compared with the use of the solid, and in addition the volatility of the catalytically active salt is reduced. Acid–base and redox equilibria involving oxide ions (e.g., $NO_3^- \rightarrow NO_2^+ + O^{2-}$) are alternatives to proton-transfer equilibria, and it may well be possible to create specifically designed fused-salt catalysts because of the ability of ionic melts to offer low-energy electrostatic environments for ionic transition states. The well-known enhancement of the nucleophilic reactivity of anions towards acceptor molecules in dipolar aprotic solvents may be even greater in ionic melts due to the field symmetry around the anion. However, it is important to remember that catalytic reactions involving a volatile species must take place at the gas–melt interface and that strongly polarizable molecules and polyatomic ions are likely to be sorbed at the surface so that the ionic environment at the site of the reaction may not be that of the bulk melt. This and other physicochemical factors which determine the reactivity of ionic melts have been succinctly analyzed by Ubbelohde.[77]

A topical problem in which the use of ionic melts may figure significantly is that of hydrocarbon cracking. Halide melts at relatively high temperatures have been used since 1959 for this purpose and offer considerable advantages over solid-state catalysts which are expensive to prepare and require regeneration. For example, the hydrocracking of coal and coal extracts to produce high-octane gasoline, using a molten zinc chloride catalyst at pressures up to 3500 psi H_2 and temperatures up to 430°C, has been carried out in a continuous system with a conversion efficiency of up to 85% for a residence time of 60 min.[78] For countries with substantial reserves of solid fossil hydrocarbon fuels such processes may rapidly become economically more favorable than petroleum-based production.

While they are not environmentally unobjectionable, molten-salt waste-disposal systems must also be mentioned. Currently developed in the form of a sewage-treatment plant,[79] the molten-salt reactor apparently offers considerable advantages over traditional grate and fluidized bed incinerators. For those organic wastes which are not easily biodegradable, such reactors may find a place in our future technology, particularly if they can be developed to produce end products other than carbon dioxide and water. The use of liquid high-temperature electrolytes in the processing of carbonaceous raw materials and recycled wastes will be a necessary part of the future environmental planning.

9. Conclusion

The field of high-temperature ionics has been placed on a sound physicochemical foundation over the past 30 years, and in many ways the Faraday Discussion of 1961 on ionic melts[80] appropriately marks a halfway house. We have passed from a period of exploring the existing to an era of creating the new, and our technology is moving from the exploitation of what is available to the design of what is required.

The development of the future science and technology of high-temperature electrolytes will inevitably be profoundly influenced by the state of our global energy economy. Reliance on existing power sources will restrict the use of the energy-intensive technology of high-temperature electrolyte processing but, on the other hand, nuclear fusion or efficient solar conversion may allow greater lavishness in the use of energy and expansion into some of the areas this paper has touched on. The final session of this conference is therefore very appropriately placed, and its conclusions must color the spectacles with which the high-temperature electrochemist views the future of his subject.

The contributions of John Bockris to this field in the past 30 years have been immense, not only in terms of his own research, but equally by the enthusiasm and optimism with which he has encouraged others. I do not doubt that with his help the next 30 years will be equally exciting.

References

1. *Faraday's Diaries*, Vol. II, G. Bell and Sons Ltd., London (1932).
2. P. Drossbach, *Elektrochemie Geschmolzener Salze*, Springer-Verlag, Berlin (1938).

3. R. Lorenz, *Die Elektrolyse Geschmolzener Salze*, Knapp, Leipzig (1923).
4. A. J. Allmand and H. J. T. Ellingham, *Appled Electrochemistry*, Edward Arnold, London (1924).
5. C. L. Mantell, *Electrochemical Engineering*, 4th ed., McGraw-Hill, New York (1960).
6. F. D. Richardson, *Physical Chemistry of Melts in Metallurgy*, Vols. 1 and 2, Academic Press, New York (1974).
7. J. O'M. Bockris and A. K. N. Reddy, *Modern Electrochemistry*, Plenum Press, New York (1970).
8. G. Mamantov, ed., *Molten Salts, Characterization and Analysis*, Marcel Dekker, New York (1969).
9. S. Petrucci, ed., *Ionic Interactions*, Vols. 1 and 2, Academic Press, New York (1971).
10. B. R. Sundheim, ed., *Fused Salts*, McGraw-Hill, New York (1964).
11. M. Blander, ed., *Molten Salt Chemistry*, Wiley-Interscience, New York (1964).
12. H. Bloom, *The Chemistry of Molten Salts*, W. A. Benjamin, New York (1967).
13. J. Lumsden, *Thermodynamics of Molten Salt Mixtures*, Academic Press, New York (1966).
14. J. O'M Bockris, Some newer concepts in electrochemical science, *Chem. N.Z.*, 37(2) 51 (1973).
15. J. H. R. Clarke and G. J. Hills, *Chemistry in Britain*, 1973, 9(1), 12.
16. W. Van Gool, ed., *Fast Ion Transport in Solids, Solid State Batteries and Devices*, North Holland, Amsterdam, (1973).
17. Sodium Sulphur Batteries for Electric Vehicles, Electricity Council Research Centre Report (1974).
18. B. C. H. Steele and G. J. Dudley, *M.T.P. International Review of Science (Series II), Solid State Chemistry*, Butterworths, London (1974).
19. B. Cleaver, S. I. Smedley, and P. N. Spencer, *J. Chem. Soc. Faraday Trans. 1* 68, 1720 (1972).
20. J. E. Bannard, A. F. M. Barton, and G. J. Hills, *High Temp. High Pressures 3*, 65 (1971).
21. G. E. Blomgren, *Ann. N.Y. Acad. Sci. 79*, 781 (1960).
22. F. H. Stillinger, J. G. Kirkwood, and P. J. Wojtowicz, *J. Chem. Phys. 32*, 1837 (1960).
23. C. M. Carlson, H. Eyring, and T. Ree, *Proc. Natl. Acad. Sci., U.S.A. 46*, 333 (1960).
24. K. Furukawa, *Disc. Faraday Soc. 32*, 53 (1961).
25. D. A. McQuarrie, *J. Phys. Chem. 66*, 1508 (1962).
26. I. G. Murgulescu and G. H. Vasu, *Rev. Roumaine Chim. 11*, 681 (1966).
27. A. R. Ubbelohde, *Rev. Roumaine Chim. 17*, 357 (1972).
28. J. G. Kirkwood, *J. Chem. Phys. 14*, 180 (1956); S. A. Rice and A. R. Allnatt, *J. Chem. Phys. 34*, 2144, 2156 (1961).
29. S. A. Rice, *Trans. Faraday Soc. 58*, 499 (1962); B. Berne and S. A. Rice, *J. Chem. Phys. 40*, 1347 (1964).
30. G. Morrison and J. E. Lind, *J. Chem. Phys. 49*, 5310 (1968); T. Grindley and J. E. Lind, *J. Chem. Phys. 56*, 3602 (1972).
31. B. Cleaver, S. I. Smedley, and P. N. Spencer, in: *Atomic Transport in Solids and Liquids* (A. Lodding and T. Lagerwall, eds.), Verlag der Z. für Naturforschung, Tübingen (1971).
32. J. Nazhizadeh and S. A. Rice, *J. Chem. Phys. 36*, 2710 (1962).
33. G. Morrison and J. E. Lind, *J. Phys. Chem. 72*, 3001 (1968); K. Ichikawa and M. Shimoji, *Trans. Faraday Soc. 66*, 843 (1970).
34. S. A. Rice, *Phys. Rev. 112*, 804 (1958).
35. B. J. Alder and J. H. Hildebrand, *Ind. Eng. Chem. Fund. 12*(3), 387 (1973).
36. W. A. Wassam and F. K. Fong, *J. Chem. Phys.* (in press); W. A. Wassam, Quantum Statistical Mechanical Approach to Molecular Relaxation Processes, Thesis, Purdue Univ., Lafayette, Indiana, Diss. Abs. Int. B, 1975, 36(2), 749–50.
37. A. F. M. Barton and R. J. Speedy, *J. Chem. Soc., Faraday Trans. 1*, 70, 506 (1974).
38. J. D. Bernal, *Nature 183*, 141 (1959); *185*, 68 (1960); *188*, 910 (1960).
39. J. W. Tomlinson, *Rev. Pure Appl. Chem. 18*, 187 (1968).
40. M. J. Rice and W. L. Roth, *J. Solid State Chem. 4*, 294 (1972).
41. W. H. Flygare and R. A. Huggins, *J. Phys. Chem. Solids, 34*, 1199 (1973).
42. S. Pizzini and G. Bianchi, *Chim. Ind. (Rome) 54*, 224 (1972).
43. L. V. Woodcock and K. Singer, *Trans. Faraday Soc. 67*, 12 (1971).
44. I. R. McDonald and K. Singer, *Chem. Br. 9*, 54 (1973).
45. T. Emi and J. O'M. Bockris, *J. Phys. Chem. 74*, 159 (1970).
46. A. Rahman, R. H. Fowler, and A. H. Narten, *J. Chem. Phys. 57*, 3010 (1972).
47. S. I. Smedley and L. V. Woodcock, *J. Chem. Soc., Faraday Trans. 2, 70*, 955 (1974).
48. S. A. Rice, *Trans. Faraday Soc. 58*, 499 (1962).
49. O. J. Kleppa, *Annu. Rev. Phys. Chem. 16*, 187 (1965); L. S. Hersh, A. Navrotsky, and O. J. Kleppa, *J. Chem. Phys. 42*, 1309 (1965); L. S. Hersh and O. J. Kleppa, *J. Chem. Phys. 42*, 3752 (1965); see also reference 13.

50. H. Reiss, S. W. Mayer, and J. L. Katz, *J. Chem. Phys.* 1961, *35*, 820 (1961); M. Blander, *Adv. Chem. Phys. 11*, 83 (1967).

51. See for example A. Klemm in reference 11 and B. R. Sundheim in reference 10.

52. R. W. Laity, in: *Encyclopaedia of Electrochemistry* (C. A. Hempel, ed.), p. 653, Reinhold, New York (1964).

53. P. F. Hart, A. W. D. Hills, and J. W. Tomlinson, *Advances in Extractive Metallurgy*, p. 624, Institution of Mining and Metallurgy, London (1968).

54. E. H. Amstein, W. D. Davis, and C. Hillyer, in: *Advances in Extractive Metallurgy and Refining*, Paper 10, Institution of Mining and Metallurgy, London (1972).

55. A. F. M. Barton, *Rev. Pure Appl. Chem. 21*, 49 (1971).

56. L. Grantham and S. J. Yosim, *J. Phys. Chem. 67*, 2506 (1963); *J. Chem. Phys. 45*, 1192 (1966).

57. B. Cleaver and S. I. Smedley, *Trans. Faraday Soc. 67*, 1115 (1971).

58. A. J. Darnell and W. A. McCallum, *J. Chem. Phys. 55*, 116 (1971).

59. J. E. Bannard and G. Treiber, *High Temp High Pressures*, *5*(2), 177 (1973).

60. G. Treiber and K. Tödheide, *Ber. Bunsenges. Phys. Chem. 77*, 541 (1973).

61. S. D. Hamann and M. Linton, *Trans. Faraday Soc. 62*, 2234 (1966).

62. W. Holzapfel and E. U. Franck, *Ber. Bunsenges. Phys. Chem. 70*, 1105 (1966).

63. S. D. Hamann and M. Linton, *Trans. Faraday Soc. 65*, 2186 (1969).

64. E. U. Franck, *Pure Appl. Chem. 24*, 13 (1970); *J. Solution Chem. 2*, 339 (1973).

65. D. Severin, Thesis (1969–1971), Institute for Physical Chemistry, Karlsruhe, quoted in references 60 and 64.

66. M. Buback and E. U. Franck, *Ber. Bunsenges. Phys. Chem. 77*, 1074, (1973).

67. W. L. Marshall, *Rev. Pure Appl. Chem. 18*, 167 (1968).

68. S. K. Fellows and J. W. Tomlinson (to be published).

69. J. D. Mackenzie, Chapter 22, this volume.

70. R. J. Heus and J. J. Egan, *J. Phys. Chem. 77*, 1989 (1973).

71. J. Braunstein, *Inorg. Chim. Acta 2*, 19 (1968).

72. S. Sourirajan and G. C. Kennedy, *Am. J. Sci. 260*, 115 (1962).

73. E. U. Franck, *Ber. Bunsenges. Phys. Chem. 73*, 135 (1969).

74. W. Sundermeyer, *Angew. Chem. Int. Ed. 4*(3), 222 (1965).

75. J. E. Gordon, Applications of fused salts in organic chemistry, in: *Techniques and Methods of Organic and Organometallic Chemistry*, (D. B. Denney, ed.), Vol. I, Marcel Dekker, New York (1969).

76. A. F. M. Barton, *J. Chem. Ed. 48*, 156 (1971).

77. A. R. Ubbelohde, *Chem. Ind.* 313 (1968).

78. C. W. Zielke *et al.*, *Ind. Eng. Chem.*, *Proc. Res. Dev. 8*(4), 546 (1969).

79. L. Lessing, *Fortune*, July 1973, p. 138.

80. The Structure and Properties of Ionic Melts, *Disc. Faraday Soc. No. 32* (1961).

81. B. C. H. Steele, *M.T.P. International Review of Science*, *Series One*, *Inorganic Chemistry*, Vol. 10, Butterworths, London (1972).

24

Summing Up: High-Temperature Electrolytes

S. H. White

During the past 30 years or so, electrochemistry has extended in many directions, resulting in the variety of topics illustrated by the sessions held at this conference. High-temperature electrolytes have received such attention over this same period of time that it is now becoming necessary to subdivide this subject in a manner similar to that adopted in aqueous electrochemistry.[1] It is important that the molten-salt researcher be aware of almost all the facets of electrochemistry, perhaps with the exception of bioelectrochemistry, if he is to achieve a measure of success in his particular area of research. Add to this the tremendous diversity of molten-salt solvent/solution combinations and it is apparent that fruitful research in the field of high-temperature electrolytes is extremely challenging. The significance of Prof. Bockris's contributions to the field of high-temperature electrolytes should be seen in this context.

Profs. Mackenzie and Tomlinson have emphasized, quite rightly, the contributions made and to be made in the area of pure electrolytes and their simple mixtures. However, I feel that the chemistry and electrochemistry of solutions (up to about 1 M) are equally important, and I shall confine my remarks mainly to this area.

It had been the hope that high-temperature electrolytes would offer such structural simplifications over aqueous and organic solutions that theories of the liquid state could be tested readily using data obtained for them. This has been confirmed for a limited number of selected melts (as illustrated in the papers of Prof. Tomlinson and Mackenzie). However, the great diversity of melts (consider for example the chlorides, which cover the range from more or less completely ionic potassium chloride to polymeric aluminum chloride) has prevented any far reaching generalizations. The experimental studies complementing these theoretical investigations have led to the accumulation of a considerable amount of fundamental data (e.g., density, viscosity, transport properties, thermodynamic properties, etc.). Janz and coworkers over the past 15 years have played an important role in collecting, collating, and assessing this material, together with information on the chemistry, the purification of solvents, and the sparse results of electrodic studies such as double-layer capacity, overall electrode reactions, etc.[2] Much more work is required in the future to improve and to extend this collation to areas of less common (academically!) multicomponent systems of technological importance. The need for such technologically important data is well illustrated by Nolan Richards's earlier remarks concerning the economics of anodic overvoltage in the aluminum cell.

The earlier suggestion that molten salts and their solutions would provide media in which the chemistry can be regarded as simple has been dispelled in recent years. Jordan, Zambonin and coworkers[3] have, for example, shown that the oxide chemistry in alkali-metal nitrate solutions is extremely complex, dependent on acid–base interactions between oxide species and nitrate/nitrite anions, and unless precautions are taken, may

---

S. H. White • Department of Metallurgy, Imperial College, London SW 7, England.

involve the melt containers as well.[3,4] The availability of modern electroanalytical techniques has done much to focus attention on, and to elucidate, this complex chemistry.

Our own studies (carried out together with Mr. G. Warren and Mr. T. Mukhergee) on the mechanisms of refractory metal deposition from chloride solvents have illustrated the complex relationship between temperature, solvent structure/composition, solute oxidation state, and the overall electrode process. For example, it would be advantageous to win metal from solutions containing the metal ions in their lowest stable valency state. Simple alkali-metal chloride solvents tend to stabilize higher valency states and require higher operating temperatures. It has been suggested that the acidic aluminum chloride- or zinc chloride-based melts which tend to stabilize low oxidation states and have low melting points would be suitable media for such metalwinning processes. Recent studies[5] on the preparation of polynuclear refractory metal compounds have shown the necessity of an important synthetic step, which involves the use of aluminum chloride-based melts. This must be recognized when interpreting recent experimental observations by ourselves and by Rubel and Gross.[6] A passivation phenomenon may intervene during the reduction of lower oxidation states in these media and may indeed also occur for reductions from high oxidation states in simple halide media at low temperatures.

The intervention of surface phenomena which can lead to erroneous conclusions concerning diffusion coefficients has been illustrated recently by Hills et al.[7,8] The reported capacitance measurements[9] (in the anodic branch) at mercury electrodes in molten nitrates may well be influenced by the presence of surface phenomena resulting from the presence of residual chloride ions in these melts.[10]

These illustrations of recent complications in high-temperature electrolyte solution chemistry act not only as cautioning remarks but also as a pointer to future developments. In the former context, it should be emphasized that advanced electrochemical techniques have improved the molten-salt chemist's ability to discern chemical and electrochemical phenomena in molten-salt solutions to the extent that a detailed chemical or electrochemical understanding is now possible.

As to the future, the chemical and electrochemical behavior of electrolyte solutions must be an area for continued study, especially for transition-metal ion solutions. How the solution chemistry influences and possibly controls the electrochemical process will need to be elucidated before molten electrolyte solutions can fulfill their technological promise. The influence of adsorption/nucleation[11] phenomena is equally important and may prove exciting to theoretical electrochemists.

Other areas which, from a technological point of view, require study are liquid–liquid partition involving immiscible high-temperature liquids such as silicate/halide couples, and solid–gas–molten salt interactions. Both of these areas are now being investigated from the point of view of metal reclamation in our research group.[12] This type of investigation, of course, may emphasize the lack of understanding and the paucity of data for high-temperature oxyanionic liquids such as phosphates, silicates, borates, etc. A revived interest may be aroused in the electrochemistry of such melts, particularly from the electroanalytical viewpoint which is already becoming important to concepts of on-line control of technological processes.[13]

The development of new materials such as β-alumina, titanium disulfide, etc., and their relevance to energy production and conservation has already been partially touched upon by Prof. Mackenzie. It is felt that this will be important especially when related to the interaction of refractory materials with molten salts. The corrosion of refractory materials in contact with high-temperature electrolytes is an expensive technological problem,[14] and studies in this area are, to say the least, of economic interest. A fruitful area may be the development of solid refractories whose electrochemical properties are controlled to enable selective removal of undesirable inclusions from metals such as steel, aluminum, etc.

As an alternative to new materials, the use of conventional, readily available, energy-conserving materials treated in such a way as to produce desirable properties should be considered. The surface treatment of steel by carburization and nitridation has long been established, but alternatives may be necessary because of health and pollution hazards (cyanides are used). Similar processes which introduce alloying elements into a surface, thus modifying its properties, are being considered and need to be further investigated[15,16] The good throwing power and concurrent heat treatment make processes based on molten electrolytes potentially important, although the electrochemical behavior for such alloy–melt interfaces is not well documented.

To conclude, the recent recognition of an energy/materials crisis which, if left unchecked, will leave a derelict world for future generations, will be an important stimulus in spurring on present and past lines of research in high-temperature electrolytes with an emphasis on solvent systems of lower temperatures. To this end, the investigation of organic-based salts (perhaps tailored by the synthetic organic chemist to give maximum stability, low melting point, high conductivity, and electrochemical stability) is required.

References

1. D. Inman, J. E. Bowling, D. G. Lovering, and S. H. White, *Electrochemistry, Specialist Report No. 4* 78 (1972).
2. G. J. Janz, *Molten Salts Handbook*, Academic Press, New York (1967); G. J. Janz, F. W. Dampier, G. R. Lakshminarayanan, P. K. Lorenz, and R. P. T. Tomkins, *Molten Salts*, Vol. 1, National Standards Reference Data Series, National Bureau of Standards 15 (1968).
3. P. G. Zambonin and J. Jordan, *J. Am. Chem. Soc. 89*, 6365 (1967); *91*, 2225 (1969); and subsequent papers by P. G. Zambonin.
4. J. D. Burke and D. H. Kerridge, *Electrochim. Acta 19*, 256 (1974).
5. See for example W. C. Dorman and R. E. McCarley, *Inorg. Chem. 13*, 491 (1974); also C. R. Boston, in: *Advances in Molten Salt Chemistry* (J. Braunstein, G. Mamantov, and G. P. Smith, eds.), Vol. 1., Plenum Press 1971.
6. G. Rubel and M. Gross, *Corros. Sci. 15*, 261 (1975).
7. G. J. Hills, D. J. Schiffrin, and J. Thompson, *J. Electrochemical Soc. 120*, 157 (1973).
8. G. J. Hills, D. J. Schiffrin, and J. Thompson, *Electrochim. Acta 19*, 657, 671 (1974).
9. P. Fellner and K. Matiasovsky, *Chem. Zvesti 26*, 36 (1972); G. J. Hills and P. D. Power, *Trans. Faraday Soc. 64*, 1629 (1968).
10. S. H. White, unpublished work (1975).
11. D. Inman, D. G. Lovering, and R. Narayan, *Trans. Faraday Soc. 64*, 2487 (1968); D. Inman, R. S. Sethi, and R. Spencer, *J. Electroanal. Chem. 29*, 137 (1971).
12. The Extraction of Metals from Low Grade Sources, work supported at Imperial College by the Wolfson Foundation 1974–1975.
13. D. Inman and S. H. White, unpublished work (1974–1975).
14. For example L. D. Lucas, M. Olette, and P. Kozakevitch in: *Physical Chemistry of Process Metallurgy: the Richardson Conference* (J. H. E. Jeffes and R. J. Tait eds.), p. 187, I.M.M. London (1974).
15. N. C. Cooke *Sci. Am.* 38 (1969).
16. S. H. White and N. Godshall, unpublished work (1975).

25

Electrochemistry for a Better World—Present State of the Art

A. T. Kuhn

The role of electrochemistry in helping to keep our world as a clean and pleasant place in which to live has earlier been described at some length in the writings of John O'M. Bockris,[1,2] and it is not the purpose of this chapter to reiterate what is there, but to endeavor to collect the various applications of electrochemistry and arrange them in an ordered framework. With this, and with the up-dating of the literature, we shall concern ourselves.

If we try to create a logical framework for applications of electrochemistry to environmental control, we may use a diagram such as the one in Figure 1. We are not dealing with analytical applications of electrochemistry, rich though they may be with new ideas and uses reported almost daily. The figure shows how both liquid and gaseous effluent streams may be treated, the latter by scrubbing. Of course, as has been written before, scrubbing of hot gases—such as flue-stack gases—creates a problem because of cooling, which prevents their attainment of the high altitudes necessary to disperse the pollutants. However, some interesting ideas based on use of molten-salt or other nonaqueous scrubbing liquids are now being mooted and will be discussed later. Let us, however, consider the left-hand side of the diagram in which we find listed the so-called physical methods: (1) electrodialysis, (2) electroflotation, (3) electrophoresis, and (4) miscellaneous. In none of these is there a faradaic reaction taking place or at least not one which is of primary importance. In none of these methods is there a chemical change effected. However—and we shall see that this is equally important—these methods "recast" the problem of handling a particular effluent and are used to present the matter in a more tractable form. Some years ago, the disposal, as waste, of skimmed milk was widely deplored in the popular press, because of its nutritional value. The answer lay in the fact that the water content was so high that transport costs of this vast excess of aqueous matter far outweighed the cash value of the liquid as a food. This is the type of problem which is all too often crucial and yet unappreciated by the layman. We shall now consider in further detail the methods listed above. The first is used for separation of a liquid waste into two streams, one of which may be pure water or at least free from nonionic contaminants. Flotation is a method of segregating two phases. Electrophoresis is used to reduce the water content of sludges. In this range of applications it will be seen the methods are wholly complementary.

1. Electrodialysis

This method has been described in several sources as a general technique[3] and as one specifically applied to the problems of effluent control.[1,2] Largely under U.S. impetus,

A. T. Kuhn • Department of Chemistry and Applied Chemistry, University of Salford, Salford M5 4WT, Lancashire, England.

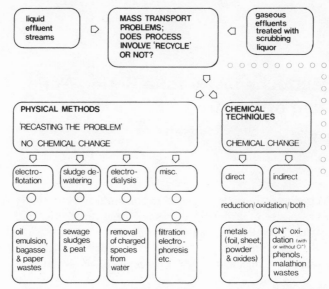

Fig. 1. Concepts in the electrochemical treatment of effluents.

much work has been done to improve the performance of electrodialysis (ED) stacks. In the main this related to improving stack design and reducing the thickness of the diffusion layer in the individual compartments, so permitting higher current without membrane polarization. Much of this work, which made use of inserted turbulence promoters, etc., has been cited elsewhere.[4,12] Progress has also been made in the development of better and (slightly) cheaper membranes. The Nafion range in particular is virtually indestructible in any aqueous environment and runs for months, if not years, in strong chromic acid, for example. Then too, there is a slowly growing body of information as to conditions which will give rise to "membrane-fouling," a situation which, as its name implies, is not conducive to good operation. In spite of these advances in the subject, one is forced to the conclusion that the technique has not found the many applications once foreseen for it. Of course most if not all the plants described[3] are still functioning well. However, the technique of reverse osmosis (itself employing a membrane) has made very substantial inroads into possible fields of application, and it is generally considered to be a less troublesome method (not that there is complete overlap) so for many applications ED remains the only answer. The literature contains some suggestions as to the future of the method. A paper by Solt[5] endorses the suggestion of earlier workers that the nature of the process is such that it is specially suited to the removal of low (1000 ppm or less) concentrations of salts from brackish waters to render these potable, and it quotes operating details of Italian, Middle-Eastern, and U.S. installations. Lancy[6] considers the applications of the method specifically to treatment and recovery of solutions used in the metal-finishing industry and quotes examples of plants in which etch solutions (such as those used for printed circuit boards, etc.) are treated and the acid is recovered. Ciancia[7] quotes a study[8] sponsored by the U.S. Environmental Protection Agency in which simulated copper cyanide rinse water was treated to recover both chemicals and water.

There is little one can add at this stage, and whether this chapter serves as an epitaph for the technique, or merely to mark a "resting point" in its career, one cannot say. Under the earlier force of the various U.S. government agencies a number of projects still continue. Among the publications we should cite are a study on membrane fouling,[9] operation of a test field rig,[10] and a larger (325,000 gal/day) plant in Kentucky.[11] A bibliography of applications of electrodialysis to saline-water treatment is found in reference 13, an engineering–theoretical study in reference 14, and a patent for *in situ* formation of membranes in reference 15. Commercially, processes are described for recovery of Ni ions from washwaters[16] for manufacture of strong brines from sea water,[36] and for treatment of cyanide washwaters by removing the CN and returning it to the process.[37] The removal of heavy metal ions by the method is described,[38] as well as the operation of a smaller stack (for Coast Guard use),[39] and a larger one.[40] Several attempts have been made to raise the operating temperature of ED stacks, and newer membranes allow this. Such work has been described,[41] and the use of inorganic membranes for electrodialysis has been discussed.[42] There are many patents relating to the subject,[42a] some dealing with problems of metal removal.

2. Electroflotation

In contrast to the previous method, electroflotation appears to be a fast-growing method and one offering immediate savings in operating and capital costs over air flotation in many situations. The basic principles of the idea were outlined,[1,2] and two articles[35] provide a near-exhaustive review of the literature and applications of the method. It is seen that the majority of the work in this area is from the U.S.S.R. and covers not only the more fundamental aspects, such as bubble size and its distribution, the effect of pH changes, and the overall relationship with the potential of zero charge (p.z.c.) of the electrode surface, but also the more practical aspects such as height of liquid and the effect of this and other practically important parameters. The early question asked regarding electroflotation vis-à-vis its conventional analog was "Are the bubbles different, not in size, but in their charge from nonelectrochemically formed ones?"[57] Though such bubbles are indeed much smaller than those normally obtained in other ways, it is not felt that they possess any unusual charge attributes. Rather, the present suggestion is that an effect is wrought on the pulp or material to be floated. Whether this effect is a direct electrode one or accomplished by the chemically different nature of the bubbles (H_2 and/or O_2) is still an unresolved question. The Soviets quote the example of the usually hard-to-float mineral cassiterite which, in an electroflotation cell becomes partially reduced at its surface, less easily wettable, and hence easier to float. How this reduction is achieved must be the main study goal of future scientists in the field. Such a line of thought leads us to postulate a processing mode in which the substrate is passed through an electrolytic cell, not in order to achieve complete conversion of any kind, but merely to modify its surface atoms in such a way as to render it more easily floatable in subsequent air-flotation cells. In a sense, we have seen one example of this quoted by an earlier speaker (Chapter 19), who discussed flotation using *in situ* electrolytically modified xanthates, and similar work from the U.S.S.R., using kerosines[49] and sodium sulfides,[50] likewise modified *in situ* by electrolysis, has been reported. The recent announcement of a new line of flotation tanks from the Powel-Duffryn/St. Gobain organization with optional air or electrolytic flotation fitment, although not identical with this concept, would seem to bring it so much closer.

The successful commercial application of electroflotation has made available a substantial amount of valuable data relating to the method. The author is grateful to the officers of the three major suppliers in the United Kingdom of this process who have provided the information below.

The Simon-Hartley Group, under their Electraflote label, has installed substantial

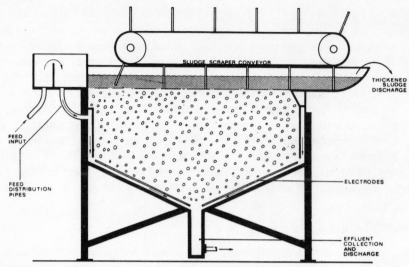

Fig. 2. Schematic flow diagram of Simon–Hartley flotation cell showing how effluent liquors are "conditioned" by passage across electrode surface prior to flotation. (Courtesy Messrs. Simon Hartley Ltd.)

amounts of plant. Their cells incorporate a design which forces the effluent to flow across the electrodes before reaching the bubble-generation zone where the actual flotation takes place. This, if one wishes, could be regarded as the electrochemical "conditioning" to which we have earlier referred. Figure 2 shows the schematic layout, while Figure 3 shows a design drawing. The data in Table 1 are from the same company.

Another very successful U.K. plant designer and contractor is the firm Morgett Electrochemicals Ltd. (Heswall, Cheshire) with plants of $48\ m^3$ h to their credit (14,000 gal/h).

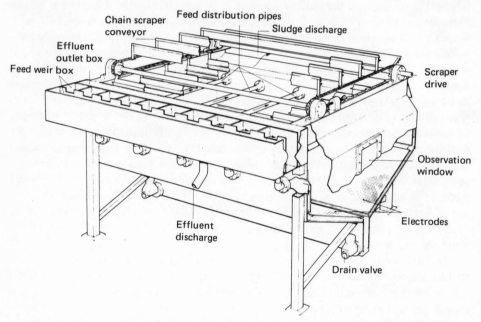

Fig. 3. Engineering drawing of Simon-Hartley cell. (Courtesy Messrs. Simon Hartley Ltd.)

Table 1. Test Results from Various Electroflotation Plants

Site type of feed (AS = Activated sludge)	Solid concentration			Power kWh/t (dry solids)	H$_2$SO$_4$ (ml/m^3)
	Feed	Floated sludge (%)	Effluent (mg/liter)		
1 Domestic + industrial AS	0.372	2.57	98	112	119
2 Domestic AS	0.790	2.91	57	238	Nil
3 Industrial AS	0.386	4.17	80	196	Nil
4 Domestic + industrial AS	0.770	3.60	239	74	263
5 Domestic + industrial AS	0.954	4.00	60	84	183
6 Domestic + industrial AS	0.276	4.66	90	78	76
7 Paper mill effluent	0.148	5.40	300	91	Nil
8 Resin recovery	0.145	31.60	134	339	Nil

The actual operating costs of such plants (to be distinguished from any estimated predictions) have ranged from $0.05 to $0.1/m^3 liquid. This wide range covers separations ranging from the simple one of fat removal or paper-fiber separation, through abattoir wastes, sludge thickening, and metal hydroxide flotation, to the difficult separations such as latex mixtures or flexographic printing inks in washwaters. Some of these require addition of flocculating agents while others do not. Intermediate cases show decreased flotation times with such additions, though not necessarily better cost effectiveness. An interesting concept raised by Morgett is the possibility of selective flotation, and it quotes an example of abattoir wastes where low pH (without additives) achieves fat removal alone. Changes of flotation condition can then be used to bring out the other suspended material. Figures 4 and 5 show Morgett plants of two different sizes, while Figure 6 shows an electrode array for large and small tank units. Morgett is now manufacturing and supplying electrode arrays to other contractors, and these units, based on stainless steel and Ti with PbO$_2$ coatings welded together and partially encapsulated in epoxy resins, provide a rugged and very easy to install unit. The last well-known U.K. manufacturer is Pollution Technical Services (Abingdon, Oxfordshire), members of the larger William Press Group. These manufacturers again have many successful operating plants to their credit, and details of some of these give insight into the type of process effluent for which the method is well suited.

1. *Plastic emulsion paint effluent*: 1.5 m^3/h. Effluent stream inlet, 25,000–35,000 ppm solids. Outflow, less than 400 ppm. Total electrical load (pumps, etc., included), 8 kW. Total operating costs (excluding amortization), $0.08/m^3.

2. *Latex-bearing wastes from foam-backed carpet manufacture*: 1 m^3/h. Inflow, 5500–8500 (peak) mg/liter. Effluent less than 400 ppm. Total Power, 7 kW. Costs $0.16/m^3 (as above).

3. *Abattoir wastes*: 5.5 m^3/h. Three fat traps followed by balancing sump from which effluent is pumped forward into a chemical dosing tank where salt is added at a controlled rate. Gravity flow to electrolytic tank thence to upflow clarifier. Sludge collection in sump for subsequent removal by contractors. Inlet, 2000 ppm av., 3000, peak. Effluent, less than 4530 ppm. Total power, 9 kW, $0.08/m^3.

Fig. 4. Electroflotation plants prior to installation (Courtesy Messrs. Morgett Electrochemicals Ltd.)

Fig. 5. Electroflotation plants as installed (Courtesy Messrs. Morgett Electrochemicals Ltd.)

Fig. 6. Electrode assembly: stainless-steel strip cathodes, lead-dioxide-coated titanium strip anodes with resin-encapsulated end connections. (Courtesy Messrs. Morgett Electrochemicals Ltd.).

4. *Scouring liquors from yarn washing*: Electroflotation with chemical dosing. 1.6 m^3/h. 3.5 kW total electrical demand. 8000–12,000 ppm inflow grease and oil. Less than 3000 ppm effluent $0.21/m^3$.

Other similar plants include wool-scouring liquors[52] (effluent better than 200 ppm), grease fat and oil from drum reconditioning plant (effluent less than 100 ppm), removal of suspended solids from dye-works effluent, and protein recovery from dairy wastes.[51]

Pechniney St. Gobain, who do not quote data, maintain that the smaller the plant, the larger the advantage of the electroflotation route. Thus for 0.25 m^2 tanks they quote a 50% saving in running costs, dropping to a 5% cost advantage on 5 m^2 tanks. Given their own figure of 7 $m^3/m^2/h$ tank surface, these figures can be converted to actual capacities.

European manufacturers seem, if anything, to prefer platinized titanium electrodes, while U.K. manufacturers go for lead dioxide-coated electrodes in the main. All non-Soviet plants use systems in which the oxygen and hydrogen rise together through the tank. This contrasts with what little is known of Soviet practice, and reference is made[35] to one description of an industrial electroflotation unit from the U.S.S.R. (grape-juice) in which the effluent gases are separated, the one being vented through a side chimney, the other used for the flotation. Most electroflotation plants are installed in the open air, and this removes dangers due to the explosive gas mixtures which they provide. Photographs of a Swissair plant show that hoods have been used for gas extraction in this indoor installation.[53] Some patents are cited in reference 70.

This description summarizes the commercial and technological situation, but leaves us curious as to the situation in the U.S.S.R. from whence a continuous flow of publications on the subject emanate. Most of these have been cited,[35] but they continue to appear. Recently, we have seen a study by Mamakov *et al.*[54] on the stability of foams formed during the electroflotation process, while Matov et al.[55] have returned to the problem of bubble size and the effect of the curvature of the electrode surface. Mamakov has further[56] considered factors affecting the size of the gas bubbles while Matov[57] has studied the charge on bubbles during their actual evolution from the electrode surface. On the practical side, there have been further interesting publications. Romanov[58,62,63] has studied the extraction of Mn and chromite as well as quartz ores using the method, while Matov[59] has investigated its application to albumen suspensions. Mamakov et al.[60] have studied the effect of the nature of the gas on the efficiency of flotation of various minerals. Hydrogen is better for cassiterite (as previously reported) while oxygen is the preferred gas for pyrites flotation. Malko[61] has published nomograms for the selection of optimal electroflotation conditions.

Other papers include work related to effluents from the leather industry,[64] the treatment of manganese-containing slurries,[65] and coke oven plants,[66] while a more general appraisal is made by Mamakov,[67] who also considered the breakaway of particles from hydrogen bubbles during flotation.[68] A particle counter devised for electroflotation studies is described.[69] The application of electroflotation processing to wastes generated during electrochemical machining is especially advantageous since there is already a large d.c. source incorporated in the machine, and this, constituting one of the largest cost elements in an electroflotation plant, affords opportunity for significant savings. One example is the work of Mitikov,[71] and this is of added interest (in the light of the fore-going) since the use of additional gases other than those electrolytically generated is specifically referred to. A modern review describing the latest work is (111).

The author must again express his gratitude to Mr. Eugene Gros, Managing Director of Messrs. Scientific Information Consultants Ltd. (661 Finchley Road, London NW2) for making available to him the translations of the above mentioned papers.

3. Miscellaneous Methods

In this section one can do little more than list a variety of methods, some of them once used on a practical scale and since abandoned, some of them now being developed, mainly in the U.S., and recipients of the lavish forecasts of the type once used to describe the fuel cell, a device which seems to be receding rapidly into obscurity. [Many people will not agree with this statement. Eds.]

3.1. Electrophoresis and Electro-osmosis

These two electrokinetic processes are sometimes confused, but in the first case the process is a movement of solids and in the second a movement of liquid (usually water). Both have been suggested as possible methods for treatment of process or effluent streams, and indeed some use of the methods has actually been made. In the following, the author follows and is indebted to Sunderland,[26] who is the authority in the field, and a recent review is due to Hiroshi.[44]

Electro-osmosis was used in the last century and much of the practical work is associated with the name of Graf von Schwerin, who favored the method. His ideas related to the dewatering of peat pulp are cited in references 17 and 18, and a later monograph[19] reviews the whole field at the end of the 1920s. There are applications to the drying of cellulose [pulp, cotton, cloth, etc.[20]], the electro-osmotic dewatering of finely divided limestone,[21] and further dewatering applications.[22] The method was used in treatment of clay in the Midland potteries prior to the war and the treatment of kaolin in Germany.[23]

There are postwar patents applied to dewatering of sludges, etc.,[24] and US Government reports.[25]

Looking back, we can see first of all a range of applications where the power consumption was too high and the process too expensive. Then too, a lack of suitable electrode materials led to problems of cost, replacement of corroded electrodes, and contamination of the process stream by electrode corrosion products. Sunderland has suggested that these objections can now be overcome and that the U.K. Electricity Council is optimistic that an economic electro-osmotic process can be developed, possibly to operate in conjunction with electroflotation. The latter process produces a sludge or floc containing up to 5% solids. Electro-osmosis could take over at this point and further reduce the water content, either to the stage where combustion becomes self-supporting or even further. Electrophoresis has been used to collect zinc cyanide precipitated by adding Zn^{2+} ions to a CN solution.[48]

3.2. Electrosorptive Methods

The removal of ionic impurities by adsorbing these on electrodes at one particular potential and then releasing them by changing the potential at a subsequent time has been examined in a number of papers and reports. Again it is suggested this might be an attractive economic method for achieving such separations. Mostly large-area carbon electrodes are used in this application. Another has been suggested that might be cheaper than electrodialysis.[27] Newman et al. have described the process[28] and at least one patent[43] has been filed.

3.3. Other Techniques

Electrofiltration—the use of a potential to prevent filters from clogging—has been described[29] using electrochemically controlled ion-exchange[30] as a possible method for demineralization. Electrodecantation[31] and piezodialysis have also been discussed.[32,45] The possibility of regenerating active carbon beds is also an interesting idea.[31] There has also been discussion[34,46] of the effect of surface potential (of a boiler or heat-exchanger) on the rate of scale formation and the possibility of scale prevention by alteration of that potential. Electrosorption in which a d.c. field is imposed across a two-phase mixture of oil and water was discussed (bimetallic coalescers, etc.)[1] and has since been discussed by Panchenkov[47] and Mason et al.,[90] the latter throwing doubts on the originally proposed mechanism of Fowkes et al. Whether all or any of the many ideas enumerated here ever become commercially important, we must wait and see.

4. Chemical Approaches to Treatment of Effluents

In contrast to the preceding section where the discussion centered on physical methods for effluent treatment, we are here concerned with chemical methods, that is to say processes which are wholly or mainly faradaic in their nature.

Referring back to Figure 1, we see there is the option of direct or indirect processing. In the former, the electrochemical reaction occurring at the electrode is one in which the effluent itself is reacted. In the latter case, an intermediate species is oxidized/reduced at the electrode, and the active species so produced reacts homogeneously with the pollutant in solution. Common examples of indirect processing in this context include NaCl, which forms NaOCl as used to destroy cyanide etc. The merits of indirect electrochemical processing as a whole have been recently discussed,[72] and the points will not be reiterated here. The salient question is whether the system is one which is recycled (in which case NaCl or any other species will be continuously recirculated) or whether the process stream goes to waste (in which case an additional "carrier" such as NaCl can neither be afforded nor tolerated). The other interesting application of "indirect" treatment of effluents is based on the use of zinc to recover iron, and this will be discussed below.

Apart from the question of direct or indirect processing, we must consider the whole question of reactor design and the efficiencies and capital costs associated with various types of electrochemical reactors. Thus emphasis has now shifted from a search for reactors either with very thin effective diffusion layer thicknesses or those with high space-time yields to a search for designs at once cheap and reliable and easy to maintain. Since this subject has been recently discussed in the literature,[4, 73] we shall not repeat the matter here, except to say firstly that the design of Lopez, using fluidized inert beads contacted with wire mesh electrodes[74, 75] seems a good one, and also to update the references above. Thus recent studies of relevance here relate to mass transfer to a rotating cylinder[76] under turbulent conditions (still favored by many workers, especially when it is desired to recover metals from solution in powder form) and further studies of mass-transfer enhanced by simultaneous evolution of gas bubbles at the electrode surface.[77]

An interesting cell using very high specific surface areas achieved with carbon-fiber electrodes was discussed at a recent meeting[78] while another concept which may find a use is the parallel-plate cell with turbulence promotion achieved by inert plastic meshes (Netlon).[73] However, Houghton and Kuhn[73] have used such cells in metal deposition work in a mode where the metal grew right through the meshes, and the cell was operated until short-circuiting began to take place. The metal can then easily be removed by withdrawal of the Netlon. Depending on the deposition conditions, this deposit may be powdery, in which case a quick shake frees the metal from the plastic. Under other conditions, a really coherent growth is seen. In this case, where the plastic is firmly imbedded, either the metal can be melted to free the plastic (which is so cheap that it can be destroyed in this way) or the composite structure can be used as a dissolving anode in an electrochemical process such as electroplating or similar applications. The mass transfer behavior of water in packed beds at low Reynolds numbers[79] should be of interest in modeling the behavior of three-dimensional electrodes.

In summary, it is seen that the emphasis has shifted in the past decade from the initial question—can we deposit metals cathodically from effluent solutions or destroy anionic or uncharged species?—to the more sophisticated approach—how efficiently can we deposit metals from solutions of their ions at low (1000 or 100 ppm) concentrations and how can we do this most cost-effectively and reliably? The benefits of a 10^{-6}-cm diffusion layer are less impressive if this is produced only by means of components rotated at high speeds. Simple cells with turbulence internally introduced by obstructions and baffles, having high pressure drops at relatively low flow velocities, appear to be preferable to those in which a very high flow rate is required to produce the same reduction in diffusion layer thickness. The former approach has the advantages that the remainder of the pipework circuit can be designed to have low hydraulic losses and so reduce energy consumption. Such an approach, coupled with the use of pumps rather than rotating disks/cylinders to create turbulence, seems to offer the best hope of success, and there are distinct signs that the pendulum has begun to swing away from cell designs with ultrahigh space-time yields because the high current densities (across the diaphragm) result in high ohmic drops there and power costs per coulomb are far higher than those obtaining in a lower-rated cell. Furthermore the physical nature of the recovered metal (e.g., powder, sheet, beads) is as important as the efficiency with which this is achieved.

4.1. Cathodic Deposition of Metals

As we have seen, the emphasis has now shifted from striking demonstrations of the ability to deposit with 99% faradaic efficiency from extremely dilute solutions. The problem is now seen to be one of what metals can be deposited from solution. Surfleet[80] has broken down the metals into three classes as follows:

1. Straightforward cathodic deposition: Au, Ag, Cu, Sn, Cd, Zn, Hg, As, Sb
2. Difficult deposition: Ni, Co, Cr
3. Deposition not possible: Pb, Al, Ti (and refractory metals)

As he rightly states, much depends on the nature of the solution, and presence or absence of certain anions or other metal ions can promote or demote a metal from one category to another.* The focus is then on the effects of anions on the one hand, and the presence of a second metal on the other. In time, there seems no doubt there will be the accumulation of knowledge as to the deposition behavior of a metal from a range of solutions. At the present, however, it is still a mixture of intuitive guesswork or trade "know-how" often most reluctantly disclosed, if indeed at all. Studies such as those of Houghton[81] on the deposition of antimony are the type of paper on which one has to depend, along with odd comments such as those of Surfleet,[80] who states that the smallest traces of Sb (1 ppm) inhibit deposition of Zn, and quotes other examples. Thus Ni can be easily deposited from a Watts-type bath but only with great difficulty from a Ni cyanide solution. [A recent paper by Bishop,[82] with a more detailed examination of the system, shows nickel cyanide to be highly unusual among the metal cyanides as a whole.] Cobalt behaves in much the same way. Nitrate and nitrite anions, as well as persulfates, give poor results, certainly from a current efficiency standpoint since they are themselves reducible. Lead can of course be deposited from a fluoroborate, but only rarely does this occur. The electrolytic recovery of Ni from dilute solutions was the topic of a paper delivered recently by Barker and Plunkett,[83] and is also described by Trivedi.[104]

Other references to metal recovery include British patent 1,359,809 which uses ion exchange as an intermediate stage for Ni recovery, British patent 1,362,601 for Cu removal, and U.S. patent 3,804,733. The conversion of Cr(VI) to the trivalent state (when it can be precipitated) using electrolytically generated ferrous hydroxides is described by Onstott and Gregory,[84] while other approaches to Cr(VI) reduction or regeneration are described in U.S. patent 3,728,238, and references 105 and 106. A very detailed and interesting study of treatment of pickle liquors from a brass rolling mill in the U.S. has been described.[85] A plant was installed using sulfuric acid and hydrogen peroxide pickle with electrochemical regeneration of the acid and metal being a valuable by-product. Costs indicate this method to have shown a great saving over conventional methods and simple dumping. Of course hydrogen peroxide itself (in that case it was brought in) could be electrochemically manufactured on site, especially in a conducting medium such as sulfuric acid. A related idea, using electrolytically formed H_2O_2 is found in U.S. patent 3,788,967. Various patents relating to the RCI (bipolar bed)[1,2] are found as U.S. patent 3,692,661 and British patent 1,279,650. Silver recovery cells are described in British patent 1,259,374 and U.S. patent 3,767,558. Nalco has patented electrolytic recovery of heavy metals (U.S. patent 3,748,240). A dual-bed cell for metal recovery is described in German patent 2,218,124 and other cells are described in reference 97.

*Recovery of metals as sulfides is discussed in British patent 1,362,943.

4.2. Modes of Electrodeposition

Granted that various electrochemical reactors can provide the efficient mass transport which ensures recovery of metals down to a low level, there are several schools of thought as to the best type. The use of the rotating cylinder provides metals in a powdered form and by fitting such a device with a cone-shaped lower opening, such metals can be readily recovered. However, the use of high-speed rotating equipment is undesirable if it can possibly be avoided. Sporadic trials continue with the fluidized-bed cell, and these are now taking place in Africa for possible copper recovery. However, the problem of obtaining suitable beads with which to feed the cell as the older ones, now grown in size are withdrawn, remains embarrassing if not actually insoluble. The cell devised by Lopez, using fluidized inert beads such as glass against metal grids, appears the better answer.[74, 75] For ultra-efficient recovery of metals, both Newman[86] and later Hills and Chu have considered use of porous electrodes in which a metal can be slowly cathodically deposited and either later or simultaneously anodically stripped out to give a more concentrated solution.

The behavior of electrodeposition systems at the very lowest concentration ($<$10 ppm) metal is not fully understood. While one would expect the metal-removal process to decelerate as the concentration decreases, it is surprising to find that in many cases it ceases altogether and that no amount of electrolysis will lower the level of concentration further. This has been the subject of comment both by Houghton,[81] who found that if solutions containing less than the limit value of Sb were fed to the cell, redissolution of earlier deposited metal would take place, and Jennings, who shows (confirming earlier studies)[87] that oxygen is deleterious in attempts to reach very low concentrations of Cu or Ag. Such problems require further investigation. The indirect recovery of metals is not only possible, but has been demonstrated at the Electricity Council Laboratories on a large scale. Zinc was used to reduce ferric solutions, the dissolved Zn^{2+} ions being reduced (as deposited metal) at a cathode in a well-stirred tank. Tseung has quoted experiments on the deposition of metals at high current densities which raise an interesting point. Traditionally the conventional wisdom is that metal deposition should not occur at a rate greater than, or even close to, the diffusion-limiting current for a given system. However, Tseung shows that by exceeding this limit a very rough deposit is obtained, and that this in turn sets up a much higher diffusion limit because of the many rugosities which penetrate the diffusion layer.

Yet another mode of metal deposition is in conjunction with some type of solvent-extraction system, and this can be especially attractive for feeds with low levels of metal ion concentration. Japanese patent 73:43,733 describes the extraction of Cu, Ni, or Co at the 5-ppm level with naphthenic or versatic acids. However, cell voltages seem high and efficiencies low. A very detailed examination of solvent extraction coupled with electrolytic metal deposition of copper is the paper by Barthel and Heinisch,[88] and Surkov[91] describes Zn recovery from benzotriazole production wastes. The treatment of Fe pickle liquors is described by Jangg,[92] Kerti,[93] and in a U.S. government report.[94] Recovery of scrap uranium is mentioned,[95] copper pickle treatment is described,[96] as well as various cells for metal recovery.[97] Jorne[98] describes removal of alkali metals from Hg amalgams in a propylene carbonate electrolyte. Though aimed at the Cl_2 industry, the idea might have applicability to effluent treatment. A full-scale Zn recovery plant is understood to be operating at a British Steel factory in Wales.

4.3. Destruction of Anionic and Uncharged Species

Little scientific literature has appeared here although the patents appearing suggest that the importance of this area in the destruction of cyanides, cyanates, and the thio

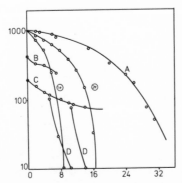

Fig. 7. Anodic destruction of cyanide (as KCN): 80°C, 1 M KOH; current density 10 mA/cm^2; anode area, 10 cm^2; undivided cell, 800 ml electrolyte vol. A, bright Pt anode, pulsed hourly. B, mild steel anode pulsed hourly. C, mild steel anode, not pulsed. D (2 runs), mild steel anode pulsed hourly. Lines 1e and 2e indicate theoretical calculated (100% efficient) destruction for this experiment assuming 1 and 2 electron mechanisms respectively. Axes: horizontal, time in hours; vertical, ppm CN$^-$.

analogs continues. Our understanding of the cyanide system improves. Kuhn[89] has pointed out that, especially at higher temperatures, the nonelectrochemical alkaline hydrolysis rate of cyanide destruction is appreciable and requires a correction which is frequently absent in reported studies of the system. Other authors[89] show how blowing a gas such as air or N$_2$ through the cell causes the loss of HCN from the system, again introducing a nonelectrochemical effect. Then too the cathodic reduction of the CNO species formed at the anode is slowly being realized as a process which affects not only the many experiments carried out in undivided cells but also the efficiency of the industrial cells so constructed. Indeed this is now suggested to be the main reason for the inefficiency of such cells at concentrations below 1000 ppm, although Figure 7 does not support that contention.

The question of pulsing, in the context of CN destruction, has been raised by Tseung, Kuhn, and Marquis. The former used Pt electrodes, the latter a variety of materials. The purpose of the pulse appears to be a reactivation of the electrode very similar to the idea advanced a decade earlier by Bockris for fuel cells. And of course the slow pulsing of electrodes in metal deposition is nothing new. It is thus doubtful that any form of patent protection could be envisaged for such an idea. Figure 7 is a composite constructed from many series of runs and shows how the rate of CN removal varies with a steady current imposed on the one hand and an hourly pulse on the other. Further information on CN oxidation is found in references 112, 113.

4.4. Miscellaneous Studies

Either outside the structure we have previously built, or possibly making use of two or more electrochemical effects simultaneously, we have a plethora of ideas which range from the obvious to the surprising. Anodic oxidation, as Bockris *et al.* have shown, can oxidize most compounds from methane to cellulose. It is hardly surprising then that the application to the general field of pathogens has been considered. Stoner *et al.*[99] report the use of pulsed Pt electrodes, although our understanding of the literature would suggest that unpulsed PbO$_2$ anodes would be equally good, if not better.

The anodic dissolution of silver to inject controlled amounts of the metal into drinking water is also an idea which has been brought to commercial fruition, although some authorities ask why copper is not equally good. Possibly it gives rise to unacceptable color effects. U.S. patent 3,192,146 is an example of this idea. While we have earlier[1] discussed the regeneration of etch baths by removal of Cu and recreation of the acid, many extensions of the idea are now evident, mainly in respect to regeneration of persulfate etches (German patent 1,264,920, U.S. patents 3,072,545 and 3,470,044, British patents 1,141,407 and 1,191,034), while citrate etch baths can be regenerated too (British patent 1,141,758, French patent 1,517,482, and U.S. patent 3,425,920). A stream of patents and other literature emerges with the simple message (not stating how or why) that electrolysis sterilizes. A German patent (2,003,426) describes a small unit to hang in a domestic flush cistern. The regeneration of color photographic bleach solutions such as ferricyanides is discussed,[100] as well as the treatment of acid mine water by passage through a packed-bed reactor.[101] An idea for removal of anions such as sulfate or phosphate is found in U.S. patent 3,766,034. Pellets of Pb or Al or other metals are coated with ion-exchange membrane. Anions which percolate through then form the insoluble sulfate or phosphate on the metal. The controlled electrodissolution of lead to precipitate anions as the lead salt does not sound like a very attractive idea (U.S. patent 3,766,037). Sterilization of water with a pulsed system (but the authors, in admitting that the presence of Cl^- is required, give a clue to the *modus operandi* of this and similar ideas) is described in German patent 2,311,504.

The electrochemical treatment of gases continues to be a subject of interest, although little has been heard of late regarding the large Stone and Webster plant[1,2] in Florida. The reduction of CO_2, using metal phthalocyanins and other electrodes[102] has potential, not only in the context of submarine engineering, but also as a route to the chemical feedstocks or liquid fuels in the future. The more prosaic subject of sulfur-dioxide treatment and its electrochemical oxidation is discussed.[103]

The organolead compounds, which are so much more toxic than the lead itself, can be treated both anodically and cathodically (U.S. patents 3,799,851-3 and 3,696,009).

The last broad area to be mentioned concerns the use of electrochemically formed sodium hypochlorite to treat effluent streams or to prevent contamination or fouling by marine or freshwater growths. The range of applications of this technique, together with a detailed description of commercially available cells and their designs, has been recently published,[107] and we shall not therefore cover this ground again. However, further studies of electrolytic hypochlorite cells are in progress at the University of Salford, partly to study the performance of such cells under reasonably well-defined conditions and partly to measure the corrosion rates of various anode materials for such cells.[114,115] A recent U.S. government report considers the use of such cells for sterilization of sewer overflows.[108]

The fraternity of plant designers operating in the field of effluent treatment is more troubled than most by problems of electrode fouling. Indeed, as we have previously indicated,[1,2,107] this has been at times a limiting factor. Apart from the solution suggested at various times for coating of the electrode with a membrane, current reversal is a frequently employed device. With composite electrodes of the coated types (either with noble metals or lead dioxide), this procedure leads to difficulties. One idea for avoiding total current reversal, and the polarity change which ensues, is to split the cathodes into two groupings and to reverse polarity only partially and not below the hydrogen-evolution potential. According to the patent, the system has been working for a considerable period of time.[109] Another approach[110] is to blow air in periodically, and this is stated to be a satisfactory solution. A description of the PEPCON cell (Pacific Engineering) in which this is shown is found in reference 107.

References

1. J. O'M. Bockris and B. E. Conway, eds., *Modern Aspects of Electrochemistry* Vol. 8, Plenum Press, New York (1972).

2. J. O'M. Bockris, ed., *Electrochemistry of Cleaner Environments*, Plenum Press, New York (1972).
3. A. T. Kuhn (ed.), *Industrial Electrochemical Processes,* Elsevier, Amsterdam (1971).
4. R. W. Houghton and A. T. Kuhn, *J. Appl. Electrochem. 4*, 173 (1974).
5. G. Solt, *Effluent Water Treat. J. 12*, 293 (1972).
6. L. E. Lancy, Interfinish Surface, Proc. 8th Congr. Int. Union for Electrodeposition and Surface Finish, p. 435, Foerster Verlag, Zurich (1972).
7. J. Ciancia, *Plating 60*, 1037 (1973).
8. Environmental Protective Administration Investigation of Treatment of CN Waste by ED, U.S. Gov't. EPA Grant Project 12010 DFS to RAI Corp.
9. Report PB 218-325/9, *U.S. Gov. Res. Dev. Rep. 10*, 65 (1973).
10. Report PB 204-624, *U.S. Gov. Res. Dev. Rep.* Feb. 10th, 51 (1972).
11. Report PB 215-183/5, *U.S. Gov. Res. Dev. Rep. 8*, 60 (1973).
12. Report PB 215-176/9, *U.S. Gov. Res. Dev. Rep. 8*, 60 (1973).
13. Report PB 209-026, *U.S. Gov. Res. Dev. Rep.* July 10th, 55 (1972).
14. Report PB 210-415, *U.S. Gov. Res. Dev. Rep.* Aug. 25th, 66 (1972).
15. US Pat. 3,725,235.
16. D. S. Trivedi, *Ion Exch. Membr. 1*, 37 (1972); *Jpn. Kokai 74*, 1974 (Pat.), 10, 555, *Chem. Abstr. 81*, 4601s.
17. US Pats. 720,186; 894,070.
18. G. V. Schwerin, *Z. Elektrochem. 9*, 739 (1903).
19. J. Reitstoetter, *Elektrophorese, Elektro-osmose und Elektrodialyse*, Steinkopf Verlag, Dresden (1931).
20. Brit. Pat. 351,170.
21. Brit. Pat 662,568.
22. Brit. Pats. 344,547; 383,199, US Pat 1,156,715.
23. US Pats. 1,133,967 and 1,174,946; also S. R. Hind, *Trans. Ceram. Soc. 24*, 73 (1924).
24. Brit. Pat. 1,170,413, US Pat. 3,664,940 and 3,520,793, Brit. Pat. 1,182,019.
25. N71-36511 (STAR 09-23), *U.S. Gov. Res. Dev. Rep.* Feb. 25, 61.
26. G. Sunderland, Lecture delivered to Symposium on Electrochemical Techniques for Environmental Control, Salford University, Capenhurst, Chester, U.K. (1973).
27. Report PB 207-364, *U.S. Gov. Res. Dev. Rep.* May 10, 74 (1972); PB 200-056, *U.S. Gov. Res. Dev. Rep.* Aug 10, 92 (1971); PB 206-354, *U.S. Gov. Res. Dev. Rep.* March 25th, 45 (1972); PB 207-007, *U.S. Gov. Res. Dev. Rep.* Apr. 25th, 79 (1972).
28. J. Newman, *J. Electrochem. Soc. 118*, 510 (1971).
29. J. O'M. Bockris and Z. Nagy, *Electrochemistry for Ecologists*, p. 111, Plenum Press, New York (1974).
30. Report PB 207-019, *U.S. Gov. Res. Dev. Rep.* Apr. 25th, 79 (1972).
31. AD Report 744-142, *U.S. Gov. Res. Dev. Rep.* Aug. 25th, 69 (1972); *Plaste u. Kautschuk* (East Germany) *18* 1, 35–38 (1971); *Ger. Offen.* 2,351,445, *Chem. Abstr. 81*, 29321.
32. Report PB 203-833, *U.S. Gov. Res. Dev. Rep.* Dec. 25th, 51 (1971); PB 212-124, *U.S. Gov. Res. Dev. Rep.* Nov. 25th, 70 (1972).
33. Ger. Offen. Pat. 2,351,445 (*Chem. Abstr. 81*, 29321).
34. Report PB206-352 *USGRDR* March 25, 1972, p. 45.
35. A. T. Kuhn, *Chem. Processing* (London) *20*, pp. 9–12 (June 1974); *20*, pp. 5–7 (July 1974).
36. Reports PB 229-628/3GA and PB 229-629/1GA, *U.S. Gov. Res. Dev. Rep.* May 17th, 64 (1974).
37. Report PB 231-263/5GA, *U.S. Gov. Res. Dev. Rep.* June 28, 60 (1974).
38. Report PB 235-398/5GA, *U.S. Gov. Res. Dev. Rep.* Dec. 27th, 74 (1974).
39. AD 786-398/8GA, *U.S. Gov. Res. Dev. Rep.* Dec. 13th, 69 (1974).
40. Report PB 225-189/OGA, *U.S. Gov. Res. Dev. Rep.* Jan. 25th, 71 (1974).
41. Reports PB 226-825/8GA, PB 226-827/4GA, PB 226-826/6GA, *U.S. Gov. Res. Dev. Rep.* March 22nd, 68 (1974); Report PB 227-280/5GA, *U.S. Gov. Res. Dev. Rep.* (*Apr. 5th*) 59 (1974).
42. US Pat. 3,463,713.
42a. Br. Pat. 1,122,636; Br. Pat. 1,298,814; Br. pat. 1,347,896.
43. US Pat. 3,244,612.
44. Y. Hiroshi; *Kagaku Kogyo 22*(3), 121 (1971); available in translation as AD 782-075/6GA *U.S. Gov. Res. Dev. Rep.* Sept. 20th, 55 (1974).
45. Reports PB 236-613/6GA and 236-614/4GA, *U.S. Gov. Res. Dev. Rep.* Dec. 27th, 75 (1974).
46. AD 787-172/6GA, *U.S. Gov. Res. Dev. Rep.* Dec. 27th, 71 (1974).
47. G. M. Panchenkov. *Khim. Tekhnol. Topl. Masel 17*, 31 (1973).

48. *Jpn. Pat. Kokai, 73*, 14, 573.
49. A. L. Sagradyan, *Tr. Arm. Nauchno-Issled Proektn. Inst. Tsvetn. Metall. 1972*(1), 75 (1972); *Chem. Abstr. 82*, 9304.
50. A. L. Sagradyan, *Tr. Arm. Nauchno-Issled Proektn. Inst. Tsvetn. Metall. 1972*(1), 83–94 (1972); *Chem. Abstr. 82*, 9305.
51. D. C. Lewin and C. F. Forster, *Effluent Water Treat. J. 14*, 142 (1974).
52. B. R. Evans, *Effluent Water Treat. J.* Feb. *14*, 85–88 (1974).
53. H. P. Roth, *Wasser Energiewirtschaft (Baden)* 6 (1974).
54. A. A. Mamakov, *Electron. Obrab Matov 1972*(2), 63 (1972).
55. B. M. Matov, *Electron. Obrab Matov 1972*(5), 48 (1972).
56. A. A. Mamakov, *Electron. Obrab Matov 1973*(5), 66 (1973).
57. B. M. Matov, *Electron. Obrab Matov 1973*(5), 71 (1973).
58. A. M. Romanov, *Electron. Obrab Matov 1973*(1), 54 (1973).
59. B. M. Matov, *Electron. Obrab Matov 1973*(3), 61 (1973).
60. A. A. Mamakov, *Electron. Obrab Matov 1973*(4), 46 (1973).
61. V. F. Malko, *Electron. Obrab Matov 1973*(3), 33 (1973).
62. A. A. Romanov, *Electron. Obrab Matov 1974*(3), 50 (1974).
63. A. A. Romanov, *Electron. Obrab Matov 1974*(2), 57 (1974).
64. V. B. Chebanov, *Electron. Obrab Matov 1972*(6), 81 (1972).
65. V. A. Glembotskii, *Electron. Obrab Matov 1975*(1), 51 (1975).
66. A. A. Mamakov, *Electron. Obrab Matov 1974*(6), 40 (1974).
67. A. A. Mamakov, *Electron. Obrab Matov 1974*(5), 47 (1974).
68. A. A. Mamakov, *Electron. Obrab Matov 1974*(1), 58 (1974).
69. I. S. Lavrov, *Electron. Obrab Matov 1974*(4), 82 (1974).
70. US Pats. 3,766,035-039, 3,793,178, 3,783,114, 3,769,186; Brit. Pats. 1,338,476, 1,347,041.
71. Kh. Mitikov, *Mashinostroene 21* (8), 351 (1972); *Chem. Abstr. 78*, 230 66a.
72. R. Clarke, A. T. Kuhn, and E. Okoh, *Chem. Br. 11*, 59 (1974).
73. R. W. Houghton and A. T. Kuhn, Society for The Advancement of Electrochemical Science and Technology *Centenary Celebration Volume* (1975), Karaikudi, India.
74. C. L. Lopez-Cacicedo, paper presented at Institute of Metal Finishing Conference, London (June 4–6, 1975).
75. C. L. Lopez-Cacicedio, paper presented at AIME Meeting, Manchester (April 5, 1975).
76. V. V. Djordik, *Elektron. Obrab. Mater. 1973*(3), 56 (1973).
77. L. J. J. Janssen and J. G. Hoogland, *Electrochim. Acta 18*, 534 (1973).
78. S. Das Gupta, private communication (1975).
79. LBL-1831/GA June, 1973, *U.S. Gov. Res. Dev. Rep.* Dec. 25th, 89 (1973).
80. B. R. Surfleet, paper presented at Symp. on Electrochemical Methods for Environmental Control, University of Salford (1973).
81. R. W. Houghton and A. T. Kuhn, *J. Appl. Electrochem. 4*, 69 (1974).
82. E. Bishop, paper presented at Symposium on Electrochemistry and Environmental Control, Polytechnic of the South Bank, (April 1975).
83. B. D. Barker and B. A. Plunkett, paper presented at Institute of Metal Finishing Conference, London (June 4–6, 1975).
84. E. I. Onstott and W. S. Gregory, *Environ. Sci. Technol. 7*, 333 (1973).
85. EPA-12010-DPF 11/71 (PB 215-697/4) *USGRDR*, April 25 1973 p. 148.
86. D. N. Bennion and J. Newman, *J. Electrochem. Soc. 120*(3), 109C, Abstr. No 236 (1973).
87. D. Jennings, A. T. Kuhn *et al.*, *Electrochim. Acta* (1975) *20*, 903.
88. G. Barthel and R. F. Heinisch, *Tech. Mitt. Krupp Werksber. 32* 29 (1974).
89. A. T. Kuhn, "Operflaeche" (Zurich) in press.
90. S. G. Mason *et al.*, *J. Colloid. Inteface Sci. 48*, 172 (1974).
91. Yu. Surkov, *Izv. Vyssh. Uchebn. Zaved. Khim. Khim. Tekhnol. 16*, 1700 (1973); *Chem. Abstr. 81*, 15339d.
92. G. Jangg, *Electrodeposition Surf. Treat. 1*, 139 (1972).
93. F. Kerti, private communication.
94. Report PB 201-651, *U.S. Gov. Res. Dev. Rep.* Sept 25th, 70 (1971).
95. US Report MCW-1494, Mallinckrodt Works, (1966).
96. PB 215-697/4 *USGRDR* April 25 1973, p. 148. Brit. Pat. 601; US Pat. 3,804,733.
97. US Pats., 3,718,552, 3,725,266, 3,718,559, 3,803,016, 3,804,739; Brit. Pat. 1,364,167.
98. J. Jorne, *J. Electrochem. Soc. 120*, 255C, (1973) Abstr. 295.
99. G. Stoner *et al.*, *J. Electrochem. Soc. 122*, 106C, Abstr. 360 (1975).

100. *Japn. Kokai* 74;27,058; 27,059 (*Chem. Abstr. 81*, 97734a); AD 772-807/4GA, *USGRDR* Mar. 22, 62 (1974).

101. Report PB 208-820 *U.S. Gov. Res. Dev. Rep.* June 25th, 85 (1972). PB 232-764, *USGRDR* Aug. 23, 54 (1974)

102. *J. Chem. Soc. Chem. Commun. 5*, 159 (1974); AD Report 654,146 (Ionics Inc.) (1967).

103. US Pat. 3,655,547 (Lockheed): Brit. Pat. 1,184,768 (Ionics); UCRL Trans. 10525 *U.S. Gov. Res. Dev. Rep.* Sept. 10th, 91 (1971); Czech. Pat. 153,372 (*Chem. Abstr. 82* 23760); see also T. R. Kozlowski *et al., J. Inorg. Nucl. Chem. 32*, 401 (1970).

104. D. S. Trivedi, *Ion Exch. Membr. 1*, 37 (1972).

105. E. M. Golubchik, *Zh. Prikl. Khim. (Lenin) 45*, 2083 (1972).

106. R. Clarke *et al., J. Appl. Chem.* Biotechnol *26*, 407(1976).

107. A. T. Kuhn, *Processing* (March & April 1975) *21* (3) 6; *21* (4) 10 (Pts I & II); also A. T. Kuhn and R. B. Lartey, *Chem. Ing. Tech. 47*, 129 (1975).

108. Report PB 211-243 (Ionics), *U.S. Gov. Res. Dev. Rep.* Oct. 10th, 46 (1972).

109. French Demande 2,196,201 (St. Gobain), *Chem. Abstr. 81*, 98689.

110. US Pat. 3,394,067 (Pacific Engineering).

111. P. Hogan and P. Wills, *J. Cambome School Mines*, Nov (1976).

112. C. Campbell-Wilson, *J. Cambome School Mines*, 1977 (in press).

113. A. T. Kuhn, *J. Cambome School Mines*, 1977 (in press).

114. R. Lartey, "Corrosion" NACE (in press).

115. R. Lartey and A. T. Kuhn, submitted for publication.

26

Electrochemistry for a Better World

A. J. Appleby

1. Introduction

It would be rash to promise a "better world" in the future, using electrochemistry or not, at a time when we are beset by the simultaneous pressures of an expanding population and dwindling material resources. It should, however, be our task to examine ways by which the world of the future might be prevented from becoming worse. I have chosen to concentrate mainly upon the problems that will dominate our future, namely energy production and use, and the associated areas of energy and raw materials conservation. The impact of electrochemistry on these areas is potentially very great, and in some cases, dominant. Since the lead-time for the development of new technologies to the point of major use in any field is at least 30 years (e.g., nuclear fission), I have decided to restrict the scope of this survey to society in the world of the twenty-first century. Many parts of the world, Europe and Japan included, will then be in the post-fossil-fuel age. The only fossil fuel available in major quantities in the U.S. will be coal, but environmental, capital, and transportation difficulties will prevent its wide use in the energy economy.

Nonfossil energy will be extremely capital-intensive. Costs, calculated using conventional accounting methods, will vary from about $3.60 (on-site nuclear electricity, perhaps thermochemical hydrogen using ultra-high-temperature reactors) to about $6.00 (delivered electricity, solar-produced hydrogen) per GJ (1970 dollars).[1] Mean costs of abundant primary energy (oil) in the U.S. in 1970 were ~40¢/GJ; they are now ~$1.30 ($ 1970), and they represent almost 10% of the GNP. Unless vast economies in the use of energy are effected, costs may be as high as 40% of the GNP by the twenty-first century.[1] This will be an impossible situation, and it seems clear that survival of society will depend on a change from the present transportation-based, throwaway philosophy to a low-key, energy-conserving, energy-production-based structure. It is assumed below that electrochemistry will contribute to the creation of such a society. A broad outline of future energy production, transportation, telecommunications, energy storage and solar conversion, and the future life sciences is given, in terms of an extrapolation of present electrochemical technology. No future major breakthroughs are assumed. Present and postulated future energy consumption for the U.S. by sector is given in references 1–3.

2. The Energy Industries

Society in the twenty-first century will rely on nuclear (fission and ultimately fusion), solar (direct or indirect, the latter including ocean temperature gradients,[4] wind and wave power), and possibly geothermal energy sources. Energy will be in the form of electricity for immediate local use and in the form of synthetic hydrogen for storage, use in transportation and in the chemical industry (see appropriate sections below), and for

A. J. Appleby • Laboratoire d'Electrolyse, C.N.R.S. Bellevue, 92-Bellevue, France.

pipeline transmission to remote locations.[5] The latter aspect of the hydrogen economy will be particularly important in the future, due to the geographic mismatch of population centers and areas of power production (offshore nuclear, the oceans, deserts). As we will see below, the majority of this hydrogen will be produced by electrolysis.

In areas where fossil-fuel energy, in the form of coal, is still available, the fuel cell will find wide application. Such fuel cells will be based on the acid system (initially phosphoric acid, 150–180°C; perhaps eventually using higher temperatures and electrolytes based on the molten half-salts of polymeric fluorosulfonic acids) and on the molten carbonate system. The latter will become increasingly important for stationary fuel cell applications, since lower-temperature systems are incapable of directly using CO (derived from partial oxidation of coal) as fuel. It seems certain that the technological and cost problems associated with the molten carbonate fuel cell will be much less important than those for the development of high-temperature solid oxide (doped ZrO_2) systems. The latter will consequently find little application.

Production of hydrogen from water requires a free energy and a heat input which varies according to the path used. If the direct decomposition of water (in particular by electrolysis) is the path chosen, free energy input in the form of electrical work is considerable (about 56.7 kcal/mole at 25°, for liquid water). This work input alone is not sufficient to dissociate water, as a further heat input to bridge the gap between the above free-energy value and the heat of dissociation (about 67.8 kcal/mole for liquid water at 25°C) is necessary. In practice, it is well known that this difference is more than provided by an irreversible thermodynamic affinity, representing an additional work term, that must be added to the reversible free energy value. In electrolysis, this affinity is represented by the combined losses due to ion and electron transport (ohmic drop), molecular transport (mass transport concentration gradients), and finite kinetic rates (overpotential). A reasonable research goal is to reduce these affinity terms so the total work required corresponds to the heat of dissociation of water, with the possibility of reducing the work requirement still further, provided that a cheap source of heat at the electrolysis temperature is available. The real heat requirements for the process must be taken into account, allowing for the PdV term at the proposed gas production pressure. The latter factor is of particular importance since physical compression is much less efficient than direct formation by electrolysis at the pressure required for hydrogen storage. In the future, this will be on the order of 1000 psi (68 bar).[5] A plot of the ΔG and ΔH terms for water vapor between ambient temperatures and 1750 K, and for liquid water up to 450 K, is shown in Figure 1. $\Delta G°$ and $\Delta H°$ terms are shown in solid lines; the dashed lines are the corresponding values calculated for mean gas mixtures in practical electrolyzers, supposing that H_2 and O_2 are produced at a pressure of 100 bar. It can be seen that for low-temperature electrolyzers, using liquid water feedstock, thermoneutral operation will occur at about 1.55–1.6 V, whereas for high-temperature systems (1200 K) operating on water vapor, the corresponding value is about 1.4–1.45 V. In addition, high-temperature electrolyzers will require a heat source at about 300°C for formation and compression of water vapor. Enough heat will be recoverable from the hydrogen and oxygen produced to heat the compressed water vapor to the working temperature.

Some present electrolyzers, or electrolyzer concepts, are shown in Figure 2.[6–10] Current pressure electrolyzers (30 bar), shown in Figure 2A, use low-surface-area woven electrodes, thick asbestos separators (5 mm) and operate in 25% KOH at 80°C at 200 mA/cm^2 and 1.9 V. The electrodes are "activated" (surface-catalyzed) nickel. An advanced concept,[7, 8] which should be capable of economic operation at 400 mA/cm^2 at the thermoneutral potential at 70 bar (1.57 V) and 130°C, would incorporate a thin,

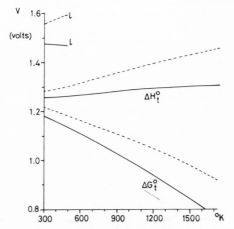

Fig. 1. Thermodynamics of water electrolysis as a function of temperature. Solid lines: ΔH_t^0, ΔG_t^0 values; dashed lines: ΔH_t^0, $\Delta G_t^0 + PdV$ calculated at gas pressures of 70 bar for practical gas mixtures. Dotted line: Open circuit potential for gas mixtures experimentally used in doped zirconia electrolyzers operating at 1.2 V, 2 A/cm^2 (80% H_2O, 20% H_2). Lines marked l: ΔH_t^0 and $\Delta H_t^0 + PdV$ (H_2, O_2 at 70 bar) for liquid water.

temperature- and KOH-resistant separator (perhaps of titanate material) to reduce IR drop, and will have efficient high-surface-area electrodes with nonnoble catalysts. A third concept, also working at 130–150°C, but using a stable fluorinated sulfonic acid ion-exchange membrane, is shown in Figure 2C. Acid-resistant materials and noble-metal anode catalysts, of high capital cost, require operation at 2 A/cm^2 for economical use; 1.6 V operation is projected.[9] A major breakthrough would be the development of a stable *anionic* membrane, which would allow cheap hardware and catalysts and permit the use of pressurized hydrogen and ambient air in a fuel-cell mode. The final type of

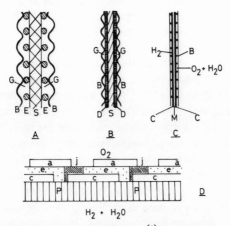

Fig. 2. Electrolyzers. (A) Zdansky–Lonza pressure type:[6] B,B, bipolar plates; E,E, gauze electrode; G,G, gas electrolyte space; S, asbestos separator (1.9 V at 200 mA/cm^2, 80°C). (B) Advanced version: D,D, high-surface catalyzed electrodes,[7,8] S, high-temperature separator. Projected performance: 1.6 V at 400 mA/cm^2, 130°C. (C) Membrane type:[9] M, anionic (fluorosulfonic acid) membrane; B, bipolar plate; C,C, catalyst layers. Projected performance: 1.6 V at 2 A/cm^2, 150°C. (D) ZrO$_2$–Y$_2$O$_3$ solid-state system:[10] P,P, porous support; C,C, cathodes (Ni); e, solid electrolyte; j,j, oxide intercell junction, a,a, oxide (In$_2$O$_3$) anodes. Projected performance:[11] 0.35 V polarization at 2 A/cm^2, 900°C.

electrolyzer shown is the solid oxide type (doped ZrO_2, 900°C), a proposed form[10] of which is illustrated in Figure 2D.

The present systems suffer from thermal cycling and materials (oxygen electrode, intercell junction) problems. Manufacture will be difficult, since each cell unit will have a maximum length of about 8 mm to avoid excessive IR drop. High current densities will therefore be required for economic operation. Current experimental units, using dilute mixtures of hydrogen in water vapor as cathode gas (O C potential 0.85 V at 1200 K, see Figure 1), work at about 1.2 V and 2 A/cm²,[11] indicating a total polarization of about 350 mV at this current density. This polarization is mostly IR drop in the oxygen electrodes (In_2O_3 and similar oxides) and the intercell junctions and is likely to be worse (on a long-term basis) in practical systems. The latter will therefore work around the thermoneutral potential for the mixture of compressed gases (1.4 V), which will in any case be necessary unless cheap heat is available at the electrolyzer at 900°C, implying reactor core temperatures >1000°C. Such reactors will not be available until at least 2025,[12] and when they are available the most advantageous route for hydrogen manufacture may be thermochemical (see below).

Low-temperature electrolysis will be capable of supplying hydrogen at an efficiency of about 29% (based on high heating value of hydrogen at ambient pressure and taking into account electrical generating efficiency and conversion losses) using conventional nuclear heat sources (e.g., the PWR or BWR), or 38% (using HTR heat sources). For the latter, electrical generation efficiency has been taken as 42%, and in all cases electrolyzers are presumed to work under thermoneutral conditions at the required hydrogen pressure. Since the high-temperature electrolyzer will be only at most 1–2% more efficient when supplied with electricity from the above reactors, taking into account heat from the steam loop required to produce superheated water vapor, it is unlikely to find much practical use in future.

An alternative suggested method for hydrogen manufacture is via thermochemical cycles.[13–18] The approach used is to find a sequence of reactions whose individual ΔG values are as close as possible to zero, and which use heat at different temperatures, as in a heat engine. Generally 3–5 steps are required; alternatively, the number of steps may be reduced if electrochemically driven reactions[16] (negative ΔG) are incorporated. A list of processes, taken from published work, is given in Table 1.[1–9,18] Process 2 is particularly adapted to combined thermochemical–electrochemical (for the 125°C step) operation. Process efficiencies (based on hydrogen high heating value), calculated using practical heat-transfer and recuperation data,[14,17] but neglecting reaction and separation affinities due to finite kinetic rates,[19] are given in Table 2. When these affinities are taken into account, efficiencies drop dramatically, especially for multistep processes (Table 2). Unless nuclear heat will be available at very high temperatures (>1250°C), which are difficult to use advantageously in electricity generation, electrolysis is likely to be more efficient and cheaper* than direct thermochemical cycles. Low-temperature electrolysis techniques will dominate in hydrogen production from solar, wind, and ocean energy sources, and they will be advantageously used in combined thermochemical–electrochemical (2-step) cycles if and when very-high-temperature nuclear reactors become available.

Finally, before leaving the subject of energy production, continuous electrochemical extraction methods may render the molten-salt breeder reactor feasible.

---

*Heavy chemical plant costs and materials problems, together with the inventory of chemicals and their turnover,[20] must be taken into account in assessing estimated costs. By comparison, electrochemistry is a known technology.

Table 1. Thermochemical Cycles[18]

	Cycle	Temp. (°C)
1. De Beni	$C + H_2O \longrightarrow CO + H_2$	725
	$CO + 2Fe_3O_4 \longrightarrow C + 3Fe_2O_3$	225
	$3Fe_2O_3 \longrightarrow 2Fe_3O_4 + \frac{1}{2}O_2$	1425
2. IGT	$Cd + H_2O \longrightarrow CdO + H_2$	125
	$CdO \longrightarrow Cd + \frac{1}{2}O_2$	1225
3. IGT	$Fe_3O_4 + 2H_2O + 3SO_2 \longrightarrow 3Fe(SO_4)_3 + 2H_2$	125
	$3FeSO_4 \longrightarrow 3/2Fe_2O_3 + 3/2SO_2 + 3/2SO_3$	725
	$3/2Fe_2O_3 + \frac{1}{2}SO_2 \longrightarrow Fe_3O_4 + \frac{1}{2}SO_3$	925
	$2SO_3 \longrightarrow 2SO_2 + O_2$	925
4. IGT	$2CrCl_2 + 2HCl \longrightarrow 2CrCl_3 + H_2$	325
	$2CrCl_3 \longrightarrow 2CrCl_2 + Cl_2$	875
	$H_2O + Cl_2 \longrightarrow 2HCl + \frac{1}{2}O_2$	850
5. Euratom, No. 7	$3FeCl_2 + 4H_2O \longrightarrow Fe_3O_4 + 6HCl + H_2$	600
	$Fe_3O_4 + \frac{1}{4}O_2 \longrightarrow 3/2Fe_2O_3$	400
	$3/2Fe_2O_3 + 9HCl \longrightarrow 3FeCl_3 + 9/2H_2O$	100
	$3FeCl_3 \longrightarrow 3FeCl_2 + 3/2Cl_2$	400
	$3/2H_2O + 3/2Cl_2 \longrightarrow 3HCl + 3/4O_2$	800
6. Euratom, No. 1	$Hg + 2HBr \longrightarrow HgBr_2 + H_2$	250
	$HgBr_2 + Ca(OH)_2 \longrightarrow CaBr_2 + HgO + H_2O$	200
	$CaBr_2 + 2H_2O \longrightarrow Ca(OH)_2 + 2HBr$	725
	$HgO \longrightarrow Hg + \frac{1}{2}O_2$	600
7. Euratom, No. 9	$3FeCl_2 + 4H_2O \longrightarrow Fe_3O_4 + 6HCl + H_2$	650
	$Fe_3O_4 + 3/2Cl_2 + 6HCl \longrightarrow 3FeCl_3 + 3H_2O + \frac{1}{2}O_2$	120
	$3FeCl_3 \longrightarrow 3FeCl_2 + 3/2Cl_2$	420
8. G. E. "Beulah"	$2Cu + 2HCl \longrightarrow 2CuCl + H_2$	100
	$4CuCl \longrightarrow 2CuCl_2 + 2Cu$	100
	$2CuCl_2 \longrightarrow 2CuCl + Cl_2$	600
	$Cl_2 + Mg(OH)_2 \longrightarrow MgCl_2 + H_2O + \frac{1}{2}O_2$	80
	$MgCl_2 + 2H_2O \longrightarrow Mg(OH)_2 + 2HCl$	350
9. G. E. "Agnes"	$3FeCl_2 + 4H_2O \longrightarrow Fe_3O_4 + 6HCl + H_2$	550
	$Fe_3O_4 + 8HCl \longrightarrow FeCl_2 + 2FeCl_3 + 4H_2O$	110
	$2FeCl_3 \longrightarrow 2FeCl_2 + Cl_2$	300
	$Cl_2 + Mg(OH)_2 \longrightarrow MgCl_2 + H_2O + \frac{1}{2}O_2$	80
	$MgCl_2 + 2H_2O \longrightarrow Mg(OH)_2 + 2HCl$	350

Table 2. Comparison of High-Heating-Value Efficiencies of Thermochemical Cycles
(Table 1)

Thermochemical cycle	Higher heat input temp. (°C)	Eff., reversible (%)	Eff., real (%)
1	1425	69	49
2	1225	66	55
3	925	57	37
4	875	46	34
5	800	36	28
6	725	39	30
7	650	41	34
8	600	27	23
9	550	29	24

3. Transportation

The transportation sector will be the largest consumer of primary energy in the postulated future economy, using about 40% of total high-grade energy (compared with a present value of about 25% in the U.S.). Normalized consumption (i.e., per GNP unit) for transport will be ~60% of that today. The increased relative importance of transportation in the high-grade energy economy compared with today results from the fact that energy in this sector is required in the form of work, whereas in other areas much of present energy requirements can be met by the use of waste or other low-grade (e.g., solar) heat. The saving of about 40% of normalized energy will be achieved by replacing all present intercity traffic by rail (or bus) up to a distance of 400 miles, and by the use of more efficient power sources in urban vehicles. The numbers of such vehicles will be minimized by reducing the necessity for commuting (more rational distribution of employment location, replacement of commuting by telecommunications).

The gasoline-powered vehicle as currently used has a thermal efficiency of under 10% in urban traffic.[21,22] Based on nonfossil fuel derived from primary nuclear energy or solar power (electricity or equivalent chemical free energy), an even lower overall efficiency will be expected. Since such artificial fuels will have untaxed constant dollar costs perhaps three times higher than those for present-day oil-based fuels ($10/barrel,* $1.60/GJ), economic pressure alone will force a change to higher-efficiency (and quieter, nonpolluting) energy conversion for the urban environment. In addition to saving direct costs to vehicle owners, such a change will result in considerable saving in investment in primary energy sources, thus avoiding increased strain on the future GNP and on the environment.

For short-distance urban or rural travel, the battery- or fuel-cell-powered vehicle will predominate. Battery-driven vehicles may be expected to have efficiencies in the 40–75% range, depending on the system used, and considering electricity as primary energy. Hydrogen fuel-cell vehicles may be expected to have an ultimate realistic efficiency of 45–50% (based on the electricity required to produce hydrogen, or on the high heating value of hydrogen). However, if hydrogen piped over long distances is the source of a community's primary energy, battery vehicles will be at a disadvantage due to the on-site conversion efficiency of hydrogen to electricity (60% in a fuel cell, based on the free energy of hydrogen). Logistic considerations will therefore largely determine the choice of energy system with vehicle mission and user preference (refuelability vs. rechargeability) being further factors.

*Current dollars.

Fuel cells may find some place in heavy intercity road vehicles in circumstances where rail is impractical. Rail systems will in all probability return to their nineteenth-century importance. All major lines will be electrified, and branch lines serving smaller communities will ultimately use battery-driven railcars recharged on the main-line system.[23] A similar battery-overhead-electric hybrid system, using trolley-coaches,[24] is attractive for mass transport in larger cities.

Electrochemical systems will therefore account for the major part of power sources for terrestrial transportation in the future economy. An additional, although related, area in which electrochemistry will be of prime future importance is that of telecommunications, which will replace marginally necessary people movement in the future economy.[25] Telecommunications satellite links, which will largely replace long-distance business travel, will require long-life electrochemical energy-storage systems. Similarly, telephone, videophone, and computer links will require electrochemical back-up systems of longer lifetime and greater reliability than those currently available.

3.1. Energy Sources for Transportation

The types of batteries (or fuel cells, or battery–fuel cell hybrids) that will be used in the future transportation sector will depend on the following essential factors:

1. Capital cost (per kW or per kWh, whichever is the limiting factor characteristic of the system) per unit of operating lifetime.
2. Efficiency
3. Energy and power density
4. Convenience

The first two factors serve to determine overall transportation cost, while the second two factors determine the user acceptance (or practicality) of the system. In general, the user will require a sufficient power output to give an acceptable performance, together with a maximum range that is greater than an average day's driving, so that flexibility in recharging or refueling is allowed for. The latter is particularly important if immobilization of the vehicle for several hours for recharging is necessary, as will be the case for battery systems. In addition, deep cycling of batteries will be generally avoided to conserve system lifetime. Overall emphasis will be on systems of low cost and high energy density, with, if possible, high efficiency. Refuelable (fuel-cell) systems will have a high convenience factor—more so than exchangeable charged batteries, which will incur insignificantly higher overall capital costs.

3.2. Batteries

Most of the possible permutations of low-cost anodic and cathodic active materials and electrolytes have by this time been examined. It is unreasonable to expect major breakthroughs in certain systems which have excellent energy density but are unpromising in the areas of power density and chargeability. This is particularly true for organoelectrolyte lithium batteries, which are improbable for the vehicle power-source application.* If we first consider aqueous secondary battery systems, present indications are that only PbO_2 (acid) and $NiOOH$ (base) positives are satisfactory from the viewpoint of cost and chargeability. MnO_2 (alkaline) systems show poor recharge characteristics.[26] High-performance, ambient-temperature, secondary oxygen electrodes currently require different substrates for evolution and reduction of oxygen in alkaline solution, unless noble metals

*Added in proof: See, however, Ref. 52.

are used. Silver-based (alkaline) systems may be excluded on cost grounds. Halogen-based systems (in particular chlorine) may show promise, provided that materials and self-discharge problems can be overcome. Other systems may be excluded on the grounds of poor rechargeability and/or cost. Among the possible negative electrodes involving dissolving metal couples, only Pb (acid), Fe (alkali), and Zn (alkali or buffer) systems seem to be viable from the cost and chargeability viewpoint. Reversible hydrogen electrodes (using hydrogen at ambient pressure) are marginally possible, but they will require major improvements in catalysis and in the storage of hydrogen. High-pressure hydrogen systems are promising, however, provided certain cost and construction difficulties can be solved. Dissolved redox couples have energy densities that are too low for the transportation application.

A list of candidate low-temperature systems, indicating their relative strong points and problems, is given in Table 3. Only systems with an energy density at the 1-h rate greater than 65 Wh/kg need be considered for the traction application. O_2/Fe can be immediately dismissed on efficiency grounds. Zinc-alkaline systems at present suffer from the poor cycle life of the zinc electrode due to the high solubility of its reaction products, but rapid progress is currently being made to remedy this problem.[27] The Zn-compressed O_2 system[28] has some attractive features, but the cost of the O_2 electrode catalyst and the pressure vessel probably exclude it from consideration. The Zn-ambient air system again has the fundamental problem of a low-cost, high-performance, secondary oxygen electrode and in addition suffers from certain engineering difficulties (electrolyte circulation, CO_2 removal). It is best considered from the viewpoint of a metal fuel cell (see below). Ni-compressed H_2 batteries have a number of interesting features including excellent overcharge and inversion protection and very high charge acceptance.[29] They may find use as hybrid peaking batteries, incorporating regenerative braking, for a compressed hydrogen fuel-cell-powered vehicle, rather than as a primary power source. In such an application, the cost disadvantage resulting from the pressure vessel is eliminated. The Zn-Cl$_2$ system, using refrigerated hydrate chlorine storage,[30] is at present something of an unknown. If the materials and cycling problems associated with it can be

Table 3. Ambient-Temperature Secondary Cells for Traction

Couple	Relative cost	Wh/kg (1-h rate)	Peaking capability[a]	Cycle life	Efficiency	Problems
Pb/PbO$_2$	1	30	+	~500	~60	Poor energy density
Ni/Zn	3	65	++	500+[b]		Zn cycle life
Ni/Fe	3	40	+	1000+	~50	Poor energy density
Ni/H$_2$	>3	65	++	2000+	~65	Cost (H$_2$ electrode, pressure vessel)
O$_2$/Zn (a)	high	120	++	500+[b]	50	Zn cycle life; cost (O$_2$ electrode, pressure vessel)
O$_2$/Zn (b)	?1–2[b]	100	–	500+[b]	~40	Zn cycle life, O$_2$ electrode, efficiency, carbonation
O$_2$/Fe	?1–2	70	–	1000+	<30	Very poor efficiency, O$_2$ electrode, carbonation
Cl$_2$/Zn	?	?110	++	?	65	Materials

[a] ++ = Very good; + = good; – = poor.
[b] Future.

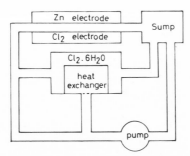

Fig. 3. Schematic view of $Zn\text{-}Cl \cdot 6H_2O$ battery.[30]

solved, it will be a very interesting rival to high-temperature systems (see below) in applications involving large power sources. The system is illustrated schematically in Figure 3.

High-temperature batteries are still in an early stage of development. Problems to be faced are immobilization of the constituents, structural changes on cycling, self-discharge, and corrosion. Systems involving all-liquid-active materials should theoretically show infinite cycle life, but containment problems are considerable. Two such systems (sodium–sulfur and lithium–sulfur) are shown in Figure 4. In the lithium–sulfur system,[31] containment is achieved by porous current collectors which adsorb the reactants and by the use of a porous separator impregnated with molten salt (halide eutectic, e.g., LiCl–KCl). While very high energy densities (>200 Wh/kg) and power densities are possible, realization of a practical system is far off since very-high-capacity electrodes (1 Ah/cm^2), involving large volume changes on cycling, are needed to exploit the possibilities of the system. Other problems are self-discharge and corrosion at the high operating temperature used (>400°C). The sodium–sulfur[23,32] system solves the containment problem by the use of a solid Na ion-conducting membrane of sodium polyaluminate (so-called β-alumina). This allows high Ah capacities without the difficulty of compensating for volume change. Since β-alumina with a suitably high conductivity and long lifetime (over 5000 cycles, 20,000 h continuous cycling) are now available on a laboratory scale,[33] the sodium–sulfur system must be considered the present favorite in the field of high-temperature batteries. Practical systems, which will be of low cost (material costs $2/kWh, compared with

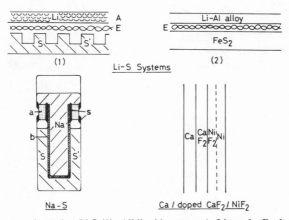

Fig. 4. High-temperature batteries. Li-S (1): All liquid system; A, Li anode; E, electrolyte + separator; S,S', comb-type S + carbon cathode (400°C). Li-S (2): Solid–liquid–solid system; E, electrolyte + separator (400°C). Na-S (concept): a, α-alumina; b, β-alumina; S, seals; S, S', S + carbon cathode (300–350°C). CaF_2: all solid type.

about \$10 for Pb-PbO$_2$) and high performance (200 Wh/kg at the C/10 rate, 150 Wh/kg at the C rate), are currently being developed. They seem to be assured of wide application in hybrid (railcar, trolley coach) and heavy-duty traction applications.

Alternative battery concepts involve use of solid active materials, either with liquid or solid electrolyte. Both types will necessarily require thin cells with high electrode area, since capacity in Ah/cm^2 will be low. Two concepts are shown in Figure 4. The Li-Al alloy, chloride eutectic, FeS$_2$ system[34] is an extrapolation of the all-liquid Li-S system described above. If electrode deterioration on cycling is within acceptable limits, it may be a good contender for transportation applications. It should be capable of 150 Wh/kg. Other possible concepts that may eventually offer similar energy density use chlorine storage on high-surface-area carbon, using either Al[35] or Na-β-alumina[36] cathodes, and chloraluminate electrolytes. Such an approach to a secondary halogen electrogen is necessary since high-rate gaseous diffusion electrodes for halogens in molten salts are probably not feasible owing to materials (attack and wetting) problems. All-solid concepts (e.g., Ca-doped CaF$_2$-NiF$_2$,[37] shown schematically in Figure 4) are intriguing, but they will require much further research into ionic conduction in the solid state, preparation of thin films of doped ionic solids, and prevention of irreversible changes on cycling, before practical systems can be constructed.

This brief review of the extrapolation of present technology shows therefore that the Ni-Zn system is the most probable ambient-temperature secondary battery for future use in smaller, short-range vehicles (electric scooters, utility cars), with the Zn-Cl$_2$ hydrate system a possible candidate for heavier applications, if its cycling ability can be proven. The sodium-β-alumina-sulfur battery (300-350°C), which will offer a 30-40% higher energy density than the ambient-temperature Zn-chlorine hydrate system, will probably be the most widely used electrically rechargeable energy source for heavy-duty terrestrial transportation applications in the foreseeable future.

3.3. Fuel Cells

Fuel cells, either of hydrogen-air or metal-air type, are very attractive from the viewpoint of convenience, in that the system can be rapidly replenished by refilling. Introduction of a successful system within the next few years will probably result in wide acceptance of the electric urban car. To give good performance ($\sim$160 km between charges; acceleration corresponding to the Stockholm cycle; cruising speed $\sim$80 km/hr; maximum speed 105 km/hr) about 100 Wh/kg (3-h rate), 80 W/kg (maximum power), is required if system weight is not to exceed 25% of overall vehicle weight. The only ambient-temperature secondary battery that may meet these requirements is Zn-Cl$_2$. Characteristics of some fuel cells are given in Table 4. While the expected energy densities of Li- and Al-based systems are high, their efficiencies* remove them from consideration. In addition, since proposed cells have replaceable metal electrodes of flat plate type, refueling difficulties may be considerable. The most attractive system is the tubular electrode Zn power-KOH slurry fuel cell, whose reaction product is a suspension of colloidal zincate in KOH.[21] Such a system will permit recharging from a built-in small electrolysis unit, or refueling with reprocessed slurry. Fuel-cell lifetime and capital cost will correspond to vehicle requirements. A schematic view of the system is given in Figure 5.

Fuel cells using synthetic fuels (methanol, methane) directly are not considered here, as their engineering and catalytic problems are orders of magnitude greater than those for hydrogen cells. Based on primary energy (e.g., nuclear electricity), their overall efficien-

*See footnote[a] to Table 4.

Table 4. Vehicle Fuel Cells (80 W/kg)

Fuel	Wh/kg	Efficiency (%)
Al	200	25[a]
Li[b]	270	?25
Zn	110+	~40
H_2	see text	60

[a] Present value. See however data reported by Despic (Chapter 1, this volume).
[b] Extrapolated data based on Lockheed[63] Li-H_2O system.

cies are likely to be very poor. The latter is also true for hydrazine cells (10% overall efficiency). While fuel cells burning synthetic gases derived from coal will be important in stretching out fossil-fuel reserves (where they exist), it is unlikely that they will be used for transportation: because reforming and heat recovery are required for high efficiency, economics and water provision favor large units. Pure hydrogen fuel cells will be adequate for vehicle propulsion, provided that an efficient storage system can be devised. Liquid hydrogen storage makes little sense, since liquefaction represents an irreversible loss of about 50% of the energy content of the gas. Reversible hydrides (approx. 80–100 kg/kg H_2, including low-pressure container)[38] are possible, but they will probably be costly and will require slow recharging, unless a high-pressure container is used. In the latter case, there is no advantage over a pressure storage system. Storage in a tubular system (possibly electroformed) constituting the car chassis, at the proposed hydrogen cross-country pipeline pressure [1000 psi, 68 bar[5]] seems more logical. A vehicle concept is shown in Figure 6. It has a chassis of high-strength tubes (yield stress 45 hectobar, specific gravity 8.5), 10 cm diameter, 2 mm wall thickness, storing 1.6 kg of hydrogen at 68 bar with a safety factor of 2.5. Overall weight of the car chassis is 160 kg. The tubes contain approx. 20 kg (15+ kW) of Ni-H_2 batteries for peaking and regenerative braking. The main fuel cell will be a 10-kW unit, weighing 125 kg. Nonnoble catalysts will be used. Range (90% depth of discharge) will be 200 km, and maximum cruising speed 85 km/h. System energy density is 207 Wh/kg (H_2 storage not included) and 110 Wh/kg (H_2 storage included). The Ni-H_2 battery is shown in Figure 7. This type of hybrid hydrogen fuel cell will replace the zinc slurry cell in a full hydrogen economy.

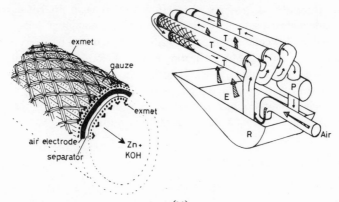

Fig. 5. Schematic drawing of Zn powder fuel cell.[21] T,T, electrode tubes (shown in detail at left); R, reservoir for Zn + KOH electrolyte (E).

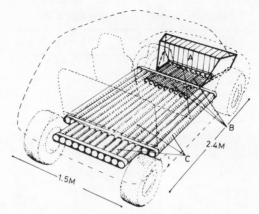

Fig. 6. H$_2$-fuel-cell car concept. A: 10 Kw fuel cell; B: Ni–H$_2$ peaking batteries; C: chassis tubes (diam = 10 cm, wall thickness 2 mm) storing 1.6 kg of H$_2$ at 68 bar (1000 Psi).

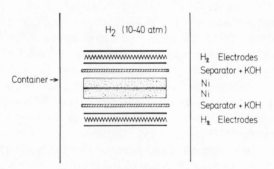

Fig. 7. Ni–H$_2$ battery (schematic).

4. Telecommunications—An Alternative to Transportation

The necessity for routine business travel (day-to-day commuting, business trips) will be substantially reduced in the future by the increased use of telecommunications systems.[25] A majority of people at the present time work at jobs that do not require a physically concerted effort. These people could work at home (or in neighborhood centers) just as efficiently (or perhaps more efficiently) than in present central locations provided that adequate telecommunications (videophones, computer linkages, copying services) are available. A second advantage of such a system would be an economy of investment in workspace, and in reduced heating energy. Electrochemistry will have a major part to play in the functioning of such systems. Since the application will require energy storage systems with very long and trouble-free lifetimes (10+ years), an ideal battery candidate for telecommunications use (either terrestrially or in satellites) is the nickel–hydrogen system. Na-S-β alumina may be used for standby power.

5. Energy Storage: Solar Energy

The primary problem of the future society will be that of matching energy production and demand. If energy production is concentrated in remote locations, the most favorable energy vector and storage medium will be piped hydrogen. Production of hydrogen will be essentially electrolytic, as discussed above, unless very-high-temperature (>1200°C) fission reactors become economically realistic.

Energy transport over distances of less than about 400 miles will be best carried out using electricity.[5] For efficient use of the primary production plant, which in most cases will be a fission (eventually fusion) reactor, off-peak energy storage will be necessary. Electrochemical storage systems are free from the geographical and geological restrictions imposed by pumped air or water systems, and can additionally be placed in the grid so that investment in equipment is optimized. Such electrochemical storage systems will have to be competitive on a cost basis with current storage methods (with correction for credits resulting from grid investment savings, since electrochemical storage is independent of locational considerations). Break-even investment costs on the order of $10–$20/kWh have been suggested for electrochemical storage systems,[35] depending on efficiency (minimum required ~60%), battery cycle life (minimum ~5 years), and saving in investment costs in the grid structure.

Some battery systems, including the high-temperature concepts given earlier (in particular, sodium–sulfur) and perhaps ambient-temperature systems such as Zn–chlorine hydrate, may reach these cost and performance goals. Indeed, the problem of energy storage in a nuclear economy will be so crucial that it will be necessary for electrochemists to meet the required goals to render the whole concept feasible on a large scale. One can anticipate that electrochemical energy storage systems, in particular those using the Na–S couple, will account for 10% of the future power capacity of electricity grid networks supplied by nuclear fission reactors.

Local generation and storage of power will have an important place in the future, particularly in rural areas and in developing countries. In such cases, the source of energy will be solar radiation or wind work. Since such energy is very time dependent, storage is a necessary adjunct to these systems. Unless very large storage units are envisaged, ambient-temperature batteries will be the most logical choice (e.g., for single households, farms, etc.). Low-temperature batteries for traction applications (e.g., Ni–Zn, see above) may be suitable for household energy storage, but the liberation from the system weight and volume constraint imposed by the requirements of transportation energy storage give a further dimension to the possibilities of electrochemical systems. Low-cost dissolved redox couples (probably using ion-exchange membranes to separate anodic and cathodic constituents) become a possibility.[39] Such systems offer very low energy densities (maximum perhaps 30 Wh/kg) owing to the volume of water required for solution of the redox couple, but in compensation they have the potential advantage of heat storage. Such redox systems may be used most advantageously as a domestic energy system in an adaptation of the electrochemical solar cell described by Gerischer.[40] Such a system, which avoids the costly solid-state p–n junction of conventional solar cells, is shown diagrammatically in Figure 8. An n-type semiconductor surface will be capable of injecting holes into an electrochemical redox system on illumination with light of photon energy greater than the band gap. The electrons liberated at the semiconductor surface diffuse along the conduction band in the opposite direction, and are collected at a transparent counter electrode, reducing the redox species produced by the holes at the semiconductor electrode. A p-type semiconductor will work in the opposite direction, reducing the redox system at the semiconductor surface by electron injection. If the electrolyte containing the redox system is circulated, a very high overall energy efficiency may be expected, since about 10% of the incident light can be collected as electrical energy and 90% as heat. The latter may be used as a heat source for a heat-pump in winter, and in an absorption air-conditioning system in summer.

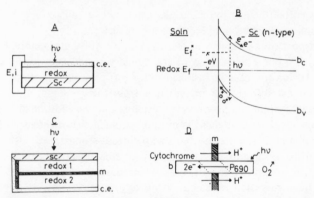

Fig. 8. Electrochemical solar-cell concepts. A: photovoltaic system; B: principle, C: storage cell; D: biological storage cell analog.

The electrochemical photocell of such a total energy system can be used to charge either conventional batteries or a battery based on the redox system used for ion transfer. An alternative and extremely intriguing possibility is a solar–chemical energy storage system, an analog of biological photosynthesis, as shown in Figure 8. As illustrated, the system is assumed to use an extremely thin n- (or p-) type semiconductor, so that irradiation can be carried out from the back side. The semiconductor is shorted to a counter electrode, the ionic path being maintained by an ion-exchange membrane, so that the reduced and oxidized species of the redox couple remain separated for use in an electrical energy storage system. Any such system will require research into suitable semiconducting materials that are stable under the conditions proposed. Band-gap energy must correspond to that for the visible region ($\sim$1.5 eV); previous experiments on electrochemical photosynthesis have used semiconductors whose bandgaps are in the ultraviolet (TiO_2[41]). A possible approach is via organic semiconductors of structures similar to those found in nature. Figure 8 gives the biological analog of such a system—it shows the photochemical light system II,[42] in which light adsorption at P690 centers (gap energy 690 nm) injects holes into water, producing oxygen. The corresponding electrons are donated via a semiconducting molecule to cytochrome b redox couples, the circuit being completed by proton transfer through molecular groups that represent the analog of an ion-exchange membrane. The reduced cytochrome b serves to produce adenosine triphosphate, and ultimately reduced CO_2 (sugars), after further photochemical energy storage via a further process [light system I[43]].

6. Industrial Processes

Industry in the U.S. consumed 41% of total primary energy in 1970.[1,2] Energy savings in this sector will be effected in the future in two ways:

1. Use of low-grade waste heat (from nuclear plants) and solar heat, which may save as much as 35% of total present industrial energy requirements.
2. Manufacture of products that are longer lasting and less energy-intense in fabrication.

Electrochemistry can be of great help in the latter sector. Many of today's manufactured products have a limited lifetime because of corrosion, which is essentially an electrochemical phenomenon. At the present time, corrosion represents perhaps 1% of total energy consumption, expressed as energy for recycling and replacement. Emphasis in the future will be on the use of much longer-lasting products, which will not only directly save energy, but will reduce the rate at which nonrenewable natural resources are consumed. The latter factor will be the ultimate limiting process for the development of mankind.[44]

An analysis of energy consumption in the industrial sector[1] shows that in the future rearranged economy devoted to energy production and replacement as a primary industry (rather than to transportation, as is currently the case), primary energy (e.g., nuclear electricity, hydrogen) requirements per GNP unit will be only 34% of those of today. Built-in obsolescence will have disappeared, thus saving precious mineral resources. Emphasis will be on repair, rather than production.

Electrochemistry will be very important in effecting many of these energy savings. Traditional electrosynthetic methods will be improved to save electrical energy and product separation work. Chlorine production cells will use an ion-exchange membrane for the direct preparation of pure caustic, using an air-diffusion counterelectrode (40% energy saving compared with present methods). A schematic view of such a cell is shown in Figure 9.[45] The aluminum production process will be improved, perhaps by the use of low-melting chloraluminate electrolytes. Electroforming will become a current technique. Resistant metals like titanium will see much greater use than at present, since they promise virtually infinite lifetimes. They will be produced and electroformed by electrolysis, e.g., from molten fluorides.[46]

Since primary energy in the future economy will consist either of nuclear electricity or (electrolytic) hydrogen, it is clear that electrolytic processes (either direct, or indirect using hydrogen) will replace many of the thermal processes currently used in the metallurgical and chemical industries. In metallurgy, electrochemical processes can virtually eliminate the energy required for separation of constituents of ores and the energy losses currently associated with rejection of slag material. In the future, iron ores will be reduced by hydrogen, the spongy iron thus produced being separated from mineral matter in electrochemical reactors. Among more classical separation methods, electroflotation[47] will become current. Sulfide ores will be electrochemically reduced to the metal sponge and sulfur, using polysulfide or chloride melts (cf. Figure 4). Electrochemical methods of converting S to SO_2 and SO_3 (molten sulfate electrolytes) may become realistic, electricity for on-site use being a by-product. Emphasis will be on sequential coupled electrochemical processes, so that excessive electrical consumption is avoided and valuable products are formed at both working and counterelectrodes. Alternatively, depolarized counterelectrodes will be used, e.g., in oxidative processes in aqueous solution, and reduction of atmospheric oxygen will be preferred to hydrogen evolution at the counterelectrode: In reductive processes, oxidation of a suitable waste product, e.g., SO_2 or even sewage waste,[48] may be feasible.

Electrochemical processes will be equally current in the field of organic synthesis. Potentially, such processes allow higher yields, a wider free-energy range than is available with homogeneous chemical methods (~6 eV or 140 kcal/mole in nonaqueous solvents), fewer by-products, simpler separations, and fewer reaction steps. Homogeneous chemical

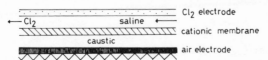

Fig. 9. Cl_2–caustic cell with cationic membrane and air diffusion electrode.

processes involving electroformed intermediates will become important, e.g., in the chlorination of hydrocarbons. New types of electrolytes (low-temperature molten salts, e.g., fluorosulfonates and alkali-metal fluorosulfonate half-salts) will become important (see earlier), since the higher working temperatures they permit will allow much higher reaction rates. Eventually, the future economy will require the synthesis of carbonaceous material from CO_2 (either from the atmosphere or from limestone). This will be best conducted electrochemically, in adaptations of high-temperature molten carbonate or possibly solid-state doped ZrO_2 or other conducting oxide fuel cells. Reversible reduction of CO_2 to CO will thereby be effected, using either oxygen-evolution or hydrogen-depolarized counterelectrodes. The CO can be chemically combined with either OH^- or H_2 to produce formates or methanol; these will ultimately serve as chemical feedstocks. Enzymatic formation of protein from such synthetic methanol via yeasts or bacteria is possible (see next section). Other biochemical reactions involving electrochemical regeneration of enzymes may become important.

Materials recuperation and recycling will become an important part of future industrial planning. Electrochemistry is particularly adapted to the treatment of dilute wastes, since it avoids the problem of irreversible losses of separation work. Low-grade dilute industrial waste containing such metals as Ag, Cu, Zn, and Hg will be monitored by electrochemical analysis techniques (polarography) and the constituents will be extracted by electrodeposition. Leachings from low-grade ores will be similarly treated.

The environmental impact of electrochemical processes will be much less than that of traditional thermal processes. In addition, electrochemical techniques will be of value in maintaining future environmental quality. Oxygen produced as a by-product in hydrogen manufacture will be used in water aeration (sewage waste, natural waters) so that correct biological conditions are present in waterways near heavily populated areas. Water derived from the electrochemical or chemical combustion of pure hydrogen fuel will reduce the need for reservoir storage. Electrochemical methods will also be of importance in the treatment of nuclear waste.

Due to the ability of electrochemical redox systems in contact with semiconductors to reproduce a solid-state semiconductor p–n junction, as discussed in the previous section, electrochemical light sources of very high efficiency may one day become available. Such systems will be less costly and less energy-intensive to manufacture than all-solid-state devices. Presently lighting, at an average efficiency of about 5%, accounts for about 25% of the electricity budget in the U.S., and growth is reportedly about 12% per annum.[49] Very large energy savings will be possible based on solid-state/electrochemical devices of perhaps 50% efficiency. Theoretically, electrochemical methods may be used for laser pumping, but the energy densities (W/cm^2) will have to be much higher than those currently reported for such an application.

In summary, it seems certain that electrochemical processes of all types will have an increasing part to play in future industry. This will occur partly because much primary industrial energy will be nuclear electricity, supplied on site in "nuclear–industrial complexes" designed as heat–electricity total energy systems,[50] and partly because of the flexibility, efficiency, and lack of environmental impact of electrochemical processes. Future engineers will require intensive training in electrochemical techniques, and electrochemistry should be part of all general education, so that a broad base can be built for its future development.

7. Electrochemistry and the Life Sciences

Medical applications of electrochemistry will be of certain, although not predominant, importance in the future. Some of these areas (e.g., cardiac stimulation, bone growth, even acupuncture) have been discussed in other papers at this meeting.[51] Only applications of fuel cells or electrolyzers involving biological materials will be discussed here.

Implanted fuel cells using compounds in blood ultrafiltrate as fuel [glucose,[53] glucosamines if polysaccharide-breaking enzymes are used[54]] give the possibility of a permanent power source for assist or control devices necessitated by a patient's state of health. Such devices will only be capable of low power levels ($\sim$100 $\mu W/cm^2$), but in small units they will be able to serve as power sources for cardiac pacemakers,[53] urinary stimulators,[54] automatic insulin dosers, and similar devices. For the insulin-dosing application, blood glucose concentration will be measured polarographically at an electrode. Such devices will use selective cathodes fed by dissolved oxygen in blood ultrafiltrate (intercellular fluid).

A far more ambitious undertaking is a biological fuel cell for powering a prosthetic heart or (more probably) natural heart assist device. Such a unit must produce a mean power output of 4–5 W (10 W peak). While a Ni–H_2 battery of, say, 60 Ah capacity (1 kg, lifetime 10 years) would be a suitable power source if daily charging via a percutaneous high-frequency coupling is acceptable, a better solution would be the use of a hybrid 5-W biological fuel cell. Such a system will require a large area of electrode–blood ultrafiltrate membrane (perhaps 1 m^2), with a high loading of an efficient glucose oxidation catalyst [Rh, Pt–Ru[55]] in contact with low-oxygen-content venous blood at high pressure (pulmonary artery). Forced convection derived from blood pressure oscillation would be used to provide efficient mass transport.[56] Since oxygen supply from the arterial systems will require a complex and bulky mass-transport membrane system, it is proposed that ambient air be fed to the cathodes via a tissue-compatible percutaneous tube. Cathodes would be of Teflon-bonded, air-diffusion type. Periodic pulsing of the system, by shorting or by using the Ni–H_2 peaking battery, would prevent electrode poisoning. Such a technique has been suggested in connection with direct organic fuel cells and in electrosynthesis.[57]

An electrolysis concept that will be of future value in medicine is that of urea removal by oxidation in renal dialysis systems.[58] If reaction rates are sufficiently high, the process may be carried out concurrently with dialysis,[58] the alternative possibility being oxidative regeneration of a urea adsorbant (e.g., zirconium phosphate) outside periods of dialysis. Such a concept brings an efficient, really portable artificial kidney unit a stage nearer.

In the previous section, reference was made to a possible role of electrochemistry in food (protein) production. Such protein may be used either as animal food or as a supplement to human diet, especially in developing regions with climatic problems. The importance of this subject is underlined by the fact that it currently requires approximately 10 kcal of primary energy to produce 1 kcal of food using Western intensive agricultural methods.[59] If such methods are required in the future to feed the anticipated world population for the year 2000, primary energy requirements will be *double* those of present projections.[1] In addition, primary energy value added costs alone will account for sums on the order of $100–$150 (constant dollars) for annual per capita food costs.[1] Such figures are greater than the annual per capita GNP of many developing countries.[1] Such a situation will be clearly intolerable, and any "green revolution" in the future will have to take into account the energy efficiency of food production. CO production from hydrogen produced by electrolysis and CO_2, e.g., from waste or refuse, could be carried out at practically 100% efficiency. Further conversion to methanol would be at high efficiency, and it is reasonable to assume that production of high-grade protein could be achieved at a rate of 1 kcal for 3 kcal of primary energy input, or three times the present figure. An easier route to protein production may involve direct consumption of electrolytic hydrogen, atmospheric nitrogen, and atmospheric CO_2 by suitable bacteria.[60]

Further research is needed to establish the bacterial species and precise conditions required for protein synthesis via such a route. The importance of such an approach to the problem of feeding the increased future world population cannot be overestimated.

Finally, it is important to state the contribution electrochemistry can make towards the understanding of biological processes. This has already been alluded to above in the discussion of electrochemical solar-energy storage systems. Biological systems involving energy storage are clearly electrochemical couples using shorted anodes and cathodes (via macromolecules) and selective proton transfer (selective ion-exchange membranes). In biological systems, both electron and proton transfer are often combined in the same molecule or macromolecule. On this basis, analogies between electrochemistry and biochemistry become clear, especially in photosynthesis and in the respiratory cycle. The problems remain elusive. Energy conversion in muscle contraction is perhaps the most interesting problem. This operation cannot be antithermodynamic, and must involve known phenomena (e.g., electro-osmosis, piezoelectricity). As *concerted* energy and separate reaction sites are involved, biochemical energy conversion must proceed by electrochemical reactions.[61] Understanding of these processes, which will involve basic comprehension of the mechanisms that control all human activity, provide the greatest challenge of the future of scientific research.

8. Conclusions

Society of the post-fossil-fuel age will be energy- and materials-conservative, and production and replacement of energy-producing facilities will be the major industry. Electrochemistry will have a major part to play in this society. Transportation will be the largest consumer of primary energy, despite the fact that personal transportation will not be used for intercity travel. Personal transportation will be entirely electric. Small utility vehicles will probably use nickel–zinc batteries, whereas higher-energy-density systems ($Zn-Cl_2$, Na–β-alumina–S) will be used in heavy traction applications where recharging is necessary (hybrid battery–electric railcars, hybrid trolley coaches, vehicles on fixed daily schedules, e.g., for delivery from train to consumer). Personal cars will be fuel-cell powered, first with powered metal cells (Zn), and finally, in a full hydrogen economy, by hybrid (Ni–H_2 battery) pressurized hydrogen systems. Primary energy will be in the form of electricity (nuclear, direct or indirect solar, perhaps geothermal) and hydrogen. The latter will be entirely electrolytic unless very-high-temperature nuclear reactors (>1200°C) become feasible. Up to 10% of the installed capacity of electricity grids for domestic and commercial supply will be in the form of peak-shaving batteries (e.g., Na–β-alumina–S). Communities supplied with hydrogen piped over long distances will use fuel cells, which will also be useful in stretching out fossil-fuel (coal) reserves, where they still exist. Waste heat and low-grade solar heat will be extensively used. Electrochemical solar cells will become generally used for on-site generation (coupled with battery storage, probably redox type), since they will be low cost and will permit simultaneous heat recovery. Industrial and metallurgical (production and forming, recycling) processes may be largely electrochemical. Certain medical applications of electrochemistry (implanted power sources and sensors, artificial kidneys) may become important. Finally, electrochemistry may ultimately be of great value in feeding the expanding world population in an energy-conserving way, until stability, and eventual population reduction, is assured. This will point the way to a "better world."

Acknowledgments

In appreciation of many stimulating discussions with John Bockris, from which many of the ideas given above originated.

References

1. A. J. Appleby, *Energy Policy 4*, 87 (1976).
2. Committee on Interior and Insular Affairs, Conservation of Energy, p. 47, U.S. Senate, Ser. No. 92-18, Washington, D.C., (1972).
3. Stanford Research Institute, Patterns of Energy Consumption in the U.S., Menlo Park, California, November 1971. Office of Science and Technology, Washington, D.C. (January 1972).
4. C. Zener, *Phys. Today*, p. 48 (January 1973).
5. D. Gregory, A Hydrogen Energy System, American Gas Association, Cat. No. L21173 (1973); see also D. Gregory, D. Y. C. Ng, and G. M. Long, in: *The Electrochemistry of Cleaner Environments* (J. O'M. Bockris, ed.), Plenum, New York (1972).
6. D. H. Smith, in: *Industrial Electrolytic Processes* (A. T. Kuhn, ed.), Chapter 4, Elsevier, Amsterdam (1971).
7. R. L. Costa and P. G. Grimes, *Chem. Eng. Progr. Symp. Ser., No. 71 63*, 45 (1967).
8. J. Jacquelin and A. J. Appleby, *Rev. Gen. Electr. 85*, 551 (1976).
9. L. J. Nuttall and W. A. Titterington, presented at the Conference on Electrolytic Production of Hydrogen, City University, London (February 26-27, 1975).
10. C. C. Sun, E. W. Hawk, and E. F. Sverdrup, *J. Electrochem. Soc. 119*, 1433 (1972).
11. D. Gregory, A Hydrogen Energy System, American Gas Association, Cat. No. L21173 p. III 36 (1973).
12. D. Gregory, private Communication.
13. G. DeBeni and C. Marchetti, *EuroSpectra 9*, 46 (1970).
14. J. E. Funk, Proc. ACS Symposium on Nonfossil Fuels, p. 79, Boston, Massachusetts (1972).
15. R. H. Wentorf and R. E. Hanneman, *Science 185*, 311 (1974).
16. R. Savage *et al.*, eds., A Hydrogen Energy Carrier, p. 31. NASA/ASEE Engineering Systems Design Institute, Houston, Texas (1973).
17. J. B. Pangborn and J. C. Sharer, Proc. Miami Hydrogen Energy Conf., p. S11-35 (March 1974).
18. D. Gregory and J. B. Pangborn, Proc. 9th IECEC Conf., p. 749014, San Francisco, California (1974).
19. A. J. Appleby, *Nature 253*, 257 (1975).
20. F. Joly, Proc. Miami Hydrogen Energy Conf., p. S5-19 (March 1974).
21. A. J. Appleby, J-P. Pompon, and M. Jacquier, presented at 3rd. International Electric Vehicle Symposium, Washington, D.C. (February 1974).
22. J. T. Salihi, *IEEE Spectrum, 12*(3), 62 (1975).
23. I. Dugdale and J. L. Sudworth, *Power Sources 2, 1968* (D. H. Collins, ed.), Proc. 6th International Symp. on Power Sources, Pergamon Press Oxford (1970).
24. E. A. Lisle, H. Brown *et al.*, in: *Technology of Efficient Energy Utilization*, Rep. NATO Sci. Comm. Conf. (E. G. Kovach, ed.), Les Arcs, France, October 1973, p. 60; Pergamon Press, Oxford (1974).
25. E. A. Lisle, H. Brown *et al.*, in: *Technology of Efficient Energy Utilization*, Rep. NATO Sci. Comm. Conf. (E. G. Kovach, ed.), Les Arcs. France, October 1973, p. 62; Pergamon Press, Oxford (1974).
26. K. V. Kordesch and A. Kozawa, in: *Batteries, Vol. 1, Manganese Dioxide* (K. V. Kordesch, ed.), Chapters 2 and 3, Marcel Dekker, New York (1974).
27. A. J. Appleby, J. Bouet, G. Feuillade, and R. Gadessaud, extended abstracts, I.S.E. Conf. on Primary and Secondary Batteries, Marcoussis, France, (May, 1975); W. Vielstick, University of Bonn, private communication.
28. M. Klein, Proc 7th IECEC Conf., p. 729017, San Diego, California (1972).
29. J. F. Stockel, G. Van Ommering, L. Swette, and L. Gaines, Proc. 7th IECEC Conf., p. 729019, San Diego, California (1972).

30. P. C. Symons, Extended Abstracts, Conference on Electrolytes for Power Sources, Society for Electrochemistry, Brighton, England, (December 13–14, 1973).

31. M. L. Kyle, H. Shimotake, R. K. Steunenberg, F. J. Martino, R. Rubischko, and E. J. Cairns, Proc. 6th IECEC Conf., p. 719013, Boston, Massachusetts (1971).

32. S. Gratch, J. V. Petrocelli, R. P. Tischer, R. W. Minck, and T. J. Whalen, Proc. 7th IECEC Conf., p. 729008, San Diego, California (1972).

33. Y. Lazennec, extended abstracts, I.S.E. Conf. on Primary and Secondary Batteries, Marcoussis, France (May 1975).

34. E. C. Gay, W. W. Schertz, F. J. Martino, and K. E. Anderson, Proc. 9th IECEC Conf., p. 749122, San Francisco, California (1974).

35. D. A. Crouch, Jr., and J. Werth, Proc. IEEE PES Winter Meeting, New York, New York (1973).

36. J. Werth, U.S. Pat. 3,847,667 (November 12, 1974).

37. W. Baukal, Battelle, Frankfurt/Main, private communication.

38. R. H. Wiswall and J. J. Reilly, Proc. 7th IECEC Conf., p. 729210, San Diego, California (1972).

39. L. H. Thaller, Proc. 9th IECEC Conf., p. 749142, San Francisco, California (1974).

40. H. Gerischer, *J. Electroanal. Chem. 58*, 263 (1975); see also K. H. Hauffe, Chapter 12, this volume.

41. A. Fujishima and K. Honda, *Bull. Chem. Soc. Jpn. 44*, 1148 (1971); *Nature 238*, 37 (1972).

42. R. Hill and F. Bendall, *Nature 186*, 136 (1960).

43. D. O. Hall and K. K. Rao, *Photosynthesis*, Studies in Biology No. 37, Inst. of Biology; Edward Arnold, London (1972).

44. D. H. Meadows, D. L. Meadows, J. Randers, and W. W. Behrens, *The Limits to Growth*, p. 161 *et seq*, Potomac Associates, Washington D.C. (1972).

45. Energy Workshop on Electrochemistry, Kenan Laboratories of Chemistry, University of North Carolina, Chapel Hill, N.C.; National Science Foundation, June 1974. NTIS No. UNC-100, Springfield, Virginia (1974).

46. S. Senderoff and G. W. Mellors, *J. Electrochem. Soc. 112*, 840 (1965), *113*, 66 (1966).

47. A. T. Kuhn, Chapter 25, this volume.

48. J. O'M. Bockris, A. J. Appleby *et al.*, in: *Technology of Efficient Energy Utilization*, Rep. NATO Sci. Comm. Conf. (E. G. Kovach, ed.), Les Arcs, France, October 1973, p. 35, Pergamon Press, Oxford (1974).

49. R. Bleekrode, F. Oostvoels *et al.*, in: *Technology of Efficient Energy Utilization*, Rep. NATO Sci Comm. Conf. (E. G. Kovach, ed.), Les Arcs, France, October 1973, p. 21, Pergamon Press, Oxford (1974).

50. Oak Ridge National Laboratory, Nuclear Energy Centers, Industrial and Agro-Industrial Complexes, ORNL 4290, United States Atomic Energy Commission, Washington D.C., (November 1968).

51. S. Srinivasan, G. L. Cahan, Jr., and G. E. Stoner, Chapter 4, this volume.

52. L. H. Gaines, R. W. Francis, G. H. Newman, and B. M. L. Rao, Proc. 11th IECEC Conf., State Line, Nevada (Sept. 1976).

53. S. K. Wolfson, Jr., S. L. Gofberg, P. Prusiner, and L. Nanis, *Trans. Am. Soc. Artif. Intern. Organs 14*, 198 (1968).

54. S. J. Yao, A. J. Appleby, A. Geisel, H. R. Cash, Jr., and S. K. Wolfson, *Nature 224*, 921 (1969).

55. A. J. Appleby and C. Van Drunen, *J. Electrochem. Soc. 118*, 95 (1971).

56. A. J. Appleby, D. Y. C. Ng, and H. Weinstein, *J. Appl. Electrochem. 1*, 79 (1971).

57. J. O'M. Bockris, B. J. Piersma, E. Gileadi, and B. D. Cahan, *J. Electroanal. Chem. 7*, 487 (1964).

58. S. J. Yao, S. K. Wolfson, Jr., B. K. Ahn, and C. C. Liu, *Nature 241*, 471 (1973).

59. J. S. Steinhart and C. E. S. Steinhart, *Science 184*, 307 (1974).

60. C. Marchetti, *Chem. Econ. Eng. Rev.*, p. 7, (January 1973).

61. J. O'M. Bockris, *Nature 224*, 775 (1969).

62. A. Despić, Chapter 15 this volume.

63. H. J. Halberstadt, Proc. 8th IECEC Conf., p. 739008, Philadelphia, Pennsylvania (1973).

27

Electrochemistry for a Better World—Past and Future Summary

M. A. Slifkin

The two excellent expositions which I have been asked to summarize, together with the ensuing discussions, are themselves summaries, in a sense, of what has been presented at this symposium in earlier sessions. One obvious point which must be made is that there appears to be a wide area of disagreement as to what will contribute to a better world. This is not perhaps surprising as futurology is a somewhat less exact science than electrochemistry. In his very interesting review of the past contribution of electrochemistry to a better world, which included accounts of such fascinating techniques as electroflotation and the electrolytic destruction of cyanide, Dr. Kuhn managed to inject a note of controversy with the assertion that the fluidized-bed cells were not really a viable proposition as the great economy effected in initial capital expenditure would be more than offset in the higher running cost due in part to the large ohmic losses. This is a view certainly not shared by workers with this technique present here to day.

Dr. Appleby's encyclopedic account of the future went even more into controversial topics. Dr. Appleby appears to be less optimistic about the availability of resources and energy in the future than other speakers at this symposium, particularly as he feels the switch from one economy to another takes decades rather than years. Dr. Appleby sees a world after the year 2000 in which we will be geared to energy sources based on hydrogen and/or electrochemical sources. None of the participants see much future for fossil fuels, or, perhaps more surprisingly, nuclear fission and fusion.

Dr. Appleby envisions the production of hydrogen from the electrolysis of water and feels that the thermochemical splitting of water will not compete in the future with this method. However, other workers in this field, particularly in the United States, appear to hold just the opposite view. This is perhaps just as well as it means that research into both types of hydrogen production will continue.

Hydrogen, it is suggested, will be used primarily in the hydrogen fuel cell which, it is predicted, will be sufficiently developed in about 30 years time to take over from the zinc-powder fuel cell, which it is hoped will be a feasible source of energy in the slightly nearer future. This again is a contested prediction and views have been put forward here expressing doubts as to whether such fuel cells will ever be developed into alternative energy sources. However, this is not necessary to the development of a hydrogen economy. Hydrogen could be used as a fuel in itself. It can drive a hot-air engine such as the Stirling engine simply by burning it at the appropriate place, or hydrogen could be readily converted to methane thus enabling our present highly developed internal combustion technology to be utilized based on methane rather than petroleum.

M. A. Slifkin • Department of Pure and Applied Physics, University of Salford, M5 4WT, Lancashire England.

The idea of using hydrogen to produce methane suggests to me one way in which we might obtain our energy in the year 2000 and that is a do-it-yourself method. Each family would have a fermentation pit in its own backyard in which all its own refuse and waste matter would be fermented to produce methane gas. While this is not necessarily a particularly efficient process if costed out by normal commercial methods, nevertheless, as each family would be responsible for its own labor, this still might prove the best method available. In addition, each household would have its own solar hot-water heater on the roof. The technique for this is well developed; over 100,000 such units are currently in use in Israel. Dr. Appleby has pointed out that there is a bright future for electrochemical cells, particularly the sodium sulfur battery, which promises great improvements in energy/weight ratio and storage capacity over the lead–acid accumulator. Such batteries would help to overcome the bugbear of the usually predicted future sources of energy such as tides, wind, and sun, by allowing storage of the energy for use during peak demand periods.

The very interesting idea has been put forward of the use of organic light-to-electricity converters based on the biological process occurring in oxidative phosphorylation and the use of such systems for the storage of energy. A somewhat different system has begun to arouse interest and that is a very interesting bacterium called *Holobacterium holobium*. This bacterium, which lives in very brackish water (4 M NaCl), has the unique ability to synthesize a purple membrane when the oxygen level drops, to enable it to phosphorylate by utilizing sunlight directly. The purple chromophore appears to be a form of rhodopsin, the visual pigment in the eye. On the absorption of sunlight a proton is picked up outside the cell and passed very efficiently through to the inside whence it returns via specific sites producing ATP in the process. The passage of the proton to the inside of the cell appears to be due to a conformational change of the primary photoreceptor. This process of a light-driven proton pump appears to be a very efficient one. One can somewhat speculatively see this as the basis of a direct solar–electrical transducer of great efficiency wherein the pigment will be suitably adsorbed and aligned on an artificial membrane immersed in a suitable electrolyte, so that the action of sunlight will produce a proton gradient across the membrane and the consequent generation of electricity. Alternatively large shallow pools of brackish water could be utilized to the same end.

It has been possible to refer to only a few brief items, but it is clear from these that electrochemistry and electrochemists have a lot to offer to make this world tolerably inhabitable in the next 100 years. Certainly there is enthusiasm and what is more important: ideas. Unlike many other people I view the future with equanimity if that enthusiasm and breadth of outlook can only be diffused to a much wider circle of people.

28

Synopsis: Thirty Years Back and Thirty Years Forward

J. O'M. Bockris

1. Introduction

The present symposium gives me much pleasure, and I am happy to be able to attempt a summary of what has been recounted at it as well as sketching some aspects of the position, both in the past and (as I see it) in the future, for fields in electrochemical science in which my co-workers and I have made contributions.

Before I begin this task I should say something about how I regard the work which we have done—and which we hope to do for a number of years to come. The group's work has the following characteristics:

1. It has been one of a few (Frumkin, Gerischer, Vetter are names one thinks of as leaders of some of the others) the main lines of the work of which have taken electrode-process chemistry from a stultified stage in the 1940s to the active subject which it now is. But we must not claim too much because, although, with our German colleagues, the Electrochemistry Group which began at Imperial College led the evolution of electrode-process chemistry in Western Europe and the United States in the 1950s and 1960s, there is no doubt that we began in the late 1940s by standing on the shoulders of those who had published from the Electrochemistry Laboratory of the Academy of Sciences of the U.S.S.R. in Moscow.

2. Our electrochemistry has mostly been aimed at an eventually practical end result. The phrase "fundamental applied research" describes exactly what we do and did, particularly in the American period of the group's work. It consists of taking fields which need development, electrocatalysis is a central example, and studying the fundamentals behind these fields until the applied side could be lifted from an art to a science.

3. A third characteristic of our work has concerned the human and interactional aspects. The group has always been "hot" in this respect, both in the considerable number of positive associations among human beings in it, but also, it must be admitted, in an abnormal number of difficult personality interactions. Scientists work emotionally, not as the cold fish so often pictured by the public, and it is frequently the rivalries and hostilities, the fears as well as the loves and inspirations, which made the workers in London, Philadelphia, and Adelaide maybe, seem more creative and active than researchers in many parallel groups. *Emotions* drive people. Their suppression would lead to a set of uncreative zombies.

4. A fourth characteristic of our work is one which may surprise some readers. A close reading of the literature was less encouraged than in some laboratories. We begin our projects slowly, reading and note-making, and calculating, for months before the

J. O'M. Bockris • School of Physical Sciences, The Flinders University of South Australia, Bedford Park, South Australia, 5042.

project is defined and experimental plans are made. But by and large we try to think each project from emphasizing the basic equations in physics upon which all phenomena are represented. This is so, even if we do not know (as yet) the terminology of the field. Such an approach is one in which "lateral thinking" is used. One tries to jump from the position of the time not to the next step but to a quite different track, that of a future time.

We all have intellectual fathers, and I have gained by my acquaintance with and admiration for, particularly, A. N. Frumkin and Henry Eyring. At an earlier stage, W. E. Scott, at the Brighton Technical College, taught me a bit about basic chemistry, and more about courage, and the days at Imperial College (1943–1945) with my supervisor, H. J. T. Ellingham, provided me with my negative reaction to the over-application of thermodynamic principles, but also bred in me the following positive reaction:* "Is it new?" was the *first* thing that mattered in all research considerations, and by "new" was not meant some tinkering with the structure of an established field, some extension, some screw tightening; new was meant as New, Really New, and woe betide the ignoramus who suggested anything which could be put down with a contemptuous "But that's been done *before*."

This symposium is meant to look back, and that I have just been doing. But it is also meant to look forward, and that I must now do. It is a simpler task, for one may be free of fact and soar into that blue sky, the existence of which, daily, charges one up for renewed soaring. It seems to me simple to sketch the electrochemistry of the future. Electrochemistry is a core science of the world which must evolve from the disastrous pile-up which now begins to be visible of the old world of fossil fuels, pollution, resource exhaustion and, indeed, disease. Of course, we shall run out of materials, in particular the fossil fuels, and we unfortunately may have to go through some decades in which the attempt will be made to prolong, what will indeed seem a Golden Age, with coal. But emerging beyond it, not far beyond 2000, and certainly not after 2050,[1] we shall be in a different world: clean, zero growth, and built for the long term (Figure 1). It is this world in which electrochemical technology will be central. Electrochemistry is more than a branch of chemistry. It is a different way of looking at reactions. By using the *electrical* properties of reactions, not their *thermal* properties, as now, it will be possible to store and convert energy; to apply it rationally to many bio-situations; to use it in electrocatalysis to lower the energy of activation and hence the wasteful overpotential in a situation; to apply it to the light–matter hydrogen interface; to make substances last longer; to have an extra element in control of the organic chemical situation with the aid of the electrochemical interface; to recycle and extract at the steady state of zero growth, and to do this cleanly and with far less, if any, pollution[2] than at present.

The future world will be associated with a great increase in *electrochemical* technology. That is what our group and our work is about.

2. Energy Conversion and Storage

Until about 1961 the group at the University of Pennsylvania did little in the direction of energy conversion, but the initiative in this area was handed to us by Manfred

---

*Another lesson Ellingham taught me was to try to make every sentence true, and lucid, *out* of its context: and he taught me to *contemplate*, to look at a thought backward, sideways, and from 99 other directions. His extreme stress on semantics encouraged me to think penetratingly. A never-to-be-forgotten example was that half-hour at his home in my thesis-writing days (typical appointment time: 8 a.m., Sundays) where we contemplated the weighty matter of whether overpotential was *on, at* (or maybe *in*?) an electrode?

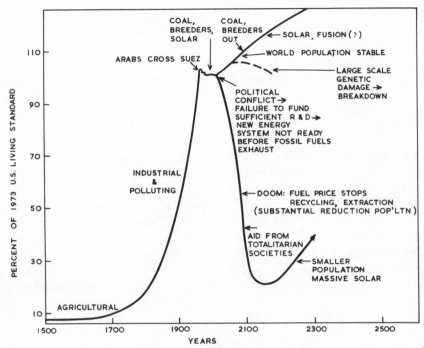

Fig. 1. The past and possible future living standards relative to the 1973 U.S. living standard.[1]

Altman, founder of the Energy Conversion Institute at the University of Pennsylvania, and Ernst Cohn, who led Electrochemical Space Power in the National Aeronautical and Space Agency (NASA). The impetus of this work lasted from about 1962 until 1971 and formed the principal funding of the group at about $100,000 per year (some one quarter of our funds). The NASA connection gave some contact with the Washington scientific policy thinking and made us feel that we were adding a grain or two to the beachhead into the Solar System and perhaps further. Ernst Cohn's personality, too, driving, acrid, critical, striving, optimistic, was a help in his visits to us, always creating days of great intellectual anxiety followed by evenings of excellent intellectual enjoyment (discussion of extraelectrochemical areas not excluded); they formed one of the pleasantest memories of the group's whole course to date.

When I speak of energy conversion, readers may imagine that the group made fuel cells, but such was not the case. Cohn was sceptical of our abilities to do something commercially practical, and we were meant to arm his understanding of what was going on at the electrode–solution interfaces, in respect to the large contracts which he was giving out to organizations such as that at Pratt & Whitney. Hence he let us work widely, and we were encouraged for a few enjoyable years to pick unripened plums from the electrochemical tree and force-grow them until they dropped. All we had to do was to convince Ernst Cohn that when they did drop it might help him to understand the functions being examined by his more practically oriented contractors.

Let me, then, mention from those halcyon days a few of the achievements. One was the theoretical work carried out by Devanathan and Muller, and resulting in the well-known "BDM theory" of "the structure of change interfaces."[3] The paper was written with a plentitude of time and relaxed attention which it would be difficult to imagine now because of the diminished funding and over-preoccupation forced on academics with administrative tasks, and an over-degree of lecturing to students who sometimes don't want to learn. It was carried out in the same rich and leisurely Rolls-Royce approach used for writing the book by Bockris and Reddy. Devanathan was a postdoctoral fellow from Ceylon, but had worked with Roger Parsons in the early days in London. Klaus Muller came to us from Kurt Schwabe's laboratory in Dresden. Both collaborators were of the highest quality. Devanathan, a small, rather stout man with twinkling eyes, had already made many original contributions to the double-layer field and about one third of the paper was an elaboration at a more quantitative level of some ideas he had published in 1954.

The BDM paper represents a contribution to the theory of the double layer analogous to those made by Helmholtz and by Gouy in their day, because it introduces a new concept into double-layer theory (Figure 2): the effect of the water-molecule layer. There were precursors of the formulation published in 1963, for example, the suggestion of Lange and Miscenko[4] of 1930 about two contributions to the double-layer electrode field, and the contributions made (independently) by both MacDonald[5] in 1954 and MacDonald and Barlow[6] in 1962, together with those of Mott and Watts-Tobin.[7] The formulation made by BDM was an independent mathematical formulation of models which involved imaging in the double layer and the effect of two-position water molecules

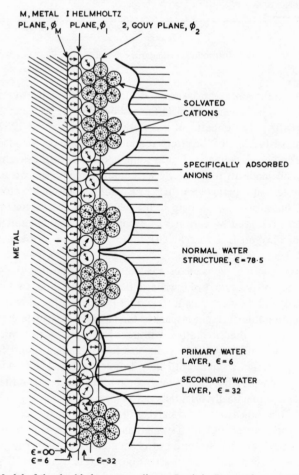

Fig. 2. Model of the double layer according to Bockris, Devanathan, and Müller.[3]

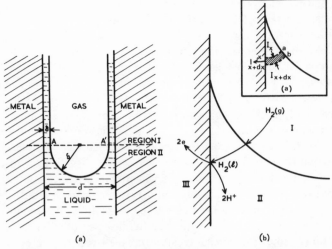

Fig. 3. (a) The "thin-film" model for porous electrodes. (b) The "finite contact angle meniscus" model for porous electrodes developed by Bockris and Cahan.[13]

in the electrode, a solution of half a dozen different problems of the electrical double layer of the time, together with the first solid formulation of the molecular theory of the adsorption of organic molecules at the surface of electrodes. Its repercussions have been widespread,[8] have led to a change in the concepts of specific adsorption[9] and of the possibility of calculating individual components to the electrode potential,[10] for example, components due to solvent orientation, and the electron-overlap potential,[11] and has in general substantially changed the model of the structure of the interface.

A good deal of the NASA work was electrocatalytic in nature and this work will be mentioned in a separate section below.

Much other mechanism work was carried out with NASA contracts on energy conversion and among these must be mentioned some of the earliest, and perhaps most complete, analyses of the problem of porous electrodes. Here the mathematics of Hurwitz and Srinivasan[12] were applied in taking into account current line distribution in thin layers of the electrode pores, and Boris Cahan[13] then came up with the concept of a very small contact angle, the experimental measurement of which he pioneered at the three-phase-boundary between gas, liquid, and metal (Figure 3), developing for the first time complex equations which were computerized and the solutions graphed for easy application.

The energy conversion work was never finished because the pulse which was behind it was the NASA Moon Mission, and when this had been accomplished in 1969, the drive fell off from the fuel-cell program which began to be reduced, and pointed in another direction, that of the development of high-energy batteries. Here again, some achievements were made, and particularly those of Zoltan Nagy, who had already received some electrochemical education before he came to the United States as a result of the revolution in Hungary. He worked in the Electrochemistry Laboratory in Pennsylvania on the distribution of the current within the pores of zinc cells. What had been intended, had it been possible to continue work on fuel cells for a sufficient time, was a more realistic mathematical formulation of the "total equation" for the design of fuel cells, so that electrode practical size, boundary pore length, surface condition, electrolyte concentration, and catalyst loading could all be optimized with the design criteria mathematicized at a standard equal to that, say, used in turbine engineering. Because Cohn had to push the work in the direction of batteries as of 1969, we tried to attain something a bit similar by an investigation using optical methods, of the current distribution down the pores of zinc electrodes. We were interested in the question of why the electrode lasted for a limited time,[14] and what happened to the unused zinc,[15] apart from questions which we had examined elsewhere concerning the growth of dendrites and the cessation caused in this growth by the addition of traces of tetraalkylammonium salts.[16]

I have dwelt fairly fully on the past of energy conversion and storage. The reason is that the NASA support was one of the greater things which occurred to the Electrochemistry Group in its middle period, and the results show what can be done when scientists have sufficient funds to work with and lessened anxiety. I now turn to the future of energy conversion and storage and make a statement which is sociopolitical. It took me too long, until I was perhaps over 35 years of age, to recognize that the achievements of scientific advances, which seemed inevitably to be a benefit to the community, did not have much chance of getting ahead until they fulfilled two other criteria. Firstly, they have to offer some people the possibility of *profit*, the Great Motivator. This is now so obvious that I can hardly believe that it took me so long to *understand* it properly and accept it, apart from the superficial intellectual recognition which I had of it years earlier. I add here that I have found mature and intelligent scientists aged 50 and more, who still do not take into account this simple principle: If a funding source cannot see a money gain for itself in the enterprise, it won't be funded.*

The second great criterion, however, is that the proposed advances must be compatible with the views of the establishment in the country concerned, for, of course, all research and advances depend upon a turning on of the spigot down the appropriate pipe, and the spigot is in the hands of the reigning establishment, who control it through (so-called) government committees on which are carefully chosen to sit the senior servants of the masters of the greater financial groups, the banks, the oil, steel, and automotive concerns. Here, I jump forward to make clear that I do not point only at the influence of giant international corporations. The establishment principle must apply as much within the communist world as within the capitalistic one, although the shape of what the single giant international company, which is the Soviet government, may want may be different from the things motivating the Western corporations (the personal financial gains of their directors). However, it did take me a long time, until I was more than 40 years of age, to lose my naive way of thinking that, were I to be able to direct or inspire some kind of scientific advances of definite and undisputed value to the community, it could escape being smothered by the power beneath the surface which is that of the men who control the economic base of a country, and which in turn manipulate the apparent political "leaders" on the surface. What could better illustrate the limitations of scientists' power than the absence of a storm of work on fuel cells in the 1970s? How is it possible that an era where energy shortage and its permanence and increasing seriousness are obvious does not see this storm on energy conversion devices which would double the time during which we could use our present resources? It is a fine example of the negative influence of vested interests in modern times, and the fact that there are a few bulwarks that stand muddied but straight, in the swampy murk swirling largely in the wrong direction, is a tribute to the history-bending influence of individuals.† A dou-

---

*A decisive part in my education here came in my many contacts with the communist world, where energetic attempts are made to suppress the profit motive. (It is of course as unsuppressible as man and religion.) I saw the relative lack-lusterness of everything.
†For example, Podolny of the United Aircraft Company.

bling of efficiency, elimination of noise, a reduction to zero of pollution, all these are unambiguous achievements of electrochemical energy conversion, and yet there are still only a few organizations, albeit some giant ones, (Pratt & Whitney, Siemens, Exxon, and the leviathanic Soviet government groups) who work upon this field.

However, although it is possible that the tide is held up in its direction, it cannot be changed and there is no doubt that, even if we should not need development of high-energy-density batteries, because of the advent of a hydrogen economy, we shall need electrochemical energy conversion, and therefore the prospects before us in this area are limitless and will occupy electrochemical engineers for decades until the multimegawatt fuel cell is commonplace, and the cost of energy correspondingly decreased. After we are through our last resources of fossil fuels in 3-4 decades, it will be the fuel cell which will convert the hydrogen-borne energy to mechanical energy via electric motors.[17] The change to other transducers does not seem to be a serious one because economics is thoroughly on the side of the electrochemistry, and if hydrogen is used to drive cars, there will be a trend towards transduction to mechanical energy via fuel cells.

In this respect, I would like to draw attention particularly to the presentation made by John Appleby at this meeting in which he portrayed the best design of an electric vehicle yet shown. Thus, Appleby overcame the problem of liquefication and cryogenic storage by using high-pressure storage and hydrogen within the tubular structure of the car and fuel cells for driving it under normal situations. At the same time he used a hybrid concept, in which a nickel-hydrogen battery would be used for the acceleration situation.

The future of the electrochemical energy conversion and storage seems to be assured, for, let us suppose that the present apparently bright future for a hydrogen economy should become obliterated, then it will have to be replaced by electrochemical energy storage according to the predictions of Kordesh.[18] These were partly presented in the meeting by the calm and solid picture given by MacBreen. Kordesh predicts that in the 1980s and 1990s the alkali metal–air battery will have the best chance of success, along with alkali metal–halide (or air?) storers. We do not yet know which of these alternatives, the large-scale development of the fuel cell with hydrogen, or the development of new high-density batteries, will be the more important, but what we do know is that a pollution-free, zero growth, energy-short civilization will not get far without one or both.

3. Bioelectrochemistry

The contributions of electrochemistry to biology have received description from members of our group both in the introduction of Bockris and Reddy,[19] where the beginnings from the original paper of Galvani on the movement of frogs legs were described, and also in the erudite paper of Srinivasan at this meeting, who, with Philip Sawyer at the Down State Medical Center in Brooklyn, has made remarkable contributions to the clotting of blood under electrochemical control.[20]

My own summary of the past centers upon one idea which was described in a paper in *Nature* in 1969. Perhaps it should rather be called a suspicion. In describing it here, I would like to begin with a general statement: wherever, at least in biology, one finds electrical potentials, the conclusion in the past has been that these potentials originate in a Nernstian manner.

My introduction to newer thoughts began a long time ago, perhaps about 1950, when I understood, partly under the tutelage of Sam Hoar, the meaning of the work of Wagner and Traud. These gentlemen, as electrochemists know well, published a seminal paper[22] in 1938 in which they suggested that the basic concepts of corrosion which had for so long held sway, and were effectively Nernstian, could give way to a view more in consonance with the kinetic concepts of Erdey-Gruz and Volmer, and instead of having to have two components in the corroding situation, so that there would be a potential difference developed according to the electrochemical series, one could draw two Tafel lines and let them intercept. Each would represent the rate–potential relation of a process, and if the processes were opposite in respect to the direction of a passage of electrons across the interface, then one would see the possibility of a chemical reaction being carried out electrochemically. No net electron would be involved, and yet there would be electrochemical control by potentials across the interface.

Wagner and Traud put only a rudimentary bit of mathematics into their paper, which was probably indeed not the first to mention the idea (something similar was earlier implied by Hoar and Evans).[23] I got carried further forward by Philip George at the University of Pennsylvania, who stressed to me the difference in the nature of biological reactions: their tendency to have heats of activation which are low, their corresponding high rate constants, and, of course, their dependence upon enzymes.

In the meantime, and still in the 1950s, I was acquainted during consulting work at the DuPont Company with the quasi-electrochemical mechanism of the Kroll reaction in which, it seemed, the exchange of electrons occurred across the magnesium drops, the electrons transferring to titanium ions in solution form titanium metal while the magnesium dissolved, to partner the lost chlorides.

In 1965 I read the paper of Del Ducca and Fucsoe[24] in the *International Journal of Science and Technology*. Here was described what I had been comtemplating, the existence of the fuel cell–biological cell. I recall clearly the moment at which the idea tumbled upon me. I was sitting at the Midway Airport in Chicago, perhaps in 1958, my elbows on my knees contemplating the floor with my head in my hands and I noticed my regular breathing rhythm. It occurred to me that the lungs are in fact like porous electrodes and food is the organic fuel: mammals, then, get their energy in fuel cell mode! The thought was amplified in 1967 when Srinivasan and I wrote a note to *Nature* pointing out the difficulty of explaining the high efficiency of the biological energy conversion quantities (40%) with anything but an electrochemical kind of mechanism.[25]

At the house of the irrepressibly enzymatic Felix Gutmann in Philadelphia, about 1965 I met Freeman Cope, and he had expressed in a diagram in one of his papers a concept of the anodic and cathodic current in cells, the concepts I had in mind since 1958.

It seemed too late to author an original contribution myself ("Is it new?") but I tried a high lift shot by writing a paper in *Nature* (would my former supervisor have "considered it advisable"?) which synthesized Wagner and Traud, Del Ducca and Fucsoe, and Freeman Cope (Figure 4). I pointed out that biological systems had a large surface to volume ratio, and little bulk liquid within them, which alone made likely a surface reaction concept of biological reactions. I gave some examples with the membrane (a semiconductor) as the joining "wire" with pores as the "tubes between the two electrodes."

I had put myself out on a limb with my jousts with electrochemistry into biology and the establishment in the field, still deeply buried in the quietude of a residual part of the Nernstian hiatus, tried not to recognize my existence. One of their own, however, the well-known Britton-Chance, asked me to address his large group in 1971 and summarize the results of what I had achieved in basic electrochemistry in the last 20 years (the majority of which had been spent at the University of Pennsylvania). My lecture to the biologists in Chance's group went down like a ton of lead bricks, and I don't think that

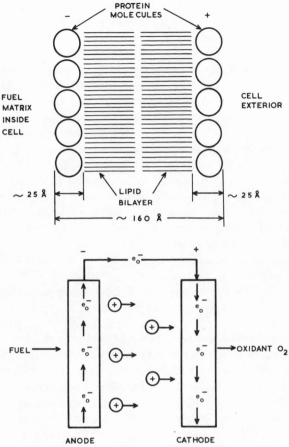

Fig. 4. Resemblance of the gross structure of a membrane around a biological cell (magnification 180,000×) to an electrochemical cell.[25]

anyone understood 5% of what I tried to say. However, in the discussion which took place between a few of us afterwards, I showed them by means of scrap-paper drawn diagrams how a process could occur at steady state, electrochemically, without the apparent use-up of electrons. I ended with the words "thus all your assumptions about equilibrium thermodynamics may well be completely wrong," and at this point I observed that "Britt" had removed himself from the edge of the group and was creeping away in mimed fashion. "This is the point at which I had better go," he muttered, and disappeared from the room.

The field of electrophysiology depends upon the outcome of whether the essentially Nernstian or Volmerian concepts prevail. This large area, containing more workers than are in electrochemistry itself, is boxed in with books, traditions, assumptions, and a vocabulary of its own. It appears to electrochemists who look at it from outside to be like a house of cards.*

---

*I owe this description to Lazzaro Mandell, who made it one day in Philadelphia after we had discussed the interpretation of his own discovery of Tafel's relation at the mucosa membrane.[26]

To comment upon the future of electrochemistry in biology, so stimulatingly outlined by Dr. Eckert in the London symposium, is to look into an area which may have more meaning than the space-oriented studies which the group contributed to in the NASA time.

Thus, writing in 1969, Reddy and I already predicted that the greater part of the electrochemical future would be in biology. "A large, immediately developable area lies in the electrochemical aspects of molecular biology and in the development of circuitry which will join the brain and its probable electrochemical mechanisms to artificial limbs with their electrochemical functions and, perhaps, even to circuits not kept within the body."

These prognostications have found substance in hard science, and they are summarised in the paper of Becker and Pilla.[27] I hope, too, that the brief account which Gutmann, Bonciocat, and I[28] have tried to put forward in the little book entitled *A Primer in Electrochemistry* will be of value as an introduction. I refer, for example, to the proof of the electrochemical nature of bone growth, to the electrochemical nature of thrombosis and blood clotting, and to the advances made in the electric interpretations of the action potentials by Lowenhaupt,[29] which have been interpreted for so many years in a Nernstian costume, and for which a Nobel prize was given to Hodgkin and Huxley many years ago.

However, allow me to dare to step out of this modern scientific framework, electrodic electrochemistry in biology, and attempt to lead away from it to matters which now are on the edge of science. May the knowledge gained from science, in a context of millions of years, be regarded as a perturbation? Could it be that science is essentially a *popularizer* and *practical* effectuator of basic knowledge, which, in a different guise, was available to earlier civilizations? Thus, some of the concepts to which we now crabwisely approach, seem to have been known for thousands of years to the brains of abnormally perceptive persons in the deep past. If this statement causes not only a raising of the eyebrows of the conservative, and a throwing up of hands of the more reactive,* then one should read the works of H. S. Burr who already in 1935 wrote a paper with Northrop entitled "An Electro-Dynamic Theory of Life." Burr worked for decades upon the thesis that electrochemical currents were determinative in living functions, and one of his key papers was published in the scientifically immaculate *Proceedings of the National Academy of Science* in 1943.[30] In it he describes the electrochemical correlations which govern the growth of pure and higher strains of corn.

There is a tendency in the present culture to consider data which do not arrive through the senses and to bring them out of myth into normal consciousness. The key work for those still skeptical is that by Köstler[31] entitled *The Roots of Coincidence.* There a study is made of the resemblance between some new higher concepts of quantum physics (e.g., time flowing backwards) and the trend toward the scientification of the approach to nonsensory knowledge in which random number generators[32] are faced by persons who are asked to forsee the future a few seconds ahead. Prediction several orders of magnitude above chance can be achieved by several percent of the population. Just as a great revision has had to be made in the concepts of chemistry by the advent of quantum concepts of the 1920s,† so a corresponding revision, a deeper one, may have to be made in the concepts which bring us to join the ideas of materialism to ideas which have been present in Eastern philosophy for millenia. Such concepts are available in language

---

*Not to say a curling of the lip of the critical; or a muttered exclamation of contempt from the older scientist some of whom may read but narrowly.

†What quantum concepts say is: The *senses cannot indicate* the working of individual particles small enough to be the building blocks of matter. There is already here a hint of Maya.

which may be understood by scientists in the works of biologist Lyall Watson,[33] apart from those of the populist philosopher, Köstler.

What has all this to do with the future of electrochemistry in biology? I am not sure, and if I were, I would be doing what I had promised some colleagues long ago, working hard upon the ideas that the electric currents arising from electrochemical reactions in biological systems would enable us to get a bridgehead between rational science (sense data knowledge) and knowledge sometimes called "direct." For those still skeptical, let me point out the *scientific fact* of the connections between events in nature and their dependence upon the heavenly constellations.[34]* These facts, so recently unburied from legend by Ganquelin, prove that some biological functions vary with the positions of planets and moon. A view one may take of this is that it is simply beyond our degree of perception and that its integration into the world picture must wait for a revolution in thinking wider then that wrought by the quantum theory and deeper than that of relativistic physics. This is probably right and attempts, like those made by those involved in the heavily funded official Russian government research into ESP,[35] to try to explain the new phenomena in terms of the development of present scientific concepts, may be bound to failure. Yet, were one to understand the body and its functions more widely in bioelectrochemical terms, one might be able to link material concepts and such unlikely but established events as the dependence of some biological kinetics on external magnetic fields.[36]

4. Electrocatalysis

In a disguised form, namely in the study of the dependence of the exchange-current density for the same reaction upon the metal, electrocatalysis has been studied since the time of Bowden and Rideal.[37] However, it was Grubb[38] who, in 1963, introduced the term "electrocatalysis."

Our own group was much involved in electrocatalysis from 1963 to 1971, and one of the major contributions we made in the American period was to give some experimental reality to the term.

Thus, the first contribution was in the form of a review which I wrote with Wroblowa and published in the Proceedings of the AGARD meeting at Cannes in 1964.[39] This article, which I liken to my own review entitled "Overpotential" in the *Chemical Review* of 1948,[40] or to David Grahame's review on the double layer of 1947,[41] provided a nucleus for growth in the subject, which was taken up extensively in Russia by Bagotzky,[42] Schumilova,[43] and others; and in the United States by Paul Stonehart.[44]

In looking back and trying to pick up the plums from the electrocatalysis tree, it is necessary to ask: "Electrocatalysis of what?" It would be nice to think that we are able to study the electrocatalysis of any reaction, the dictates being only made by the scientific desiderata. However, in spite of the relative freedom which Ernst Cohn gave us, this attitude is not possible. We had to perform work which could help the only two reactions which were of interest in the 1960s for *application*.

The first of these was the important oxygen-reduction reaction:

$$O_2 + 4H^+ + 4e_0^- \longrightarrow 2H_2O$$

The importance of this reduction reaction cannot be overemphasized, for not only is it the half of every fuel cell, but it also has importance in other parts of both technological and biological electrochemistry.

---

*Gauquelin[36] was a biophysicist and statistician who stressed the scientific method. He appeared not to know of the nonscientific earlier body of knowledge, which he rediscovered.

The other reaction which had interest in the 1960s was the electrochemical oxidation of hydrocarbons, for example:

$$C_2H_4 + 4H_2O \longrightarrow 2CO_2 + 12H^+ + 12e_0^-$$

Although there is no longer interest in the electrochemical conversion of oil or natural gas to electricity, because of the timescale of the exhaustion of these products, there is interest in the application of this type of reaction to the oxidation of coal and its products where success of electrochemical conversion would as much as double the available energy.

The contributions of the group here were on two fronts. First, questions in electro-catalysis are secondary to that of the mechanism of the reaction concerned. We asked our colleagues in industry how it was that they could attempt research with their empirical methods without knowing what the rate-determining step in the reaction is likely to be. The simplistic answer is that their bosses wanted "quick results" of a patentable kind and despised the idea of having to take a few years to find out what was going on.

It soon became clear, for the hydrocarbon oxidation particularly, with the work of Kuhn and Wroblowa in 1967[45] (preceded by the excellent work of Wroblowa and Piersma)[46] that only the noble metals would do. Slightly faster rate constants could be obtained by using some alloys of noble metals with, e.g., copper, but it was never demonstrated that there would be a longer lifetime of the situation concerned. The reason for the necessity of using the noble metals alone is easy to discern: the oxidation potential of the hydrocarbons is above 1.00 V, and it is more positive than the dissolution potential of most of the ordinary metals. This puts a crimp upon ideas one might have of using something cheaper *among the metals*, and so we decided to go towards metal oxides.

At this time, about 1966, knowledge of the conducting properties of metallic oxides was not widespread, and I asked Darko Sepa, who was being supervised by Alexander Damjanovic, to research the literature for conducting oxides. He returned with the tungsten bronzes, of which I had never heard at the time, and the name of which I could not understand.*

We soon learned how to synthesize bronzes electrochemically, and Sepa started with an apparent success,[47] showing that the bronzes did catalyze oxygen, when in single crystal form, at much the same rate as platinum (Figure 5). This claim caused a ripple through the fuel-cell world, and an attempt was made to repeat the work at the General Electric Company. Although success could be obtained with Sepa's crystal, crystals prepared by others did not give the same results, although a catalytic activity, less than that of platinum, was shown.

The suggestion that, somehow, the results were due to trace of platinum in our crystals did not seem at first feasible, because, although we had used a platinum counter-electrode in their preparation, when we analyzed how much platinum there was on the surface of the bronzes, it was so tiny that, assuming that the rate constant of the small particles present would be the same as bulk platinum, we could not get sufficient effects.[48]

Later on this problem was solved by John McHardy in a creative way.[49] He showed that the particles on the surface of the bronzes were platinum, but that the platinum was associated with a surrounding band of alumina, which had come from the alumina crucible used in preparation. A hypothesis was made that the alumina got there at the same point

---

*It appears to have been gratuitously granted because compounds of the formula $M_{o.x}WO_3$ glint. They can therefore be called "bronzes" although, of course, their formulae bear no resemblance whatsoever to bronzes in the normal sense.

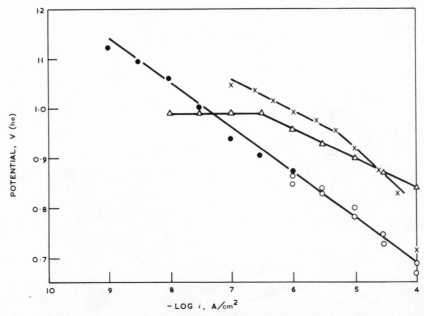

Fig. 5. Tafel lines for oxygen reduction in acid solution. ×, sodium tungsten bronze; △, oxide-free platinum; •, oxide-covered platinum; ○, oxide-covered platinum.[47]

as the platinum because it might be associated as colloid and counterion in the solution. Several of the results, the dependence of the rate of the oxygen dissolution reaction upon the amount of platinum present, could be fitted to a model in which their synergistic effect to platinum traces was to be explained by the platinum's catalysis of the first step in the oxygen-reduction reaction, a complicated Temkin-associated dependence of the coverage of radicals upon the platinum, and the "spillover" of those radicals to be catalyzed on the bronzes.[50] Alone, the bronzes could not work, for they did not have the ability to catalyze the first step (Figure 6).

There does not seem as yet to have been a major development of this fertile viewpoint. A difficulty in practice may be that the platinum, and its vital alumina setting, would not be stable; another difficulty would arise when one powdered the catalyst for use in a porous electrode. Thus, McHardy established by the use of ion-probe microscopy that alkali metal dissolved easily and retreated into the interior of the bronze. The part of this which would be denuded of its alkali metal component would become nonconducting. Although this would be an acceptable situation if the interface occurred only once, because the nonconducting path would then be so small, yet over the repetition of the interface for many millions of times, as with the small particles, a substantial resistance would add up. The practical value of the work on the bronzes, therefore, remains questionable, but the inherent heightened activity of tiny traces of platinum (which has been confirmed by other workers)[51] remains an original to Damjanovic, Sepa and McHardy.

Does electrocatalysis have a bright future? In the old days, and that means until about 1968 or 1969, we used always to think that "more research" would solve everything. It was perhaps Vietnam which taught us that quantity is not enough, that there is something more insidious about the difficulties of a research problem when it advances in complexity. When positive entropy increases, driving force decreases, and the difficulty of obtaining a solution to a problem like electrocatalysis is that there are so many factors

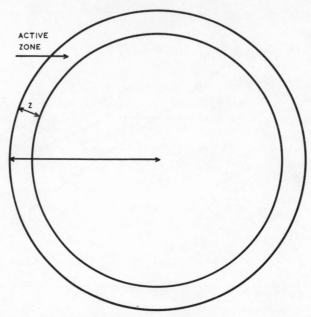

Fig. 6. Model proposed by Bockris and McHardy to explain the platinum-enhanced activity of sodium tungsten bronzes in oxygen evolution. In this model an "active" zone of platinum, width z, covers an inclusion of alumina, radius r.[50]

which affect the issue that one pales a bit before the research tasks involved. Just to mention a few gaps in knowledge before we could hope to make electrocatalysis rational:

1. We know very little about the adsorption of ions at *solid*-solution interfaces.[52]
2. What of the effect of adsorption at neighboring sites on the properties of catalytic sites?
3. What about adsorbed water and anions?
4. How can we obtain, about 1000 times more readily than at present, information on rate-determining steps and reaction paths?[53]
5. Catalysts with sites of different metals, and synergistic effects, may have a relatively short life.

There are two brighter notes on which to end: (a) The research which would answer some of the above questions has hardly begun, and so a degree of pessimism on the future of electrocatalysis is a reflection of an ignorance which may be expunged. The rate of this change will be a function of the amount of money government and industry put into its solution. (b) Nature has evolved enzymes and when these are better understood we may have learned much about electrocatalysis.

There seems to be an inverse relationship between the usefulness of a subject and the work done upon its background basic research. When no application can be foreseen there is enthusiasm among fundamental scientists for carrying out the work and even occasionally a bit of government money for doing it. But when one comes to subjects like polymers, friction, catalysis, fuel cells—subjects of vast practical importance, where advances could result in community benefit within a decade—the government funding is absent and scientists in universities lack enthusiasm. At any rate, in the days of plentiful research funding by NASA, many very complex research and development tasks were solved. Thus, if realization of the relationship between electrocatalysis and cheap energy comes to the money determiners before we are too poor because of rising energy costs, it may be that sufficient funding will do the trick in electrocatalysis research.

5. Electrode Processes

Although it is far from my thoughts to express criticism of the organizers of the London symposium, it does seem odd that we put the basic topic of the symposium in the middle, instead of at the beginning, although I suppose one could justify it.

The achievements of the groups in London, Philadelphia, and Adelaide have been considerable in the basic electrode-process field, and, one may submit, within the first few of the contributions to fundamental electrode process studies of the electrochemical research groups in the world. One of the ways in which I can present the work that has been done, or at least to skim over its surface, is simply to line it up with individuals concerned, roughly in order of dates.

Brian Conway's independent contributions to electrochemistry have been enormous. Even those which he made with me could form the subject of a long review, so let me pick out only the two headlines here of Conway's: his original paper (part of his Ph.D. thesis) in which he showed that 10^{-10} moles of impurities would affect an electrode surface and change the kinetics thereat (Figure 7), and his paper on the metal deposition problem (written at Pennsylvania), in which the potential energy–distance method was applied to prove the likely steps which would determine the rate at which a metal could cross the double layer, diffuse to building sites, and incorporate itself into a lattice.[54]

Roger Parsons would certainly be said to be one of the two or three most well-known people who obtained his Ph.D. in the Electrochemistry Group at Imperial College, and his own many later and authoritative contributions have overwhelmed those which he made with me. However, one of the papers which he wrote independently of my authorship, but shortly after he got his Ph.D., is what I call, within the group, "the 1951 paper," by which I mean his remarkably comprehensive analysis[55] of mechanism determination according to the quasi-thermodynamic approach. I have been surprised at the lack of resonance by others to this major contribution which Parsons made at such an early time in the development of electrode kinetics, but I suppose that we may have to await the attainment of some of the conditions laid down by A. K. N. Reddy in his address here on the future of the field before we are able to apply Parson's ideas, for they would need much experimental work.

E. C. Potter, who is now well-known in Australia, particularly in government scientific circles, was a founder-member of the group in London, and he should really be quoted here in terms of his great forte of solidity and reliability. However, I venture to suggest that the most original thing he did with me was the application, for the first time, of the concept of the water dipole potential as having an effect in electrode kinetics, and its addition of a $\Delta\chi$ term to the potential,[56] which enables us to claim priority over the paper of McDonald of 1954, where similar things were discussed although in a much wider context.

R. G. H. Watson has been an original contributor in several ways, and that which he made in our own group was the mechanism of hydrogen evolution from alkaline solutions.[57] However, it is never possible for me to mention Watson's work without saying that he made his greatest contribution with Bacon. The Bacon fuel cell would more justly have been named the Bacon–Watson cell because of the period which Watson spent with Bacon after 1952, in being the main electrochemical contributor to Bacon's effort on electrochemical energy conversion.

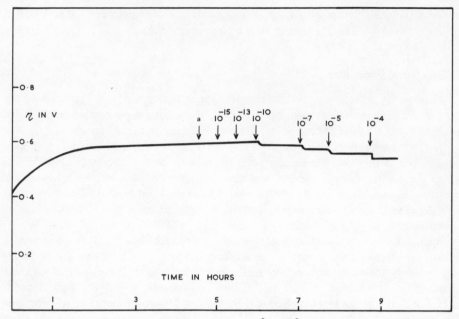

Fig. 7. Poisoning of Ni by As_2O_3 in 0.1 N aq. HCl at 10^{-2} A/cm^2. (a) 50 drops of 0.1 N HCl added as blank. Arrows show conc. of As_2O_3 in mole/liter.[54]

I. A. Ammar[58] is now a fortunate professor in the University of Riyadh, in Saudi Arabia. He was the first to give the theory for the Tafel line as effected by the change in the electrokinetic potential near the potential of zero charge with potential, i.e., he showed that the effect of the electrode potential upon the $\phi_{2\text{-}b}$ potential would make the Tafel slope different from what it would be on simple mechanism considerations. Such work was formalized by Parsons.[59] There is a degree of ambiguity about it, and about its separation from the "activationless discharge" studies of Krishtalik,[60] but at any rate, whatever the cause of a second straight line of slope less than $2RT/F$, e.g., for hydrogen on metals, Ammar was the first to elucidate in equation form the effect in the variation of the zero potential with potential on the Tafel slope.

D. L. Hill, who spent many years later doing electrochemistry in Hong Kong, contributed in the early 1950s to the original work in the purification of molten salts. It was his initiative which enabled us to get so far with elimination of water, which was a puzzle in the early days in London, in our finally successful attempts to deposit pure titanium from molten salts.[61]

Lucy Oldfield worked in the group in London when Roger Parsons was still there, and her original contribution is a simple one: the discovery of what I have called "the 0.84 potential," which is, in fact, a mixed potential formed by some oxidation and reduction reactions of hydrogen peroxide.[62]

Parry-Jones has claim to originality in the fact that he took the Rehbinder pendulum method of measuring "hardness" and used it so that it responded to friction.[63] There was some dispute between Frumkin and Rehbinder on the one side, and ourselves on the other, about the meaning of the results, but under the conditions in which Parry-Jones worked, it was the rolling friction of the ball of the Kater's pendulum upon the substrate which determined the decay of the motion, and it was this which was potential dependent, the method leading to a method for determining the potential of zero charge.

A. K. M. S. Huq was a bridge person between London and Philadelphia, and his name is celebrated in books of electrochemistry for his first experimental determination of the

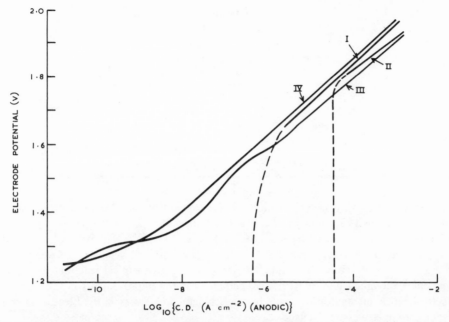

Fig. 8. Typical Tafel lines for oxygen evolution as a function of pre-electrolysis conditions. I, no pre-electrolysis; II, cathodic pre-electrolysis at 10^{-2} A/cm² for 24 h; III, only anodic pre-electrolysis at 10^{-2} A/cm² for 48 h; IV, cathodic pre-electrolysis 24 h; anodic 48 h at 10^{-2} A/cm².[65]

reversible potential of the oxygen electrode, something which had earlier only been calculated[64] (Figure 8).

Erik Blomgren was a giant among us and had so many endearing personal characteristics that a book should be written about him. Along with my first wife, Linda Bockris, he gave me the greatest help in the formation of the Pennsylvania Group. His scientific contributions, those which should have a lasting place in some list of "firsts," were his work on the relation between the shape of an organic molecule and its ability to adsorb upon mercury[65] and iron, and how this can be related to the tailoring of corrosion inhibitors.

Mauser will be remembered by those in Philadelphia in 1956–1957 who lived with him through his frequent emotional ups and downs. In the literature he does not get his fair mention, because his paper was published in the less well-known *Canadian Journal of Chemistry*,[66] rather than one of the American or British Journals. It is the first full analysis [before that of Vetter[67]] of hydrogen electrode kinetics, taking into account both the backwards and forwards steps of the large number of possible paths, and evaluating with mathematical rigor and objectivity the precise conditions for a "rate-determining step."

Wolfgang Mehl was from Göttingen and was possessed of a very pleasant part of "the Old Spirit"—the plentiful use of creative leisure in the academic class. His work was interspersed with games of tennis in the afternoon and absences for (thinking) vacations at unannounced times. It was partly because of this, he said, that his very intelligent contributions were so substantial, and it was he who made the first evaluation of the concentration of metal adions on the surface of electrodes (Table 1).[68] Later, particularly after the introduction of the potentiodynamic method in which the current and potential are both varied simultaneously, thus making the analysis of the significance so much more difficult, the work of the electrochemistry group was criticized for not being sufficiently

Table 1. Values of Surface Adion
Concentration on Ag Electrodes
from Various Workers[68]

	C_{ad}^o (mole/cm^2, $\times$ 10^{-11})
Mehl and Bockris	90
Gerischer	15
Lorenz	3
Despic and Bockris	
Nonactivated	3
Activated	160

oriented towards "fast-sweep" methods. It is noteworthy that Mehl's contributions to this field came in 1957, followed by those of Inman in 1959, in the nanosecond range, being preceded only by hydrogen studies by Knott, Breiter,[69] and Will,[70] and by the original paper of Bowden and Rideal.[37] Mehl's galvanostatic sweep analysis was the first to evaluate the rate constant of the *fast* step in an electrochemical reaction. It served as a stimulus for much else, particularly, perhaps, for the improvements in the theory of electrochemical crystallization made by Fleischmann and Thirsk in the Electrochemistry Laboratory in Newcastle, England.

Michio Enyo, with whom we have had so many contacts more recently in our connections with Japanese electrochemists from the Australian position of the group, was the author of an original contribution towards the theory of metal deposition, in one of which (with Kita) he derived the potential dependence of the beginning of rotation of spirals in crystal growth.[71]

Nanosecond measurements are still rare in electrochemistry. They were made for the first time in 1959 in Philadelphia by Douglas Inman[72] (inspired by the instrumentally masterful Blomgren), who also contributed originally with the first quarter-time potential method within the molten-salt situation to obtain the concentration of complexes formed by halide ligands in nitrate groups.[73] He was the first person to measure the lifetime of complexes in molten salts and thus destroy the myth that such entities might not have any operational reality.

Despic, the source (with Branko Lovrececk) of so many of our Yugoslav co-workers, was particularly original in his high-current-density (fast-sweep!) measurements with metal deposition, in which he was the first to observe the ending of the Tafel line (Figure 9), corresponding to the potential (as we now realize) at which the electron energy level is lifted above the ground state and rate of the acceptor levels in solution.[74]

David Koch, who must now be in terms of numbers one of the greater leaders of electrochemical teams within the Western world, was the first to use the primary isotope effect in electrochemical systems and to discuss the effect of water on chemisorbed bonds.[75]

Jim Barton, who now works in Paris, did not realize the importance of his own contributions at the time he made them. He built up an involved piece of equipment for the observation of the growth of dendrites, but the original part of his contribution was the first absolute calculation of a crystal growth rate.[76] An absolute equation for the dendritic case, in Barton's model, turned out to be quite easy, and I am not aware of any other such absolute calculation of crystal growth rate at this time (Figure 10).

Devanathan was such an original man in so many ways, both within and without the

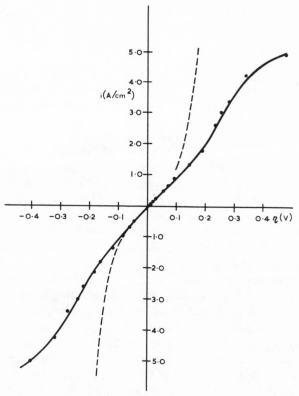

Fig. 9. Typical i–η relation with an activated Ag electrode in 0.364 N AgClO$_4$ + 1 N HClO$_4$ solution. Dashed line, Butler–Volmer curve with β kept constant. Note the non-Tafelian behavior of the experimental i–η curve.[74]

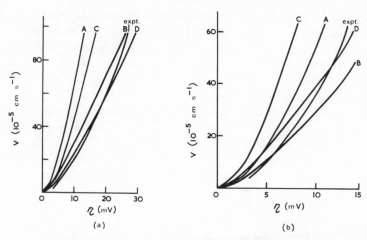

Fig. 10. (a) and (b): Comparison of some calculated (A, B, C, and D) and observed velocity–overpotential curves for the electrolytic growth of dendrites in ionic solutions.[76]

electrochemistry group, that it seems a pity only to mention one of his many contributions: This must, of course, be the theory of the double layer, including the isotherm with imaging in the metal and the contributions from the adsorbed water-molecule layer.[3]

Sascha Damjanovic, who supervised part of the group in the U.S.A. for a decade, made considerable impact with his comprehensive formulation of the theory of diffusion-controlled transport of metal with nucleation,[77] but his was the mind which kept us clearly on the solid-state track in many aspects involving the theory of metals.

Klaus Müller winds himself in and out of our tales in respect to contributions to the theory of double-layer structure with Devanathan; to the more advanced theory made with Gileadi, after Devanathan had gone back to Ceylon for the third time, and also in respect to the evaluation of the faradaic admittance, with Gileadi. *Nearly* always happy, particularly when working very closely to Zlata Kovac, Klaus was at his best when it came to long and involved numerical calculations with computers, then, of course, still novel. Among research students who published papers in worthy journals before his Ph.D. (a full paper in the *Proceedings of the Royal Society* and one in the *Journal of Chemical Physics*), his situation must make something of a record.

Halina Wroblowa was the second mother of the group. There is no doubt that the contributions made in the setting up of the group by Linda Bockris were unexceeded, but the sophisticated intellectual contributions which the hypercritical Halina was able to bring, lifted many discussions to a high level. As to original contributions published, one recalls particularly that on the oxidation of ethylene, the interpretation of the negative pressure coefficient, and the concept of rate-determining water discharge.[78]

Anderson was the first person to put forward a quantitative theory of contact (or specific) adsorption.[79] He showed that the vague suggestions of David Grahame in terms of chemical bonding were not tenable. He utilized the observations of Devanathan and Müller that anions were nonhydrated and cations were hydrated in the double layer to work out a theory which related the dispersive interaction energy of the large anions with the metal and their consequent ability to displace water molecules and to cover the electrode in the inner Helmholtz plane.

Zlata Kovac has recently won many awards in the hard school of I.B.M. research, and may now be counted among the more well-known women scientists in the U.S.A. She contributed more slowly at the beginning, but at any rate [considerably helped by Wroblowa in many weekend threesomes in the Pocono Mountains near Philadelphia, proving the efficacy of work in idyllic circumstances on hot days near swimming pools[80]] she proved that the position of the hump in capacitance-charge studies was identical with that at which the inflection on the specific adsorption vs. metal charge curve occurred, thus confirming a prediction of the BDM model.

Srinivasan must be considered, publication-wise, as one of the most productive persons in our group. He contributed most originally to the calculation of the separation factor[81] and also to the theory of porous electrodes, together with his writing of the book on fuel cells.[82] In terms of rapidity of repartee and penetrating observation, discussion with Srinivasan was the fastest and brightest of all.

Leonard Nanis, who was a supervisor for three years in the early 1960s, had his most original contribution in collaboration with Beck and McBreen. Together they established a simple way (Figure 11) of proving that the concentration of hydrogen in a metal depended upon the local stress exponentially, so that when the stress exceeded a certain critical amount, the solubility increased suddenly and greatly. This contribution,[83] the originality of which was for a time disputed by Oriani and Darken,[84,85] has had re-

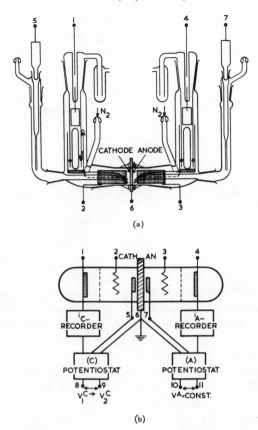

Fig. 11. (a) The electrochemical cell used to study the effect of stress on the solubility of hydrogen through metals. (b) The electrical circuit of the cell.[174]

percussions in basic metallurgical work, and its significance is fully reviewed by Subramanyan in the first volume of the MTP series.[86]

Bhimasena Rao has been mentioned here in respect to his contribution to the first automatic ellipsometer. This leads us directly to A. K. N. Reddy and his contributions to the group, which took place in Philadelphia. I certainly have to place Reddy along with Blomgren, Conway, Damjanovic, Devanathan, Gileadi, Parsons, Srinivasan, Wroblowa, and a very few others, for his superb contributions to the group, which were in many directions other than the best known ones in the major textbook. Stress upon this tends to underrate the originality of Reddy in his suggestions, and it should be well known that he was the first to introduce potential control and computerized calculations into ellipsometry[87] in electrochemistry, although there again we must acknowledge the contributions of Jerry Kruger, who worked along with Reddy in those days and who contributed the programming (Figure 12).

Marvin Genshaw, whose contribution to the group included supervision of a number of students, is best quoted for his theory of the dependence of the current at the rotating disk and ring upon the angular velocity of the rotation,[88] developed with the prolific Damjanovic. This technique has been developed extensively, and improved, by Russian workers.[89]

In noting the contributions of Eliezer Gileadi to the group, one comes to the level of the greatest names of those who worked in it. Now one of the most active electrochemists

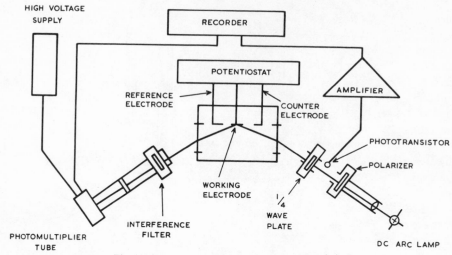

Fig. 12. Generalized ellipsometry experiment.[52]

in the West, Gileadi made many contributions as a supervisor of the group in Philadelphia, one being the formulation of the equations for the faradaic admittance of electrodes as a function of frequency when the electrode is polarizable.[90] A qualitative theory had been given at an earlier time, and indeed the idea harked back to a visit which I made to David Grahame's laboratory in Amherst, Massachusetts, in 1953. Grahame was very worried that as he always plotted the resistive component of his bridge, he found it depending greatly upon frequency. He then threw the results away because he could not explain them. Gileadi, working with the indefatigable Klaus Müller, was able to give a mathematical framework to the ideas advanced by Conway, Mehl, and Young much earlier.[91] Further, Gileadi was able, again with Müller,[92] to take the molecular theory of the adsorption of organic molecules, which had been originated by Devanathan and Müller, to a higher degree of approximation. This is not always recognized, one finds, by some of our Russian colleagues.[93,94]

Shyam Argade will be remembered as a worker who determined,[95] under one of our NASA contracts, the potentials of zero charge of various metals, and he stressed this concept to a Western audience, after its having been put forward originally by Frumkin,[96] and so frequently by Antropov.[97] However, Shyam Argade's most important contribution, that for which he makes this list, was not in the determination of the potential of zero charges, nor in the origination of a new method to do this, but in the paper which he published[10] in 1969, in which he showed how to calculate the absolute metal–solution potential difference. This gigantic *pons asinorum* of electrochemistry had long been a stumbling block to progress, and a number of graves of the reputations of electrochemists who have tried to *determine experimentally* this difficult quantity lie around the world. Argade showed that the determination of the outer potential of metals, together with a knowledge of the metal–metal PD in cells and an *estimate* of the amount by which the surface potential difference was changed upon bringing an aqueous and a metal phase into contact, could give rise to a *calculation* of the absolute metal–solution potential difference.[10] The principle of the approach was later, essentially, confirmed by Trasatti,[11] although the latter made important numerical corrections to it: he improved the value of the absolute heat of solvation of the hydrogen ion which had been used by Argade, and also that of the Fermi energy. Recently, M. A. Habib,[98] working in

Adelaide, has shown that these concepts can give rise to the calculations of "$\chi_{electron\ overlap}$," that elusive and little-mentioned, but vital, quantity (Figure 13).

Boris Cahan's contributions cannot be judged in a simple way. He gave many positive benefits to the group—as well as detracting from it in his many fights with me—but this account is not a history of the group, rather an attempt to look back on 30 years of its work, and therefore I would like here just to pull out the really original thing about Cahan's contributions: the measurement of contact angle in a porous electrode situation, and the ability, thereby, to form up a basic and complete theory of the distribution of current in a situation near to that of pore-type porous electrodes.[99] Many would say, however, that Cahan's constructive and directive suggestions on apparatus construction and experimental methodology gave a vital backbone to the execution of many ideas for new techniques which I had, which were ill-formed and impractical.

Ljerca Duic was the first to check simultaneously, on the same systems, the electrochemical radio-tracer method of determining the adsorption of organic materials on metals. With Gileadi it was possible to find from her work an alternative approach towards the determination of radicals on surfaces.[100]

Rowshan Mannan,[101] who later contributed two second-generation members in the shape of Habib and Khan to the group in its Adelaide days, was also often active during weekends near the swimming pool in the Pocono Mountains in Pennsylvania. The theory to which her work gave rise is a proof of what had often been stated, but never established, that the heat of activation for reactions not involving chemibonding would be independent of the Fermi level of the metal (Figure 14).

Vlasta Brusic comes next, not, I think, because of her work on the mechanism of the passivation of iron,[102] which was good enough but could be compared in originality with those of others, but because she was the first (along with her co-authors, Genshaw and Cahan) to build a *successful* automatic ellipsometer.[103] One automatic ellipsometer had been built by Bhimasena Rao and A. K. N. Reddy, as early as 1966,[104] but it never functioned well. The apparatus built by Brusic with the (technical and emotional) assistance

Table 2. Summary of the Internal Pressure Calculations as a Function of the Mechanism of the Hydrogen-Evolution Reaction.[105]

Mechanism	Fugacity (f_{H_2} atm)	Remarks
Fast discharge–slow combination	$\exp\left(-\dfrac{2\eta F}{RT}\right)$	Nernst equation valid, could be embrittling
Slow discharge–fast combination	1	Nonembrittling
Slow discharge–fast electrochemical desorption	$\exp\left(\dfrac{2\eta F}{RT}\right)$	Nonembrittling
Fast discharge–slow electrochemical desorption	$\exp\left(-\dfrac{2\eta F}{RT}\right)$	Nernst equation valid, could be embrittling
Coupled discharge–combination	$10^{1.5} \times \exp\left(-\dfrac{\eta F}{2RT}\right)$	$\theta \ll 1$, could be embrittling
Coupled discharge–electrochemical desorption	$\exp\left(-\dfrac{2\eta^* F}{RT}\right)$	Only predictable exactly if η^* known experimentally

of Cahan, and the encouraging supervision of Marvin Genshaw, was the machine which later on became the model built by industry.

P. K. Subramanyan has a claim to this list by virtue of his proof of the pressure theory of the breakdown of metals: namely the calculation based upon the work of Beck, Bockris, Nanis, and McBreen, which gave rise to a calculation of the pressure inside the voids. He showed that the corresponding expected gas pressure, based upon overpotential studies (mechanism dependent), would be the same (Table 2).[105]

John Appleby, who now circulates between the large French government electrochemistry laboratory at Bellevue, near Paris, and the Laboratoire d' Electricité de France, came to the group at the end of its American period, and stayed only a few months. During this time, however, he proved fertile in interaction with Rajat Sen, and together they debunked much of the Levich (cripplingly simplistic) theory of the long-distance activation of electrochemical particles by polaronic interactions with surrounding librators.[106] In a typically quiet way, Appleby showed that the basic equations of the

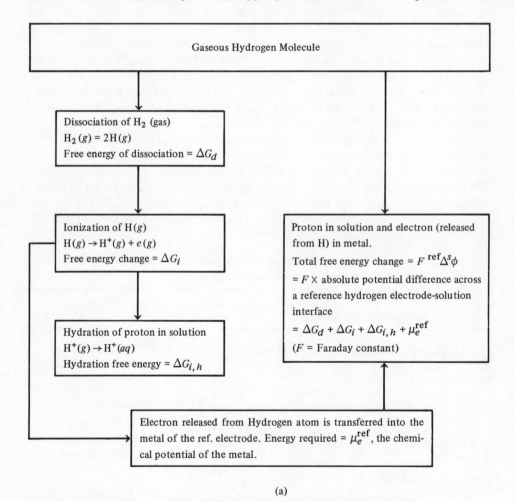

(a)

Fig. 13. (a) Procedure to estimate the absolute potential difference across a reference hydrogen electrode–solution interface. (b) Estimation of the electron overlap potential of a metal in contact with a solution and the absolute potential difference across a metal–solution interface.[98]

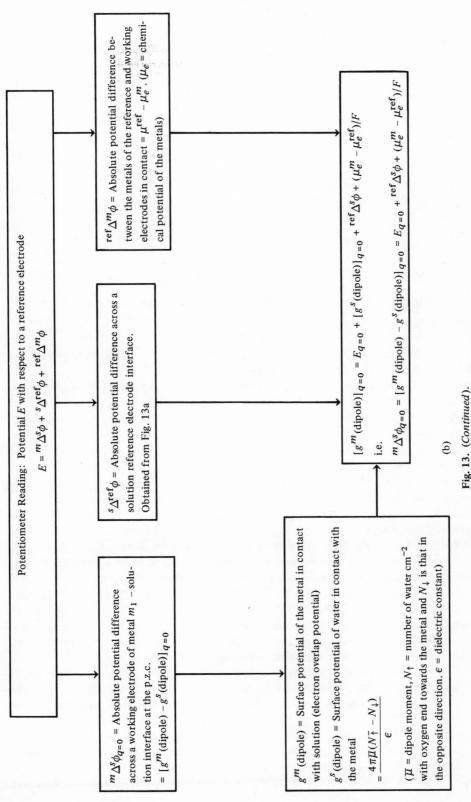

Potentiometer Reading: Potential E with respect to a reference electrode

$$E = {}^m\Delta^s\phi + {}^s\Delta^{\text{ref}}\phi + {}^{\text{ref}}\Delta^m\phi$$

$^{\text{ref}}\Delta^m\phi$ = Absolute potential difference between the metals of the reference and working electrodes in contact = $\mu^{\text{ref}} - \mu_e^m$. ($\mu_e$ = chemical potential of the metals)

$^s\Delta^{\text{ref}}\phi$ = Absolute potential difference across a solution reference electrode interface. Obtained from Fig. 13a.

$^m\Delta^s\phi_{q=0}$ = Absolute potential difference across a working electrode of metal m_1 – solution interface at the p.z.c.

$= [g^m(\text{dipole}) - g^s(\text{dipole})]_{q=0}$

$g^m(\text{dipole})$ = Surface potential of the metal in contact with solution (electron overlap potential)

$g^s(\text{dipole})$ = Surface potential of water in contact with the metal

$$= \frac{4\pi\bar{\mu}(N_\uparrow - N_\downarrow)}{\epsilon}$$

($\bar{\mu}$ = dipole moment, $N_\uparrow$ = number of water cm^{-2} with oxygen end towards the metal and $N_\downarrow$ is that in the opposite direction. ϵ = dielectric constant)

$[g^m(\text{dipole})]_{q=0} = E_{q=0} + [g^s(\text{dipole})]_{q=0} + {}^{\text{ref}}\Delta^s\phi + (\mu_e^m - \mu_e^{\text{ref}})/F$

i.e.

$^m\Delta^s\phi_{q=0} = [g^m(\text{dipole}) - g^s(\text{dipole})]_{q=0} = E_{q=0} + {}^{\text{ref}}\Delta^s\phi + (\mu_e^m - \mu_e^{\text{ref}})/F$

(b)

Fig. 13. *(Continued).*

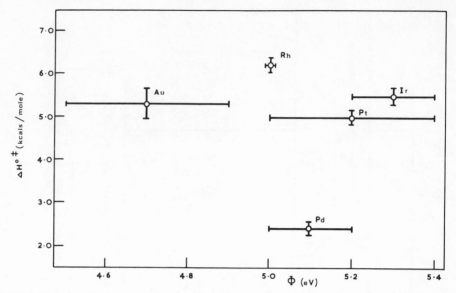

Fig. 14. Heat of activation involving chemibonding is independent of the workfunction ϕ of the metal.[101]

so-called quantum model, which Levich had been putting forward for many years, could be deduced in a few lines upon classical grounds.[106]

Won-Kie Paik, a silent man who seldom spoke unless it was to say something important, quietly watched our efforts in ellipsometry for more than a year and saw how much they were dependent upon the crutch of auxiliary measurements, e.g., coulombic measurements of the thickness or of refractive index. He started off along the same lines as the others, partly tutored by Brusic and by Cahan, but one day he drifted into my room and said in his undertonic way: "I have solved the ellipsometry problem." After that we wrote a paper[107] (I hasten to make clear that my part in it was largely correcting the English and making it understandable to others, first to myself) which changed the direction of ellipsometry in electrochemistry and made it possible to dispense with the auxiliary methods. In claiming so much for Paik, we must acknowledge the help we had from Jerry Kruger and Mrs. McBee in the refereeing and improvement of the paper. They had, in fact, considered the same idea at an earlier time but thought it unattainable. Paik, however, had found out the particular condition at which it worked.

Ron Fredlein, the last import from Australia to the group in the U.S.A., attained a goal which many electrochemists had attempted: he turned the electrocapillary thermodynamics from something which applied only to mercury to something which applied to the rest of the metals by making a device which bent, on application of a potential, to a degree proportional to the surface tension (Figure 15).[108]

John McHardy has already featured in this account for his contributions to synergistic effects in catalysis at electrodes and for his application of spillover theory.[49,50] The latter, perhaps, explains the inherent effect of tiny traces of noble metal material on electrodes.

Bill O'Grady represented the first student in our group who was of the "new type," a jeans wearer with beard, frequent expressions of dissidence with the surrounding established order, and an advanced version of the concept of flexitime. Bill turned out, correspondingly, to be an extremely original worker, and his first application of Mössbauer

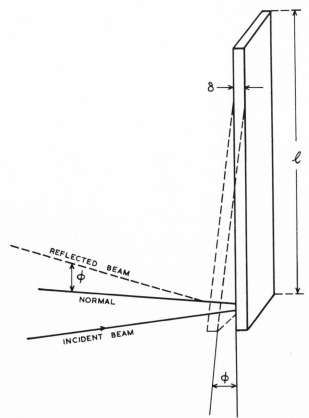

Fig. 15. The measurement of the surface tension of a metal electrode according to Bockris and Fredlein. The method is based on the bending of a thin beam of electrode material with potential.[175]

spectroscopy to electrochemistry[109] gave definitive information on the mechanism of passivation. It indicated, according to O'Grady, that the earlier work of Kruger[110] and of Cohen[111] on electron diffraction had been spoiled by heating; they had, in fact, been examining the lattice of the dehydrated film. The essence of what O'Grady said was that "the morphology of the passive layer created by water molecules is the essence for the reason for the passive behavior." This has now been confirmed by Winston Revie and Bruce Baker in Adelaide, using Auger spectroscopy (Figure 16).[112,113]

One of the well-known names which shall be mentioned here is certainly the principal organizer of this meeting, Felix Gutmann, the *enzyme of electrochemistry*. But Felix's more recent contributions and inspirations of so many organizational efforts are backed by solid, original contributions of his own in basic electrochemistry. Apart from that seminal gift which he gave to us (with Breyer) in a.c. polarography many years ago,[114] I mention electrochemical noise, about which he was the first to publish.[115] Incidentally, although Felix is really an *honorary* member of the group, he can be justified for inclusion for he has worked in the group's laboratories in Philadelphia from time to time and made contributions as a consultant to some of our work.

Shahed Khan[116] is a man of rare gifts, a brilliant student. It was he who succeeded Rhajat Sen in making contributions to quantum mechanics in Adelaide. He is the author of a paper which I believe will be well known in a few years in showing, for the first time,

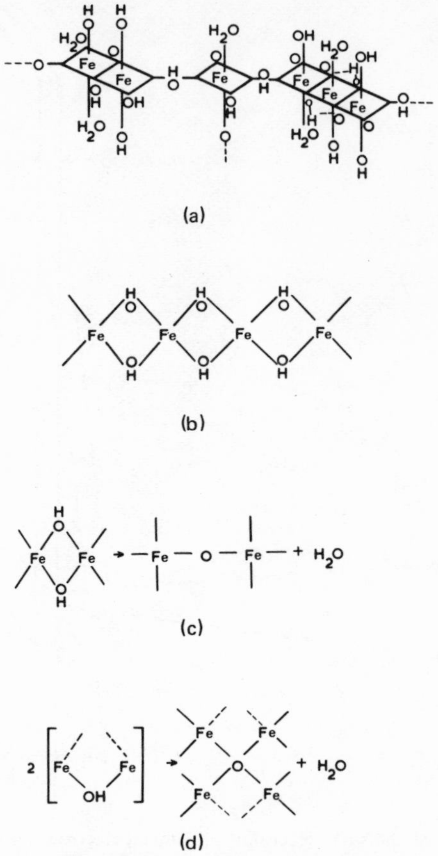

(a)

(b)

(c)

(d)

Fig. 16. (a) and (b). Postulated structures of the passive layer on iron before dehydration. (c) and (d). Postulated structures of the passive layer on iron after dehydration.[109, 112, 113]

the relation between the quantal aspects of spectroscopic and electrochemical transitions (Figure 17).

Having spent so much time upon the glorious past, it seems to have the worst implication—to give a relatively short treatment of the glorious future. However, an outline of the future course of research in electrode processes has been given well by Amulya Reddy at this meeting, so I am going to refer to his paper for ideas which, though they are utterly his own, I find are similar to those I would have liked to express.

I want to add only one major point to what has been said in Reddy's paper and that is that electrochemical kinetics must be thoroughly quantized.[117] It remains a lacuna in this respect. This is all the more curious because Gurney's paper of 1932[118] could be regarded as one of the first two or three treatments of chemical topics in terms of quantum ideas. Electrochemistry, again, had the lead, but lost it, as occurred two centuries earlier when electrochemistry was so much ahead of "physical chemistry."

Further, the electrochemical system (looked at quantally, a distribution of electronic states for which the laws are well known in the metal, and a problem in time-dependent perturbation theory for the frequency of transition to acceptor states in the solution) is a particularly suitable problem for an exercise in quantum kinetics.[119]

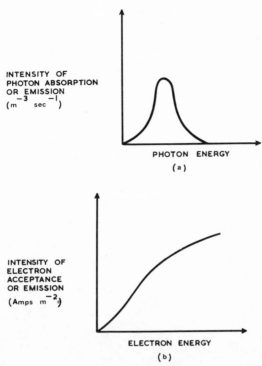

INTENSITY OF
PHOTON ABSORPTION
OR EMISSION
$(m^{-3} sec^{-1})$

PHOTON ENERGY
(a)

INTENSITY OF
ELECTRON
ACCEPTANCE
OR EMISSION
$(Amps\ m^{-2})$

ELECTRON ENERGY
(b)

Fig. 17. The intensity of absorption and acceptance of photons and electrons as a function of their respective energies. (a) Intensity of photon absorption or emission with photon energy in the spectroscopic case. (b) Intensity of electron acceptance or emission with electron energy in the electrochemical case.[116]

It is in this direction that one major pulse of electrochemical work at the Flinders University in South Australia is now proceeding. Its direction is continuous with the contributions made in Philadelphia to the quantum mechanics of proton transfer at interfaces by Dennis Matthews. It harks back to the well-known contributions to the mechanism of proton motion in aqueous solution made by Brian Conway in Philadelphia, utilizing the London-based experiments of Hedda Lynton (who, of course, is really Hanna Rosenberg).

6. Semiconductors

The contribution made by our group in this field is not great in quantity, but it has had the honor of being the seat of some original contributions, because of the presence in the group for some three years of Mino Green. His contributions[120] preceded those of Gerischer and Beck,[121] to whom, erroneously, much of the original theoretical work in semiconductor electrochemistry is usually ascribed.*

In respect to the future of work on semiconductor electrochemistry, it is easy to make optimistic predictions. One thinks of an increase in range of properties on the solid

---

*In fact, the original publication in semiconductor electrochemistry was made by Garrett and Brattain[122] in the *Bell Telephone Journal* for 1954. The paper was wrong in its fundamental equations, Nernstian oriented, but nevertheless it did describe the semiconductor in an electrochemical situation and was part of the program at Bell which led to the transistor, for which the authors received a Nobel Prize.

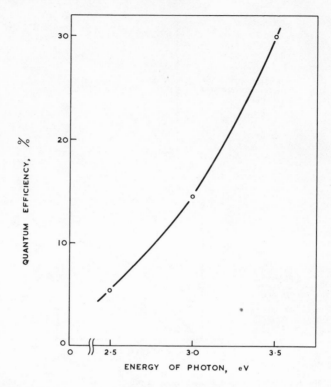

Fig. 18. The predicted quantum efficiency of a p-GaAs electrode as a function of the energy of the illuminating photons at −1.0 V with respect to the flat band potential.[123]

side of the double layer and the possibility of adjusting the energy levels in these to an almost infinite variety, so that any desired condition can be attained. However, one comes down to earth with a bump when one realizes that the maximum transfer rates of electrons and holes in most semiconductors are small, so that current densities which can be passed across the interfaces concerned are less than 1 mA/cm^2. From the point of view of applications, semiconductor systems have limitations.

In looking towards the future of semiconductor electrochemistry, one must not omit considering light effects. Interfaces involving semiconductors are particularly subject to the effect of light, and it is easy to comprehend, quantally, why there is more chance of getting the electrons raised to the conduction band to live long enough to exist at the surface and do something useful. It is because of the energy gap, a *forbidden* zone, and hence the falling back of the electrons to lower levels is less easy than it is with metals.

We are at the frontier here at Flinders with an attempt by K. Uosaki and J. McCann to make hydrogen by photo-driven cells, but the possibilities in respect to efficiency are not yet well defined (Figure 18).[123]

7. Corrosion

Contributions to the fundamentals of corrosion processes have been a major part of the work carried out by the electrochemistry group. The work has extended over many years: the fundamental formulation of the electrode-kinetic situation, the application of optical methods to the examination of the passive film, and above all the formulation of the properties of metals as a function of the hydrogen content. However, it is only wished

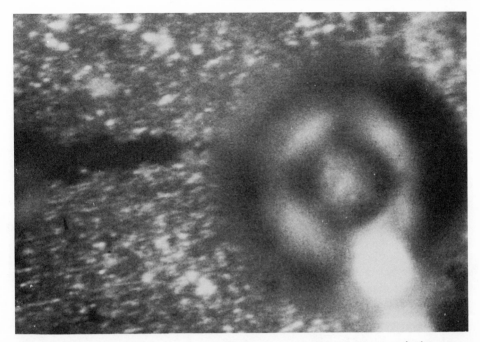

Fig. 19. A hydrogen bubble evolving from the tip of a stress corrosion crack.[165]

here to point out one historical fact and that is that the 1954 paper which I authored in Volume 2 of *Modern Aspects of Electrochemistry* contains the first formulation of the equations which describe corrosion in electrode-kinetic terms.[124] There is no doubt that I derived my ideas qualitatively from those of Wagner* and Traud. However, these authors had not obtained the relations of the mixed potentials in terms of exchange-current densities. During the writing of the article on electrode kinetics in the first volume of *Modern Aspects*, I was in the fortunate position of being in my Visiting Professor year at Pennsylvania. I had only one research student, the illustrious Shamshul Huq, who first observed the reversible oxygen electrode potential. I was thus free to think alone, and I became fascinated by the thought that I could derive a corrosion rate entirely theoretically, even though only for an ideal system. There always has been a flaw in these equations in that they assume an equality of the cathodic and anodic areas. Attempts to eliminate this flaw could be of value, although the complications of the real systems are great and stress in corrosion theory has passed largely from the steady corrosion of ideally imagined flat surfaces to the theory of the spreading of cracks (Figure 19).

In taking a look into the future of work on corrosion, one is amazed at conferences by the large number of workers and the small amount of fundamental work. Thus, even elementary processes, such as the degree of aggregation of hydrogen towards the head of an advancing crack, are still relatively unresearched.

Einar Mattsson has given at this meeting a most imaginative and helpful survey of some advances now going on in corrosion, but it seems unlikely that I shall again contribute fundamentally to the corrosion area, although I have founded the Corrosion Laboratory at the Flinders University, largely for practical help to industry in nondestructive testing.

*Although my personal acquaintance with Carl Wagner dates back to 1951, and my contacts with him at M.I.T. in the 1950s were fairly frequent, I did not have contact with him in the development of the equations to which I refer.

Mattsson's description of the use of light and sound for the examination of a corroding specimen is fascinating. One can let the imagination fly towards the evolution of metal structures with decreased dislocation concentrations and utilize the formula derived by Beck *et al.* to relate the local concentration of hydrogen to stress. In compressive stresses, this concentration is reduced, and there should be some possibilities of utilizing this fact for reducing stress corrosion cracking.

Of all the areas in science and technology, it is in the field of new sources of energy that the need for funding is the greatest, for here we could terminate our civilization with a couple of generations if we do not act in time.[1] However, one may think that a second most important area—so long as one interprets the term widely—is that of corrosion. Corrosion is the essential wasting function of the material world. It bears to materials a relation analogous to that of cancer in biology. Although it will not be eliminated, its reduction is within the capabilities of our present knowledge, were this known to and applied by engineers. There can be few subjects where the ratio of likely commercial gain to money invested in research looks so large. Here again we are at a turning point. What steel company wishes, really to stop corrosion, i.e., to cut sales? All this will be different in the future in a resource-exhausting world. The diminution of corrosion, and its reversal by electrochemical regeneration, will have to become a regular part of the steady-state system which we must attain within the next decades.

8. Electro-organic Chemistry

The main contributions to this area which we have made are associated with the name of Wroblowa, as also with those of Johnson,[125] Piersma,[126] Kuhn[127] and others. We were the first to evaluate the mechanism of the oxidation of hydrocarbons and to study the adsorption situation, including the distinction between reversible and irreversible adsorption, as a necessary prerequisite to the study and evaluation of the mechanism.

A curious rift has developed in this field in which there appear to be those who study it without considering the adsorption of intermediate radicals on the surface of the electrode! This is something like attempting to put on Hamlet without the Prince, for many of the rate determining steps in the oxidation of organic compounds are not electron-transfer reactions across the double layer, but chemical combinations on the surface. It is possible that there are cases in nonaqueous solutions where the adsorbed concentration is small and where the mechanism of the electro-organic reaction approaches that of a redox reaction, but it seems that a treatment of electro-organic chemistry without taking into account the adsorption and the surface kinetics of radical intermediates is sadly lacking in contact with the rest of chemical kinetics and the electrochemical (and electro-catalytic) evidence.

As to the future, here, it will have two aspects. First, there is the mechanistic side, and I refer once more to Reddy's paper for an imaginative sketch of how we may tackle complex mechanisms in the future. The other aspect—how much we shall go in for electro-organic syntheses, rather than the thermal syntheses of the present time—depends on how much the idea of electrogenerative reactions is developed. Organic reactions which occur spontaneously in the thermal mode must have negative changes in free energy. It should be possible to form with them an electrochemical system which would produce the necessary product *and* electricity, easily collected and stored instead of the waste heat, which is the main product with most thermal reactions: and the collection and storage of which is impractical depending upon the economics of our energy situation in the future, it may be possible to utilize electricity to drive some reactions when this is needed.

The electrochemistry of (organic) charge-transfer complexes, a field recently opened up by Felix Gutmann,[174] also offers considerable prospects e.g. for the stabilization of otherwise highly unstable systems such as metal azides.

9. Electrowinning and the Deposition of Aluminum

It has been my fortune to collaborate for some 20 years with Noland Richards, the Research Director of the Reynolds Metals Company in Sheffield, Alabama, and this experience has led me to the view that in Dr. Richards one finds an example of the optimal union of fundamental knowledge with a talent for fruitful application and organization of personnel. These years of work with Richards in the aluminum industry have taught me much, and there is no doubt that the balance of knowledge exchange between the industry and myself has been in my favor. I learned two great lessons:

1. Real systems are very complicated beside the simple ones with which the fundamental scientist deals. One becomes humbled by reality in dealing with the actualities of an industrial plant. Thus, for several years, Richards and I worked (in collaboration particularly with Barry Welch) to lower the overpotential in the anodic reaction at carbon. However, when we succeeded,[128] we found we had increased the rate at which the electrodes were thermally consumed, so that although it appeared desirable to reduce the cost of the electrical energy per unit weight of Al produced, it turned out that our efforts were, overall, in the wrong direction. Feedback and systems analysis—these are what one lacks in insight at first, when applying fundamental knowledge to real systems.

2. I learned the reality of the influence of invested capital. I had learned about this intellectually when I was 15, but I *realized* it properly for the first time at the Reynolds Metals Company where I saw the hundreds of millions of dollars invested in the plant, the vast halls full of cells of a certain type, which *had* to be amortized, i.e., the plant had to be used until the bank loans which had bought them on certain assumptions had been paid off. After that, I was humble with my suggestions, as a consultant, as to what might be changed. A suggestion which a consultant can make to industry has to be very brilliant if it is to show the company a way whereby it can throw away an unamortized plant and build a new kind of plant with a new method, and yet still come out with a better net profit *and* pay off the bank.

As to the future of the winning of aluminum, there is one sense whereby the present situation can be described as stupid. Here is this old process of Hall[129] and Herriault,[130] with its complexity and high cost, set up the world over, firmly entrenched. On the other hand, there are several suggestions—this is not the place to outline them—for cheaper ways of preparing aluminum than the old process developed in the last century. But the vested capital holds the old way firm. When a new plant has to be built, and the vested capital argument can be forgotten, one comes to the vested knowledge argument: The engineers the company employs (not too young, some of them) *know* how to rebuild the old kind of plant, in a new place, and they are against building a new kind which may not run well and thus cost them their jobs, with all the anxiety of "relocation."

Do I express pessimism about aluminum? Perhaps! But the change may come in respect to the fact that we have limitless aluminum only in clay, whereas supplies of easily available bauxite are exhausting. But clay demands extraction of aluminum by chlorine, and therefore the development of a new process, and one may see a breakthrough towards a simpler and cheaper method for the extraction of aluminum, via perhaps the electrolyzing NaCl and the use of the products to form an electrogenerative synthesis of Al from $AlCl_3$ in NaCl.

10. Electrowinning of Metals other than Aluminum

This field was well ploughed in the group in London days, particularly by Douglas Inman,[131] along with Hill, Menzies, and G. J. Hills.[62] The work was in full swing at the end of the London period and resulted in, e.g., the evaluation of the conditions for obtaining dendritic titanium deposits, to be distinguished from the fine particles obtained when there is a preliminary deposition of alkali metal and the titanium comes from a secondary reaction in solution.

Interesting work in this area was carried out by Inman in the early part of the American period: For example, he made pioneer chromopotentiometric measurements which gave information on complex concentration in the solution.[132]

The future in this field appears to be a valuable one because of the increasing viability of recycling, and one wonders that more research money is not yet being devoted to it.

11. Mineral Processing

There are excellent minds who devote themselves to the electrochemistry of this subject in Australia, and David Koch, whose mind is among these, is a former member of our group. This is a complex area, and the contributions to it of members in the group—while they were in the group—have been minimal so I shall not say much about it except in respect to one basic matter.

This concerns whether the interfaces between such bodies as the cupric sulfide ores, and the surrounding solutions, should be regarded in a Nernstian or a Sternian way. Roughly, this means whether the interfaces are nonpolarizable, and thus subject to Nernst's Law, or largely polarizable. There seems to be a mix-up in the theory here, and part of the time the workers concerned are dealing with the interface as though it were in thermodynamic equilibrium, whereas the rest of the time it is supposed to be subject to Stern's equations.

I think that there is much future in theoretical work in this field, and some was outlined by Tom Healey at the London meeting. Pollution control should give rise to large-scale changes in the extraction of minerals (no more SO_2 evolved into the atmosphere), and therefore I suppose that one would expect the development of electrodeposition and electroextraction processes generally. However, there is much to be worked out and former group members are working on this, for example, in the National Institute of Metallurgy in Johannesburg, South Africa,[133] as well as in the CSIRO in Melbourne, Australia.[134] Here, the properties of semiconductor electrodes, the electrochemistry of which was originated by Mino Green, a former supervisor in the group, form the predominant background for the theory.

12. High-Temperature Electrolytes

The high-temperature electrochemistry carried out in the electrochemistry group represents one of its most pioneering efforts. Along with the development of high-purification techniques for solid electrodes, it made one of the most important experimental contributions to the London period. The original workers here were Stan Ignatowicz[135] and Alec Liberman,[136] both of whom were older than I at the time of our collaboration. Joe Kitchener and I were asked to enter the high-temperature field—about which we knew nothing whatsoever—in 1947 by Sir Charles Goodeve and Dennys Richardson of the British Iron and Steel Research Association. They strode into my room at Imperial College one day in 1947, and said, "Will you work on the physical chemistry of slags?" I did not know, myself, at this time, what a slag was, but they offered not only research support but also a consulting retainer which was roughly equal to my university salary, so it became really an offer which I could not refuse.

The work we did is a good example of how one's interest can be fired with material inducements. It was first in the lime–silica system where we measured the conductivity,[137] finding that it was ionic and that the presentations made, for example, by Schrenck in his theory of slags[138] (where entirely molecular presentations of the siliceous components of a slag had been made) were fallacious. We examined the transport numbers,[139] just to make sure, and also the viscosity,[140] densities,[141] partial molar volumes,[142] and much else.[143]

Here we achieved several firsts in instrumentation, for example, the use of tungsten and molybdenum as refractories,[144] the first viscometer working at 1900°C,[145] (Figure 20) and contributed a good deal to furnace design and other high-temperature-

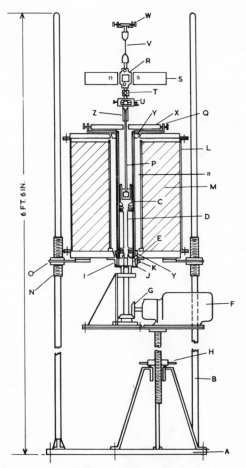

Fig. 20. A viscometer designed to work to high temperatues (~1800°C). A, base plate; B, main supports; C, 2-in. internal diameter recrystallized alumina tube; D, crucible support; E, self-centering chuck; F, ¼ hp motor; G, bevel gears; H, capstan jack; I, chuck housing; J, chuck backing plate; K, grease nipple; L, brass furnace casing; M, lamp-black furnace packing; N, threaded sleeves; O, furnace supporting rings; P, graphite center spindle; Q, D-plates housing; R, Mg–Al coil former; S, Alcomax II magnet; T, galvanometer mirror; U, total reflecting prism; V, Be–Cu suspension wire; W, inner cylinder centering mechanism; X, water-cooled D plates; Y, furnace element contact cones; Z, Tufnol collar; a, molybdenum furnace element.

technique matters. Some of this material is reported in the book *Physicochemical Measurements at High Temperature*, by Bockris, White, and Mackenzie.[146]

It was in this period that the collaboration between J.W. Tomlinson and myself began, and lasted for several fruitful years, until Tomlinson became professor at Wellington, New Zealand.

One of the most important things which came out of the work on slags was unexpected. We had plenty of evidence that the silicate particles in the slag were ionic, and therefore we came to the conclusion (first with Dennis Lowe)[140] that we had to fit into some scheme a known O—Si—O angle, and a known silicon–oxygen ratio. There is no doubt that the theory which we made, illustrated in Figure 21, has had repercussions, not only in the metallurgical understanding of slags, but—the surprising fact—also in the theory of glasses. The old Zachariasen theory on networks[147] was conclusively disproved for compositions above 10% metal oxide, and the most important work here was done by the originator of this conference, Doug Mackenzie.[148] Thus, work done in the group in the 1950s is the origin of the present frontier of the structural theory of glasses.

What is the future of this kind of work? Toshiko Emi,[149] who worked in the late American period in the group, has written an article about this in the second edition of

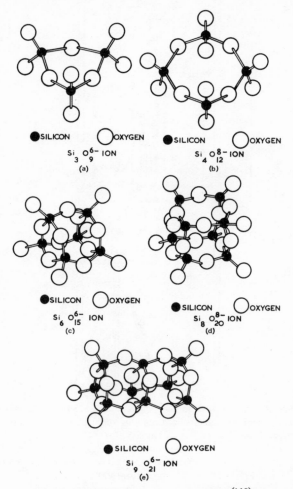

Fig. 21. Possible discrete anions in silicates.[148]

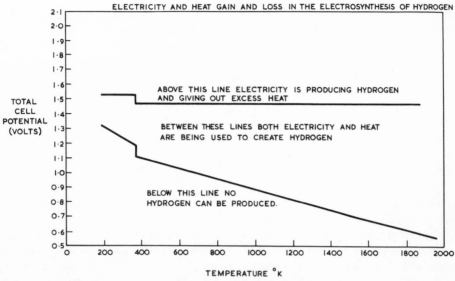

Fig. 22. The dependence of the electrochemical cell potential required to produce hydrogen from water on temperature.[152]

the MTP series, to be published in 1976. He reviews the advanced state of the electro-chemistry of the metal–slag interface, which is largely based, in its structural aspects, upon the work done in London in the 1950s, together with the contributions made by Essin[150] in the U.S.S.R. and recent Japanese work by Emi and his colleagues.[151]

Apart from these steel-oriented possibilities, one sees a need for work near to what was done in London in the high-temperature production of hydrogen. There is a case for the production of hydrogen at 1000°C,[152] perhaps 1500°C, (Figure 22). There is a possibility that the work now being carried out by Bevan (formerly of the Nuffield Research Group in the days of Richardson and Tomlinson) and Badwal in Adelaide, will enable the zirconia-yttria membranes to be made sufficiently thin, and sufficiently ionically conductive, to allow the cheap electrolysis of steam.[153]

13. Structure of Ionic Liquids

Having said a little about our contributions to what may be called "very-high-tempera-ture liquids," 1600-2400°C, I would like to go down one thousand degrees or so and say a word about our contributions to molten salts—the "ordinary" molten salts, mainly ionic liquids such as sodium chloride.

I should like to start off with an interesting paper contributed by Joe Kitchener and Alina Borucka.[154] Here, for the first time, were clear results very difficult to explain in any way but that in which ion-size holes existed in the model for the liquid. We had started out upon an applied basis. Borucka and I had gone down to "get the money" from one of the ministries, and we were supposed to determine the diffusion coefficients of carbonates. In fact, we did do sodium chloride, and the Nernst–Einstein equation was the relation we hoped to use for bridging the gap between radiotracer diffusion coeffi-cients and the conductivity. Dr. Borucka's excellent technique rapidly got us some diffu-sional coefficients, but when we tried to relate them, via the Nernst-Einstein equation, to the experimentally known conductivities for these liquids, we found that there was a fairly systematic discrepancy. I had gone off to my visiting professor year in the U.S.A. at this point and had to try to contribute to the project by means of correspondence.* Here, a meeting with Carl Wagner in Boston must be mentioned and discussion of the results with him which germinated the idea that, if two ions jumped together into paired vacancies they would *count* in the diffusion coefficients, but not be affected by an electric field, for they would have no net charge.

This concept was fed over to Borucka and Kitchener in London, and they came out with a splendid treatment of Borucka's results, which seemed to fit a hole model for simple molten salts very well, and this was the first molecular theory for deviations from the Nernst–Einstein equation.

Work was also done with Wally Hooper[155] on a more extended scale and also with S. R. Richards[156] later on the alkaline earth halides, at higher temperatures.

Earlier, in 1956, Harry Bloom had been on a visit to Pennsylvania and there was a time when Harry, Nolan Richards, and I were all at work together. This proved to be a fruitful period. Nolan Richards was a product of the Bloom laboratory, and Bloom drew our attention to the advantages of the Furth version of the hole theory—a suggestion which sparked Richards and I to develop a comprehensive paper which was published in the *Proceedings of the Royal Society* in 1957.[157] This paper was a foundation mark in the development of the fundamentals of molten salts, because it was the first one in which properties such as compressibility and expansivity, which Richards had determined with techniques he had developed with Ed Brauner,[158] could be quantitatively interpreted within some 10% *without the use of scaling parameters* or other confidence-destroying crutches beloved of later theorists.

This was pleasing for, at this time, in the mid-1950s, ideas on the structure of molten salts were very unclear and here we had come along and calculated the properties. It seemed to be going well, and particularly so when, in 1963, Leonard Nanis[159] used a formula which had first been described in the paper by Hooper to obtain the heats of activation not only of the molten salts, but also of a number of liquids, from the rare gases to the molten metals, throwing in the molten salts as well.

We seemed to have a hunk of the theory of liquids under control at this time, but a punch on the nose was delivered to us when it was pointed out that we had forgotton the entropy, and that the entropy of hole formation cancelled out with that of the heat, if one accepted the formalism of the Furth model. This led, of course, to some reddening of faces, particularly mine, but we struggled on, both with Cyril Solomons[160] and finally with the erudite and academic Emi to come through, finally, with a *fairly* good escape which was published in the *Journal of Physical Chemistry*.[161]

It must be admitted that the derivation which Emi produced was still a crude one indeed, and there is still no really satisfactory and sophisticated argument which gives an explanation of $dE_{(n)}/dT_{mp}$ with a slope at about $3.64R$, a powerful experimental generalization.

I have said that Eyring had much influence on my mode of thinking in London and Philadelphia, and for this reason I was keen to try out the Eyringian concept of constant volume and constant pressure in transport on the molten salts. Briefly, if one keeps to constant volume, one avoids the hole formation work and gets the heat to jump, but the experimental work is difficult to do and was first done in my laboratory by Barry Trickle-bank,[162] another graduate from the Bloom stable, and latterly by Nagarajan[163] with an enormous room full of equipment and measurements up to 1200 atmospheres, at

---

*Alina Borucka still possesses, she says, two boxes which contain the letters. Some of the correspondence may be, of course, somewhat admonitory.

1000°C, in 1965 a very difficult requirement. Later it was made to look easy by the more advanced work of one of the last collaborators from the London period of the work, viz., G. J. Hills, in frequent association with Ernest Yaeger, at Southampton.

What the results showed was that, indeed, the hole formation work was much greater than the work jumping. The interpretation of the $3.74R$ coefficient to the T_{mp} does depend upon neglecting the work of jumping, and there is still much to do here in refining the difference between the two types of work.

We have mentioned the work of Douglas Inman in using the quartertime potential method to elucidate the decay of simple complexes, like $CdCl^+$, and the growth of complexes like $CdCl_3^-$, as the chloride ligand is added to the cadmium nitrate molten salt. Here, Balasubramanyam[164] played a part with huge Raman equipment bought for us out of the University of Pennsylvania's funds. Later, however, C. Solomons and J. H. R. Clark[160] contributed a definitive study of the cryolite aluminum system, which was a beautiful example of high-temperature spectroscopy on a very difficult system and I think remains a standard for the field. At any rate, it gave a basis to much molten-salt complex ion determination, which has been so successfully carried forward by authorities such as George Janz and with increasing sophistication in the development of new spectroscopic methods in the laboratory at Southampton under Grahame Hills.

To some extent, we had reached a plateau by 1967 with our molten-salt work. On the experimental side, we were clear about what we wanted to do—high pressures and high-temperatures—with the objectives the raising of the pressure and temperature to about 2000 atmospheres and 1500°. The costs of such apparatus, however, if one is going to use big furnaces and high-pressure equipment, become astronomical and the AEC'S funding, which had been constant and generous for many years, was beginning to show a tendency to be cut back ("Vietnam"), so we decided not to try to press the issue of this expensive method but to attempt to generate the pressure at the end of a moving bullet.

Looking towards the future of this work on the molten state, where the main interest is less in the structure of the particles, as with the silicates, borates, and phosphates, and much more with the "liquid" aspects of the matter, I am puzzled. Usually, I have no difficulty in jumping on from where I have been. My coworkers react against my demand that they neglect the last fellow's work and get on with thinking it all through as though nothing had been done before. Here, with the liquid structures, I am not sure that I can feel great enthusiasm for techniques which have developed outside models. I suppose it is a psychological, not an intellectual, matter. I intellectualize my objection by maintaining that the job of structure determination has been done well by the modelistic approaches and is only being repeated and confirmed by the Monte Carlo and numerous other computer-simulation techniques which are now the focus of the attack. How close a calculation do you want? What degree of "scaling" will you accept? How often do you want to go on confirming? Eventually, it becomes a matter of the philosophy of the objectives of fundamental research. It is good to know that much of what is being done in the molten-salt field, particularly by Woodcock and G. J. Hills, seems to confirm aspects of the models which originated from the work of Borucka and of Richards in the 1950. However, it clearly makes it much more well based and solid and certainly 1000% more sophisticated.

14. Other Topics

The symposium in London was designed, ostensibly, to pick out topics from the work of the electrochemistry group which I led in London and Philadelphia, and now lead in Adelaide. Perhaps it is cheating if I mention things not dealt with at the meeting, and therefore I shall be brief in mentioning creamy bits from other topics which we have tackled. Further, I want to allow myself the luxury of a word about my attitude toward topics I do not know anything about—and taking them on in research. I have been brash about this. My attitude spread from the early days when Sir Charles Goodeve and Dennys Richardson asked to investigate "slags," the definition of which I did not know, although I was being asked to speak about their structure internationally only three or four years later. So it never troubled me if I did not know what the words meant, or anything about the topic. It does not trouble me now. It seems to me that what you need in research is a knowledge of basic physics, say the Feynmann lectures, and after that it is all a matter of *Thinking*. The main rule is not to take any notice of what anybody else has said in the past. It is wrong, of course, and it is better to think for oneself and "new."*

These principles were applied in the early days to metallurgical topics: the measurement of the amount of solubility of oxygen in iron and equilibria involving slags and sulfides. Here, the point was the experimental technique: Many said reliable measurements over 1650°C were exceedingly difficult—and sometimes dangerous. Courage and perseverance pulled us through all that in London—and the great help of our furnace maker, James Royce. Walter Beck[83, 164] in Philadelphia led me into interesting problems in metallurgy, steel, its breakdown, and thereby to paths with hydrogen, which are now being exploited by Harvey Flitt[165] in Adelaide.

Another aspect of our work not dealt with at the meeting was that of solutions where we contributed extensively to models in solvation.[166, 167] One of the best papers with which I have ever been associated is the paper with Hedda Linton and Brian Conway[168]— one of the products of the idyllic period at Pennsylvania when Conway, Bloom, and I and only two or three others were in a small group together. Here, particularly just before my departure for the brief period in South Africa, we used to work very late at night, pottering around the campus at Pennsylvania between twelve and two in the morning, discussing in the hot Philadelphian summer, up and down the pathways of the campus, just how those protons got across the solution so quickly. It was the usual method, numerically calculating out all the models we could think of and it came out very clearly that it was only the field-induced rotation of the water molecule which would quantitatively fit all the facts. The final coup of this work was the realization that it should not happen in ice, because there the protons are so few that the water molecules will always turn round in time to receive them. Then, the quantum mechanical tunneling of protons becomes rate determining so that the mobility of protons in ice should be greater than that in water. This was realized in a breakfast discussion between Brian Conway and myself one Sunday morning in 1955. We rushed off down to the laboratory and wrote up the note[169] to *Nature* and mailed it on the Monday.

This would be a good point to end the brief mention of the other topics, but I must throw in a last one, more because of the brilliant investigator who carried it out than the actual topic. I refer to the "other" Parsons, Donald Parsons, whom I was privileged to have working with me from 1950 to 1952. "DFP" worked on the magnetochemical

---

*After one has thought for a year or two, there is no harm in taking a peep at what other people say. It pleases the referees of papers to see their names mentioned, and it is of course part of morality among scientists to credit previous work. Further, one must not carry the "do it yourself first" method too far, or else one may miss something terribly important that one has not thought of at all. However, it is good principle to "think first alone," and I practiced it even before David Grahame told me how he always tried to avoid both reading the literature and going to meetings. (I can't agree with him about the latter practice. I think one has frequently to expose ones ideas to the slings and arrows of one's competitors.)

detection of semiquinone radicals,[170] their transient existence as intermediates in reactions in solution and we ended up with a tremendous apparatus, perhaps the largest
with which I have ever worked (because of the giant coil inside which we worked in order
to eliminate the effect of the earth's field). At the time 1952, we had the most sensitive
magnetic measurements anywhere, and it was a fitting climax in respect to the invention
of experimental techniques within the London period. The subsequent career of D. F.
Parsons* in medicine, and his eminent position in cancer research, is a great joy to me,
though it is 100% due to his own incisive work and energy.

15. Electrochemistry for a Better World

As I stated at the beginning, much of the work carried out in the electrochemistry
group over the years can be described as "fundamental applied research", and its eventual
objectives are not badly described as "electrochemistry for a better world." It is seldom
realized how much of the present world, independently of any repercussions from work
of the last 30 years, utilizes electrochemical science and technology. Electrochemistry in
everyday life is met in battery-driven or -started equipment, electrochemically produced
aluminum in buildings, fresh water from saline environments, electrochemically produced
nylon, the synthesis of antiknock agents in cars, just to mention a few.

How will electrochemistry affect life in the future? Because of the ubiquity of the
subject—it is more than a branch of chemistry, it is "another chemistry"—there are so
many ways that a list could only detract by limiting the scope. One can but point out
that future sources of energy can only be atomic or solar, that electricity will be produced from both directly, and that it, being the medium of energy, will encourage the use
of electrochemistry.[171] If we go through hydrogen as an intermediate, and that is the
medium of energy, then electrochemistry will be the more encouraged, for hydrogen is
the best electrochemical fuel. Antipollution measures and recycling are practical methods
for the future, and they are associated with many electrochemical possibilities.

However, it is no use being Panglossian about these matters. Electrochemical technology is not standing there waiting to be used.[172] The rate at which its virtues can become
useful depends on the amount and skill of the research being done (the right books attrac-
tively written). All that depends on research funding, on training the right people, on
training, in fact, some hundred new electrochemical engineers per year and perhaps 10 or
20 per year of the fundamental kind. This is where the present situation, not only in electrochemistry, but in many fields in science in the Western world, looks so dark because research funding—much more needed than in the past, because of the economic difficulties

---

*Among any group of men and women with a long history, there arise a number of anecdotes, some-
times partly apocryphal. I have intentionally refrained from relating any of these in this account
which attempts to correspond to its title. In respect to D. F. Parsons' apparatus, however, I wish to
go on record with a scene which arose from Parsons' method of making exceedingly thin quartz
fibers for the suspension of his supersensitive magnet which responded by tiny movements to the
introduction of transient paramagnetism. This was to use an enormous bow and arrow. Parsons, a
tall man, stood at one side of the room, and shot his arrow, followed by the thinning freezing
fiber, across the room to a target on the far wall. All went well until Parsons' habits of starting
about 7 A.M. collided with those of an enthusiastic chief cleaning lady, vigorously terminating her
day by entering Parsons' lab without knocking. The resulting arrow–head momentum exchange
might well have proved the end of her life had not the lady's crown been detachable, and carried on
along with the fiber toward the target. Parsons' comment is not recorded, but I am certain he would
have been extremely concerned for the fiber.

of the present—is being given increasingly less.* We have to try to do more with less, and the end of that tendency is like that of Freud's horse of Schilda. Its owners found it could live on half the hay which they were giving it, so they cut it to a quarter—the horse still worked—and then to one eighth and the end of the horse. We are facing a kind of "lemmingism," and discussion of its roots lies outside the limits of this paper and is perhaps to be sought in the cyclical concepts of history of Spengler and of Toynbee.

Finally, to come back to this meeting, may I say, lastly, a *final word*. It was wonderful to get together! The simple and straight-forward reasons were there: good information exchange and increase of knowledge among many who had similar interests, but a more important reason was to resuscitate "the Old Spirit." That is the point, I think, which a number of scientists may tend to miss. Spirit, atmosphere, loyalty, feeling! These are the things which make men *work and achieve*. Hanna Rosenberg![173]† She was *there*, and that meant a lot, not only to me.

Lastly, many people have asked me why I go ahead so strongly on this field of electrochemistry. I don't think that it is necessary to answer for members of our group because those who understand the field will understand my reasons. However, I will give my reasons for those who do not understand them. Electrodic electrochemistry treats the relation between electricity and matter. Electrochemical science also deals with that relation: how matter can be converted electrically and how electricity can interact with matter. This is a powerful and general sort of relation, and it seems to me that its study leads through a hitherto closed door into another room in chemistry. By this I mean that the chemistry which we have always studied has had as its central kinetic idea the theory of reactivity in terms of collisional processes. I think that we shall understand that there is another kind of chemistry, and that wherever a *thermal* reaction existed, there is often the possibility of an interfacial one with electrical activation. It is the applications of this concept which are so exciting for modern and for near future times.

Acknowledgments

I was helped in the diagrammation and reference finding by M. A. Habib (Leader), H. Flitt, S. U. M. Khan, J. McCann, F. Uddin and K. Uosaki.

References

1. J. O'M. Bockris, *Energy: The Solar-Hydrogen Alternative*, Australian & New Zealand Book Company, Syndey (1975).
2. J. O'M. Bockris, ed., *The Electrochemistry of Cleaner Environments*, Plenum Press, New York (1972).
3. J. O'M. Bockris, M. A. V. Devanathan, and K. Müller, *Proc. R. Soc. London A274*, 55 (1963).
4. E. Lange and K. P. Miscenko, *Z. Phys. Chem. A149*, 1 (1930).
5. J. R. Macdonald, *J. Chem. Phys. 22*, 1857 (1954).
6. J. R. Macdonald and C. A. Barlow, *J. Chem. Phys. 36*, 3062 (1962).
7. N. F. Mott and R. J. Watts-Tobin, *Electrochim. Acta 4*, 79 (1961).

---

*There is evidence for the view that the long-term (multidecade) cycles in capitalist economies (the great long ascents and declines) reflect areas of the ebb and flow of technological innovation. The time for the heaviest spending to produce innovative technology is during economic declines. Were research funding on a decade scale varied according to this principle, one might see the evening out of booms and busts.

†Alias Hedda Linton, the third author of CBL, the Conway, Bockris, and Linton paper on proton transfer. One thinks easily here of the analogy to the coulombic energy in the Schroedinger for H_2^+, with Miss Linton as the electron, of course.

8. H. Wroblowa and K. Müller, *J. Phys. Chem. 73*, 3528 (1969); T. Bejerano, Ch. Fargacs, and E. Gileadi, *J. Electroanal. Chem. Interface Electrochem. 27*, 69 (1970); R. G. Barradas and J. M. Sedlak, *Electrochim. Acta 16*, 2091 (1971); B. E. Conway and H. P. Dhar, *Electrochim. Acta 19*, 445 (1974).

9. D. J. Schiffrin, *Trans. Faraday Soc. 67*, 3318 (1971).

10. J. O'M. Bockris and S. Argade, *J. Chem. Phys. 49*, 5133 (1969).

11. S. Trasatti, *J. Electroanal. Chem. Interface Electrochem. 52*, 313 (1974).

12. S. Srinivasan, H. D. Hurwitz, and J. O'M. Bockris, *J. Chem. Phys. 46*, 3108 (1967).

13. J. O'M. Bockris and B. D. Cahan, *J. Chem. Phys. 50*, 1307 (1969).

14. J. O'M. Bockris and Z. Nagy, *J. Electrochem. Soc. 119*, 1129 (1972).

15. J. O'M. Bockris, Z. Nagy, and A. Damjanovic, *J. Electrochem. Soc. 119*, 286 (1972).

16. A. R. Despic, J. Diggle, and J. O'M. Bockris, *J. Electrochem. Soc. 115*, 507 (1968).

17. N. Viziroglu, ed., *Hydrogen Energy*, Plenum Press, New York (1975).

18. K. Kordesh, in: *Modern Aspects of Electrochemistry* J. O'M. Bockris and B. E. Conway, eds. Vol. 10, Plenum Press, New York (1975).

19. J. O'M. Bockris and A. K. N. Reddy, *Modern Electrochemistry*, Rosetta Edition, Plenum Press, New York (1973).

20. L. Duic, S. Srinivasan, and P. N. Sawyer, *J. Electrochem. Soc. 120*, 348 (1973).

21. J. O'M. Bockris, *Nature 224*, 775 (1969).

22. C. Wagner and W. Traud, *Z. Electrochem. 44*, 391 (1938).

23. U. R. Evans and T. P. Hoar, *Proc. R. Soc. London A137*, 343 (1932).

24. M. Del Ducca and K. Fucsoe, *Int. J. Sci. Technol.* (March, 1965).

25. J. O'M. Bockris and S. Srinivasan, *Nature, 215*, 197 (1967).

26. L. Mandel, in: *Modern Aspects of Electrochemistry* J. O'M. Bockris and B. E. Conway, eds. Vol. 8, Plenum Press, New York (1972).

27. R. O. Becker and A. A. Pilla, in: *Modern Aspects of Electrochemistry* J. O'M. Bockris and B. E. Conway, eds., Vol. 10, Plenum Press, New York (1975).

28. J.O'M. Bockris, N. Bonciocat, and F. Gutmann, in: *An Introduction to Electrochemical Science*, Chapter 9, Wykeham Press, London (1973).

29. B. Lowenhaupt, *J. Theor. Biol. 25*, 187 (1969).

30. H. S. Burr, *Proc. Natl. Acad. Sci. U.S.A. 29*, 163 (1943).

31. A. Köstler, *The Roots of Coincidence*, Hutchinson, London (1972).

32. A. Wissenberg, University of Adelaide, Private Communication (1975).

33. L. Watson, *The Romeo Error*, Hodder and Stoughton, London (1974).

34. M. Ganquelin, *L'influence des astres*, Dauphin, Paris (1955).

35. S. Ostrander and L. Schroeder, *Psychic Discoveries behind the Iron Curtain*, Prentice-Hall, Englewood Cliffs, New Jersey (1970).

36. A. S. Pressman, *Electro-Magnetic Fields and Life*, Plenum Press, New York (1969).

37. F. P. Bowden and E. K. Rideal, *Proc. R. Soc. London 120A*, 59 (1928).

38. W. T. Grubb, *Nature 198*, 883 (1963).

39. J. O'M. Bockris and H. Wroblowa, *J. Electroanal. Chem. 7*, 428 (1964).

40. J. O'M. Bockris, *Chem. Rev. 43*, 525 (1948).

41. D. C. Grahame, *Chem. Rev. 41*, 441 (1947).

42. V. S. Bagotzky, Yu. B. Vassiliev, and I. I. Pyshnograeva, *Electrochim. Acta 16*, 2141 (1971).

43. V. S. Bagotzky, N. A. Shumilova, G. P. Samoilov, and E. I. Krushcheva, *Electrochim. Acta 117*, 1625 (1972).

44. P. Stonehart and J. Lundqvist, *Electrochim. Acta 18*, 907 (1973).

45. A. T. Kuhn, H. Wroblowa, and J. O'M. Bockris, *Trans. Faraday Soc. 63*, 1458 (1967).

46. J. O'M. Bockris, H. Wroblowa, E. Gileadi, and B. J. Piersma, *Trans. Faraday Soc. 61*, 2531 (1965).

47. D. B. Sepa, A. Damjanovic, and J. O'M. Bockris, *Electrochim. Acta 12*, 746 (1967).

48. A. Damjanovic, D. B. Sepa, and J. O'M. Bockris, *J. Res. Inst. Catal.* (Hokkaido Univ.) *16*, 1 (1968).

49. J. McHardy and J. O'M. Bockris, *J. Electrochem. Soc. 120*, 53 (1973).

50. J. McHardy and J. O'M. Bockris, *J. Electrochem. Soc. 120*, 61 (1973).

51. L. S. Kanevsky, V. Sh. Palenkev, and V. S. Bagotsky, *Elektrokhimya 5*, 1390 (1969); *6*, 1879 (1970).

52. W. K. Paik, M. A. Genshaw, and J. O'M. Bockris, *J. Phys. Chem. 74*, 4266 (1970).

53. A. K. N. Reddy, Chapter 10, this volume.

54. J. O'M. Bockris and B. E. Conway, *Trans. Faraday Soc. 45*, 989 (1949).

55. R. Parsons and J. O'M. Bockris, *Trans. Faraday Soc. 47*, 914 (1951).

56. J. O'M. Bockris and E. C. Potter, *J. Electrochem. Soc. 99*, 169 (1952).

57. J. O'M. Bockris and R. G. H. Watson, *J. Chim. Phys. 49*, 1 (1952).

58. J. O'M. Bockris, I. A. Ammar, and A. K. M. S. Huq, *J. Phys. Chem. 61*, 879 (1957).

59. R. Parsons, in: *Advances in Electrochemistry and Electrochemical Engineering*, P. Delahay, ed., Vol. 1, Chapter 1, Interscience New York (1961).

60. L. I. Krishtalik, *Elektrokhimiya 6*, 507 (1970).

61. I. A. Menzies, D. L. Hill, G. J. Hills, L. Young, and J. O'M. Bockris, *J. Electroanal. Chem. 1*, 161 (1959–1960).

62. J. O'M. Bockris and L. Oldfield, *Trans. Faraday Soc. 51*, 249 (1953).

63. J. O'M. Bockris and R. Parry Jones, *Nature 171*, 930 (1953).

64. J. O'M. Bockris and A. K. M. S. Huq, *Proc. R. Soc. London A237*, 277 (1956).

65. E. Blomgren and J. O'M. Bockris, *J. Phys. Chem. 63*, 1475 (1959); E. Blomgren, J. O'M. Bockris, and C. Jesh, *J. Phys. Chem. 65*, 2000 (1961).

66. J. O'M. Bockris and H. Mauser, *Can. J. Chem. 37*, 475 (1959).

67. K. J. Vetter, *Electrochemical Kinetics*, Academic Press, New York (1967).

68. J. O'M. Bockris, W. Mehl, B. E. Conway, and L. Young, *J. Chem. Phys. 25*, 776 (1956).

69. M. W. Breiter, *Ber. Bunsenges. Phy. Chem. 69*, 612 (1965).

70. F. G. Will, *J. Electrochem. Soc. 112*, 451 (1965).

71. H. Kita, M. Enyo, and J. O'M. Bockris, *Can. J. Chem. 39*, 1670 (1961).

72. E. Blomgren, D. Inman, and J. O'M. Bockris, *Rev. Sci. Instrum. 32*(1), 11 (1961).

73. D. Inman, G. J. Hills, L. Young, and J. O'M. Bockris, *Ann. N.Y. Acad. Sci. 79*, 803 (1960).

74. A. R. Despic and J. O'M. Bockris, *J. Chem. Phys. 32*, 389 (1960).

75. J. O'M. Bockris and D. F. A. Koch, *J. Phys. Chem. 65*, 1941 (1961).

76. J. L. Barton and J. O'M. Bockris, *Proc. R. Soc. London A268*, 485 (1962).

77. A. Damjanovic and J. O'M. Bockris, *J. Electrochem. Soc. 110*, 1035 (1963).

78. H. Wroblowa, B. J. Piersma, and J. O'M. Bockris, *J. Electroanal. Chem. 6*, 401 (1963).

79. T. Anderson and J. O'M. Bockris, *Electrochim. Acta 9*, 347 (1964); J. O'M. Bockris, R. K. Sen, and K. L. Mittal, *J. Res. Inst. Catal.* (Hokkaido Univ.) *20*, 153 (1972); J. O'M. Bockris, B. E. Conway, and R. K. Sen, *Nature 240*, 197 (1972); J. O'M. Bockris, R. K. Sen, and B. E. Conway, *Nature 240*, 143 (1972); J. O'M. Bockris, S. U. M. Khan, and D. B. Matthews, *J. Res. Instr. Catal.* (Hokkaido Univ.) *22*, 1 (1974).

80. H. Wroblowa, Z. Kovac, and J. O'M. Bockris, *Trans. Faraday Soc. 61*, 1523 (1965).

81. J. O'M. Bockris and S. Srinivasan, *J. Electrochem. Soc. 111*, 853 (1964).

82. J. O'M. Bockris and S. Srinivasan, *Fuel cells; Their Electrochemistry*, McGraw-Hill, New York (1969).

83. W. Beck, J. O'M. Bockris, J. McBreen, and L. Nanis, *Proc. R. Soc. London A290*, 220 (1966).

84. J. C. M. Li, R. A. Oriani, and L. S. Darken, *Z. Phys. Chem. 49*, 271 (1966); J. O'M. Bockris and P. K. Subramanyan, *Corros. Sci. 10*, 435 (1970).

85. J. O'M. Bockris and P. K. Subramanyan, *Corros. Sci., 10*, 435 (1970).

86. P. K. Subramanyan, *MTP International Review of Science*, Vol. 1, Butterworth, London (1972); P. K. Subramanyan, in: *MTP International Review of Science*, Vol. 6, (J. O'M. Bockris, ed.), Butterworth, London (1973).

87. A. K. N. Reddy, M. A. V. Devanathan, and J. O'M. Bockris, *Proc. R. Soc. London A279*, 327 (1964).

88. A. Damjanovic, M. Genshaw, and J. O'M. Bockris, *J. Phys. Chem. 70*, 3761 (1961).

89. N. Schumilova and V. S. Bagotsky, meeting of the Electrochemical Society, Montreal, Canada (1968).

90. J. O'M. Bockris, E. Gileadi, and K. Müller, *J. Chem. Phys. 44*, 1445 (1966).

91. J. O'M. Bockris, W. Mehl, and B. E. Conway, *Proceedings of the 5th International Conference on Electrochemistry* (Acad. Sci. U.S.S.R.), Moscow (1959).

92. J. O'M. Bockris, E. Gileadi, and K. Müller, *Electrochim. Acta 12*, 1301 (1967).

93. B. B. Damaskin and A. N. Frumkin, *J. Electroanal. Chem. 34*, 191 (1972).

94. J. O'M. Bockris, *J. Electroanal. Chem. 34*, 201 (1972).

95. J. O'M. Bockris, S. Argade, and E. Gileadi, *Electrochim. Acta 14*, 1259 (1969).

96. A. N. Frumkin, *Z. Phys. Chem. 111*, 190 (1924); *116*, 466 (1925).

97. L. I. Antropov, *Kinetics of Electrode Processes and Null Points of Metals*, Council of Scientific and Industrial Research, New Delhi (1959).

98. J. O'M. Bockris and M. A. Habib, *J. Electroanal. Chem. Interfacial Electrochem. 68*, 367 (1976).
99. J. O'M. Bockris and B. D. Cahan, *J. Chem. Phys. 50*, 1307 (1969).
100. E. Gileadi, L. Duic, and J. O'M. Bockris, *Electrochim. Acta 13*, 1915 (1968).
101. J. O'M. Bockris, R. J. Mannan, and A. Damjanovic, *J. Chem. Phys. 48*, 1989 (1968).
102. M. A. Genshaw, V. Brusic, H. Wroblowa, and J. O'M. Bockris, *Electrochim. Acta 16*, 1859 (1971); J. O'M. Bockris, M. A. Genshaw, and V. Brusic, *Symp. Faraday Soc. 4*, 177 (1970).
103. V. Brusic, M. A. Genshaw, and B. D. Cahan, *J. Appl. Opt. 9*, 1634 (1970).
104. A. K. N. Reddy and B. Rao, *Can. J. Chem. 47*(14), 2693 (1969).
105. J. O'M. Bockris and P. K. Subramanyan, *Electrochim. Acta 16*, 2169 (1971).
106. A. J. Appleby, J. O'M. Bockris, R. K. Sen, and B. E. Conway, in: *MTP International Review of Science, Physical Chemistry Series One* (J. O'M. Bockris, ed.), Vol. 6 Butterworth, London (1973).
107. W. K. Paik and J. O'M. Bockris, *Surface Sci. 27*, 191 (1971); W. K. Paik, in: *MTP International Review of Science*, (J. O'M. Bockris, ed.), Vol. 6, Butterworth, London (1973).
108. R. A. Fredlein, A. Damjanovic, and J. O'M. Bockris, *Surface Sci. 25*, 261 (1971); R. A. Fredlein and J. O'M. Bockris, *Surface Sci. 46*, 641 (1974).
109. W. E. O'Grady and J. O'M. Bockris, *Chem. Phys. Lett. 5*, 116 (1970); W. E. O'Grady and J. O'M. Bockris, *Surface Sci. 38*, 249 (1973).
110. J. Kruger and J. P. Calvert, *J. Electrochem. Soc. 114*, 43 (1967).
111. N. Sato and M. Cohen, *J. Electrochem. Soc. 111*, 512 (1964).
112. W. Revie, B. G. Baker, and J. O'M. Bockris, *J. Electrochem. Soc. 122*, 1460 (1975).
113. W. Revie, J. O'M. Bockris, and B. G. Baker, *J. Electrochem. Soc. 122*, 1460 (1975).
114. B. Breyer and F. Gutmann, *Trans. Faraday Soc. 43*, 785 (1947).
115. F. Gutmann, *Rev. Mod. Phys. 20*, 457 (1948); *Electrochim. Acta 5*, 87 (1961); *7*, 55 (1962); D. Vasilescu, F. Gutmann *et al.*, *Electrochim. Acta 19*, 181 (1974).
116. J. O'M. Bockris and S. U. M. Khan, *J. Inst. Catalysis*, University of Hokkaido, in press, 1976.
117. J. O'M. Bockris, *Energy Convers. 8*, 141 (1968).
118. R. W. Gurney, *Proc. R. Soc. London A134*, 137 (1931).
119. S. Golden, *Quantum Statistical Foundations of Chemical Kinetics*, Oxford Univ. Press, London (1969).
120. M. Green, in: *Modern Aspects of Electrochemistry* (J. O'M. Bockris and B. E. Conway, eds.), Vol. 2, Plenum Press, New York (1959); H. Gerischer and F. Beck, *Z. Phys. Chem. Neue Folge 13*, 389 (1957).
121. H. Gerischer and F. Beck, *Z. Phys. Chem. Neue Folge 23*, 113 (1960).
122. W. H. Brattain and C. G. B. Garrett, *Bell Syst. Tech. J. 34*, 129 (1955).
123. J. O'M. Bockris and K. Uosaki, Solar Energy Conference at Dharan, Saudi Arabia (1975).
124. J. O'M. Bockris, in: *Modern Aspects of Electrochemistry*, (J. O'M. Bockris and B. E. Conway, eds.), Vol. 1, Butterworths, London (1954).
125. J. W. Johnson, H. Wroblowa, and J. O'M. Bockris, *Electrochim. Acta 9*, 639 (1964).
126. J. O'M. Bockris, H. Wroblowa, E. Gileadi, and B. J. Piersma, *Trans. Faraday Soc. 61*, 2531 (1965).
127. A. T. Kuhn, H. Wroblowa, and J. O'M. Bockris, *Trans. Faraday Soc. 63*, 1458 (1967).
128. N. E. Richards and B. E. Welch, *Electrochemistry Proc. of the 1st Australian Conference*, p. 90 (1962).
129. C. M. Hall, Eng. Pats. 5669; 5670 (April 2, 1889).
130. P. Herriault, Fr. Pat. 175711 (April 23, 1886).
131. D. Inman, G. J. Hills, L. Young, and J. O'M. Bockris, *Trans. Faraday Soc. 55*, 1904 (1959).
132. D. Inman and J. O'M. Bockris, *J. Electroanal. Chem. 3*, 126 (1962).
133. A. Saunders, National Institute of Metallurgy, Johannesburg, Report No. 1778 (1975); J. Diggle, several publications concerning electrochemical behavior of galena (in press, 1976).
134. D. F. A. Koch, in: *Modern Aspects of Electrochemistry* (J. O'M. Bockris and B. E. Conway, eds.), Vol. 10, Chapter 4, Plenum Press, New York (1975).
135. J. O'M. Bockris, J. A. Kitchener, S. Ignatowicz, and J. W. Tomlinson, *Disc. Faraday Soc. 4*, 265 (1948).
136. J. A. Kitchener, J. O'M. Bockris, and A. Liberman, *Disc. Faraday Soc. 1*, 49 (1948).
137. J. O'M. Bockris, J. A. Kitchener, S. Ignatowicz, and J. W. Tomlinson, *Trans. Faraday Soc. 48*, 75 (1952).
138. H. Schrenck, *The Physical Chemistry of Steel Making*, BRISA London (1946).
139. J. O'M. Bockris, J. A. Kitchener, and A. E. Davies, *J. Chem. Phys. 19*, 225 (1951).

140. J. O'M. Bockris and D. C. Lowe, *Proc. R. Soc. London A226*, 423 (1954).

141. J. W. Tomlinson, M. S. R. Heynes, and J. O'M. Bockris, *Trans. Faraday Soc. 54*, 1822 (1958).

142. J. W. Tomlinson, J. L. White, and J. O'M. Bockris, *Trans. Faraday Soc. 52*, 229 (1956).

143. J. O'M. Bockris, Symposium on the Vitreous State, University of Sheffield p. 48, (May 1953).

144. J. W. Tomilson, D. C. Lowe, G. W. Mellors, and J. O'M. Bockris, *J. Sci. Instrum. 31*, 107 (1954); J. O'M. Bockris and D. F. Parsons, *J. Sci. Instrum. 30*, 343 (1953).

145. J. O'M. Bockris and D. C. Lowe, *J. Sci. Instrum. 30*, 403 (1953).

146. J. O'M. Bockris, J. White, and J. Mackenzie, *Physicochemical Measurements at High Temperatures*, Butterworths, London (1959).

147. W. H. Zachariasen, *J. Am. Chem. Soc. 54*, 3842 (1932).

148. J. O'M. Bockris, J. D. Mackenzie, and J. A. Kitchener, *Trans. Faraday Soc. 51*, 1734 (1955).

149. T. Emi, in: *MTP International Review of Science Series* (J. O'M. Bockris, ed.), Vol. 6, Butterworths, London (1973).

150. C. A. Toporitzev, S. G. Melomyd, O. A. Essin, and B. A. Robnyschkin, *Izv. Vuz. Chem. Metall. 12*, 18 (1971).

151. T. Emi, H. Nakato, and K. Suzuki, *Tetsu To Hagane*, T-60 (1974).

152. J. O'M. Bockris, *Energy Convers. 14*, 81 (1975).

153. R. E. Carten *et al.* General Electric Co. press release (December 26, 1962).

154. A. Borucka, J. O'M. Bockris, and J. A. Kitchener, *J. Chem. Phys. 24*, 1282 (1956); *Proc. R. Soc. London A241*, 554 (1957).

155. J. O'M. Bockris and G. W. Hooper, *Disc. Faraday Soc. 32*, 218 (1961).

156. S. R. Richards, L. Nanis, and J. O'M. Bockris, *J. Phys. Chem. 69*, 1627 (1965).

157. J. O'M. Bockris and N. E. Richards, *Proc. R. Soc. London A241*, 44 (1957).

158. N. E. Richards, E. J. Brauner, and J. O'M. Bockris, *Br. J. Appl. Phys. 6*, 387 (1955).

159. L. Nanis and J. O'M. Bockris, *J. Phys. Chem. 67*, 2865 (1963).

160. C. Solomons, J. H. R. Clarke, and J. O'M. Bockris, *J. Chem. Phys. 49*, 445 (1968).

161. T. Emi and J. O'M. Bockris, *J. Phys. Chem. 74*, 159 (1970).

162. S. B. Tricklebank, L. Nanis, and J. O'M. Bockris, *J. Phys. Chem. 68*, 58 (1964); *Rev. Sci. Instrum. 35*, 807 (1964).

163. M. K. Nagarajan, L. Nanis, and J. O'M. Bockris, *J. Phys. Chem. 68*, 2726 (1964); M. K. Nagarajan and J. O'M. Bockris, *J. Phys. Chem. 70*, 1854 (1966).

164. W. Beck, J. O'M. Bockris, M. A. Genshaw, and P. K. Subramanyan, *Metall. Trans. 2*, 883 (1971); J. O'M. Bockris, W. Beck, M. A. Genshaw, P. K. Subramanyan, and F. S. William, *Acta Metall. 19*, 1209 (1971). M. Tanaka, K. Balasubramanyam, and J. O'M. Bockris, Raman Spectrum of the Col Cl_2-KCl system: *Eléctrochim. Acta. 8*, 621 (1963).

165. H. J. Flitt, R. W. Revie, and J. O'M. Bockris, *Aust. Corros. Eng.* In press, 1976.

166. J. O'M. Bockris and P. P. S. Saluja, *J. Electrochem. Soc. 119*, 1060 (1972); *J. Phys. Chem. 76*, 2140 (1972).

167. J. O'M. Bockris and H. Egan, *Trans. Faraday Soc. 44*, 151 (1948).

168. B. E. Conway, J. O'M. Bockris, and H. Linton, *J. Chem. Phys. 24*, 834 (1956).

169. B. E. Conway and J. O'M. Bockris, *J. Chem. Phys. 28*, 354 (1958); *31*, 1133 (1959).

170. H. G. Effemy, D. F. Parsons, and J. O'M. Bockris, *J. Sci. Instrum. 32*, 99 (1955).

171. J. O'M. Bockris and J. Appleby, *Environment 1*, 29 (1972).

172. J. O'M. Bockris and Z. Nagy, *Electrochemistry for Ecologists*, Plenum Press, New York (1974).

173. J. O'M. Bockris, R. Parsons, and H. Rosenberg, *Trans. Faraday Soc. 47*, 766 (1951).

174. F. Gutmann, *Adv. Biochem. Biophys. 9*, 15 (1974); J. P. Farges and F. Gutmann, in *Modern Aspects of Electrochemistry*, (J. O'M. Bockris and B. E. Conway, eds.), Vol. 12, Chapter 5, Plenum Press, New York (1977).

Appendix

Bockris and Collaborators: Principal Fields of Studies with Periods of Activity

Field	Period	Most recent paper
Ion-solvent interaction	1946–1952; 1955–1959; 1969–1972	With Saluja, 1972
Structure molten salts	1956–1971	With Emi, 1971
Structure liquid silicates	1948–1960	With Kojunen, 1960
Theory of double layers	1951–	With Habib, 1976
Adsorption of organic molecules	1959–1968	With Wroblowa, 1968
Electronic spectra of surfaces	1971–	With Revie, 1968
Ellipsometric studies	1963–1972	With Paik, 1972
Impedance of interfaces	1956–1968	With Gileadi, 1968
Quantum electrochemistry	1971–	With Khan, 1976
Photoelectrode kinetics	1973–	With Uosaki, 1976
Hydrogen electrode kinetics	1946–1966	With Mannan, 1968
Electrocatalysis	1963–1973	With MacHardy, 1973
Metal deposition	1956–1973	With Razumney, 1973
Corrosion	1946–1948; 1954; 1967–	With Flitt, 1976
H in metal	1964–	With Flitt, 1976
Bioelectrochemistry	1967–1969; 1974–1976	With Loewenhaupt, 1976
Hydrogen economy	1971–	Bockris, 1975
Energy supply theory	1973–	Bockris, 1976
Physical chemistry—high temperatures	1948–1953	With Tomlinson and White, 1956

Participants

Dr. L. Alcazer, University of Technology, Lisbon, Portugal

Dr. Ibraheem Ammar, University of Riyadh, Saudi Arabia

Dr. A. J. Appleby, Laboratoire d'Electrolyse de C.N.R.S., Bellevue, Paris, France

Dr. J. Augustynski, Chimie et Electrochimie Appliquees, Geneva, Switzerland

Dr. M. Baizer, Monsanto Chemical Company, St. Louis, Mo. 63166, U.S.A.

Dr. B. S. Baker, Energy Research Corp., Bethel, Conn., U.S.A.

Prof. E. Barendrecht, Technical University, Eindhoven, Netherlands

Dr. H. Binder, Battelle Institute, Frankfurt/Main, West Germany

Prof. J. O'M. Bockris, School of Physical Sciences, Flinders University, Bedford Park, S.A., Australia

Dr. L. Bourgeois, Solvay & Cie, Brussels, Belgium

Dr. C. W. Bradford, Johnson Matthey & Co. Ltd., Wembley, Middlesex, England

Prof. E. B. Budevski, Central Laboratory for Electrochem. Power Sources, Bulgarian Academy of Science, Sofia, Bulgaria

Miss Chateau-Gosselin (student), University of Brussels, Belgium

Prof. S. G. Christov, Institute of Physical Chemistry, Bulgarian Academy of Science, Sofia, Bulgaria

Prof. F. Colom, Instituto de Quimica Fisics "Rocasolano," Madrid, Spain

Prof. B. E. Conway, Department of Chemistry, University of Ottawa, Ontario, Canada

Dr. A. Damjanovic, Physical Chemistry Branch, Xerox Research Laboratories, Webster, N.Y. 14580, U.S.A.

Dr. Das Gupta, Binla Institute of Technology, India

Dr. G. M. Deriasz, ICI Mond Div., Northwich, Cheshire, England

Prof. A. Despic, Department of Physical Chemistry, University of Belgrade, Belgrade, Yugoslavia

Dr. M. R. V. Devanathan, Tea Research Institute of Ceylon, St. Coombes Estate, Talawakele, Ceylon

Prof. D. M. Drazic, Department of Physical Chemistry, University of Belgrade, Belgrade, Yugoslavia

Dr. J. R. Duncan, University of Nottingham, Nottingham, England

Prof. L. E. Eberson, Chemical Center, University of Lund, Sweden

Dr. G. M. Eckert, Sydney Hospital, Sydney, N.S.W., Australia

Dr. H. Egan, U.K. Government Chemist, London SE1, England

Mr. J. H. Entwisle, ICI Mond Div., Northwich, Cheshire, England

Prof. M. Enyo, Research Institute for Catalysis, University of Hokkaido, Sapporo, Japan

Dr. Cl. Feneau, Metallurgie Hoboken-Overpelt, Hoboken, Belgium

Prof. M. Fleischmann, University of Southampton, Southampton, SO9 5NH, England

Dr. A. Franks, National Physical Laboratories, Teddington, Middlesex, England

Dr. D. Fresnel, France

Dr. M. V. Ginatta, Politechnico di Torino, Italy

Dr. M. Gleiser, Apt. 104, 2510 Le Conte, Berkeley, Calif. 94709, U.S.A.

Dr. Mino Green, Imperial College, London, England

Prof. J. Guion, Université de Nice, France

Dr. F. Gutmann, Macquarie University, Sydney, N.S.W., Australia

Dr. K. Hass, Niederkassel, West Germany

Professor K. Hauffe, University of Göttingen, West Germany

Dr. T. Healy, University of Melbourne, Parkville, Victoria 3052, Australia

Dr. W. N. Heiland, Institut fur Plasma Physik, 8046 Garching-Munchen, Germany

Dr. D. L. Hill, Department of Chemistry, State University College of Brockport, Brockport, N.Y. 14420, U.S.A.

Dr. T. P. Hoar, Metallurgy Department, University of Cambridge, England

Dr. G. Hodes, Weizmann Inst. of Science, Rehovoth, Israel

Dr. D. Inman, Nuffield Research Group, Imperial College of Science, London SW7, England

Dr. R. O. James, University of Melbourne, Victoria, Australia

Dr. D. R. Jenkins, Shell Research Ltd., Thornton Research Center, Chester, England

Dr. J. W. Johnson, University of Missouri, Department of Chemical Engineering, Rolla, Mo. 65401, U.S.A.

Dr. R. Keller, Alusuisse Neuhausen am Rheinfall, Switzerland

Prof. W. Kemula, Institute of Physical Chemistry, The Academy of Sciences, Warsaw, Poland

Prof. H. Keyzer, California State University, Los Angeles, Calif., U.S.A.

Dr. D. F. A. Koch, CSIRO Div. Mineral Chemistry, Port Melbourne, Victoria, Australia

Dr. Zlata Kovac, IBM Inc., Box 218, P.O., Yorktown Heights, N.Y. 10598, U.S.A.

Dr. A. T. Kuhn, Department of Chemistry and Applied Chemistry, University of Salford, Salford, Lancashire, England

Mr. G. Lecayon, Centre Nucléaire de Saclay, Gif-sur-Yvette 91190, France

Dr. S. A. Losi, Sel-Rex-Corp. Italy.

Dr. J. McBreen, Research Laboratories, General Motors Technical Center, Warren, Mich., U.S.A.

Prof. J. Ross Macdonald, University of North Carolina, Chapel Hill, N.C., U.S.A.

Dr. J. McHardy, Pratt & Whitney Aircraft, Materials Engineering & Research Division, Middletown, Conn. 06457, U.S.A.

Dr. J. K. Mackay, Oxy Metal Indust. Intermatl. S.A. 1211 Geneva, Switzerland

Prof. J. D. Mackenzie, School of Engineering & Applied Science, University of California, Los Angeles, Calif. 90024, U.S.A.

Prof. Y. Matsuda, Yamaguchi University, Japan

Dr. Einar Mattsson, Swedish Corrosion Institute, S-114 28 Stockholm, Sweden

Prof. Ian Menzies, Department of Materials Technology, University of Technology, Loughborough, Leicestershire, England

Mr. G. Mitchell, Consulting Engineer, North Sydney, N.S.W., Australia

Dr. A. Le Moël, Centre Nucléaire de Saclay, Gif-sur-Yvette 91190, France

Prof. K. Müller, Schlossgasse 55, D-7889 Grenzach, Germany

Dr. D. C. Olson, Shell Develop. Co. Research Lab., Wood River, Ill., U.S.A.

Dr. J. Padova, Chemistry Division, Oak Ridge National Laboratory, Oak Ridge, Tenn. 37830, U.S.A.

Prof. G. J. Patriarche, Free University of Brussels, Belgium

Dr. N. Pentland, Physics Department, Brighton Polytechnic, Moulescombe, England

Dr. M. E. Peover, National Physical Laboratory, Teddington, Middlesex, England

Dr. B. J. Piersma, Houghton College, Houghton, N.Y., U.S.A.

Dr. B. P. Piggin, IBM (U.K.) Ltd., Millbrook, Southampton, England

Dr. E. C. Potter, CSIRO Mineral Research Laboratory, North Ryde, N.S.W., Australia

Mrs. D. Prince, 51 Blvd. de Latour-Manbourg, 75 Paris 7ène, France

Dr. R. G. Raicheff, Institute of Chemical Technology, Sofia, Bulgaria

Dr. J. J. Rameau, Ecole National Superieure de l'Electrochimie, St. Martin d'Heres, France

Prof. A. K. N. Reddy, Department of Inorganic and Physical Chemistry, Indian Institute of Science, Bangalore 12, India

Dr. Nolan Richards, Reynolds Metals Co., Listerhill, Sheffield, Ala., U.S.A.

Miss Madeleine de Riepen, University of Göttingen, West Germany

Dr. C. R. Riou, ISSEC, Ferney-Voltaire, France

Dr. T. Russow, Höchst A. G. 623 Frankfurt/Main, West Germany

Dr. P. P. S. Saluja, University of Banaras, India

Dr. G. Sandstede, Battelle Institute, Frankfurt/Main, West Germany

Dr. M. A. Sattar, Chemistry Department, University of Southampton, West Germany

Dr. Ing. W. Schmidt-Hatting, Swiss Aluminium Co., Chippis, Switzerland

Prof. E. Sheldon, Lowell Technological Institute, Nuclear Center, Lowell, Mass. 01854, U.S.A.

Dr. M. A. Slifkin, University of Salford, Salford, Lancashire, England

Dr. P. M. Spaziante, Ontonzio di Nora, Impianti Elettrochim. Milano, Italy

Dr. R. Spencer, U.K.

Dr. P. Spinelli, Politechnico di Torino, Torino, Italy

Dr. S. Srinivasan, Brookhaven National Laboratory Upton, N.Y., U.S.A.

Dr. F. von Sturm, Siemens A. G., Erlangen, West Germany

Dr. Abdul Ghani Hamza Suliman, University of Riyadh, Saudi Arabia

Dr. Fouad Sulimani, University of Riyadh, Saudi Arabia

Dr. A. G. Sussex, Materials Research Laboratory, Ascot Vale, Victoria, Australia

Dr. B. V. Tilac, France.

Dr. P. Tissot, Chimie et Electrochimie Appliquées, Geneva, Switzerland

Prof. J. W. Tomlinson, Chemistry Department, Victoria University, Wellington, New Zealand

Prof. A. R. Ubbelohde, Chemical Engineering Department, Imperial College, London, England

Dr. S. Valcher, Lab. di Polarografia e Elettrochimica CNR, Padova, Italy

Prof. D. Vasilescu, Université de Nice, France

Dr. T. Vereecken, Free University of Brussels, Belgium

Dr. B. Verger, Alsthom DRE, Massy 91301 France

Mr. N. Viré, France

Dr. W. H. Visscher, Chemical Engineering Department, University of Technology, Eindhoven, Netherlands

Dr. R. G. H. Watson, Chemical Defence Establishment, Porton Downs, Salisbury, England

Dr. S. H. White, Imperial College, London, England

Dr. D. E. Williams, Imperial College, London, England

Dr. I. O. Wilson, BICC Research Laboratories, London, England

Prof. R. F. P. Winand, Chemistry Department, Free University of Brussels, Belgium

Index

445